# Strained
# Organic Molecules

This is Volume 38 of
ORGANIC CHEMISTRY
A series of monographs
Editor: HARRY H. WASSERMAN

A complete list of the books in this series appears at the end of the volume.

# Strained
# Organic Molecules

## Arthur Greenberg

Department of Chemical Engineering and Chemistry
New Jersey Institute of Technology
Newark, New Jersey

## Joel F. Liebman

Department of Chemistry
University of Maryland–Baltimore County
Catonsville, Maryland

ACADEMIC PRESS   New York   San Francisco   London   1978
A Subsidiary of Harcourt Brace Jovanovich, Publishers

ACADEMIC PRESS, INC.
111 Fifth Avenue, New York, New York 10003

*United Kingdom Edition published by*
ACADEMIC PRESS, INC. (LONDON) LTD.
24/28 Oval Road, London NW1    7DX

Library of Congress Cataloging in Publication Data

Greenberg, Arthur.
    Strained organic molecules.

    (Organic chemistry series ; )
    Includes bibliographical references and index.
    1. Chemistry, Physical organic.  2.  Molecular
structure.  I. Liebman, Joel F., joint author.
II.  Title.  III.  Series:  Organic chemistry, a series
of monographs ;
QD476.G66        547'.1'3              77–11212
ISBN  0–12–299550–3

# Contents

## CHAPTER 4

### Polycycles: Aesthetics, Rearrangements, and Topology

## CHAPTER 5

### Kinetic and Thermodynamic Stability

## CHAPTER 6

### A Potpourri of Pathologies

# Preface

Strained molecules have been known for about a century. For example, spiropentane was first correctly reported in the 1890's, although countless others had been claimed erroneously even earlier. However, interest in these species has accelerated greatly only during the last two decades. It is hard to account precisely for this phenomenon, but we believe it is predominantly due to the development of new synthetic methods, which, together with advances in instrumentation, have allowed experimental access to species in increasingly minute concentrations and with increasingly brief lifetimes under more exotic conditions. In addition, the commonplace use of spectroscopy, thermochemistry, and/or both quantitative and descriptive quantum chemistry provides continuing research impetus. The "old rules" no longer seem to hold—for example, we often blithely draw structures containing penta- and hexacoordinate carbons, as well as tetracoodinate carbons having "inverted tetrahedra," and hideously "bent and battered" benzene rings. Moreover, we seriously consider the potential stability of molecules such as 1,2-cyclobutadiene. For example, 1,3-cyclobutadiene, the more stable isomer, has been "tamed" in the form of its tri-*tert*-butyl derivative, and researchers courageously, if not happily, have proceeded to studies aimed at the observation of the isomeric tetrahedranes.

This book attempts to consider the vast field of strained organic molecules. We attempt to achieve a level of sophistication and coverage intermediate between that of an advanced organic chemistry text and that of highly specialized monographs and review articles. Developed, in part, from our review article [*Chem. Rev.* **76,** 311 (1976)], this current volume is more pedagogical and, naturally, more up-to-date. It also includes extensive discussions of thermochemistry and its customary (if not implicit) assumptions, transition metal reactions with strained molecules, and strained heterocycles, all of which are topics largely neglected in the review article. Although there are many unifying themes, the book is arranged such that any chapter (and indeed, almost any section) can be read independently.

There are many respectable reasons for the painstaking synthesis of esoteric, highly strained molecules. It is true that they usually hold little potential as components of "materials," pharmaceuticals, and other "useful" substances. However, many of them have been postulated as intermediates in a variety of reaction mechanisms, and their study may serve to either validate or disprove such mechanisms. Also, strained organic molecules may be useful as high-energy synthetic intermediates. Moreover, the synthesis and study of strained molecules allow chemists to explore the limits of stability in organic chemistry: Molecules strained seemingly to their limits often manifest unprecedented chemical behavior. Additionally, such studies permit calibration of theory and experiment in extreme cases.

Dodecahedrane, sought in earnest for approximately the last fifteen years, nicely illustrates the rationale for the laborious synthesis of a strained molecule. This molecule is anticipated to (a) possess the physical properties of a noble gas; (b) be an excellent substrate for studies of long-range substituent effects, and (c) offer potential as a template for synthesis. In addition, a determination of its standard heat or enthalpy of formation might allow proper choice of a molecular mechanics calculational package for hydrocarbons, since there are considerable differences in the "competing" heats of formation for it. There is also the strong motivation of "climbing the mountain" in order to be first in the synthesis of an exotic molecule.

The history of the benzene valence isomers (Chapter 5,B), once suggested as possible structures for benzene, illustrates the role of serendipity. These isomers have played a prime, and certainly unanticipated, role in exemplifying and clarifying orbital symmetry and catalysis. Moreover, the recent exciting studies of the storage of light energy as latent heat have extensively utilized the benzene valence isomer, "Dewar benzene." This same group of isomers has also played a major role in showing the increased stability afforded strained molecules by trifluoromethyl groups. The only long-range predictions about future discoveries we wish to make are that they will be many in number.

We have in this book attempted to blur the divisions between alicyclic and heterocyclic chemistries, as well as between organic and inorganic chemistries. Here such divisions appear artificial. Moreover, in order to provide quantitation wherever possible, we have melded quantum chemical and molecular mechanics calculations, experimental thermochemical data, and the semiquantitative bond-additivity and group increment methods. Although we use the terminology associated with the theory of the conservation of orbital symmetry and related approaches, an introduction to this theory has not been included since there are many excellent monographs on the subject (see references in Chapter 5).

We suspect that a chemist transported from the year 1958 to 1978 would probably hardly believe the structures of molecules observed, isolated, and (worse!) actively sought and theoretically investigated. As such, we hesitate to predict the nature of structures featured in the *Journal of the American Chemical Society* **120** (or even **110**). Hopefully, some of the discussions in this book will provide indications and suggestions of areas for future exploration.

> It is not thy duty to complete the work
> but neither are thou free to desist of it.
>
> Ethics of the Fathers (The Talmud) 2:21

We wish to thank Professors I. J. Borowitz, B. B. Jarvis, A. Liberles, A. P. Marchand, P. H. Mazzocchi, R. M. Pollack, M. Pomerantz, S. W. Staley, V. P. Vitullo, D. L. Whalen, and most of all Professor P. v. R. Schleyer (Nürnberg–Erlangen) and Dr. E. J. Domalski (National Bureau of Standards) for the helpful comments or criticisms of this manuscript and/or the review article on which it is partially based. We also express our gratitude to those individuals, already cited in the references, for providing us with unpublished data and often partially formulated conclusions. One of us (A. G.) completed most of his work on this book while a faculty member at Frostberg State College, Frostberg, Maryland. We thank Ms. Deborah Van Vechten and Dr. C. C. Van Vechten for their editorial assistance, Mr. Stuart I. Yaniger for indexing, and Ms. Michelle Amyot for typing. Above all, we acknowledge the patience, understanding and love of our wives, Susan Joan Greenberg and Deborah Van Vechten, to whom we gratefully dedicate this book.

Arthur Greenberg
Joel F. Liebman

# Strained
# Organic Molecules

# CHAPTER 1

# Energy and Entropy

## 1.A  Reference States and Reference Books

The study of organic chemistry, as both art and science, consists in large part of understanding regularities in structure, stability, and reactivity.[1-4] Although stability and reactivity are fundamentally inseparable concepts, stability is a simpler concept than reactivity since we need not ask about reaction conditions or about the other reagents. The energetics of a molecule are also defined in terms of other species. Stability is always relative: we can only speak of "stabilization energy" as manifested by resonance, conjugation, aromaticity or delocalization, steric attraction, or by the ubiquitous coulombic attraction of oppositely charged species. (We limit this last effect to ions interacting either with one another or with some nearby solvent molecule.) "Destabilization energy" likewise implies comparisons, wherein we now talk about negative resonance and antiaromaticity, steric repulsion, or the ubiquitous coulombic repulsion of similarly charged species. (We limit this last effect to ions interacting with one another.) The fact that stability is always relative leads to the idea of reference states. One could say that propene is more stable than cyclopropane because the former can be made from the

---

[1] G. W. Wheland, "Advanced Organic Chemistry," 3rd Ed. Wiley, New York, 1960.

[2] C. K. Ingold, "Structure and Mechanism in Organic Chemistry," 2nd Ed. Cornell Univ. Press, Ithaca, New York, 1969.

[3] K. B. Wiberg, *in* "Determination of Organic Structures by Physical Methods" (F. C. Nachod and J. J. Zuckerman, eds.), Vol. 3, pp. 207–245. Academic Press, New York, 1971.

[4] W. J. LeNoble, "Highlights of Organic Chemistry." Dekker, New York, 1974.

1

latter by thermal isomerization (see Chapter 2). But how do we compare octane and "iso-octane" (2,2,4-trimethylpentane)? Heating octane yields a variety of products; we remind the reader of the almost inexorable complexity of petroleum refining.[5a,b]

The customary reference states from thermodynamics[6-17] are the so-called standard states of the elements,[18] i.e., the most stable form of the elements under standard temperature and pressure (STP), 25°C, and 1 atm are arbitrarily given a heat of formation, $\Delta H_f^0$, of 0 kcal/mole. Accordingly, hydrocarbons, the major class of compounds discussed in this book, are compared with graphite and hydrogen gas. This comparison provides insufficient intuitive understanding of the relationships between structure and energetics. (Any other single reference state, such as atomic carbon and atomic hydrogen, or, say, methane and benzene, would be of equally little use because the new results are merely linear combinations of the old.)

It is, then, necessary to know the heat of reaction from the compound of interest to the chosen reference states. This is rarely done directly. Rather,

[5a] F. D. Rossini, B. J. Mair, and A. J. Streiff, "Hydrocarbons from Petroleum." Reinhold, New York, 1953.

[5b] "Refining Petroleum for Chemicals," Adv. Chem. Ser. No. 97. Am. Chem. Soc., Washington, D.C., 1970.

[6] Some books on thermodynamics and thermochemistry that we have found useful constitute the following 11 references.

[7] S. W. Benson, "Thermochemical Kinetics: Methods for the Estimation of Thermochemical Data and Rate Parameters," 2nd Ed. Wiley, New York, 1976.

[8] H. A. Bent, "The Second Law: An Introduction to Classical and Statistical Mechanics." Oxford Univ. Press, London and New York, 1965.

[9] A. Bondi, "Physical Properties of Molecular Crystals, Liquids and Glasses." Wiley, New York, 1968.

[10] J. D. Cox and G. Pilcher, "Thermochemistry of Organic and Organometallic Compounds." Academic Press, New York, 1970.

[11] F. D. Rossini, ed., "Experimental Thermochemistry," Vol. 1. Wiley (Interscience), New York, 1956; H. A. Skinner, ed., "Experimental Thermochemistry," Vol. 2. Wiley (Interscience), New York, 1962.

[12] G. J. Janz, "Thermodynamic Properties of Organic Compounds: Estimation Methods, Principles and Practice." Academic Press, New York, 1967.

[13] W. Kauzmann, "Thermal Properties of Matter, Vol. II. Thermodynamics and Statistics: With Applications to Gases." Benjamin, New York, 1967.

[14] I. M. Klotz and R. M. Rosenberg, "Chemical Thermodynamics: Basic Theory and Methods," 3rd Ed. Benjamin, New York, 1974.

[15] C. T. Mortimer, "Reaction Heats and Bond Strengths." Pergamon, New York, 1962.

[16] R. C. Reid and T. K. Sherwood, "The Properties of Gases and Liquids: Their Estimation and Correlation," 2nd Ed. McGraw-Hill, New York, 1966.

[17] D. R. Stull, E. F. Westrum, Jr., and G. C. Sinke, "The Chemical Thermodynamics of Organic Compounds." Wiley, New York, 1969.

[18] The reader is advised to be careful to distinguish C for atomic carbon and for graphite.

both the compound and the reference states are transformed into the same final products and the difference in the energy released is computed. Hess' law guarantees that this difference is independent of the final products.[19] The most commonly used reaction is combustion, in which, at least for hydrocarbons, the final products are simply $CO_2$ and $H_2O$. Equally valid reactions[20] include combustion in fluorine, hydrolysis of an ester to an alcohol and acid of known heats of formation, and reduction of an olefin to the related but more thermochemically characterized saturated hydrocarbon.

Furthermore, care must be taken not to confuse inter- and intramolecular sources of stabilization and destabilization. It is thus necessary for the substance to act as an ideal gas at STP. Unfortunately, most compounds of interest are in a condensed phase (i.e., liquid or solid) under these conditions. It is therefore necessary to determine the heat of vaporization or sublimation; the latter procedure often is done stepwise as fusion and then vaporization. Extensive compendia for heats of formation, vaporization, and sublimation exist.[10,17,21-28d] But for many compounds of interest, these data are missing,

[19] H. A. Bent (Ref. 8) presents two very good examples totally unrelated to the compounds discussed in this book.

[20] See Refs. 10, 11, 15, and 17 for numerous examples.

[21] J. D. Cox and G. Pilcher, (Ref. 10). For reasons motivated by convenience, we have chosen this reference as the major source of thermochemical data in our book.

[22] "Selected Values of Properties of Hydrocarbons," Am. Pet. Inst. Res. Project 44, Carnegie Inst. Technol., Pittsburgh, Pennsylvania, 1953.

[23] S. W. Benson (Ref. 7), appendices and tables within.

[24a] D. D. Wagman, W. H. Evans, V. B. Parker, I. Halow, S. M. Bailey, and R. H. Schumm, *Natl. Bur. Stand. (U.S.), Tech. Note* 270-3 (1975). This is a corrected version of the 1968 edition and supersedes both Technical Notes 270-1 and 270-2.

[24b] For Technical Notes 270-4 through 270-7, see D. D. Wagman, W. H. Evans, V. B. Parker, I. Halow, S. M. Bailey, and H. Schumm, *Natl. Bur. Stand. (U.S.), Tech. Note* 270-4 (1969); D. D. Wagman, W. H. Evans, V. B. Parker, I. Halow, S. M. Bailey, R. H. Schumm, and K. L. Churney, *Tech. Note* 270-5 (1971); V. B. Parker, D. D. Wagman, and W. H. Evans, *Tech. Note* 270-6 (1971); R. H. Schumm, D. D. Wagman, S. Bailey, W. H. Evans, and V. B. Parker, *Tech. Note* 270-7 (1973). These are compendia of "Selected Values of Chemical Thermodynamic Properties."

[25] The following are updates of the JANAF Thermochemical Tables: D. R. Stull and H. Prophet, eds., *Natl. Stand. Ref. Data Ser., Natl. Bur. Stand.* NSRDS-NBS 37 (1971); M. W. Chase, J. L. Curnutt, A. T. Hu, H. Prophet, A. N. Syverud, and L. C. Walker, *J. Phys. Chem. Ref. Data* 3, 311 (1974); M. W. Chase, J. L. Curnutt, H. Prophet, R. A. McDonald, and A. N. Syverud, *J. Phys. Chem. Ref. Data* 4, 1 (1975).

[26] H. M. Rosenstock, K. Draxl, B. W. Steiner, and J. T. Herron, *J. Phys. Chem. Ref. Data, Suppl.* 1 (1977). This is the major compendia on the "Energetics of Gaseous Ions."

[27] The following is an annual review of thermochemical data: "The Bulletin of Thermodynamics and Thermochemistry" (E. F. Westrum, Jr., ed.) Univ. of Michigan Publ. Distrib. Serv., Ann Arbor, **Michigan**.

[28a] The following are compendia of organic thermochemical data:

whereas for numerous others, "merely" the heat of vaporization or sublimation is lacking. Fortunately, many methods enable the heats of formation of gas-phase molecules to be estimated with varying degrees of reliability.[29]

Unfortunately, even when all the data are available, conceptual problems may still arise to haunt us. Cyclopropane, $(CH_2)_3$ or $C_3H_6$, is an archetypally destabilized molecule (see Chapter 2), whereas benzene, $(CH)_6$ or $C_6H_6$, is a correspondingly archetypally stabilized molecule.[1-4] Under the above conditions of STP and ideality of gas, the experimental heats of formation of cyclopropane and benzene are 12.7 and 19.8 kcal/mole, respectively. Reminding the reader that the heat of formation of a compound is defined as the heat absorbed or released upon formation from the elemental standard states, we deduce that both substances are unstable relative to graphite and hydrogen. Indeed, the hypothetical reaction

$$C_6H_6(g) \longrightarrow C_3H_6(g) + 3C(graph.)$$

is exothermic by 7.1 kcal/mole, leading us to conclude that cyclopropane is the more stable compound. Our conventional use of the archetypes requires the exact opposite. What is the problem?

As noted earlier, cyclopropane is less stable than its isomer, propane. Similarly, benzene is the most stable $C_6H_6$ isomer and, *a fortiori*, the most stable $(CH)_6$ valence isomer. (Benzene and its valence isomers will be discussed further in Section 5.B.) Must we then limit our discussions to members of single sets of isomers? If so, cyclopropane and benzene can never be compared unless we say that benzene and (cyclopropane + 3 graphitic carbons) are isomeric. While such a seeming redefinition of "isomeric" can prove useful (see Section 1.D, for example), this returns us to the undesirable conclusion that cyclopropane is more stable than benzene.

One resolution of this problem is to somehow use the molecule itself as part of its own reference state. Molecules may be hierarchically recognized[30] as collections (1) of atoms, e.g., C and H; (2) of bonds, e.g., C—C, C=C and C—H; (3) of groups or "superatoms,"[31] e.g., $CH_3$—, —$CH_2$—, —CH< and

---

[28b] S. W. Benson, F. R. Cruickshank, D. M. Golden, G. R. Haugen, H. E. O'Neal, A. S. Rodgers, R. Shaw, and R. Walsh, *Chem. Rev.* **69**, 279 (1969).

[28c] E. J. Domalski, *J. Phys. Chem. Ref. Data* **1**, 221 (1972).

[28d] H. K. Eigenmann, D. M. Golden, and S. W. Benson, *J. Phys. Chem.* **77**, 1687 (1973).

[29] We wish to emphasize that this is a markedly incomplete listing. Dr. Eugene J. Domalski, data evaluator and "thermochemical archivist" for the National Bureau of Standards, has mentioned to us that he has a compendium of approximation methods that exceed 500 in number.

[30] S. W. Benson and J. H. Buss, *J. Chem. Phys.* **29**, 546 (1958).

[31] L. M. Masinter, N. S. Sridharan, J. Lederberg, and D. H. Smith, *J. Am. Chem. Soc.* **96**, 7702 (1974).

$>$C$<$ ; and (4) of rings, e.g., a "$C_3$ ring."[32] The first member of the hierarchy is the traditional basis of chemistry: excepting studies in nuclear chemistry, it is tacitly assumed that the number and type of atoms are unchanged in any given reaction. As such, this level leads us back to such unclarifying reference states as atomic hydrogen and atomic carbon, or to the more customary hydrogen gas and graphite. The second member of the hierarchy is the basis of the method of bond energies, i.e., of bond additivity schemes.[33] Let us quantitatively illustrate this sort of reasoning. The model describes cyclopropane as composed of three C—C bonds and six C—H bonds. If cyclopropane were to lack any stabilization or destabilization, then the total bond energy (or, as more customarily called, the atomization energy) would equal the sum of the individual bond energies. While we admit that more sophisticated paradigms or models exist for these quantities, we can simply set the C—C bond energy as equal to that of ethane, and the C—H bond energy to be that of the first homolysis of methane (the process $CH_4 \rightarrow CH_3 + H$). Numerically,[34–38b]

$$D(C—C) = 2\,\Delta H_f^0(CH_3) - \Delta H_f^0(C_2H_6) = 2(34) - (-20) = 88 \text{ kcal/mole} \quad (1)$$

and

$$D(C—H) = \Delta H_f^0(CH_3) + \Delta H_f^0(H) - \Delta H_f^0(CH_4) =$$
$$34 + 52 - (-18) = 104 \text{ kcal/mole} \quad (2)$$

Accordingly, the atomization energy of cyclopropane, $\Delta H_{at}$, is $3(88) + 6(104) = 888$ kcal/mole. The heats of formation and of atomization of an arbitrary hydrocarbon $C_cH_h$ are interrelated as shown in Eq.(3). From the

$$\Delta H_{at} = c\,\Delta H_f^0[C(g)] + h\,\Delta H_f^0[H(g)] - \Delta H_f^0(C_cH_h\,(g)] =$$
$$171c + 52h - \Delta H_f^0[C_cH_h(g)] \quad (3)$$

experimental heat of gaseous cyclopropane, we obtain a value of 811 kcal/mole. Accordingly, cyclopropane appears destabilized by $888 - 811 = 77$ kcal/

$$D(C=C) = 2\,\Delta H_f^0(CH_2) - \Delta H_f^0(C_2H_4) = 2(86) - 12 = 160 \text{ kcal/mole} \quad (4)$$

[32] S. W. Benson (Ref. 7), pp. 31–33.

[33] The following five references are some sources of bond energy data we have found useful.

[34] J. A. Kerr, *Chem. Rev.* **66**, 266 (1966).

[35] B. deB. Darwent, *Natl. Stand. Ref. Data Ser., Natl. Bur. Stand.* **NSRDS-NBS 31** (1970).

[36] D. M. Golden and S. W. Benson, *Chem. Rev.* **69**, 125 (1969).

[37] K. W. Egger and A. T. Cocks, *Helv. Chim. Acta* **56**, 1516, 1537 (1973).

[38a] R. T. Sanderson, "Chemical Bonds and Bond Energy," 2nd Ed. Academic Press, New York, 1976.

[38b] R. T. Sanderson, "Chemical Bonds and Organic Compounds," Sun and Sand Publ. Co., Scottsdale, Arizona, 1976.

mole. In order to perform comparable calculations on benzene, one needs $D(C{=}C)$. This value may analogously be taken as shown in Eq.(4). Proceeding as in Eq.(4), we find the calculated heat of atomization of benzene to be 1368 kcal/mole, while the experimental value is 1328 kcal/mole. That is, benzene is destabilized by 40 kcal/mole relative to our reference states. Because benzene is less destabilized relative to our reference states than cyclopropane, this scheme is sufficient to reaffirm our intuitive logic that benzene is more stable than cyclopropane. (Note that if we consider destabilization per carbon atom or C—C bond, benzene is even more stable.) However, it is distressing that we do not recover the customary 36 kcal/mole stabilization energy for benzene nor the ca. 25 kcal/mole destabilization energy for cyclopropane.

This failure documents a very important point—the numerical results are apparently dependent on the reference state. Indeed, we might wonder whether methane is destabilized by our model. It is tempting to answer, "Of course not!" since methane is one of our reference states. However, it is not the entire methane molecule but rather its $CH_3$—H bond that is one of our reference states. We would calculate the atomization energy of methane as $4(104) = 416$ kcal/mole. The experimental value, as determined by Eq. (3), is 397 kcal/mole. Accordingly, methane is experimentally destabilized by 19 kcal/mole! Rather remarkably, both $C_2H_6$ and $C_2H_4$ are likewise destabilized by 38 kcal/mole; or the same 19 kcal/mole per carbon. Correcting for this "obviously" invalid destabilization in our model compounds, we find cyclopropane to be destabilized by $77 - 3(19) = 20$ kcal/mole and benzene to be stabilized by $-40 + 6(19) = 74$ kcal/mole. The results remain in quantitative disagreement with experience, but at least cyclopropane and benzene are on the appropriate "sides" of the reference-state stability scale.

A major component of the above ambiguities arises from the seeming absence of unique C—H, C—C and C$=$C bond energies. Indeed, there are none. For example, we are not surprised that the C—H bond strength in methane exceeds that of toluene (104 vs. 85 kcal/mole) because of the resonance stabilization of the resulting benzyl radical.[39] However, it is distressing to note that the C—H bonds in benzene and ethane are respectively 5 kcal/mole greater than and less than those of methane, or, more precisely, $D(C_6H_5$—H) and $D(C_2H_5$—H) are 5 kcal/mole larger and smaller, respectively than $D(CH_3$—H). Despite all these problems, we still would like to have reference states in a bond-additivity scheme.

One notably successful method is to use so-called "isodesmic reactions."[40] These reactions "are examples of chemical changes in which there is retention

[39] D. Griller and K. U. Ingold, *Acc. Chem. Res.* **9**, 13 (1976).

[40] W. J. Hehre, R. Ditchfield, L. Radom, and J. A. Pople, *J. Am. Chem. Soc.* **92**, 4796 (1970). We wish to note additionally that most articles by these authors, subsequent to this paper, have made use of this approach.

of the number of bonds of a given formal type [e.g., C—H, C—C, C=C] but with a change in their formal relation to one another."[40] In this scheme, $CH_4$, $C_2H_6$, and $C_2H_4$ return as reference states, but in a somewhat more consistent, and legitimized manner. Consider, in reactions (5) and (6) for cyclopropane and benzene, respectively, that the same number of C—H, C—C, and C=C bonds

$$(CH_2)_3 + 3CH_4 \longrightarrow 3C_2H_6 \tag{5}$$

$$C_6H_6 + 6CH_4 \longrightarrow 3C_2H_6 + 3C_2H_4 \tag{6}$$

appear on both sides of the equation; thus the precise values for the associated energies were never needed. Numerous quantum and thermochemical errors cancel out, and so, "the energies of isodesmic reactions measure deviations from the additivity of bond energies."[40] Interestingly, the energetics of reactions (5) and (6) are nearly equal to what we computed earlier for these two hydrocarbons. More precisely, these reference states reproduce the above destabilization (read *strain*) energy for cyclopropane and stabilization (read *resonance*) energies for benzene.

There is yet another, and even more conceptually satisfying, bond-energy scheme by George *et al.*[41a] This is called a "homodesmotic reaction," for which "(a) there are equal numbers of C atoms in their various states of hybridization in reactants and products and (b) there are equal number of C atoms with zero, one, two, and three atoms attached in reactants and products." Corresponding to reactions (5) and (6) are (5′) and (6′). By matching the local neighborhood and hybridization of each atom, again, there are

$$(CH_2)_3 + 3C_2H_6 \longrightarrow 3CH_3CH_2CH_3 \tag{5′}$$

$$C_6H_6 + 3C_2H_4 \longrightarrow 3CH_2CHCHCH_2 \tag{6′}$$

numerous cancellations, and so the intrinsic strain or resonance energies may be determined directly. Indeed by using propane, isobutane, and neopentane $[C(CH_3)_bH_{4-n}: n = 2,3,4]$ and propene and isobutene $[(CH_3)_nH_{2-n}$-$C=CH_2: n = 1,2]$, qualitative and quantitative understanding was achieved for a wide variety of alicyclic hydrocarbons. Specifically, the destabilization energy of cyclopropane was 26.5 kcal/mole. These authors analogously used[41b] butadiene, 2-vinylbutadiene, and 2,3-divinylbutadiene to achieve corresponding success for aromatics; specifically, one may obtain a stabilization energy of ca. 20 kcal/mole for benzene. (No thermochemical data exist for the two vinylbutadienes and so direct comparison with the admittedly sparse experimental data for the aromatics of interest is generally precluded.) We wish to cite the related, unpublished, and in large part preempted by George *et al.*[41a,b] models both of Dill[42] for substituted alicyclics on his "group

[41a] P. George, M. Trachtman, C. W. Bock, and A. M. Brett, *Tetrahedron* **32**, 317 (1976).

[41b] P. George, M. Trachtman, C. W. Bock, and A. M. Brett, *Theor. Chim. Acta* **38**, 121 (1975).

[42] Dr. James D. Dill, personal communication. This approach was described in his Princeton University final oral research proposal (October, 1975).

separation reaction" and of Liebman and co-workers[43] for general hydrocarbons in the "maximal covering subgraph" and "minimal reference state" approaches.

The reader will note that we have departed from the idea of individual bonds and are now talking about groups of atoms and bonds. In particular, for saturated alicyclic hydrocarbons, we have the $[C—(C)(H)_3]$, $[C—(C)_2(H)_2]$, $[C—(C)_3(H)]$, and $[C—(C)_4]$ groups.[27,29] These may alternatively be recognized as primary, methyl, or $CH_3$; secondary, methylene, or $CH_2$; tertiary, methyne, or $CH$; and quaternary, or C groups. The reader will note: (a) there is no assumption that the C—H and/or C—C bonds within these groups are equivalent; (b) these groups are not synonymous with the methyl radical, methyne, methyne, and atomic carbon species; and (c) in the absence of any stabilizing or destabilizing effects, the energy of the molecule equals the sum of the group energies. (Numerous other properties are similarly assumed to be additive but are of less interest to us in our discussion of strained species.)

Let us return to cyclopropane, which we recognize as being composed of three $[C—(C)_2(H)_2]$ groups. For ease of notation, this will be written as $CH_2$. Note that this notation and use would be inappropriate for such other $CH_2$-containing species as $CH_2Cl_2$, $CH_2NO_2^-$, and $(C_6H_5)_3PCH_2$, because the $CH_2$ group is not connected to two other carbons. To understand cyclopropane, it is necessary to know the energy or enthalpy of an unstrained $CH_2$ group. There are at least five methods of obtaining this quantity. The first method is to analyze the energies of a collection of alkanes and to thereby deduce all at once the energy values for all the above groups. This may be done in the framework of both rigorous molecular mechanics[44-58] (see Section 1.B) and

[43] N. P. Adams, D. J. Houck, R. J. Kolish, I. Kramer, J. F. Liebman, and F. P. Wilgis, *Rocky Mt. Reg. Meet., Am. Chem. Soc., June, 1976* Pap. No. 170; also unpublished results of D. J. Houck, D. J. Schamp, and J. F. Liebman.

[44] The following 14 references to "molecular mechanics" are a selection of major studies in this field. The reader is referred to other works of these authors where further discussion and other literature citations are made.

[45] F. H. Westheimer, *in* "Steric Effects in Organic Chemistry" (M. S. Newman, ed.), pp. 523–555. Wiley, New York, 1956.

[46] K. B. Wiberg, *J. Am. Chem. Soc.* **87**, 1070 (1965).

[47] J. B. Hendrickson and R. K. Boeckman, Jr., *J. Am. Chem. Soc.* **89**, 7036, 7043, 7047 (1967).

[48] E. J. Jacob, H. B. Thompson, and L. S. Bartell, *J. Chem. Phys.* **47**, 3736 (1967).

[49] J. E. Williams, Jr., P. J. Stang, and P. von R. Schleyer, *Ann. Rev. Phys. Chem.* **19**, pp. 531–558. (1968).

[50] S.-J. Chang, D. McNally, S. Shary-Tehrany, M. J. Hickey, and R. H. Boyd, *J. Am. Chem. Soc.* **92**, 3109 (1970).

[51] N. L. Allinger, M. T. Tribble, M. A. Miller, and D. W. Wertz, *J. Am. Chem. Soc.* **93**, 1637 (1971).

statistical mechanics (see Section 1.C), and we note that many of the "big names" of thermochemistry have developed their own sets of values. The second approach is to consider the energy difference between the gaseous straight chain, or $n$-, alkane with $M$ methylene groups and with $M - 1$ as $M$ gets larger. We would like to believe that the "$n$" for normal has more than just historical or aesthetic significance. The "by-product" of this analysis is the enthalpy of the $CH_3$ group [see Eqs. (7) and (7')]. The third alternative,

$$\Delta H_f^0[C—(C)(H)_3] = \tfrac{1}{2}\{CH_3(CH_2)_M CH_3 - M[C—(C)_2(H)_2]\} \qquad (7)$$

$$\Delta H_f^0(g, CH_3) = \tfrac{1}{2}[\Delta H_f^0(CH_3(CH_2)_M CH_3 - M \Delta H_f^0(g, CH_2)] \qquad (7')$$

a highly related scheme, is to consider those $n$-alkanes in their lowest-energy, i.e., all-*trans* or zigzag conformation (see Section 1.B).[12,59] This generates the so-called "strainless increments" as opposed to the "general." The fourth scheme is to define the energy of the $CH_2$ group as the energy difference of $CH_3CH_2CH_3$ and $CH_3CH_3$. Along with defining the energy of $CH_3$, $CH$, and $C$ as the energy $\tfrac{1}{2}CH_3CH_3$, the difference of $CH(CH_3)_3$ and $\tfrac{3}{2}CH_3CH_3$ and the difference of $C(CH_3)_4$ and $2CH_3CH_3$, this reproduces the methods of "homodesmotic reactions,"[41] "group separation,"[42] and "minimal reference states"[43] for alkanes. {Equivalently, from the enthalpy [$\Delta H_f^0(g, CH_3)$] defined as $\tfrac{1}{2}C_2H_6$, the enthalpy of $CH_2$, $CH$, and $C$ are found as the difference of the heat of formation of propane, isobutane, and neopentane minus two, three, and four $CH_3$ groups, respectively.} The fifth and final scheme is based on the energetics of cyclohexane derivatives, since most of the compounds that will be described in subsequent chapters are cyclic. The enthalpy of $CH_3$ is defined as above; of $CH_2$, as one-sixth the heat of cyclohexane; and of $CH$, defined in terms of adamantane **1** [$(CH)_4(CH_2)_6$, see Section 4.A], i.e., Eq. (8).

$$\Delta H_f^0(g, CH) = \tfrac{1}{4}[\Delta H_f^0(g, adamantane) - 6 \Delta H_f^0(g, CH_2)] \qquad (8)$$

The enthalpy of the quaternary C group is defined in terms of 1,3,5,7-tetra-methyladamantane, $(CCH_3)_4(CH_2)_6$, species **2**.

Not unexpectedly, all of these methods give approximately the same values for the heat of formation of the $CH_2$ group, but show small and potentially

[52] E.M. Engler, J. D. Andose, and P. von R. Schleyer, *J. Am. Chem. Soc.* **95**, 8005 (1973).

[53] O. Ermer and S. Lifson, *J. Am. Chem. Soc.* **95**, 4121 (1973).

[54] O. Ermer, and S. Lifson, *Struct. Bonding (Berlin)* **27**, pp. 161–212 (1974).

[55] C. Altona and D. H. Faber, *Top. Curr. Chem.* **45**, 1 (1974).

[56] N. L. Allinger, *Adv. Phys. Org. Chem.* **13**, 1 (1976).

[57] J. D. Dunitz and H. B. Bürgi, *Int. Rev. Sci. Phys. Chem.*, Ser. 2, **11**, pp. 81–120 (1975).

[58] K. Mislow, "Introduction to Stereochemistry." Benjamin, New York, 1966.

[59] E. M. Engler, J. D. Andose, and P. von R. Schleyer (Ref. 52). For convenience, we have chosen this article to be our primary reference for discussion of, and data from, molecular mechanics.

**1**, R = H
**2**, R = CH$_3$

$$\Delta H_f^o(g, C) = \tfrac{1}{4}[\Delta H_f^o(g, \mathbf{2}) - 4\,\Delta H_f^o(g, CH_3) - 6\,\Delta H_f^o(g, CH_2)] \qquad (9)$$

meaningful differences for the other groups. The set of values we shall use in our discussions arise from the "strainless increments" described in method three above; in particular, the heats of formation we have adopted for CH$_3$, CH$_2$, CH, and C are $-10.05$, $-5.13$, $-2.16$, and $-0.30$ kcal/mole,[59] respectively. We thus predict the heat of formation of gaseous cyclopropane to be $3 \times (-5.13)$ kcal/mole. This calculated value of ca. $-15$ kcal/mole is to be compared with the experimental value of $+12.7$ kcal/mole.[21] This calculation shows cyclopropane to be destabilized by 28 kcal/mole, and it is from this type of reasoning that the conventional value for the strain energy of this species arises. We feel the precise number should not be relied upon, as group-increment methods of different complexity and/or origin yield somewhat different values. However, the reader should note that regardless of the precise method of bond or group additivity we employ, we deduce that cyclopropane is highly destabilized relative to our model compounds. We would also predict that benzene is highly destabilized, in fact by $19.6 - 6(-2.16) = 32.6$ kcal/mole. Indeed it is, if we consider the CH groups in benzene to be equivalent to the CH groups in, say, isobutane. This equivalence would force us to refer to ethylene as cycloethane and to benzene as tetracyclo[6.0.$^{1,2}$ 0.$^{3,4}$ 0$^{5,6}$]hexane. While the former "synonym" in fact provides useful insights into the chemistry of cyclopropane (see Section 2.A), the latter appears absurd and seems to offer no understanding of the interesting, but conceptually evasive, concept of aromaticity.[1-4]

We now turn to the idea of rings as reference states and note that there are four interrelated ways of using them. The first is to set up group quantities for ring-containing radicals analogous to the group increment for, say, cyclopropyl [(CH$_2$)$_2$C(H)(C)].[60] The second is to employ constant corrections for a given size ring.[61,62] Of somewhat dubious value for estimating strain energies, this method is nonetheless highly useful for estimating entropy

[60] David J. Schamp and Joel F. Liebman, unpublished approach.
[61] S. W. Benson (Ref. 7), Tables A.1 and A.2.
[62] See Sect. 4.F. in this book.

corrections.[63] The third is to fuse rings when one considers polycyclic species. For example, bicyclobutane may be "synthesized"[64] by reaction (10). (The

$$\text{bicyclobutane} \rightleftharpoons 2 \text{ cyclopropane} - \text{ethane} \qquad (10)$$

synthesis for bicyclobutane is but 6 kcal/mole endothermic). If only carbocycles are considered, the first and third methods can be recast in terms of the second. A distressing problem with methods two and three as expressed is that both ignore numerous rings found within the molecule. For example, bicyclobutane has two three-membered rings. This ring-counting ambiguity will be discussed further in Section 4.F. The fourth way returns us to bond-additivity logic. For example, cyclopropanone can be "synthesized" by reactions (11) or (12).[65] In both (11) and (12), there is a preformed cyclopropane ring with one trigonally coordinated carbon, and so it is tacitly assumed

$$\text{cyclopropanone} \rightleftharpoons \text{cyclopropyl cation} + \text{acetone} - \text{isopropyl cation} \qquad (11)$$

$$\text{cyclopropanone} \rightleftharpoons \text{methylenecyclopropane} + \text{acetone} - \text{isobutylene} \qquad (12)$$

that most bond energies do not change. The predicted heats of formation of cyclopropanone are $-3$ and $0$ kcal/mole, values which are in relatively close agreement with each other and with the experimental value of $+3.8$ kcal/mole.[66] It is surprising that reactions (11) and (12) are in such close agreement. Intuitively, reaction (12) seems preferable, since one would think that the "bare" positive charge in the cyclopropyl and isopropyl cations would be extremely sensitive to the molecular environment.

As noted earlier, experimental data are often lacking. Can one substitute values that are derived from *ab initio* or semiempirical quantum-chemical calculations for thermochemistry? "Extended-basis molecule optimized" *ab initio* calculations by Newton and Schulman[64] show reaction (10) to be 12.7 kcal/mole endothermic, while one may deduce from the semiempirical MINDO/3 energies of Bingham, Dewar, and Lo[67] the nearly identical value of 12.6 kcal/mole. In these two cases, the bond-energy calculations are in error by ca. 7 kcal/mole. With regard to the absolute value for the heat of formation of bicyclobutane, the former authors provide no absolute value, while the latter are within 2 kcal/mole of the experimental number. For cyclopropanone, MINDO/3[67] calculations give a heat of formation of $-19.5$ kcal/mole. However, using data for all of the species in Eq. (12) from this

[63] S. W. Benson (Ref. 7), pp. 65–72.

[64] M. D. Newton and J. M. Schulman, *J. Am. Chem. Soc.* **94**, 767 (1972).

[65] J. F. Liebman and A. Greenberg, *J. Org. Chem.* **39**, 123 (1974).

[66] H. J. Rodrigues, J.-C. Chang, and T. F. Thomas, *J. Am. Chem. Soc.* **98**, 2027 (1976).

[67] R. C. Bingham, M. J. S. Dewar, and D. H. Lo, *J. Am. Chem. Soc.* **97**, 1294, 1302 (1975).

calculational scheme, one would predict a value of $-21.8$ kcal/mole, not $-19.5$ kcal/mole. (Remember, the experimental value is $+3.8$ kcal/mole.[66]) Looking over the calculated and experimental values for the "known" compounds in (12), it is found that the dominant error (14 kcal/mole) is found in the heat of formation of methylenecyclopropane. Comparable data are again lacking for *ab initio* values. However, we may cite two seemingly unrelated comparisons. The first is a study of cyclopropanone and its valence isomers, for which allene oxide and oxyallyl were explicitly studied.[68] Bond-additivity logic successfully reproduced both the order of stability and nearly the numerical energy differences for these species.[65] The second comparison, considerably more indirect, entails the energy of reaction (13). To the extent

cyclopropyl cation + isobutylene  $\rightleftharpoons$

$$\text{methylenecyclopropane + isopropyl cation} \quad (13)$$

that (11) and (12) are equally valid, this reaction would be essentially thermoneutral. However, using data from the 6-31G calculations of Pople, Schleyer, Hehre, and their co-workers,[69–71] this reaction may be shown to be ca. 19 kcal/mole exothermic! Earlier in this book we expressed surprise at the near equality of the energetics of (11) and (12). As (12) was believed to be more reasonable, the higher (and more nearly correct) value of 0 kcal/mole for the heat of formation of cyclopropanone is suggested. This is admittedly a Pyrrhic victory in that reactions (11) and (12) should both be thermoneutral if bond additivity in rings were to be strictly valid. From our awareness that electron-withdrawing substituents destabilize small rings and that a vacant orbital or carbocationic center is the most electron withdrawing (see Section 5.E), we are not surprised (*a posteriori*) at the failure of reaction (12) to predict the heat of formation of cyclopropanone. Substituted species will generally be ignored in our text, not because of lack of interest, but rather because of the general absence of thermochemical data. (We remind the reader that, for example, sterculic acid,[72] a naturally occurring fatty acid containing a cyclopropene ring, was isolated long before the parent carbocycle.)

[68] A. Liberles, A. Greenberg, and A. Lesk, *J. Am. Chem. Soc.* **94**, 8685 (1972).

[69] Cyclopropyl cation: L. Radom, P. C. Hariharan, J. A. Pople, and P. von R. Schleyer, *J. Am. Chem. Soc.* **95**, 6531 (1973).

[70] Isobutylene, methylenecyclopropane: W. J. Hehre and J. A. Pople, *J. Am. Chem. Soc.* **97**, 6941 (1975).

[71] Isopropyl cation: P. C. Hariharan, L. Radom, J. A. Pople, and P. von R. Schleyer, *J. Am. Chem. Soc.* **96**, 599 (1974).

[72] See the discussion in L. F. Fieser and M. Fieser, "Advanced Organic Chemistry." Reinhold, New York, 1971.

## 1.B Origin of Molecular Strain

In this section and, indeed, in most of the book, we consider the normal case, in which entropy and zero-point energy differences may be neglected in comparison to the change in enthalpy or energy (see Sections 1.C and 1.D). As such, when we speak of strain (or steric) energy, we usually view this quantity as conceptually synonymous with strain (or steric) free energy. We will initially partition this strain energy into three components,[44-58,73] all describable in terms of classical chemistry and without the use of orbitals or other bonding considerations. The first component is torsional or rotational. For the lowest energy and thus a logical reference state, all C—C—C—C fragments are assumed to be trans or zigzag[74] as opposed to gauche, cis, or some intermediate and unnamed geometry. This constitutes the idealized "strain-free" alkane geometry, although at any temperature above $0°K$ these hydrocarbons have other, higher-energy geometries or conformations as well. This may be ascribed to an entropy effect: the entropy of mixing equals $\sum_i x_i \ln x_i$, where $x_i$ is the mole fraction of the $i$th conformer.[75] The consequence of choosing the "strain-free" geometry for alkanes has little effect on the thermochemistry of these species. However, this choice of reference state is of both conceptual and numerical importance in the understanding of polycyclic hydrocarbons. We additionally think this choice of reference state is an essential component of a generalized understanding of heterocyclic and/or unsaturated species. Torsional or rotational barriers remain of considerable theoretical and experimental interest and conceptual conflict.[76a,b] It may be argued that the general preference for *trans* or staggered geometry is a special case of nonbonded repulsion[77] as in the valence shell electron pair repulsion theory.[78-80] However, in general, nonbonded repulsions and rotational barriers are considered separately when quantification is desired.[45-58] Suffice it to say, one can "synthesize" a mathematical function that gives the molecular energy in terms of the appropriate interatomic distances and/or angles by means of high-accuracy vibrational spectroscopy and quantum-chemical calcula-

---

[73] G. J. Janz, (Ref. 12), Ch. 3.

[74] S. Fitzwater and L. S. Bartell, *J. Am. Chem. Soc.* **98**, 8338 (1976).

[75] This can be seen to generate the $R \ln n$ associated with the entropy of an $n$-fold rotational barrier.

[76a] J. P. Lowe, *Prog. Phys. Org. Chem.* **6**, 1 (1968).

[76b] J. P. Lowe, *Science* **179**, 527 (1973); *J. Am. Chem. Soc.* **96**, 3759 (1974).

[77] L. S. Bartell, *J. Chem. Educ.* **45**, 754 (1968).

[78] R. J. Gillespie, *J. Chem. Educ.* **40**, 295 (1963).

[79] L. S. Bartell and V. Plato, *J. Am. Chem. Soc.* **95**, 3097 (1973).

[80] L. C. Allen, *Theor. Chim. Acta* **24**, 117 (1972).

tions.[45-58] (We note that suggestions of nonbonded attractions have been made,[81-85] but such considerations have not been incorporated into either the general mathematical or the conceptual framework of strain.)

The second contribution to strain is bond-angle distortion, a contribution long known as Baeyer strain. It seems reasonable to assert that there are "natural" bond angles such as tetrahedral (109.5°) for tetracoordinate carbon ($T_d$), trigonal planar (120°) for tricoordinate carbon ($D_{3h}$), and linear (180°) for dicoordinate carbon ($D_{\infty h}$). Indeed, it may be suggested that these angles minimize nonbonded repulsions and so represent idealized and reference geometries. However, when looking for appropriate compounds to consider, we encounter conceptual problems. Few tetracoordinate carbon compounds have precisely tetrahedral geometry—do we consider propane to be strained because of its C—C—C angle of 112°?[86] Few tricoordinate carbon compounds have precisely trigonal planar geometry—do we consider propane to be strained because of its C—C—C angle of 124°?[87] Most dicoordinate carbon containing compounds have a linear local geometry around the dicoordinate carbons—however, the reader will note few such species in our text. Quantitative as opposed to qualitative understanding seemingly is easier to acquire. For saturated alicyclic hydrocarbons, two routes have been taken, both of which start from strict regular tetrahedral geometry. One may "synthesize" a mathematical function that depends on the differences in the angles from the idealized.[45-58] Thus, for hydrocarbons this function can be written in terms of C—C—C, C—C—H, and H—C—H angle distortions.[88a,b] Alternatively, one need "merely" distort methane into the desired geometry and splice these pieces together to form the desired polycycle and then assume the strain is additive. Encouragingly, and somewhat surprisingly, this latter approach seems highly successful. For unsaturated species, both approaches are possible in principle, but we defer discussion until later in this section.

The third contribution to strain is linear bond stretching or compression. Chemical bonds often are viewed as springs; thus, there is a "natural" bond

[81] S. Wolfe, *Acc. Chem. Res.* **5**, 102 (1972).

[82] R. Hoffmann, C. C. Levin, and R. A. Moss, *J. Am. Chem. Soc.* **95**, 629 (1973).

[83a] N. D. Epiotis, *J. Am. Chem. Soc.* **95**, 3087 (1973).

[83b] N. D. Epiotis, D. Bjorkquist, L. Bjorkquist, and S. Sarkanen, *J. Am. Chem. Soc.* **95**, 7558 (1973).

[84] A. Liberles, A. Greenberg, and J. E. Eilers, *J. Chem. Educ.* **50**, 676 (1973).

[85] P. Kollman, *J. Am. Chem. Soc.* **96**, 4363 (1974).

[86] D. R. Lide, Jr., *J. Chem. Phys.* **33**, 4363 (1960).

[87] D. R. Lide, Jr., and D. Christensen, *J. Chem. Phys.* **35**, 1374 (1961).

[88a] K. B. Wiberg and G. B. Ellison, *Tetrahedron* **30**, 1573 (1974).

[88b] K. B. Wiberg, G. B. Ellison, and J. J. Wendoloski, *J. Am. Chem. Soc.* **98**, 1212 (1976).

length and force constant.[89,90a,b] (Formal, indeed isomorphic, comparison may be made with "simple harmonic oscillators" as shown in Section 1.D. However, we sense that the origin and appeal of the model arise from chemists viewing molecules as balls held together by springs.) Here again, one can "synthesize" a mathematical formula describing how the energy of the bond changes with[45-58] bond length.

Change in bond lengths may be referred to as "hidden strain," since bond-length variation is much smaller than bond-angle variation, and since customary drawings of molecular structure do not explicitly show changes in bond length from the normal, in contrast to changes in angle. For example, consider cyclopropane. While it is obvious that there must be angle compression from the normal 109.5° (or is it 112°?) to 60°, the conventional drawings of this substance as an equilateral triangle of $CH_2$ groups fail to give any information about the length of the C—C bonds. Likewise, in an exquisite study[91] of the structure of hexaaryl ethanes and numerous other species with long C—C bonds [$r_e(C—C) > 1.6$ Å], none of the authors' drawings betrays the anomalous features of these species. Two additional comments deserve note with regard to bond-length variation. First, the strain associated with distorting the angles of a C—C—C, C—C—H, or H—C—H "fragment" is usually less than that associated with bond stretching. In general, it is found that "merely" stretching bonds hardly lessens most other molecular destabilizations such as torsional or nonbonded repulsions. Second, quantum-mechanical analysis of "simple harmonic oscillators"[90a,b] shows that the average bond distance is independent of the degree of excitation. More precisely, if the interatomic potential were exactly that of an ideal spring, then the interatomic distance would be independent of the amount of external energy put into the bond.

It is important to realize that all these strain energy components—torsional of or rotational barriers and other consequences of nonbonded repulsions, deviance from standard bond angles, and bond lengths—are inseparable and interdependent. That is, they cannot generally be varied individually. For example, shutting the C—C—C angle in propane mandates an opening of either the C—C—H and/or the H—C—H angles and a lessening of the distance between the $CH_3$ groups. Likewise, increasing the C—C bond in ethane but leaving all other bonds unchanged decreases the rotational barrier.

[89] J. Goodisman, "Diatomic Interaction Potential Theory," 2 vols. Academic Press, New York, 1973.

[90a] I. N. Levine, "Quantum Chemistry," 2nd Ed., Ch. 4. Allyn & Bacon, Boston, Massachusetts, 1974.

[90b] E. Merzbacher, "Quantum Mechanics," 2nd Ed., Ch. 5. Wiley, New York, 1970.

[91] W. D. Hounshell, D. A. Dougherty, J. P. Hummel, and K. Mislow, *J. Am. Chem. Soc.* **99**, 1916 (1977).

[Extended Hückel calculations on C—C bond-distorted ethanes where all angles were taken as tetrahedral and $R(C—H) = 1.1$ Å gave the following threefold rotational barriers: $R_{C-C} = 1.3$ Å, $V_3 = 7.751$ kcal/mole; 1.5, 3.553; 1.7, 1.615; 1.9, 0.724; 2.1, 0.319; 2.3; 0.128; 2.5, 0.058; 2.7, 0.024 and $(\partial V_3/\partial R_{C-C})_{eq} = -12.5$ kcal/mole.[92]]

There is a highly complicated mathematical procedure called "molecular mechanics" or "empirical force-field calculations"[45-58] that minimizes the total energy with respect to all the coordinates. Not unexpectedly, considerable effort must be expended to find the absolute minimum energy, since there are $3N - 6$ coordinates to vary in a molecule containing $N$ atoms. (The $3N - 6$ arises from three coordinates for each atom from which three are taken for rigid translation and three for rigid rotation of the molecule.) Numerous "real" but relative minima exist. For example, consider the empirical formula of $C_4H_{10}$. Some of the minima are the *trans* and gauche rotamers of *n*-butane, the corresponding rotamers of isobutane, and the "fragments" such as propylene + methane, cyclopropane + methane, ethylene + staggered ethane, acetylene + two methanes, and even cyclobutane + hydrogen. In addition, there are numerous "false" minima associated with artificially high symmetry situations.[93]

To the best of the authors' knowledge, there is no way of predicting the number of minima and so there exists the possibility that the "true" or absolute minimum will be missed. It is desirable to insert into any calculation a "reasonable" guess of the molecular geometry for the species of interest. To illustrate the necessity for this, we merely remind the reader that the number of conventional isomers, disregarding rotamers or "fragments," of eicosane, is 366,319[94]; at ten seconds per species, the computation would exceed one month were all possibilities explored. The general agreement of molecular mechanics with experiment and the rule of thumb for the cost being "merely" quadratic in the number of atoms (per isomer) suggests that it is, in fact, rare that the true minimum will be overlooked. This, of course, postulates that the molecule will not assume some "ridiculous" structure and that the "force field" has been appropriately parameterized. Indeed, the result will generally be dependent on the parameters of the "force field" and on the "molecular mechanics" computer package.

Two distressing examples may be noted. The first deals with a highly symmetric and highly hindered molecule, tri(*tert*-butyl)methane, $[(CH_3)_3C]_3CH$, or 2,2,4,4-tetramethyl-3-(2-methyl-2-propyl)pentane.[95-98] Nonbonded repul-

[92] Prof. John P. Lowe, personal communication to Joel F. Liebman.
[93] O. Ermer, *Tetrahedron* **31**, 1849 (1975).
[94] This surprising fact is given in numerous organic chemistry textbooks.
[95] H. Lee and M. Stiles, cited as footnote 3 in Ref. 96a.
[96a] H. B. Bürgi and L. S. Bartell, *J. Am. Chem. Soc.* **94**, 5236 (1972).

sions may be expected to be nearly intolerable for the central carbon prior to distortion from the idealized geometry. Repulsions may be relieved by compressing the H—C—C(t-Bu) angle and by increasing the C—C(t-Bu) bond length. Both are realized in the calculations and in experiment. The angle is found experimentally to be reduced to 101.6° and the bond length increased to 1.611 Å.[96a,b] The former quantity is amply precedented—we recall the 60° angle in cyclopropane. The latter quantity is rather remarkable, and is indeed the longest C—C bond found in any isolable acyclic hydrocarbon.[91] (The conventional bond-stretch parameters were not sufficient for predicting this quantity.) Engler, Andose, and Schleyer (EAS)[52] computed the strain energy of this hydrocarbon using the "force field" of Allinger et al.,[51] of Boyd et al.,[50] and of their own. The results were 31.48, 40.40, and 49.61 kcal/mole, respectively. Significantly different partitioning of the strain into the earlier-mentioned components was also found. However, owing to different group increment numbers, the calculated heats of formation are in considerably better, if not remarkable, agreement: $-57.07$, $-55.80$, and $-53.08$ kcal/mole. "It would *not* be expected that the different blends of strain components and group contributions would always balance out when tested over a wide range of molecules."[52] This "rationalizes" the second example, dodecahedrane (see Section 4.D for a drawing of its structure and proposed syntheses). Suffice it to say, it has 12 five-membered rings, for which the angle strain and bond stretching should be minimal. The major strain would be expected to be torsional, since all of the hydrogens are eclipsed. For this species, the differences in both the heat of formation and the strain energy as calculated using the EAS vs. Allinger force fields exceed 45 kcal/mole! We eagerly await synthesis of this symmetric and aesthetically interesting molecule, and inquire whether experimental determination of its stability will cause at least one "force field" to go up in smoke. (The major differences between the force fields appear to be in the "hardness" of $C \cdots C$ and $H \cdots H$ interactions,[56,97a] and a recent "conciliation" has apparently been suggested.[97b,c]

We now briefly mention two more types of strain in molecules. The first arises from rotation or twisting of double bonds and will be discussed further

[96b] L. S. Bartell and H. B. Bürgi, *J. Am. Chem. Soc.* **94**, 5239 (1972).

[97a] S. Fitzwater and L. S. Bartell, *J. Am. Chem. Soc.* **98**, 5107 (1976).

[97b] L. S. Bartell, *J. Am. Chem. Soc.* **99**, 3279 (1977).

[97c] N. L. Allinger, D. Hindman, and H. Hönig, *J. Am. Chem. Soc.* **99**, 3282 (1977).

[98] For a discussion of "mono and di-*tert*-butylmethane," i.e., neopentane and 2,2,4,4-tetramethylpentane, see L. S. Bartell and W. F. Bradford, *J. Mol. Struct.* **37**, 113 (1977). Using "semioptimized" geometries, it was found that the strain energy of tetra-*tert*-butylmethane is comparable to the dissociation energy of a C—C bond (Prof. Lawrence S. Bartell, unpublished results).

in Section 3.E. This strain is much more fundamentally quantum mechanical in that we recognize that the hydrogens in ethylene are eclipsed and would "want" to be staggered. Very recently, Kao and Allinger[99] successfully fused quantum chemistry for the $\pi$ electrons and molecular mechanics for the $\sigma$ electrons. While there are competing calculational approaches for both sets of electrons (and the theoretical purist would argue that they should not be treated separately in any case), we feel this will herald a new approach to molecular structure. (Evidence for the nonseparability of $\sigma$ and $\pi$ electrons comes from many sources, and we may merely cite discussions of "through space" and "through bond" interactions.[100]) Recently, Mislow and his co-workers[101] have used molecular mechanics for "hybrid" systems such as lepidopterene (see Section 4.C), in which there is a C—C bond with an interatomic distance of 1.64 Å.[102] (The longest C—C bonds we know of are found in the crystallographically inequivalent molecules of the "[4.4.1] propellatetraene," 3, in

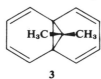

3

which the "conjoining" central bond distances are 1.780(7) and 1.836(7) Å.[103] See Section 6.B for more discussion of propellanes.) Returning to the "hybrid" systems, some of the surprisingly long bonds are systematically underestimated by molecular mechanics. Rather than reparametrize the C—C interaction term, Mislow et al.[101] have chosen to explain the seeming anomalies or inconsistencies in terms of "through bond" and "through space" effects.[100] One may conclude that, even if incompletely correct, molecular mechanics retains its utility as a major tool for the understanding of chemical phenomena.

   The second type of strain remaining to be discussed may be referred to as "electrostatic strain"[104] and is found in such species as 1,4-bicyclo[2.2.2]-octane dication[104] and cyclooctatetraene dianion.[105,106] (The former species

[99] J. Kao and N. L. Allinger, *J. Am. Chem. Soc.* **99**, 975 (1977).

[100] R. Hoffmann, *Acc. Chem. Res.* **4**, 1 (1972).

[101] D. A. Dougherty, W. D. Hounshell, H. B. Schlegel, R. A. Bell, and K. Mislow, *Tetrahedron Lett.* p. 3479 (1976).

[102] J. Gaultier, C. Hauw, and H. Bouas-Laurent, *Acta Crystallogr.*, *Sect. B* **32**, 1220 (1976).

[103] R. Bianchi, G. Morosi, A. Mugnoli, and M. Simonetta, *Acta Crystallogr.*, *Sect. B* **29**, 1196 (1973).

[104] G. A. Olah, G. Liang, P. von R. Schleyer, E. M. Engler, M. J. S. Dewar, and R. C. Bingham, *J. Am. Chem. Soc.* **95**, 6829 (1973).

[105] T. J. Katz, *J. Am. Chem. Soc.* **82**, 3784 (1960).

[106] G. R. Stevenson, I. Ocasio, and A. Bonilla, *J. Am. Chem. Soc.* **98**, 5469 (1976).

is discussed further in Section 3.B.) Inasmuch as the first types of strain have their analogues in classical mechanics, this type relates to classical electrostatics. The two like charges repel one another, and this repulsion must be balanced by the solvation and any other stabilization effects, such as aromaticity. The reader is referred to Stevenson et al.[106] for an exquisite discussion of the roles of aromaticity, solvation, and this "electrostatic strain" in determining the relative stability of cyclooctatetraene, its dianion, and the intermediate radical anion. These authors additionally discuss the strain increase that accompanies the opening of the already wide C—C—C angles of 126°[107] in cyclooctatetraene to 135° in the planar anion radical and dianion and that accompanying the compression of and extension of four C—C and C=C bonds, respectively.

We defer discussion of the quantitative aspects of this problem, not because of lack of time, space, or interest, but rather due to a sense of futility in that these results cannot be extrapolated directly to any other species for which electrostatic strain may be invoked. We note a literature model for the heat of formation of the isomeric alkanes[108a] and alkyl radicals[108b] that is derived solely from electrostatic factors, and so one could argue that more general strain considerations "reduce" to electrostatics. In the form developed by Benson and Luria, little extension[109,110a,b] seemingly has been made to the more general species that populate our text. As provable from Hellman–Feynman considerations, classical electrostatics are not violated by quantum mechanics and orbital reasoning.[111,112] However, we feel that to return to the all-electron quantum-chemical description of molecules in the name of electrical attraction and repulsion would violate the spirit of the organic chemistry lore.

The strain energy is the sum of the preceding five destabilizing terms: nonbonded repulsions, bond-angle distortions, bond stretches or compressions, rotation around or twisting of double bonds, and electrostatic strain. We remind the reader that while molecules may also have stabilizing energy contributions, these opposing effects are experimentally inseparable. Consider naphthalene, an archetypally stabilized molecule. Yet, like the earlier-discussed cyclooctatetraene dianion, the double bonds are stretched and the

[107] J. Bordner, R. G. Parker, and R. H. Stanford, Jr., Acta Crystallogr., Sect. B 28, 1069 (1972).

[108a] S. W. Benson and M. Luria, J. Am. Chem. Soc. 97, 704 (1975).

[108b] M. Luria and S. W. Benson, J. Am. Chem. Soc. 97, 3342 (1975).

[109] S. W. Benson and M. Luria, J. Am. Chem. Soc. 97, 3337 (1975).

[110a] Prof. Sidney W. Benson, personal communication to Joel F. Liebman.

[110b] S. W. Benson, Angew. Chem. Int. Ed. Engl. (in press).

[111] J. Goodisman (Ref. 89), Vol. 1, Sect. III.D.

[112] I. N. Levine (Ref. 90a), pp. 374–380.

single bonds are compressed. With near-equilateral, equiangular hexagonal geometry in each ring, it is hard to imagine that there is angle strain in this species. Yet, recalling that the "natural" C—C=C angle is 124°, we conclude that there is indeed angle distortion here. The latter strain components are not to be found in naphthalene. However, consider 1,4,5,8-tetrachloronaphthalene, seemingly as aromatic as the parent hydrocarbon. The four chlorines electrostatically repel, as do the four carbons to which they are bound. (The most obvious repulsions are between the "peri," or 1 and 8, and 4 and 5 sets of positions.) To minimize these latter repulsions, as well as those arising from merely steric or nonbonded defects, the molecule distorts to a "propeller"-like shape with the 1- and 5-chlorines above the average plane of the naphthalene ring system, while the 4- and 8-chlorines lie below[113,114] (see Section 3.H). Equivalently, there has been some rotation around the formal double bonds of the hydrocarbon.

As organic chemists, we may content ourselves with knowing merely how much a given molecule is stabilized or destabilized without concern for the individual contributions. After all, we noted in our discussion of tri(*tert*-butyl)methane that the division may depend on the precise mode of calculation or the computer program package. Nonetheless, we still are disturbed by such statements as, "the strain energy of a sample of polyvinylcyclopentane with a molecular weight of 10,000 is 677 kcal/mole" and, "the resonance energy (read *extra stabilization energy*) of a sample of polystyrene with a molecular weight of 10,000 is 3456 kcal/mole." The prefix "poly" tips us off to consider subunits, and so we would want to rephrase the above quantities in terms of strain energy or resonance energy per monomer, i.e., cyclopentane or benzene. (We remind the reader that whatever extra resonance energy styrene has over benzene, that quantity is lost on saturation of the vinyl group, either by hydrogenation or polymerization.) Indeed, the figures of 677 and 3456 were in fact computed by determining the number of "monomers" and multiplying those numbers by the respective strain energy (6.5 kcal/mole) or resonance energy (36 kcal/mole).

Even with resolution of this monomer vs. polymer dichotomy, there are still problems with this analysis. For example, do we really want to say that cyclopropane and cyclobutane are equally strained? (See Section 3.A for the appropriate strain energies but also Section 2.C for a comparison of these two species.) Is it fair to say that cubane is six times more strained than cyclobutane because the strain energy of the former is six times larger than the latter? In addition to the total strain energy (SE), customary usage (including our text) refers to strain energy per (carbon) atom ($SE_c$) and per (C—C) bond

[113] M. A. Davydova and Y. T. Struchov, *J. Struct. Chem.* (*USSR*) 2, 63 (1961); 3, 170, 202 (1962).

[114] G. Gafner and R. F. H. Herbstein, *Acta Crystallogr.* 15, 1081 (1962).

(SE$_b$). These latter two notions are highly useful in that they simultaneously solve the problem of the relative strain energies of cyclopropane and cyclobutane and return the strain energy of polymeric cyclopentane to a much more reasonable value. More precisely, either criterion shows the "normalized" strain energy of cyclopropane to exceed that of cyclobutane, and the latter to exceed the ca. 1 kcal/mole value for cyclopentane. We highly recommend the use of SE$_b$ and SE$_c$, rather than SE itself, as diagnostics when one is trying to correlate inherent thermodynamic stability with relative chemical reactivity. Fundamental questions arise even with the use of SE$_b$ and SE$_c$, but since these are more mathematical and "metachemical," we defer discussion of them until Section 4.F. Indeed, perhaps it is because quantitative data on strain energies are lacking for so many of the compounds we describe that we are willing to forego these subtleties for now.

## 1.C  Thermodynamics and Symmetry

In this book, we shall usually speak of energies, but shall be using $\Delta H$ or enthalpies. Although the former is more intuitively convenient and understandable, the latter is more experimentally accessible and thermodynamically meaningful. Interrelated by $P \Delta V$ terms, the errors are usually small under the conditions of interest: 25°C (298°K) and 1 atm pressure. Indeed, for "normal substances," the error in equating enthalpies and energies of vaporization by assuming $P \Delta V$ is zero is only 10%.[115] Additionally, we will rarely claim accuracy in the cited heats of formation better than approximately 1 kcal/mole, since both the data and our concepts are usually too imprecise and ill defined to do better. Moreover, it is not even thermodynamically valid to consider differences in enthalpies. The reader should realize that it is the Gibbs free energy $(\Delta G)$[116] that truly determines relative stability [Eq. (14)].

$$\Delta G = \Delta H - T \Delta S = -RT \ln K_p \qquad (T \text{ in } °K) \qquad (14)$$

From this equation, it is apparent that the free energy is composed of an (essentially temperature-independent) enthalpy or energy term as well as an (temperature-dependent) entropy term. In most cases near ambient temperature, the entropy term may be neglected as being comparatively small.

We may further strengthen our resolve to neglect the entropy by "polysecting" it into translational, rotational, vibrational, electronic, and symmetry-number components.[117] The translational component is generally the

---

[115] Prof. Fred Gornick, personal communication to Joel F. Liebman.

[116] This quantity is also written $\Delta F$, and in no case should be confused with the Helmholtz free energy or $\Delta A$.

[117] It will be noted that most compendia of thermochemical values automatically neglect the entropy term by giving only $\Delta H$. By contrast, thermochemistry texts normally discuss $\Delta G$, $\Delta H$, $\Delta S$ and the "polysection" of the latter.

largest by far, but as it depends solely on the mass of the molecule (logarithmically), its effect on isomer stability is zero. Indeed, even for hydrogenation reactions (e.g., for cyclopropene, cyclopropane, and propane), the change is negligible. (The ratio of the translational entropies here is $1:1.013:1.026$.) The rotational component is generally next in size but, provided that the general shape of the molecule is not *too* different for a pair of isomers or hydrogenation "relatives," again the change is rather small. (The geometry change of cycloheptane to heptane is, for example, quite large. Thus, our assumption that the change in rotational entropy is negligible appears suspect for this pair of molecules.) The vibrational component is generally still smaller. As such, if the number of C—H and C—C bonds remains more or less unchanged, again there is still change in entropy. Of course, this is an oversimplification in that this entropy component involves the precise stretching frequencies, $\nu_i$, of all the bonds or, more precisely, the normal modes.[118,119] It is well known that there is some variation[118] in these quantities—indeed, vibrational spectroscopy would largely lose its subject matter if this were not true. However, a greater error is perpetrated by neglecting zero-point energies (ca. $\frac{1}{2}\sum_i h\nu_i$, see Section 1.D), as is customarily done, than by neglecting the vibrational entropy components. As will be shown in Section 1.D, the zero-point energy term is also largely cancelled between isomeric molecules. The electronic component of the entropy may also be safely neglected unless there are low-lying excited states and/or unpaired electrons. For most of the species we are studying, neither condition is met. For those species that meet either or both conditions, it is probable that the intrinsic energies are not known reliably enough to make neglect of the electronic entropy a serious problem.

For example,[120] consider the gas-phase entropy for cycloheptane and its isomers methylcyclohexane, ethylcyclopentane, and *cis*-1,2-dimethylcyclopentane. The values given by Stull *et al.*[17] are 81.8, 82.1, 90.4, and 87.5 eu, respectively. The corresponding values for all of the related olefins are not given in this source. However, related to the last hydrocarbon are the isomeric 1,2- and 1,5-dimethylcyclopentenes. The entropies for these species are 84.0 and 86.4 eu. For comparison, the entropy of the totally saturated, hydrogenated, and extended heptane is 102.3 eu. It is thus seen that entropies remain remarkably constant.

However, symmetry-number corrections cannot be neglected[121] when the

---

[118] E. B. Wilson, Jr., J. Decius, and P. Cross, "Molecular Vibrations." McGraw-Hill, New York, 1955.

[119] J. W. Laing and R. S. Berry, *J. Am. Chem. Soc.* **98**, 660 (1976).

[120] All data for entropy comparisons in this book are from ref. 17.

[121] P. von R. Schleyer, K. R. Blanchard, and C. D. Woody, *J. Am. Chem. Soc.* **85**, 1358 (1963).

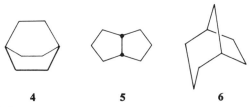

**4**        **5**        **6**

magnitudes of the enthalpy and entropy among a set of compounds are nearly comparable. Consider the set of isomeric bicyclooctanes, **4, 5,** and **6.** (See Section 3.B for a discussion of bicycloalkanes.) Experimental equilibration studies show that **5** has the greatest strain energy, while **4** and **6** have comparable strain energies. However, at 298°K, compound **5** gains stability (read $\Delta G$) because of the molecular flexibility associated with both the *cis* ring junction and two five-membered rings. Indeed, above 378°K, isomer **5** is the most stable of the three isomers. Correspondingly, isomer **6** is more stable than **4** because of the high symmetry of the latter. More precisely, **6** has but one plane of symmetry, while compound **4** has both threefold and mirror symmetry, i.e., $D_{3h}$. (This is a bit simplistic, as shown by x-ray[122] and electron-diffraction[123] and "molecular mechanics"[122–124] studies on **4,** but in fact our argument will be shown to be not impaired.) The lowering of the entropy of **4** relative to **6** is precisely quantifiable in terms of their respective symmetry numbers: the number of indistinguishable positions into which the molecule can be turned by simple rigid rotations or internal rotations. Species **4** has a symmetry number $\sigma$ of 6, while **6** has $\sigma = 1$. The entropy difference is precisely $R \ln(6) - R \ln(1)$, or 3.6 eu(gibbs), or approximately 1.1 kcal/mole at 298°K. (The sole origin of "approximately" is in the errors in the numerical value of the constant $R$.) The above-cited, more precise studies on **4** show this species to have "merely" $D_3$ symmetry. In order to lessen the repulsion due to the eclipsing hydrogens on the $CH_2CH_2$ groups (see Section 1.B), there is a slight rotation around the threefold axis of one half of the molecule. (Structures **7a** and **7b** represent top views of the $D_{3h}$ and $D_3$ form of species **4.**) However, as $\sigma$ remains equal to six, there is no effect on the entropy.

The second example involves the equilibration of the isomeric diamantanols **8** and **9.**[125] (See Section 4.A for a discussion of the parent hydrocarbon.)

**7a**          **7b**

[122] O. Ermer and J. D. Dunitz, *Helv. Chim. Acta* **52,** 1861 (1969).

[123] A. Yokozeki, K. Kuchitsu, and Y. Morino, *Bull. Chem. Soc. Jpn.* **43,** 2017 (1970).

[124] E. M. Engler, L. Chang, and P. von R. Schleyer, *Tetrahedron Lett.* p. 2525 (1972).

[125] D. E. Johnston, M. A. McKervey, and J. J. Rooney, *J. Chem. Soc., Chem. Commun.* p. 29 (1972).

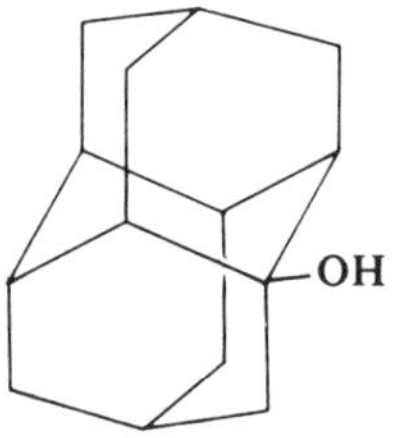

**8** 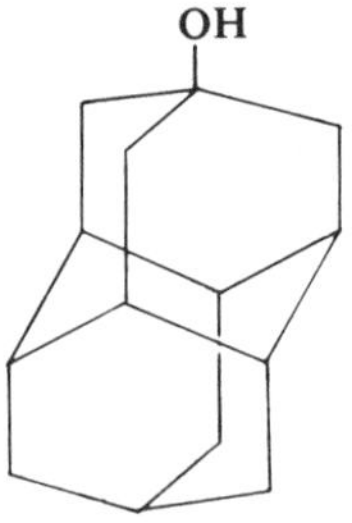                    **9**                     **10**

Strictly speaking, both species have only one plane of symmetry, and so $\sigma = 1$. However, to the extent that the C—O—H groups resemble those of simpler alcohols, we may neglect the asymmetry induced by these bent groups. (More precisely, we argue that the rotational barrier[76] is so low that the C—O—H group is effectively linear at ambient temperatures.) Even given this neglect, species **8** remains of $C_s$ symmetry ($\sigma = 1$) while **9** now has $C_{3v}$ symmetry ($\sigma = 3$). Drawing **10** is an alternate picture of diamantanol, in which both the darkened and dotted lines are C—$CH_2$—C "units." The OH in **8** is precisely at the threefold symmetry axis, where we have now "prepared" species **10**. Automatically, **9** has a lower entropy than **8** by $R \ln(3)$, or 2.2 eu. It is seen that symmetry numbers are temperature dependent essentially because the symmetry of a molecule is temperature dependent.

A particularly subtle example of this latter dependence is based on the relationship of nuclear spins and total molecular symmetry. Normally we assume that nuclear spins do not affect chemical behavior and, excepting magnetic resonance experiments, are essentially irrelevant for organic chemistry. It has been found that the precise nuclear spin of a conformationally fluxional molecule will determine, or at least severely limit, the allowed rearrangement modes.[126a–c] It thus appears that explicit consideration of the energy and entropy associated with nuclear spins should be included. Perhaps only because most molecules of interest are not fluxional have we been safe in ignoring this phenomenon.

### 1.D   Potential Energy Curves and Zero-Point Energy

In Section 1.B we discussed quantum-chemical and molecular-mechanical methods of computing the geometry of molecules. We additionally discussed the idea of potential energy curves and their natural generalization to poten-

[126a] C. Trindle and T. D. Bouman, *Int. J. Quantum Chem.* **7**, 329 (1973).

[126b] C. Trindle and T. D. Bouman, *Jerusalem Symp. Chem. Biochem. Reactiv. 6th,* pp. 51–61 (1974).

[126c] T. D. Bouman and C. Trindle, *Theor. Chim. Acta* **37**, 217 (1975).

tial energy surfaces and hypersurfaces.[127-133] These are characterized by numerous potential energy minima, which, as chemists, we may identify as the various isomers corresponding to a given atomic composition. Indeed, by recognizing chemical reactions as a specific type of isomerization (since starting and final products do have the same number and type of atoms), we may visualize all chemical interconversions as a ride on a multidimensional roller coaster. Indeed, we may argue that the rate of a reaction is related to the heights of the hills surrounding the starting point and that the energy of a reaction is related to the relative depths of the wells, i.e., to the relative maxima and minima. (We remind the reader of the difference between kinetic and thermodynamic stability, extensively exemplified in Chapter 5.) Quantification of the former relation is beyond our capabilities for all but the simplest process unless one chooses to consider only certain interconversions or mountain trails. (See the complexities arising in Section 2.B in discussing the isomerization processes of cyclopropane.) Quantification of the latter shows the assertion is false. Suppose we had a molecule precisely located at the bottom of a given well, or equivalently, described by a perfectly well-defined set of coordinates. Then the uncertainty in its position would be zero. However, as all intramolecular vibrations are accordingly frozen, the uncertainty in its velocity (or momentum) is thus also zero. This violates Heisenberg's uncertainty principle[134a,b]—that the product of the uncertainty of the position and momentum exceeds a "universal," nonzero constant [Planck's constant—[$h$, $h/2$, $\hbar$, ($h/2\pi$), or $\hbar/2$], whose precise value depends upon the reference one goes to and the accompanying derivation of this principle. Now we definitely want the uncertainty in position to be "noninfinite"—the whole idea of molecular structure depends upon knowing the atomic positions. Accordingly, the uncertainty in the momentum must be nonzero. From classical mechanics, then, we may derive that the kinetic energy is nonzero and, more importantly, so is the total energy relative to the well bottom. This minimal total energy is called the zero-point energy (ZPE or $\Delta E_z$).[135-138] That the zero-point energy is nonzero is the unique and important consequence of the above analysis.

[127] The following six references are articles on potential energy hypersurfaces we have found of particular use.

[128] J. C. Polanyi, *Acc. Chem. Res.* **5**, 161 (1972).

[129] J. J. Dannenberg, *Angew. Chem., Int. Ed. Engl.* **15**, 519 (1976).

[130] R. Daudel, *Bull. Soc. Chim. Belg.* **85**, 913 (1976).

[131] M. Simonetta and A. Gavezzoti, *Struct. Bonding* (*Berlin*) **27**, pp. 1–43 (1976).

[132] X. Chapuisat and Y. Jean, *Top. Curr. Chem.* **68**, 1 (1976).

[133] W. Gerhartz, R. D. Poshusta, and J. Michl, *J. Am. Chem. Soc.* **98**, 6427 (1976).

[134a] I. N. Levine (Ref. 90a), pp. 3–5, 157, 369–371.

[134b] E. Merzbacher (Ref. 90b), pp. 158–161.

[135] I. N. Levine, "Molecular Spectroscopy," pp. 250–251. Wiley, New York, 1975.

[136] T. L. Cottrell, *J. Chem. Soc.* p. 1448 (1948).

[137] K. S. Pitzer and E. Catalano, *J. Am. Chem. Soc.* **78**, 4844 (1956).

[138] T. L. Allen, *J. Chem. Phys.* **31**, 1037 (1959).

For conceptual and pedagogical simplicity in discussing zero-point energies, let us consider a one-dimensional problem with one minimum. The potential energy is then $V(q)$, where $q$ is some coordinate, say $r - r_e$ or $\theta - \theta_e$, where the subscript e reminds us that this is the equilibrium value. $V(q)$ can be expressed in numerous ways,[139a,b] the simplest being a Taylor series in the variable $q$ [see Eq. (15)]. The first term, $V(0)$, may safely be ignored, since it is

$$V(q) = V(0) + \frac{dV}{dq}\bigg|_{q=0} q + \frac{1}{2}\frac{d^2V}{dq^2}\bigg|_{q=0} q^2 \cdots$$

$$= \sum_{k=0}^{\infty} \frac{1}{q!}\frac{d^kV}{dq^k}\bigg|_{q=0} q^k \tag{15}$$

merely a constant. The second term, $\frac{1}{2}dV^2/dq|_{eq=0}q^2 = \frac{1}{2}kq^2$ is the first non-vanishing, nonconstant term. Let us stop here, making the tacit assumption that the third term is the dominant term in our infinite sum. This is usually true, although for the potential energy curves corresponding to the puckering of cyclobutane[140] and for hydrogen bonding[141a,b,142] this assertion is belied. Accordingly, if we neglect such cases, our potential energy curve is transformed into a parabola and renamed a "simple harmonic oscillator." (Inclusion of the higher powers of $q$ are called corrections for anharmonicity.[143,144])

For a simple harmonic oscillator, the zero-point energy is "easily" found[134a,b] to be numerically positive, i.e., destabilizing, and equal to $\frac{1}{2}hv$ where $v$ is $(k/\mu)^{1/2}$. We identify $k$ as the force constant and $\mu$ as the reduced mass. Both of these may be referred to as semiclassical (i.e., semiquantum) quantities, while $v$ is the stretching frequency deductible from spectroscopy ($v$ is often given in cm$^{-1}$ and 350 cm$^{-1} \simeq 1$ kcal/mole). In general, it is safe to assume that a molecule is composed of independent harmonic oscillators although it is unusual for each bond to correspond individually to a single oscillator. For a general, i.e., nonlinear, polyatomic molecule with $N$ atoms, there are $3N - 6$ independent vibrations (normal modes), and so it is experimentally difficult to determine all of the necessary $v$'s.

[139a] V. T. Varshni, *Rev. Mod. Phys.* **29**, 664 (1957).

[139b] D. Steele, E. R. Lippincott, and J. D. Vanderslice, *Rev. Mod. Phys.* **34**, 239 (1962).

[140] W. J. Lafferty, *in* "Critical Evaluation of Chemical and Physical Structural Information." Nat. Acad. Sci., Washington, D.C., 1974.

[141a] P. A. Kollman and L. C. Allen, *J. Am. Chem. Soc.* **92**, 6101 (1970).

[141b] P. A. Kollman and L. C. Allen, *Chem. Rev.* **72**, 283 (1972).

[142] P. Noble and R. Kortzeborn, *J. Chem. Phys.* **52**, 5375 (1970).

[143] J. Goodisman (Ref. 89), Ch. II.D.

[144] D. G. Truhlar, R. W. Olson, A. C. Jeannott, II, and J. Overend, *J. Am. Chem. Soc.* **98**, 2373 (1976).

In principle, one could determine the desired quantities via quantum-chemical calculations.[145-153] Such calculations have been done for relatively small and/or symmetric species. For example, benzene, but not the para-cyclophanes, and cyclopropane and propene, but not allylcyclopropane, have been investigated. Explicit variation of all the coordinates and/or direct calculation of all the desired matrix elements remains an arduous and expensive study. (We regrettably acknowledge that the price of an individual *ab initio* calculation goes as the 4.5th power of the number of electrons.)[154]

One might expect that when bond types do not change upon chemical reaction, the change in the zero-point energy will generally be small. This conclusion is supported by noting the near equality of the sum of the vibrational frequencies for $CH_2CCH_2$ (23320 cm$^{-1}$) and its isomer $CH_3CCH$ (23655 cm$^{-1}$), and for $(CH_2)_3$ (34700 cm$^{-1}$) and its isomer $CH_3CHCH_2$ (33768 cm$^{-1}$).[155] Converting to kcal/mole (a much more "comfortable" and customary unit of energy) and dividing by two, because $\Delta E_z = \frac{1}{2}h\nu$, the values are 33.3 and 33.8; 49.6 and 48.3 kcal/mole. It has been suggested[156] that the greater stability of "unhindered" branched hydrocarbons such as $C(CH_3)_4$ in comparison with more linear isomers such as $CH_3(CH_2)_3CH_3$ [$\Delta\ \Delta H_f(g) = 4.0$ kcal/mole] (the Rossini effect) is due, in large part, to the difference in the zero-point energy ($\Delta\ \Delta E_z = 1.7$ kcal/mole). It is impressive, however, how often the zero-point energies are so large yet cancel so nearly completely.

$$2C_2H_4 \rightleftharpoons C_2H_2 + C_2H_6$$
$$\Delta E_z\ 2(30.9) \rightleftharpoons 16.2 + 45.3\ \text{kcal/mole} \tag{16}$$

For example, consider the disproportionation reaction (16), where the zero-point energy on the left side totals 61.8, while on the right side it is 61.5

[145] The following ten references are articles on theoretically determined force constants and stretching frequencies.

[146] W. L. Bloemer and B. L. Bruner, *J. Mol. Spectrosc.* **43**, 452 (1972).

[147] K. Kozmutza and P. Pulay, *Theor. Chim. Acta* **37**, 67 (1975).

[148] P. Pulay and F. Török, *J. Mol. Struct.* **29**, 239 (1975).

[149] B. Nelander and G. Ribbegård, *J. Mol. Struct.* **20**, 325 (1970).

[150] J. S. Binkley, J. A. Pople, and W. J. Hehre, *Chem. Phys. Lett.* **36**, 1 (1975).

[151] L. S. Bartell, S. Fitzwater, and W. J. Hehre, *J. Chem. Phys.* **63**, 3042 (1975); **63**, 4750 (1975).

[152a] C. E. Blom, P. J. Slingerland, and C. Altona, *Mol. Phys.* **31** (1976).

[152b] C. E. Blom and C. Altona, *Mol. Phys.* **31**, 1377 (1976).

[153] M. J. S. Dewar and G. P. Ford, *J. Am. Chem. Soc.* **99**, 1685 (1977).

[154] This is a quantum-chemical folklore rule derived from the number of two-electron integrals ($N^4$ for $N$ electrons), with a little extra for "bookkeeping."

[155] All these zero-point energies were taken from the theoretical analysis of experimental data done in Ref. 153.

[156] F. D. Rossini, *Chem. Rev.* **27**, 1 (1940).

kcal/mole. (These values and those that follow are from Radom et al.[157]) To estimate zero-point energies, we find it very tempting to place all X—H, X—X $\sigma$, and X—X $\pi$ bonds into three distinct classes independent of the second-row element chosen for X. Fundamentally, this elemental independence is clearly in error; nonetheless, it is surprisingly accurate. For example, one may interpolate $\Delta E_z$ ($CH_3NH_2$) by averaging the values for the iso-electronic $CH_3CH_3$ (45.3 kcal/mole) and $CH_3OH$ (31.1 kcal/mole). Doing so predicts a "theoretical" value of 38.2 kcal/mole, while the experimental value is 39.2. Another example is reaction (17), where the left side totals 46.0 and

$$C_2H_2 + N_2H_4 \rightleftharpoons C_2H_6 + N_2$$
$$\Delta E_z \ 16.2 + 29.8 \rightleftharpoons 45.3 + 3.4 \text{ kcal/mole} \tag{17}$$

the right side 48.7 kcal/mole. We additionally note the near equality of the zero-point energy for the $XX'H_4$ species $C_2H_4$, $CH_3OH$, and $N_2H_4$ (30.9, 31.1, and 29.8 kcal/mole, respectively).[157] Indeed, the assertion that the zero-point energies never change in any balanced reactions seems to hold remarkably well, and is equivalent to the customary "benign neglect" given to zero-point energies in most discussions of chemical stability. For an exquisite exception, see Staley,[158] where vinylcyclopropanes and butadienes are discussed. As such, while we cannot say that no $\nu$'s is good news, the absence of experimental and theoretical values for all $3N - 6$ normal mode frequencies and zero-point energies in general need not prevent us from making comparisons, albeit somewhat imprecise, among molecules of interest. We now turn to cyclopropane, the simplest "unavoidably" strained hydrocarbon, for which a variety of models will be discussed.

[157] All of these zero-point energies were taken from the theoretical analysis of experimental data by L. Radom, W. J. Hehre, and J. A. Pople, J. Am. Chem. Soc. 93, 289 (1971).
[158] S. W. Staley, J. Am. Chem. Soc. 89, 1532 (1967).

## CHAPTER 2

# Cyclopropane and Cyclobutane

In this chapter we shall first consider cyclopropane, the conventionally simplest and unavoidably strained molecule. We use the term "conventionally," since we suspect few individuals like the names "cyclomethane," "cyclo-ethane," and "bicyclo[0.0.0]ethane" for methylene, ethylene, and acetylene. However, were the former descriptions used, the conclusion that these molecules are strained would be more plausible. The term "unavoidably" is used to mean there is no mode of stretching, angle bending, or rotation around a bond that removes the strain. In this regard, square planar methane and the eclipsed rotamer of ethane are not unavoidably strained.

## 2.A Conceptual Models for the Bonding in Cyclopropane

### 2.A.1 PROPANE VS. CYCLOPROPANE

Let us briefly compare propane and cyclopropane, the acyclic and cyclic saturated $C_3$ hydrocarbons, in order to understand the strain energy in the latter. The C—C—C angle in propane is 112°,[1] slightly larger than the idealized tetrahedral angle of 109.5°. By contrast, cyclopropane has a C—C—C angle of precisely 60°,[2,3] corresponding to the geometry of an equilateral triangle. It would seem self-evident that cyclopropane has the latter geometry until one remembers that ozone ($O_3$), formally isoelectronic

[1] D. R. Lide, Jr., *J. Chem. Phys.* **33**, 1514 (1960).
[2] W. J. Jones and B. P. Stoicheff, *Can. J. Phys.* **42**, 2259 (1960).
[3] O. Bastiansen, F. N. Fritsch, and K. Hedberg, *Acta Crystallogr.* **17**, 538 (1964).

to $(CH_2)_3$, has an O—O—O angle of $117°$.[4] We note that a cyclic form of $O_3$ has been discussed,[5-6b] and correspondingly, there has been considerable interest in the acyclic form of cyclopropane, trimethylene (see Section 2.B.2). Returning to the cyclic form, cyclopropane, it is found that the C—C bonds are shorter[1-3] and have higher vibrational stretching frequencies[7] than the corresponding quantities for propane. It is also noted[8] that the C—H bonds in cyclopropane are shorter, of higher vibrational stretching frequency or force constant, and of larger $^{13}$C—H nmr coupling constant than either the primary or secondary C—H bonds in propane. Can one say that the C—C and C—H bonds in cyclopropane are stronger than the corresponding bonds in propane? Before we give an answer, we must define "stronger." The term has two conventional, and usually parallel, meanings: higher stretching frequency or force constant, and higher dissociation or bond energy. If the former definition is accepted, we would be forced to say that cyclopropane has stronger bonds than propane. Accordingly, we would conclude that bond strength does not relate to strain energies or that cyclopropane is less strained than propane. The latter definition is harder to quantify: while spectroscopy is performed more easily than calorimetry, neither directly gives information about individual bonds in a molecule. Quantitatively, it may be shown that the dissociation energy of a C—H bond in cyclopropane is greater than the energy of the corresponding homolytic cleavage of either the primary or secondary C—H bond in propane.[9] Experimental measurements of the corresponding C—C homolyses (e.g., heating cyclopropane to form acyclic trimethylene) show that in cyclopropane the C—C bond is considerably weaker, i.e., of lower bond strength, than in propane. It should come as no surprise that it is the carbon framework of cyclopropane that is the major source of the strain in this species.[10a,b] After all, this is where the molecular geometry deviates so strongly from the normal and strainless structures.

Let us see how chemical bonding theory describes cyclopropane. While we

[4] R. H. Hughes, *J. Chem. Phys.* **24**, 131 (1956).

[5] W. A. Lathan, L. Radom, P. C. Hariharan, W. J. Hehre, and J. A. Pople, *Top. Curr. Chem.* **40**, 1 (1973).

[6a] R. J. Buenker and S. D. Peyerimhoff, *Theor. Chim. Acta* **24**, 132 (1972).

[6b] R. J. Buenker and S. D. Peyerimhoff, *Chem. Rev.* **74**, 127 (1974).

[7] T. Shimanouchi, Tables of molecular vibrational frequencies. *Natl. Stand. Ref. Data Ser., Natl. Bur. Stand.* **NSRDS-NBS 39** (1972).

[8] L. N. Ferguson, "Highlights of Alicyclic Chemistry," Part 1, Ch. 3. Franklin Publ. Co., Palisades, New Jersey, 1973.

[9] S. W. Benson, "Thermochemical Kinetics," 2nd Ed., pp. 295–299 (Tables A-11, A-12). Wiley, New York, 1976.

[10a] K. B. Wiberg and G. B. Ellison, *Tetrahedron* **30**, 1573 (1974).

[10b] K. B. Wiberg, G. B. Ellison, and J. J. Wendoloski, *J. Am. Chem. Soc.* **98**, 1212 (1976).

note that no less than 18 *ab initio* calculations on this species have been chronicled,[6a,b] the computational complexities of most of these studies are sufficient to dissuade us from asking for quantitative results. Instead, we prefer qualitative findings that are hopefully transferable to other strained species. The following five subsections describe some of these findings.

### 2.A.2  d ORBITALS AND HYBRIDIZATION

The bonding in hydrocarbons, as in most organic compounds, is, unless explicitly stated otherwise, conventionally assumed to involve only 1s orbitals on hydrogen and 2s and 2p orbitals on carbon. On the other hand, we know from the quantum-mechanical variational principle[11] that the explicit inclusion of 3d orbitals always lowers the molecule's total energy, and that this quantity, as a result, more closely approaches the experimental value. As such, when quantitative agreement with experiment is desired, the use of these higher orbitals is unavoidable.[12-14] It is well established that there is a difference between the use of these orbitals, when employed in the valence scheme, and those used as "polarization functions" to correct for inadequacies, i.e., inflexibilities, in the non-d-containing basis set.[15-17] The essentially unanswered question is how to determine, *a priori* and without "numerical experimentation," the role of d orbitals in the structure and energetics of an arbitrary molecule or pair of isomers under consideration. Equivalently, when is it safe to assume that our qualitative understanding is independent of the inclusion of d orbitals in our reasoning?

Carbon 3d and 4f orbitals have been invoked as the origin of rotational barriers in ethanelike molecules.[18a,b] While rotational barriers will be discussed more extensively (see Section 2.B), suffice it to say at this time, "It is hard to calculate the rotational barrier of ethane incorrectly."[19] The use of

[11] I. N. Levine, "Quantum Chemistry," 2nd Ed., Ch. 8. Allyn & Bacon, Boston, Massachusetts, 1974.

[12] $CH_2$: C. F. Bender, H. F. Schaeffer, III, D. R. Franceschetti, and L. C. Allen, *J. Am. Chem. Soc.* **94**, 6888 (1972).

[13] $CH_3^+$ and $CH_3^-$: R. E. Kari and I. G. Csizmadia, *J. Chem. Phys.* **50**, 1443 (1969).

[14] $C_1$ and $C_2$ Carbonium Ions: P. C. Hariharan, W. A. Lathan, and J. A. Pople, *Chem. Phys. Lett.* **14**, 385 (1972).

[15] J. I. Musher, *Angew. Chem. Int. Ed. Engl.* **8**, (1969).

[16] C. A. Coulson, *Proc. Robert A. Welsh Found. Conf. Chem. Res.* **16**, 61–120 (1973).

[17] F. Bernardi, I. G. Csizmadia, A. Mangini, H. B. Schlegel, M. H. Whangbo, and S. Wolfe, *J. Am. Chem. Soc.* **97**, 2209 (1975).

[18a] L. Pauling, *Proc. Natl. Acad. Sci. U.S.A.* **44**, 211 (1958).

[18b] L. Pauling, "The Nature of the Chemical Bond," 3rd Ed., pp. 126–128, 143. Cornell Univ. Press, Ithaca, New York, 1960.

[19] Prof. John P. Lowe, personal communication to Joel F. Liebman.

carbon 3d orbitals has been suggested in both $S_N2$[20] and $S_E2$[21] (more specifically, 1,2-carbonium ion rearrangement[21]) reactions. Since the former reaction's transition state is isoelectronic with the "hypervalent"[16] phosphoranes, it is not surprising that this argument has been repudiated by more rigorous quantum-chemical calculations.[22] This would suggest that the latter usage of 3d orbitals would be even more superfluous: there are but eight electrons around the "special" carbon in $S_E2$ reactions, while there are ten (an octet expanded) in the $S_N2$ case. Surprisingly, this conclusion is false. To see why, let us consider the simplest example of the class of reactions of interest, the automerization of $CH_3CH_2CH_2^+$ via a 1,2-methyl shift. (See Section 2.B.1 for a discussion of $C_3H_7^+$ ions and other protonated cyclopropanes.) *Ab initio* calculations show that the best "classical" structure is 3 kcal/mole more stable than the bridged structure if only s and p orbitals are considered, but 1 kcal/mole less stable if d orbitals also are included.[23] We accordingly are forced to conclude that d orbitals are qualitatively important in this case, even though we would consider the d orbitals "merely" polarization functions. There are few other cases in which d orbitals have such importance in reversing isomer stabilities. However, d orbitals always seem to stabilize strained species and seem to be essential in *quantitatively* reproducing energy differences between isomers of strained species.[24a,b]

We now consider the idea of hybridization of orbitals, admitting explicitly our qualitative biases. In particular, we will maintain the chemical "myth" of the total irrelevance of these higher orbitals. Yet even then we discover that consideration of solely s and p orbitals on carbon yields surprises. For example, it is customarily assumed that tetrahedral carbon arises from $sp^3$ hybrid orbitals. However, a recent quantum-chemical study of $CH_4$[25] showed that the presence of s orbitals was unnecessary for this species to acquire a tetrahedral geometry. Indeed, another more recent and extensive calculation[10b] showed that it is the nuclear repulsion of the hydrogens that accounts for the geometry. Not surprisingly, numerical values of the total energy and energy differences upon molecular distortion or strain depend on all the atomic orbitals[25] and "energy components."[26a,b]

[20] R. J. Gillespie, *J. Chem. Soc.* p. 1002 (1952).

[21] R. J. Gillespie, *J. Chem. Phys.* **21**, 1892 (1953).

[22] R. F. W. Bader, A. J. Duke, and R. R. Messer, *J. Am. Chem. Soc.* **95**, 7715 (1973).

[23] P. C. Hariharan, L. Radom, J. A. Pople, and P. von R. Schleyer, *J. Am. Chem. Soc.*, **96**, 599 (1974).

[24a] P. C. Hariharan and J. A. Pople, *Chem. Phys. Lett.* **16**, 217 (1972).

[24b] W. J. Hehre and J. A. Pople, *J. Am. Chem. Soc.* **97**, 6941 (1975).

[25] J. Jarvie, W. Willson, J. Doolittle, and C. Edmiston, *J. Chem. Phys.* **59**, 3020 (1973).

[26a] W. H. Fink and L. C. Allen, *J. Chem. Phys.* **46**, 2261, 2276 (1967).

[26b] L. C. Allen, *Chem. Phys. Lett.* **2**, 597 (1968).

For reference, the "energy components" are the kinetic energy of the electrons, $T$; the nuclear–electron attraction, $V_{ne}$; the electron–electron repulsion, $V_{ee}$; and the nuclear–nuclear repulsion, $V_{nn}$. Energy component analysis consists of looking at the changes in $(V_{nn} + V_{ne} + T)$ versus those in $V_{ee}$; $(V_{nn} + V_{ee} + T)$ vs. $V_{ne}$; and $(V_{nn} + V_{ee})$ vs. $(V_{ne} + T)$. The first approach[26a] contrasts one-electron terms with two-electron terms, while the second[26b] and third[27] contrast so-called "attractive" and "repulsive" terms. The virtue of any of these approaches is that the changes are much larger than the intrinsic energy difference between two rotamers or modes of distortion. The fundamental defect is that the understanding that is achieved is fundamentally *a posteriori*, i.e., arising after a calculation has been performed.[28]

To add confusion to chaos, theoretical studies designed explicitly to consider the hybridization of the carbon in methane show it to be closer to $sp^{1.7}$ (or $s^{1.5} p^{2.5}$).[29a,b] It is perhaps instructive to review the meaning of hybrid orbitals since we will soon mention the possibility of $sp^2$, $sp^3$, and even $sp^5$ hybridization in cyclopropane. Historically, at least in the organic chemical literature and tradition, $sp^n$ was limited to $n = 1, 2,$ and 3, corresponding to $1 + 1, 1 + 2,$ and $1 + 3$ bonds to the central carbon atom. There is no meaning to either $sp^{1.7}$ or $sp^5$ hybrids in this approach, since the number of bonds to carbon must be integral and cannot exeed four. However, consider the additional awareness that $sp^3$ hybrids are "synthesized" from one "part" s and three "parts" p; the $n$ in $sp^n$ is merely the p-to-s ratio. If $n = 0$, we have a pure s orbital, while $n = \infty$ is a pure p orbital. {Equivalently, an orbital is said to be an $sp^n$ hybrid if it is $100[1/(n + 1)]\%$ s and synonymously $100[n/(n + 1)]\%$ p.} From the above discussion, it might appear that orbital hybridization would be of interest solely to theoretical chemists, and indeed, the theoretical literature contains many references to it.[30–32c] However, there are additional experimental correlations of hybridization, or, as usually

[27] A. Liberles, B. O'Leary, J. E. Eilers, and D. R. Whitman, *J. Am. Chem. Soc.* **93**, 6373 (1971).

[28] Ivan Kramer and Joel F. Liebman, unpublished results.

[29a] E. Clementi and H. Popkie, *J. Am. Chem. Soc.* **94**, 4057 (1972).

[29b] T. K. Ha, *J. Mol. Struct.* **11**, 179 (1972).

[30] A. Veillard and G. Del Re, *Theor. Chim. Acta* **2**, 55 (1964).

[31a] M. Randić and Z. B. Maksić, *Theor. Chim. Acta* **3**, 59 (1965).

[31b] L. Klasinć, S. Maksić, and M. Randić, *J. Chem. Soc.* p. 755 (1966).

[31c] M. Randić and Z. B. Maksić, *Chem. Rev.* **72**, 43 (1972).

[31d] M. Randić and Z. B. Maksić, *J. Am. Chem. Soc.* **95**, 6522 (1973).

[32a] J. M. Schulman and G. J. Fisanick, *J. Am. Chem. Soc.* **92**, 6653 (1970).

[32b] M. D. Newton and J. M. Schulman, *J. Am. Chem. Soc.* **94**, 767 (1972).

[32c] M. D. Newton, *in* "Electronic Structure *ab initio Methods*" (H. F. Schaefer, III, ed.), Modern Theoretical Chemistry, Vol. 4, pp 223–268. Plenum, New York, 1977.

expressed, % s character, with $^{13}C$ nmr coupling constants.[33] In general, good qualitative correlations between % s character and bond strengths are found: the more p character in the C—C bond, the more acetylenic or olefinic; the more s character in the C—H bond, the stronger it is but the more acidic the compound.

Experiment and theory are rarely directly comparable. However, for cyclopropane, the compound of interest in this chapter, the experimental $^{13}C$—H and $^{13}C$—$^{13}C$ couplings suggest ca. 33% s character[34,a,b,35] in the C—H bonds and ca. 17% s character in the C—C bonds. Theoretical estimates[32b] give, respectively, 31% (31%) and 15% (19%) (where the two values are for "minimal basis molecular orbital" and INDO calculations). It is to be noted that the total % s character does not add up to 100% in the unparenthesized calculation. One conventionally expects that for a closed-shell organic compound, the 2s and 2p orbitals of every carbon would have two electrons each and so the sum should equal 100%. Nonetheless, qualitative logic discussing numerous experiments,[36] and this[33] and other quantum chemical calculations,[29a,b,30,37,38] belie the supposed equality. Unfortunately, in the absence of direct experiment on each of the four bonds of a given carbon, we are forced either to assume equality of the sum to 100% s or to remain ignorant. As conceptual purists, the latter alternative is more appropriate; as practicing chemists, we suggest caution but will nonetheless proceed with the former. As such, we return to simple models of the structure, energetics, and bonding of cyclopropane.

### 2.A.3  BENT BONDS AND THE COULSON–MOFFITT PICTURE

Despite all of our disclaimers, we still utilize a particularly simple model for cyclopropane. This approach[39a,b] uses methanelike $sp^3$ hybrids and joins three such units to form the alicyclic ring. In this case, the orbitals overlap outside the ring or nuclear perimeter (see Fig. 2.1), and as the maximum overlap is not along the internuclear line, the resultant bonds have been referred to as "bent."[39–41] Seemingly, even some acyclic species[42] have "bent" bonds

[33] L. N. Ferguson (Ref. 8), pp. 52–53, 101–105.

[34a] D. J. Patel, M. E. H. Howden, and J. D. Roberts, *J. Am. Chem. Soc.* **85**, 3218 (1963).

[34b] F. J. Weigert and J. D. Roberts, *J. Am. Chem. Soc.* **89**, 5962 (1967).

[35] V. S. Watts and J. H. Goldstein, *J. Chem. Phys.* **46**, 4615 (1967).

[36] M. Pomerantz and J. F. Liebman, *Tetrahedron Lett.* p. 2385 (1975).

[37] D. Peters, *Tetrahedron, Suppl.* **2**, 143 (1963).

[38] P. Kollman, *J. Am. Chem. Soc.* **96**, 4363 (1974).

[39a] C. A. Coulson and W. E. Moffitt, *J. Chem. Phys.* **15**, 151 (1947).

[39b] C. A. Coulson and W. E. Moffitt, *Philos. Mag.* **40**, 1 (1949).

[40] W. H. Flygare, *Science* **140**, 1179 (1963).

by this definition. One may argue that an atom in the molecule chooses a compromise between the maximum nuclear–electron attraction (which suggests high electron density along the bond) and maximum conformity with what we normally believe to be the intrinsic shell structure or natural hybridization.[43] One can argue whether the intrinsic or natural geometry would be tetrahedral ($sp^3$) or right-angled (pure p), but, in either case, the molecule would still be strained. An alternative approach argues for "minimum bending of orbitals"[44a,b] or "minimum bond tortuosity."[45] These studies suggest that bending intrinsically increases the kinetic energy of the electrons and so destabilizes the molecule. This kinetic energy is related to the average value of the square of the momentum. There is an experimental technique called Compton scattering that probes the momentum of the electrons in a molecule. Recently revived and reviewed,[46] this technique provides an additional mode of comparison for related molecules. In order to prove the average value of various powers of the momentum, quantum-chemical calculations have been performed on the following isoelectronic series: $CH_3X$ (X = $CH_3$, $NH_2$, OH, and F),[47a] and on $(CH_2)_2Y$ and their acyclic isomers $CH_3CHY$ (Y = $CH_2$, NH, and O).[47b] Hirst and Liebmann claim that "there are no significant differences arising from strain."[47c] This conclusion was derived from noting the near equality of the average value of the various momenta for the two series $(CH_2)_2Y$ and $CH_3CHY$. We wish to disagree because there is a marked difference between $CH_3X$ and $CH_3CHY$ (X = $CH_3$ vs. Y = $CH_2$, $NH_2$ vs. NH, OH vs. O). Our inference from the published results is that three-membered rings are more similar to two-membered (cf. cycloethane) than they are to acyclic, saturated species (see Section 2.A.6). This in fact is documented by Table 3.1 in the section on general cycloalkanes, where it is noted that the strain per carbon of cyclopropane and cycloethane are nearly equal and much higher than for any other cycloalkanes. Returning to these minimum bending or tortuosity principles, it suffices to say that despite the conceptual simplicity of these models, quantitative and qualitative applications remain surprisingly rare.

[41] D. Peters, *Tetrahedron* **19**, 1539 (1963).

[42] L. Burnelle and C. A. Coulson, *Trans. Faraday Soc.* **53**, 403 (1956).

[43] Prof. Leland C. Allen, personal communication to Joel F. Liebman.

[44a] H. Eyring, G. H. Stewart, and R. F. Smith, *Proc. Natl. Acad. Sci. U.S.A.* **44**, 259 (1958).

[44b] G. H. Steart and H. Eyring, *J. Chem. Educ.* **35**, 550 (1958).

[45] L. S. Bartell and B. Andersen, *J. Chem. Soc., Chem. Commun.* p. 789 (1973).

[46] I. R. Epstein, *Acc. Chem. Res.* **6**, 145 (1973).

[47a] D. M. Hirst and S. P. Liebmann, *Mol. Phys.* **30**, 597 (1975).

[47b] D. M. Hirst and S. P. Liebmann, *Mol. Phys.* **30**, 1693 (1975).

[47c] See Ref. 47b, p. 1693.

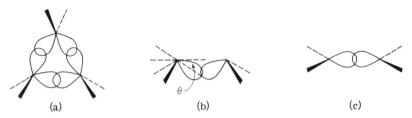

(a)                (b)                (c)

**Figure 2.1.**    (a) A bent-bond picture of cyclopropane. (b) Bent bond connecting two $CH_2$ fragments in which $\theta$ is the angle of bending. (c) A normal bond.

The earlier notion of "bent" bonds can still be quantified and extended by considering hybridizations other than sp³.[40,48] Valence bond,[39a,b,49] molecular orbital,[50–52] and maximum overlap[30,52] calculations yield an interorbital angle of approximately 104° for cyclopropane and a corresponding angle of bending of 22°. Experimental studies on cyclopropane itself seemingly have not been reported. However, the theoretical result is in encouraging agreement with high-resolution x-ray and/or neutron-diffraction studies[53] on a suitably substituted aziridine,[54] oxirane,[55a,b] norcaradiene,[56] and on the relatively simple *cis*-1,2,3-tricyanocyclopropane.[57]

For methane, interorbital and internuclear angles are equal to the natural 109.5° [i.e., $\cos^{-1}(-1/3)$] for strictly tetrahedral carbon. A better comparison is probably propane, in which the internuclear angle is 112°,[1] but we admit ignorance as to the interorbital angle. Maximum-overlap criterion calculations are conceptually simpler to understand than other studies. The greater the overlap between two orbitals, the greater the bond strength[30,58a–61d] to the

[48] W. A. Bernett, *J. Chem. Educ.* **44**, 17 (1968).

[49] C. A. Coulson and R. J. White, *Mol. Phys.* **18**, 577 (1970).

[50] E. Kochanski and J. M. Lehn, *Theor. Chim. Acta* **14**, 281 (1969).

[51] R. M. Stevens, E. Switkes, E. A. Laws, and W. N. Lipscomb, *J. Am. Chem. Soc.* **93**, 2603 (1972).

[52] C. A. Coulson and T. H. Goodwin, *J. Chem. Soc.* p. 2851 (1962); errata, p. 3161 (1963).

[53] P. Coppens, *Angew. Chem. Int. Ed. Engl.* **16**, 32 (1977).

[54] T. Ito and T. Sakurai, *Acta Crystallogr., Sect. B* **29**, 1594 (1973).

[55a] D. A. Matthews and G. D. Stucky, *J. Am. Chem. Soc.* **93**, 5954 (1971).

[55b] D. A. Matthews, G. D. Stucky, and P. Coppens, *J. Am. Chem. Soc.* **94**, 8001 (1972).

[56] C. G. Fritchie, *Acta Crystallogr.* **20**, 27 (1966).

[57] A. Hartmann and F. L. Hirschreld, *Acta Crystallogr.* **20**, 80 (1966).

[58a] L. Pauling, *J. Am. Chem. Soc.* **53**, 1367 (1931).

[58b] J. C. Slater, *Phys. Rev.* **37**, 481 (1931).

[58c] J. C. Slater, *Phys. Rev.* **38**, 325, 1109 (1931).

[58d] R. S. Mulliken, *Phys. Rev.* **41**, 49 (1932).

[59a] J. N. Murrell, *J. Chem. Phys.* **32**, 767 (1960).

[59b] T. L. Gilbert and P. G. Lykos, *J. Chem. Phys.* **34**, 2199 (1961).

[59c] A. G. Lebiewski, *Trans. Faraday Soc.* **57**, 1849 (1961).

extent that all bonds in a molecule are explicitly considered. In particular, molecular-orbital calculations[48] show that the C—C bond in cyclopropane has 20% less overlap than the C—C bond in ethane.

In the current model, we understand the strain in cyclopropane to be due to its bent bonds. An interesting complexity arises when we ask the question, "Why bend the bonds at all as opposed to using some suitable hybrid?" The conventional answer, usually implicitly assumed, is that any hybrid composed of carbon 2s and 2p orbitals may not have an interorbital angle of less than 90°. Indeed, one argues that the natural angle for p orbital bonding is 90°, and that s orbital admixture merely increases the angle. In this case, one observation and one flaw are to be noted regarding this assertion. The observation is that the spectroscopic ground state of the sp³ configuration of atomic carbon is the $^5$S.[62a,b] We remind the reader that the "S" corresponds to a spherically symmetric electron distribution,[63] and so we may ask, "what angles?" It is also important to realize a flaw, again usually implicit, in the customary construction and description of hybrid orbitals. The most general form of hybridization, or linear combination of orbitals, on a given atom allows the use of complex (i.e., $a + b\sqrt{-1}$) coefficients.[64,65] However, it may be shown that these more general orbitals cannot be pictorialized and that their "directionality" is ambiguous.[66] Coulson and White suggest a definition that obviates this problem for cyclopropane: the ideal complex orbitals for this species are nearly identical with the real (i.e., not complex, $b = 0$) orbitals earlier suggested. As we believe organic chemistry is essentially a pictorial and not a mathematical science, we will make use of only the normally used, real orbitals. Indeed, we breathe a sigh of relief that these orbitals are so nearly identical; in the probably more general case that real and complex orbitals are *not* equivalent, we must still debate which is conceptually preferable.

[60a] C. A. Coulson, "Valence," p. 76. Oxford Univ. Press, London and New York, 1961.

[60b] L. Pauling (Ref. 18b), p. 108.

[60c] A. Streitwieser, "Molecular Orbital Theory for Organic Chemists," p. 13. Wiley, New York, 1961.

[61a] J. J. Kaufman, *Int. J. Quantum Chem., Symp.* **1**, 485 (1967).

[61b] F. S. Mortimer, *Adv. Chem. Ser.* **54**, 39 (1968).

[61c] R. Manne, *Int. J. Quantum Chem.* **2**, 69 (1968).

[61d] F. Weinhold and T. K. Brunck, *J. Am. Chem. Soc.* **98**, 3745 (1976).

[62a] C. E. Moore, *Natl. Stand. Ref. Data Ser., Natl. Bur. Stand.* **NSRDS-NBS 3**, Sect. 3 (1970).

[62b] C. E. Moore, *Natl. Stand. Ref. Data Ser., Natl. Bur. Stand.* **NSRDS-NBS 35**, I (1971).

[63] I. N. Levine (Ref. 11), Ch. 11.

[64] C. A. Coulson (Ref. 60a), Ch. 8.

[65] O. Martensson and Y.Öhrn, *Theor. Chim. Acta* **9**, 133 (1967–1968).

[66] C. A. Coulson and R. J. White, *Mol. Phys.* **18**, 577 (1970).

**2.A.4** WALSH (WALSH–SUGDEN) PICTURE

Yet another picture for cyclopropane is the so-called Walsh model.[67-69] Much as ethylene may be formally synthesized from two $CH_2$ groups,[70] cyclopropane is "synthesized" from three $CH_2$ groups. These $CH_2$ groups with their associated sp² hybridization and p and sp² orbitals are combined to form cyclopropane in a manner analogous to the formation of ethylene from two such groups (see Fig. 2.2). For cyclopropane, there are six molecular

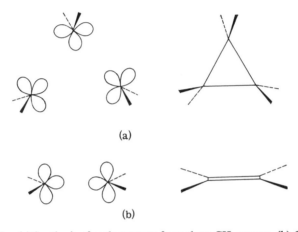

(a)

(b)

**Figure 2.2.** (a) Synthesis of cyclopropane from three $CH_2$ groups. (b) Corresponding synthesis of ethylene from two $CH_2$ groups.

(or, perhaps more descriptively, "supermolecular") orbitals that consist of two sets of three, and each set of three is likewise split into two subsets.[71] The molecule minimizes its energy by choosing the lowest-energy set. These in fact consist of one from the σ and two from the π "pools" (see Fig. 2.3). We note that in studies of general ring systems, the terms σ and π have usually been replaced: "in" and "out" cyclopropane,[71] "in" and "peripheral" cyclobutane,[72] "internal" and "external" cyclopropane and related heterocycles,[73]

[67] A. D. Walsh, *Nature (London)* **159**, 165, 712 (1947).

[68] T. N. Sugden, *Nature (London)* **160**, 367 (1947).

[69] A. D. Walsh, *Trans. Faraday Soc.* **45**, 179 (1949).

[70] R. S. Mulliken, *Phys. Rev.* **41**, 751 (1932).

[71] W. L. Jorgensen and L. Salem, "The Organic Chemist's Book of Orbitals," pp. 22–23. Academic Press, New York, 1973.

[72] R. Hoffmann and R. B. Davidson, *J. Am. Chem. Soc.* **93**, 5699 (1971).

[73] H. Basch, M. B. Robin, N. A. Kuebler, C. Baker, and D. W. Turner, *J. Chem. Phys.* **51**, 52 (1969).

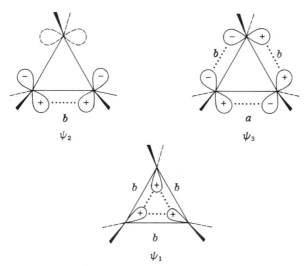

**Figure 2.3.** The three Walsh orbitals as usually drawn for cyclopropane: $b$ means bonding "pool"; $a$ means antibonding. $\psi_1$ comes from the $\sigma$ "pool," while the degenerate pair $\psi_2$ and $\psi_3$ come from the $\pi$ "pool." Strictly speaking, $\psi_2$ looks much more like a normal C—C $\sigma$ bond and $\psi_3$ looks more like a $\pi$ bond.

"inside" and "outside" cycloalkanes and general saturated heterocycles,[74] "radial" and "tangential" benzene,[75] and "horizontal" and "vertical wave" cyclohexanes.[76]

For completeness, we chronicle some of the applications of this type of approach to polycyclic ring systems: bicyclobutane,[77] **1**, spiropentane,[48,67,69] **2**, homofulvenes[78a,b] such as **3**, quadricyclane, **4**,[79] Cope-rearranging species

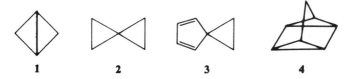

|   1   |   2   |   3   |   4   |

[74] P. E. Peterson, *J. Org. Chem.* **37**, 4180 (1972).

[75] V. O. Jonsson and E. Lindholm, *Chem. Phys. Lett.* **1**, 501 (1967).

[76] R. Hoffmann, P. D. Molliere, and E. Heilbronner, *J. Am. Chem. Soc.* **95**, 4860 (1973).

[77] M. Pomerantz and E. W. Abrahamson, *J. Am. Chem. Soc.* **88**, 3870 (1966).

[78a] R. Gleiter, E. Heilbronner, and A. de Meijere, *Helv. Chim. Acta* **54**, 1029 (1971).

[78b] P. Bischof, R. Gleiter, H. Dürr, B. Ruge, and P. Herbst, *Chem. Ber.* **109**, 1412 (1976).

[79] H. D. Martin, C. Heller, E. Haselbach, and Z. Lanyjova, *Helv. Chim. Acta* **57**, 465 (1974).

such as bullvalene, **5**,[80] and barbaryl cation, **6**,[81] and variously fused cyclopropanes with cyclohexane[82] such as diademane, **7**. It is interesting to note

that a unique set of Walsh orbitals is not always constructable.[77] (Uniqueness
is taken to mean that we are not merely redrawing a set of degenerate
orbitals.[72])

Let us now try to understand the origin of the strain of cyclopropane in the
Walsh model. The lowest-lying orbital increases the molecular stability as it
corresponds to a closed three-center/two-electron[83] (or "internal sigma"[73])
bond. Indeed, one may recall the increased $\pi$ orbital resonance energy of
cyclopropenyl cation relative to the open-chain analogue, the allyl cation.[84–86]
This, however, is admittedly deceptive and overestimates the stabilizing
influence of this orbital. In both the cyclic and the acyclic three-carbon
conjugated cations, all the $\pi$ orbital overlaps are essentially unchanged from
the normal situation found in ethylene. However, for cyclopropane, the
individual C—C bonds are bent (see Section 2.3), and so the individual $\sigma$
overlaps are reduced from those found in the normal ethane. Some restabili-
zation is expected since the C—C bonds in cyclopropane involve $sp^2$ hybridization[87] (see Fig. 2.4), but we are doubtful that this compensates.

The next, and highest-occupied, molecular orbitals are a pair of degenerate
ones. The first one (see Fig. 2.5) may be recognized as either the $\pi$ bond of a
highly distorted olefin (see Section 3.D) or as the highly distorted $\sigma$ bond
formed from two unhybridized p orbitals. In either case, there is little
stabilization arising from this orbital. The second orbital can likewise be
recognized as having two of these distorted $\sigma$ or $\pi$ bonds, but there is also a

[80] P. Bischof, R. Gleiter, E. Heilbronner, V. Hornung, and G. Schroder, *Helv. Chim. Acta* **53**, 1645 (1970).

[81] R. Hoffmann, W. D. Stohrer, and M. J. Goldstein, *Bull. Chem. Soc. Jpn.* **45**, 2513 (1972).

[82] E. Heilbronner, R. Gleiter, T. Hoshi, and A. de Meijere, *Helv. Chim. Acta* **56**, 1594 (1973).

[83] W. N. Lipscomb, "Boron Hydrides," Ch. 2. Benjamin, New York, 1963.

[84] R. Breslow and J. T. Groves, *J. Am. Chem. Soc.* **92**, 984 (1970).

[85] F. P. Lossing, *Can. J. Chem.* **50**, 3973 (1972).

[86] L. Radom, P. C. Hariharan, J. A. Pople, and P. von R. Schleyer, *J. Am. Chem. Soc.* **98**, 10 (1976).

[87] S. W. Benson (Ref. 9), pp. 63–65.

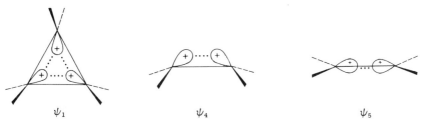

**Figure 2.4.** A closer examination of the lowest-lying Walsh orbital of cyclopropane, $\psi_1$. $\psi_4$ is the corresponding two-center orbital, a bent $sp^2$-$sp^2$ $\sigma$ bond, while $\psi_5$ is the normal $sp^2$-$sp^2$ $\sigma$ bond.

corresponding antibonding $\sigma^*$ or $\pi^*$ bond. This again yields little stabilization. There is nothing in these descriptions that forces an $sp^2$ hybridization on the carbons. This generalization of the Walsh model has been discussed[48] and yields a model equivalent to the bent-bonds picture of cyclopropane.

Still another model for describing cyclopropane uses group orbitals[88] to generate delocalized molecular orbitals. Both the C—H and C—C bonds were explicitly considered. We will consider only the latter, as we are primarily interested in the strain energy of the carbocyclic ring or framework. Three C—C bond orbitals are combined to form three new (supermolecular) orbitals. It is apparent that these three orbitals have the same nodal symmetry

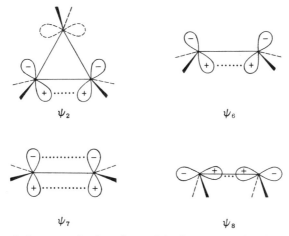

**Figure 2.5.** A closer examination of one of the degenerate pairs of Walsh orbitals, in particular, $\psi_2$. $\psi_6$ is the corresponding *strictly* two-center orbital, $\psi_7$, the $\pi$ bond of an undistorted olefin, and $\psi_8$, the $\sigma$ bond formed from two unhybridized carbon $2p$ orbitals.

[88] W. L. Jorgensen and L. Salem (Ref. 71), pp. 19–21.

(+ and − signs) as the three Walsh orbitals. This should come as no surprise: two essentially valid descriptions of a molecule should be conceptually inter-relatable. However, the group-orbital approach makes it easier to locate the source of the strain in cyclopropane. In a localized or two-center C—C bonding description, all three orbitals are bonding. However, only the lowest-lying is nodeless, or without sign changes, among all three carbon atoms. The small framework (internuclear, or C—C—C) angle increases the antibonding character of the next two orbitals, and so the molecule is de-stabilized. Analogous logic explains the relative strengths of $\sigma$ and $\pi$ bonds in olefins. Although useful for cyclopropane, however, we believe that this model will be less useful for species with lower symmetry (e.g., substituted polycyclic systems). Regrettably, we wish to emphasize that this defect is shared by most approaches in this chapter.

**2.A.5**  CYCLOPROPANE VERSUS ETHYLENE AND THE RADICAL IONS OF CYCLO-
PROPANE

Many of the erstwhile models presented here for the understanding of the bonding and energetics of cyclopropane have interrelated this species with ethylene. Extensive chemical documentation of the validity of this comparison exists,[8,89] and the reader is referred to these citations for examples. (For a dissenting view, however, see Gordon.[90]) We will concentrate here on physical properties and theoretical models. Implicit in our discussion is the assumption that what is true of the parent molecules cyclopropane and ethylene likewise applies to their substituted derivatives. (We hope the reader recognizes this assumption implicit in most treatments of molecules.) It is found that the unsubstituted (parent or archetypal) species have comparable C—H bonds as shown by bond length, force constant, ir stretching frequency, and $^{13}$C—H coupling constant.[8] The C—H bond strength in cyclopropane is less than that of ethylene but both are significantly greater than those of propane.[9] The earlier discussion of propane vs. cyclopropane (Section 2.A.1) should make the reader wary of these analyses, but nonetheless we will proceed. Turning to the C—C bonds as they are of greater interest, it comes as no surprise that the C—C bond of ethylene is considerably shorter than that of cyclopropane.[91]

Bond strengths may also be compared, but here there is an acknowledged pitfall or subtlety. Each $CH_2$ is bonded to only one other $CH_2$ in ethylene, while in cyclopropane each is bonded to two. Equivalently, standard

[89] M. Charton, *in* "The Chemistry of the Alkenes" (J. Zabricky, ed.), Part 2, pp. 511–610. Wiley (Interscience), New York, 1970.
[90] A. J. Gordon, *J. Chem. Educ.* **44**, 461 (1967).
[91] H. C. Allen and E. K. Pyler, *J. Am. Chem. Soc.* **80**, 2673 (1958).

chemical formulas draw a double bond between the $CH_2$ groups in ethylene but draw single bonds in cyclopropane. The simplest solution is to give the bond strength per $CH_2$. Table 3.1 presents these values along with the value of $1/n[\Delta H_f^0(CH_2)_n]$ for general cycloalkanes. With this definition it is found that cyclopropane is more bound than ethylene. This represents a reasonable finding once ethylene is redefined as cycloethane. This is perhaps most easily understood in terms of the early bent-bond model for this two-carbon species,[1,92,93] in which the bending is more severe than in cyclopropane.[94a,b] However, it is important to realize the near equivalence of this model with the more conventional one, in which there is one (stronger than normal) $\sigma$ bond and one $\pi$ bond.[87,95] That is, defining the two bent, or so-called "banana" bonds, as $\tau_1$ and $\tau_2$, we find (see Fig. 2.6): $\tau_1 = 2^{-1/2}(\sigma + \pi)$; $\tau_2 = 2^{-1/2}(\sigma - \pi)$; $\sigma = 2^{-1/2}(\tau_1 + \tau_2)$; $\pi = 2^{-1/2}(\tau_1 - \tau_2)$.

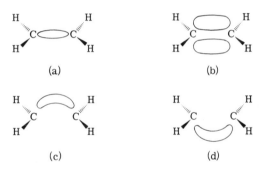

**Figure 2.6.** (a) The $\sigma$ bond of ethylene. (b) The $\pi$ bond of ethylene. (c) A "banana" bond of ethylene: $\tau_1 = 2^{-1/2}(\sigma + \pi)$. (d) A "banana" bond of ethylene: $\tau_2 = 2^{-1/2}(\sigma - \pi)$.

Bond strengths, even when defined in terms of dissociation energies, are hard to correlate with orbital energies and ionization potentials. For example, it is found that removal by ionization of a $\pi$ electron in ethylene costs less energy than removal of a $\sigma$ electron in ethane.[96] From the definition of bond strengths, $D(CH_n—CH_n) = 2\Delta H_f^0(CH_n) - \Delta H_f^0(C_2H_{2n})$ and $D(CH_n^+—CH_n) = \Delta H_f^0(CH_n^+) + \Delta H_f^0(CH_n) - \Delta H_f^0(C_2H_{2n}^+)$, we find that the change in bond strengths upon ionization may be expressed as $D(CH_n^+—CH_n) - D(CH_n—CH_n) = IP(C_2H_{2n}) - IP(CH_n)$, where IP is the ionization potential.

[92] L. Pauling, *J. Am. Chem. Soc.* **53**, 1347 (1931).
[93] C. A. Coulson and E. T. Stewart, *in* "The Chemistry of the Alkenes" (S. Patai, ed.), Vol. 1, pp. 1–147. Wiley (Interscience), New York, 1964.
[94a] G. G. Hall and J. Lennard-Jones, *Proc. Roy Soc., Ser. A* **205**, 357 (1951).
[94b] P. V. Hay, W. J. Hunt, and W. A. Goddard, III, *J. Am. Chem. Soc.* **94**, 8293 (1974).
[95] E. C. Wu and A. S. Rodgers, *J. Am. Chem. Soc.* **98**, 6112 (1976).
[96] A. J. C. Nicholson, *J. Chem. Phys.* **43**, 1171 (1965).

From the experimental IP's of $n = 3$, $C_2H_6$,[96] $CH_3$[97]; $n = 2$, $C_2H_4$,[96] $CH_2$[98]; and $n = 1$, $C_2H_2$,[96] and $CH$,[99] we find that the bond strength in ethane is halved by ionization, whereas there is but negligible change for ethylene and acetylene. One may argue that the molecular geometries of the radical cations formed by ionization and the parent neutral are quite different.[100] However, we are doubtful that this invalidates the conclusion that $\pi$ bonds are inherently weaker than $\sigma$ bonds.

Let us now consider the ionization of cyclopropane. An early ionization-potential measurement[73] gave us a value comparable to ethylene.[96] However, these authors noted that this value was an upper bound, since removal of an electron from a (doubly) degenerate orbital (see Fig. 2.2) must result in (the so-called Jahn–Teller[101,102]) distortion. More recently, the ionization potential has been remeasured[85] and found to be less than 9.93 eV and indeed less than for "cycloethane," cyclobutane, and cyclopentane. The geometry for cyclopropane radical cation has been studied by semiempirical quantum-chemical calculations.[103,104] Two nearly degenerate or equienergetic structures were found, one corresponding to a "complex of ethylene and $CH_2{}^+$"(8) and the other to "the radical cation of trimethylene"[103] (9). The former reminds us of the original Walsh model[67] of cyclopropane, a complex of ethylene and neutral $CH_2$ (10). The latter will be further defined when we discuss trimethylene (Section 2.B.2). These structures additionally mimic the again

$$
\begin{array}{ccc}
CH_2{}^+ & \overset{+}{C}H_2 & CH_2 \\
| & / & | \\
CH_2\!=\!CH_2 & \dot{C}H_2\!-\!CH_2 & CH_2\!=\!CH_2 \\
\mathbf{8} & \mathbf{9} & \mathbf{10}
\end{array}
$$

nearly equienergetic corner and edge-protonated cyclopropanes (Section 2.B.2) with their greatly elongated C—C bonds. Enforcing an artificial three-fold symmetry on cyclopropane radical cation raises the energy and indeed results in an energy maximum. In this cation, the C—C bond length is longer than the neutral by 0.02 Å. (The most "normal" structure seemingly has the highest energy.) In contrast, (*ab initio*) calculations show that the C—C bond in ethane is increased by 0.3 Å upon ionization.[100]

[97] G. Herzberg and J. Shoosmith, *Can. J. Phys.* **34**, 523 (1956).

[98] G. Herzberg, *Can. J. Phys.* **39**, 1511 (1961).

[99] G. Herzberg and J. W. C. Johns, *Astrophys. J.* **158**, 399 (1969).

[100] W. A. Lathan, W. J. Hehre, and J. A. Pople, *J. Am. Chem. Soc.* **93**, 808 (1971).

[101] R. Englman, "The Jahn–Teller Effect in Molecules and Crystals." Wiley (Interscience), New York, 1972.

[102] I. N. Levine, "Molecular Spectroscopy," pp. 313–314. Wiley, New York, 1975.

[103] E. Haselbach, *Chem. Phys. Lett.* **7**, 428 (1970).

[104] C. G. Rowland, *Chem. Phys. Lett.* **9**, 169 (1971).

Comparatively little is known about hydrocarbon radical anions. Recent experiments show that the radical anions of gas-phase ethane,[105a] ethylene,[105b] and cyclopropane[105a] are all unbound with respect to dissociation to re-form the parent hydrocarbon and an electron. There are well-defined energy levels even for unbound species (so-called "resonance levels"[106,107]), and it is found that cyclopropane radical anion is more unbound than that of ethylene. This is in accord with qualitative logic deducible from the higher, and unoccupied, orbitals for cyclopropane (e.g., Jorgensen and Salem[71]). In solution, radical anions of substituted cyclopropanes are seemingly stabilized, and it is found that numerous cyclopropanes are readily reduced by one-electron reagents.[108] The intermediates in the reductions appear to be more related to the radical anions of trimethylenes (Section 2.B.2) than to complexes of ethylene and $CH_2^-$. We suggest this combination rather than ethylene radical anion and $CH_2$ because while $C_2H_4^-$ is unbound relative to the neutral hydrocarbon,[105b] $CH_2^-$ is stable relative to such dissociation.[109] (We parenthetically note the synthesis of general $R_2C^-$ ions both chemically[110a,b] and electrochemically.[110c]) It is important to realize that radical anion chemistry is essentially complementary to that of radical cations, and as a result, many more hydrocarbons are now being studied experimentally.[111a,b] Insufficient data exist to tell whether isoelectronic comparison with cationic three-membered heterocycles (e.g., aziridinium) and $\beta$-substituted alkyl radicals (e.g., 2-aminoethyl) will prove useful in understanding cyclopropanes and their radical anions. We conclude by noting that there are other processes involving electron donation to and from olefins and cyclopropanes that include protonation and metal complex formation, and that these will be discussed in later sections of this book (2.B.1 and 5.C, respectively).

[105a] Drs. K. D. Jordan, J. A. Michejda, and P. D. Burrow, unpublished results.

[105b] P. D. Burrow and K. D. Jordan, *Chem. Phys. Lett.* **36**, 594 (1975).

[106] H. S. Taylor, *Adv. Chem. Phys.* **18**, 91 (1970).

[107] R. K. Nesbet, *Adv. Quantum Chem.* **9**, 215 (1975).

[108] S. W. Staley, *in* "Selective Organic Transformations" (B. S. Thyagarajan, ed.), Vol. 2, pp. 309–348. Wiley, New York, 1972.

[109] P. F. Zittel, G. B. Ellison, S. V. O'Neil, E. Herbst, W. C. Lineberger, and W. P. Reinhardt, *J. Am. Chem. Soc.* **98**, 3731 (1976).

[110a] G. D. Sargent, C. M. Tatum, and S. M. Kastner, *J. Am. Chem. Soc.* **94**, 7714 (1972).

[110b] G. D. Argent, C. M. Tatum, and R. P. Scott, *J. Am. Chem. Soc.* **96**, 1602 (1974).

[110c] R. W. McDonald, J. R. January, K. J. Norhani, and M. D. Hawley, *J. Am. Chem. Soc.* **99**, 1766 (1977).

[111a] Prof. Kenneth D. Jordan, personal communication to Joel F. Liebman.

[111b] Profs. Stuart W. Staley and John H. Moore, Jr., personal communication to Joel F. Liebman.

## 2.B    Selected Reactions of Cyclopropanes

### 2.B.1    PROTONATION

We shall now present the structures and energetics of protonated cyclo-propanes and, in particular, the parent $C_3H_7^+$ ion. These ions have been reviewed extensively,[8,112–119] and we largely refer the reader to these works for additional literature references. We believe a major source of the interest, and confusion, in the chemistry of these species arises from the possibility of several interconverting isomers, the reactions of cyclopropanes with electro-philes, and the classical versus nonclassical carbonium ion controversy. With regard to the first possibility, the only one to be discussed in our study, $C_3H_7^+$ has the following five isomers: **11**, 1- or *n*-propyl cation; **12**, 2- or *sec*- or iso-propyl cation; **13**, the corner-protonated cyclopropane, and **14** and **15**, the correspondingly edge- and face-protonated cyclopropanes. Recent experi-ments[120-121b] definitely show that species **12** is the most stable. However,

[112a] P. D. Bartlett, "Nonclassical Ions." Benjamin, New York, 1965.

[112b] D. Bethell and V. Gold, "Carbonium Ions: An Introduction," Ch. 7. Academic Press, New York, 1967.

[113] C. J. Collins, *Chem. Rev.* **69**, 541 (1969).

[114] F. Cacace, *Adv. Phys. Org. Chem.* **8**, 79 (1970).

[115a] J. L. Fry and G. J. Karabatsos, *in* "Carbonium Ions" (G. A. Olah and P. von R. Schleyer, eds.), Vol. 2, pp. 521–571. Wiley, New York, 1970.

[115b] J. T. Keating and P. S. Skell, *in* "Carbonium Ions" (G. A. Olah and P. von R. Schleyer, eds.), Vol. 2, pp. 573–653. Wiley, New York, 1970.

[116] C. C. Lee, *Prog. Phys. Org. Chem.* **7**, 119 (1970).

[117] D. M. Brouwer and H. Hogeveen, *Prog. Phys. Org. Chem.* **9**, 179 (1972).

[118] C. H. DePuy, *Top. Curr. Chem.* **40**, 73 (1973).

[119] M. Saunders, P. Vogel, E. L. Hagan, and J. J. Rosenfeld, *Acc. Chem. Res.* **6**, 53 (1973).

[120] S. L. Chong and J. L. Franklin, *J. Am. Chem. Soc.* **94**, 6347 (1972).

[121a] D. J. McAdoo, F. W. McLafferty, and P. F. Bente, III, *J. Am. Chem. Soc.* **94**, 2027 (1972).

[121b] P. P. Dymerski, R. N. Prinstein, and P. F. Bente, III, and F. W. McLafferty, *J. Am Chem. Soc.* **98**, 6834 (1976).

since species **11** and **12** are essentially strainless, we consider only the comparative energetics of the three protonated cyclopropanes **13, 14,** and **15.** To a first approximation, we may assume that the site of protonation should be the same as the site of interaction of cyclopropanes with other electrophiles. Carbonium ion centers are examples of suitably strong electrophiles and have been synthesized directly over the center of the cyclopropane ring or face.[122a,b,123] No particular stabilization is afforded to these cations, and so we may conclude that face protonation is energetically disfavored. However, experimental data remain insufficient to unambiguously decide whether the corner- (**13**) or edge-protonated (**14**) cyclopropane is the more stable. What does theory say? *Ab initio*[23,124] and semiempirical[125] quantum-chemical calculations confirm that **12** is the most stable $C_3H_7^+$ isomer, and that **15** is the least stable. However, these theoretical studies disagree as to the relative stability of **13** and **14.** Although we intrinsically favor *ab initio* theory (see the discussion in Pople,[126] Hehre,[127] and Dewar[128]), it is nonetheless noteworthy that the energy difference between the most stable *cyclic* and most stable *acyclic* $C_3H_7^+$ ions is predicted better by the semiempirical theory.

Let us now consider some qualitative models for the variously protonated cyclopropanes. The first recalls the bent-bond model of cyclopropane (Section 2.A.3), and explicitly protonates the region in the three-carbon framework with maximum electron density.[129] This successfully predicts that face-protonation should be energetically less favorable than either edge- or corner-protonation. The second model explicitly considers the Walsh model of cyclopropane (Section 2.A.4), and also predicts[122a,b] face-protonation to be disfavored. An interesting spin-off of this analysis is that it also explains both the relative reactivities of cyclopropane and cyclobutane with general electrophilic reagents,[130] and, in somewhat modified form, the relative proton affinities of the two cycloalkanes.[131] The third model, implicitly the most

[122a] S. A. Sherrod, R. G. Bergman, G. J. Gleicher, and D. G. Morris, *J. Am. Chem. Soc.* **92**, 3469 (1970).

[122b] S. A. Sherrod, R. G. Bergman, G. J. Gleicher, and D. G. Morris, *J. Am. Chem. Soc.* **94**, 4615 (1972).

[123] R. C. Bingham, W. F. Sliwinski, and P. von R. Schleyer, *J. Am. Chem. Soc.* **92**, 3471 (1970).

[124] L. Radom, J. A. Pople, V. Buss, and P. von R. Schleyer, *J. Am. Chem. Soc.* **94**, 311 (1972).

[125] P. K. Bischof and M. J. S. Dewar, *J. Am. Chem. Soc.* **97**, 2278 (1975).

[126] J. A. Pople, *J. Am. Chem. Soc.* **97**, 5306 (1975).

[127] W. J. Hehre, *J. Am. Chem. Soc.* **97**, 5308 (1975).

[128] M. J. S. Dewar, *J. Am. Chem. Soc.* **97**, 6591 (1975).

[129] R. Hoffmann, *J. Chem. Phys.* **40**, 2480 (1964).

[130] K. B. Wiberg, K. C. Bishop, III, and R. B. Davidson, *Tetrahedron Lett.* p. 3169 (1973).

[131] T. Pakkanen and J. L. Whitten, *J. Am. Chem. Soc.* **97**, 6337 (1975).

"popular," considers the corner-protonated ion **13** as a $\pi$ complex of ethylene and $CH_3^+$.[132a,b] In this regard, we are reminded of Walsh's original model for cyclopropane,[69] an analogous complex of ethylene and $CH_2$ (deprotonated $CH_3^+$), and a more general theory of three-membered (heterocyclic) rings of the type $(CH_2)_2X$ that explicitly considers both X-bonding and back bonding.[133] (See Chapter 5 and Fig. 5.5.)

This $\pi$ complex has also been compared with $CH_5^+$ in two ways:[124] $H_2$ replacing ethylene in its binding with $CH_3^+$, and cyclopropyl cation replacing $CH_3^+$ in its binding with $H_2$. This analysis suggests further interrelationships. First of all, we recall the isoelectronic comparisons of "small boron compounds . . . with carbocations."[134] From the literature studies of the interaction of $H_2$ with $BH_3$ and $CH_3^+$,[135,136] we anticipate the binding of ethylene with $BH_3$ to be weaker than with $CH_3^+$. (See Fehlner[137] for a review of the reactions of $BH_3$ with ethylene and other gaseous species.) It is also interesting to note that the general question of the addition of substituted boranes to substituted olefins (hydroboration[138]) has been discussed[139] in terms of intermediates isoelectronic to edge- and corner-protonated cyclopropanes. The second interrelationship makes use of the binding of $H_2$ by $CH_3^+$, $C_2H_5^+$, 2-$C_3H_7^+$, and $(CH_3)_3C^+$.[140] The experimental heat of formation of protonated cyclopropane (derivable from the values in Chong and Franklin[120] and Aue et al.[141]) is 200 kcal/mole, and of cyclopropyl cation, 238 kcal/mole. From this we deduce that the reaction of $c$-$C_3H_5^+$ with $H_2$ to form a protonated cyclopropane is exothermic by 38 kcal/mole. Since the binding energy of $H_2$ seems to increase with decreasing carbonium ion stability, we are greatly troubled by the finding that methyl and cyclopropyl cations are of comparable stability (the binding energy in $CH_3^+ \cdot H_2$ is 40 kcal/mole). We close by noting

[132a] M. J. S. Dewar, *Bull. Chim. Soc. Fr.* p. C71 (1951).

[132b] M. J. S. Dewar and A. Marchand, *Annu. Rev. Phys. Chem.* **16**, 321 (1965).

[133] M. J. S. Dewar, R. C. Haddon, A. Komornicki, and H. Rzepa, *J. Am. Chem. Soc.* **99**, 377 (1977).

[134] J. B. Collins, P. von R. Schleyer, J. S. Binkley, J. A. Pople, and L. Radom, *J. Am. Chem. Soc.* **98**, 3436 (1976).

[135] J. B. Collins, P. von R. Schleyer, J. S. Binkley, J. A. Pople, and L. Radom, *J. Am. Chem. Soc.* **98**, 3436 (1976).

[136] I. M. Pepperberg, T. A. Halgren, and W. N. Lipscomb, *J. Am. Chem. Soc.* **98**, 3442, (1976).

[137] T. P. Fehlner, *in* "Boron Hydride Chemistry" (E. L. Muetterties, ed.), pp. 175–196. Academic Press, New York, 1975.

[138] H. C. Brown, "Boranes in Organic Synthesis." Cornell Univ. Press, Ithaca, New York, 1972.

[139] D. J. Pasto, *in* "Boron Hydride Chemistry" (E. L. Muetterties, ed.), pp. 197–222. Academic Press, New York, 1975.

that the edge- and corner-protonated cyclopropanes correspond to isomeric pairs of carbonium ions found for other protonated hydrocarbons.[140,142]

### 2.B.2 THERMAL REARRANGEMENTS OF CYCLOPROPANE TO PROPYLENE AND THE TRIMETHYLENE DIRADICAL

One of the earliest but most thoroughly investigated reactions of cyclopropane is its thermal rearrangement to propylene.[143-147] Corresponding rearrangements of substituted derivatives or many-ring analogues also have been studied, and we refer the reader to more extensive reviews in the literature.[145-147] (Also see Section 5.C.1 on transition-metal catalysis of these reactions.) The rearrangement of the parent cyclopropane[148] is interesting in that it appears that only small internuclear displacements are needed for reaction (1). As such, it is most surprising that this mildly exothermic

$$
\begin{array}{ccc}
\text{H—CH} & & \text{H} \quad \text{CH} \\
\diagup \diagdown & \xrightarrow{\Delta} & \diagdown \diagup \diagdown \\
\text{CH}_2\text{—CH}_2 & & \text{CH}_2 \quad \text{CH}_2
\end{array} \qquad (1)
$$

reaction has an activation energy of ca. 65 kcal/mole.[143,144] Fortunately for our understanding, these $C_3H_6$ isomers are small enough that meaningful quantum-chemical calculations can be performed not only on these species but also on various proposed intermediates and transition states. Such a study was recently reported,[149] in which the geometry for reactant and product was optimized for all of the 21 internal coordinates (three for each atom minus six for rigid rotation and translation). Having chosen the bond angle of the migrating hydrogen atom as the reaction coordinate, "all other geometric parameters were energy minimized." The calculated transition state is characterized by long C—C and C—H bonds and indeed reminds us of both the protonated and ionized cyclopropanes described earlier (see Sections

[140] K. Hiraoka and P. Kebarle, *J. Am. Chem. Soc.* **98**, 6119 (1976).

[141] D. H. Aue, W. R. Davidson, and M. T. Bowers, *J. Am. Chem. Soc.* **98**, 6699 (1976).

[142] R. D. Smith and J. H. Futrell, *Int. J. Mass Spectrom. Ion Phys.* **20**, 347 (1976).

[143] T. S. Chambers and G. B. Kistiakowsky, *J. Am. Chem. Soc.* **56**, 399 (1934).

[144] E. W. Schlag and B. S. Rabinovitch, *J. Am. Chem. Soc.* **82**, 5996 (1960).

[145a] H. M. Frey, *Adv. Phys. Org. Chem.* **4**, 147 (1966).

[145b] H. M. Frey and R. Walsh, *Chem. Rev.* **69**, 103 (1969).

[146] S. W. Benson and H. F. O'Neal, *Natl. Stand. Ref. Data Ser., Natl. Bur. Stand.* **NSRDS-NBS 21** (1970).

[147] M. R. Wilcott, R. L. Cargill, and A. B. Sears, *Prog. Phys. Org. Chem.* **9**, 25 (1972).

[148] S. W. Benson (Ref. 9), pp. 63–65.

[149] K. Jug, *Theor. Chim. Acta* **42**, 303 (1976).

2.B.1 and 2.A.6, respectively). (We note that in principle one could inter-convert cyclopropane and propylene via these intermediates, but there is no evidence that these represent important gas-phase isomerization pathways.) The calculated energy of activation is 17 kcal/mole lower than the experimental, of which all but 1 kcal/mole may be ascribed to the theoretical errors in predicting the relative stability of cyclopropane and propylene. Three qualitative conclusions deserve to be cited. First, the reaction is initiated by a coupling of a torsional vibration or rotation of the $CH_2$ groups and an asymmetric stretch of the C—C bonds. This is in intuitive accord with the necessary atomic motions shown above for rearrangement. However, it is nonetheless distressing that the inherent $D_{3h}$ symmetry of cyclopropane has to be so violated prior to reaction. Second, this reaction is concerted, despite the high barrier. The author uses the term "concerted" to mean simultaneous bond breaking and making, as well as the absence of diradical character in the transition state. Considerable care must thus be taken when describing a reaction as concerted or not, as earlier emphasized by Lowe.[150] Third, while the earlier mechanisms[143–147] involving a trimethylene radical are seemingly in error, one must also realize the lack of flexibility inherent in describing a molecule in terms of "—" (bonds), "·" (unpaired electrons), and "---" (partial or delocalized bonds).

However, one should not summarily disregard the trimethylene diradical. Cyclopropanes undergo two rearrangements into the corresponding propylene other than isomerization. These are optical[151–155] and geometrical[152–159] isomerizations, both somewhat faster than the previous process. 1,2-Disubstituted cyclopropanes exist in two forms, *cis* and *trans*. If the two substituents are different, both isomers are chiral. However, if the two substituents are the same, the former is achiral but the latter is still chiral. Both optical and geometrical isomerization of cyclopropanes have been suggested to go through the intermediacy of trimethylene diradicals.[145–148,156–159] Let us consider the

[150] J. P. Lowe, *J. Chem. Educ.* **51**, 784 (1974).

[151] B. S. Rabinovitch, E. W. Schlag, and K. B. Wiberg, *J. Chem. Phys.* **28**, 504 (1958).

[152] R. G. Bergman, *in* "Free Radicals" (J. Kochi, ed.), Vol. 1, pp. 191–237. Wiley, New York, 1973.

[153] H. E. O'Neal and S. W. Benson, *in* "Free Radicals" (J. Kochi, ed.), Vol. 2, pp. 275–359. Wiley, New York, 1973.

[154] J. A. Berson, L. D. Pedersen, and B. K. Carpenter, *J. Am. Chem. Soc.* **98**, 122 (1976).

[155] X. Chapuisat and Y. Jean, *Top. Curr. Chem.* **68**, 1 (1976).

[156] R. J. Crawford and T. T. Lynch, *Can. J. Chem.* **46**, 1457 (1968).

[157] J. A. Berson and J. H. Balquist, *J. Am. Chem. Soc.* **90**, 7344 (1968).

[158a] W. L. Carter and R. G. Bergman, *J. Am. Chem. Soc.* **90**, 7344 (1968).

[158b] R. G. Bergman and W. L. Carter, *J. Am. Chem. Soc.* **91**, 7411 (1969).

[159] M. R. Willcott, III and V. H. Cargle, *J. Am. Chem. Soc.* **91**, 4310 (1969).

structure of these radicals and, in particular, the parent species. One might think that there should be free rotation around both C—C bonds in $\cdot CH_2$—$CH_2$—$CH_2 \cdot$. This would parallel the situation in the more common and seemingly corresponding hydrocarbon $CH_3CH_2CH_3$[160] and monoradical $CH_3CH_2CH_2 \cdot$.[161] However, owing to the possibility of the two unpaired electrons interacting in the diradical, the situation is considerably more complicated. Numerous structures may be drawn for trimethylene diradical, all identifiable as "fine-tuned" modifications of the classical $CH_2$—$CH_2$—$CH_2 \cdot$. As with the parent and closed cyclopropane, quantum-chemical calculations have been numerous[94b,155,162-168b] and suggestive. The most commonly studied structures have been the (0, 0) or $\pi$, **16**; the (0, 90), **17**; the

(90, 90), **18**; and the so-called "crab," **19**. The precise energy differences and energy barriers between these species vary with the calculational method, and the corresponding experimental values are still not definitive. However, through two ingenious studies of deuterated cyclopropane[154] and "theoretical chemical dynamics,"[155,164a-c,168a,b] the following conclusions may be deduced. The fastest process is optical isomerism, which proceeds primarily through synchronous, conrotatory[169a,b] rotation of two methylene groups. (For example, $R$-cyclopropane $\rightleftharpoons$ **18** $\rightleftharpoons$ **16** $\rightleftharpoons$ **18'** $\rightleftharpoons$ $S$-cyclopropane: in this example and the subsequent one, **18'** is the enantiomer of a substituted

[160] C. B. Kistiakowsky and W. W. Rice, *J. Chem. Phys.* **8**, 610 (1944).

[161] P. J. Krusic and J. K. Kochi, *J. Am. Chem. Soc.* **93**, 846 (1971).

[162] R. Hoffmann, *J. Am. Chem. Soc.* **90**, 1475 (1968).

[163] R. J. Buenker and S. D. Peyerimhoff, *J. Phys. Chem.* **73**, 1299 (1969).

[164a] L. Salem, *Bull. Soc. Chim. Fr.* p. 3161 (1970).

[164b] Y. Jean and L. Salem, *Chem. Commun.* p. 382 (1971).

[164c] J. A. Horsley, Y. Jean, C. Moser, L. Salem, R. M. Stevens, and J. S. Wright, *J. Am. Chem. Soc.* **94**, 279 (1972).

[165a] A. K. Q. Siu, W. M. St. John, III, and E. F. Hayes, *J. Am. Chem. Soc.* **92**, 7249 (1970).

[165b] E. F. Hayes and A. K. Q. Siu, *J. Am. Chem. Soc.* **93**, 2090 (1971).

[166] P. J. Hay, W. J. Hunt, and W. A. Goddard, III, *J. Am. Chem. Soc.* **94**, 638 (1972).

[167] N. Bodor, M. J. S. Dewar, and J. S. Wasson, *J. Am. Chem. Soc.* **94**, 9095 (1972).

[168a] Y. Jean and X. Chapuisat, *J. Am. Chem. Soc.* **96**, 6911 (1974).

[168b] X. Chapuisat and Y. Jean, *J. Am. Chem. Soc.* **97**, 6325 (1975).

[169a] R. Hoffmann and R. B. Woodward, *Acc. Chem. Res.* **1**, 17 (1968).

[169b] R. B. Woodward and R. Hoffmann, "The Conservation of Orbital Symmetry." Academic Press, New York, 1969.

derivative of **18**.) The next fastest process is seemingly geometric isomerism, which entails single methylene rotation. (For example, *cis*-cyclopropane ⇌ **18** ⇌ **17** ⇌ **18'** ⇌ *trans*-cyclopropane.) The slowest process is the isomerization to propene. In the above, we said "for example" since species **16, 17,** and **18** and intermediate $(\theta, \pm\phi)$ structures do not exhaust all modes of interconversion or automerization[170] of cyclopropanes. For example, besides synchronous conrotatory rotation of two methylene groups, there is the corresponding synchronous disrotatory process. Although seemingly "forbidden,"[170] the energy of activation is calculated[155,168a,b] to be but 2.1 kcal/mole less favorable than that of the "allowed" reaction (59.8 kcal/mole vs. 61.9 kcal/mole). Indeed, the single methylene rotation process has an activation energy of 61.3 kcal/mole. It is apparent, at least to the authors, that the intuitively understandable notions of "concerted," "allowed," and "forbidden" may depend on surprisingly minimal energy differences. One should also not discount the idea of $(\theta, \pm\phi)$ intermediates, i.e., those with neither twofold nor mirror symmetry. Indeed, the whole idea of well-defined structural intermediates appears suspect. One must seemingly consider dynamics and modes of internal energy. In particular, it is not sufficient to merely impart to a molecule the total requisite energy of activation, but rather the appropriate internal modes must be excited. A recent illustration of this principle was the use of ir frequency radiation to completely displace an equilibrium interconvertible by Cope rearrangement.[171] In Chapuisat and Jean[155] and Siu *et al.*,[165a,b] the following computed dependence on "rotational" energy was noted for the automerization of cyclopropane: $E_{rot} \leq 10$ kcal/mole, no reaction (a so-called nonreactive trajectory); 12 kcal/mole $\leq E_{rot} \leq 35.2$ kcal/mole, reaction; 36 kcal/mole $\leq E_{rot} \leq E_{act}$, no reaction again. Additionally, these authors showed that even with rotational energies exceeding the energy of activation, there was not necessarily a reaction. Of course, one cannot rigorously separate internal rotational and vibrational energies. [Analogously, as one cannot separate individual C—C bond stretching from C—C—C angle compression in cyclopropane, we now understand somewhat better how C—C stretching frequencies in cyclopropane can indeed exceed those of propane (Section 2.A.1).] It would appear that the simple ideas of molecular structure and reaction coordinates apparently have been replaced by potential-energy surfaces and trajectories. While profound understanding and even intuition may be achievable at this newer level of complexity,[155,168a,b] it is nonetheless distressing to us that such complexity is necessary to understand what seems such a simple and long-studied reaction.

A corresponding molecular-dynamics analysis has been performed on

---

[170] L. Salem, *Acc. Chem. Res.* **4,** 322 (1971).
[171] I. Glatt and A. Yoger, *J. Am. Chem. Soc.* **98,** 7087 (1976).

another route to trimethylenes and cyclopropanes, the reaction of carbenes with olefins. It has long been known that singlet carbenes generally add stereospecifically to olefins while the corresponding triplet carbenes generally do not.[172a-175b] This finding has proven invaluable in assigning spin states to carbenes and is generally known as the Skell rule or hypothesis. The conventional interpretation is that in order for triplet carbenes to react with olefins to form cyclopropanes there must be spin flipping. Since this triplet–singlet interconversion takes time, there exists the possibility of C—C bond rotation in the intermediate trimethylene with accompanying loss of stereospecificity. It is important to realize, however, that singlet carbenes cannot directly add to olefins while maintaining $C_{2v}$ symmetry. More precisely, whether one considers orbital symmetry directly,[170] "molecular orbital following,"[176a] "orbital phase continuity,"[176b] or does a complete quantum-chemical calculation,[164a-c,168a,b,177] there is an accompanying energy of activation for this mode of reaction. Several alternative transition states with non-$C_{2v}$ symmetry may be drawn: **20**, **21**, and **22**. It is a tribute to the skill

and intuition of experimental chemists[178a-c] that the relatively symmetric **22** was disqualified prior to the quantum-chemical studies. Carbenes were compared with carbonium ions (formally isoelectronic) or other species lacking two electrons. In principle, carbenes are both nucleophilic and

[172a] P. S. Skell and R. C. Woodworth, *J. Am. Chem. Soc.* **78**, 4496 (1956).

[172b] P. S. Skell and A. Y. Garner, *J. Am. Chem. Soc.* **78**, 5430 (1956).

[173] W. Kirmse, "Carbene Chemistry," 2nd Ed., Ch. 8. Academic Press, New York, 1971.

[174] D. Bethell, *in* "Organic Reactive Intermediates" (S. P. McManus, ed.), pp. 61–126. Academic Press, New York, 1973.

[175a] R. A. Moss, *in* "Carbenes" (M. Jones, Jr. and R. A. Moss, eds.), Vol. 1, pp. 153–304. Wiley (Interscience), New York, 1973.

[175b] P. P. Gaspar and G. S. Hammond, *in* "Carbenes" (M. Jones, Jr. and R. A. Moss, eds.), Vol. 2, pp. 207–362. Wiley (Interscience), New York, 1975.

[176a] H. E. Zimmerman, *Acc. Chem. Res.* **5**, 393 (1972).

[176b] W. A. Goddard, III, *J. Am. Chem. Soc.* **94**, 793 (1971).

[177] R. Hoffmann, D. M. Hayes, and P. S. Skell, *J. Phys. Chem.* **76**, 664 (1972).

[178a] P. S. Skell and A. Y. Garner, *J. Am. Chem. Soc.* **78**, 3409 (1956).

[178b] P. S. Skell and M. S. Cholod, *J. Am. Chem. Soc.* **91**, 7131 (1969).

[178c] W. R. Moore, W. R. Moser, and J. E. LaPrade, *J. Org. Chem.* **28**, 22 (1963).

electrophilic, and it is those successful interrelations that so strongly suggest general electrophilic behavior. In one respect this comparison has been unfair, in that charged species may be expected to be much more electrophilic than the corresponding neutral species (e.g., see Collins et al.[134,135] and Pepperberg et al.[136]). It is perhaps fair to say that this conclusion is also in accord with qualitative logic derived from ionization potentials and electron affinities. That is, the ionization potential of the simplest olefin and carbene, $C_2H_4$[96] and $CH_2$,[98] are comparable, while the electron affinity of the olefin[105a] is markedly less than that of the carbene.[109] We accordingly expect electron donation from the olefin to the carbene.

Of course, it is necessary to somehow transform the olefin–singlet carbene complex into the cyclopropane. Quantum-chemical calculations show[164a–c,169a,b,176a,b] a well-defined path that interconverts these species, at least for several model olefins and model carbenes. (In contrast, the corresponding olefin–triplet carbene connects only to the triplet trimethylene diradical, and Jahn–Teller logic[103,132a,b] may be used to show that triplet cyclopropane cannot have threefold symmetry.) We hesitate to use the word "trajectory" as applied earlier, because no dynamics study was performed. In particular, there is no input as to the amount of internal energy carried by the compounds of interest. We had just said that singlet carbenes react with olefins with high stereospecificity. Unfortunately, this is *only* the case with carbene–olefin reactions that are in the condensed phase, in which the solvent can "cool" or "quench" the intermediate. Gas-phase $CH_2$, synthesizable from photolyzing ketene or diazomethane, reacts with olefins not only with comparatively little stereospecificity but also with considerable isomerization of the cyclopropane. For example, the gas-phase reaction of singlet $CH_2$ (formed by photolyzing $CH_2N_2$) and *cis*-2-butene[179a–c,180] yields both *cis*- and *trans*-1,2-dimethylcyclopropane and the rearranged pentenes. We could discuss other examples of the reactions of carbenes and olefins to form cyclopropanes and rearranged products, but we have already, perhaps, strayed too far from our original interest in strained species.

We conclude with the admission that we have spoken only of the trimethylene diradical. There is seemingly also the possibility of the trimethylene zwitterion, $^+CH_2CH_2CH_2{}^-$. We recognize this as the valence-bond equivalent of the olefin–singlet carbene intermediate. In addition, suitably substituted cyclopropanes (e.g., methyl-1-cyano-2,2-diphenyl cyclopropane-1-carboxylate, **23**) not only racemize but also yield products formed by rearrangement or

[179a] H. M. Frey, *Proc. Roy. Soc., Ser. A* **250**, 409 (1955); **251**, 575 (1959).

[179b] M. C. Flowers and H. M. Frey, *J. Chem. Soc.* p. 5550 (1961).

[179c] M. C. Flowers and H. M. Frey, *Proc. Roy. Soc., Ser. A* **260**, 424 (1961).

[180] J. W. Simons and G. W. Taylor, *J. Phys. Chem.* **73**, 1274 (1969).

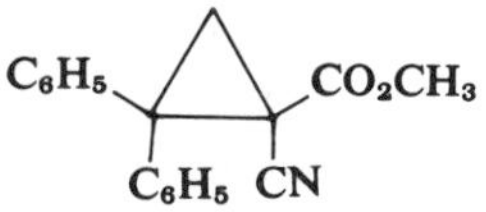

$$C_6H_5 \diagup \hspace{-0.3em} \triangle \hspace{-0.3em} \diagdown CO_2CH_3$$
$$C_6H_5 \quad CN$$

**23**

subsequent reaction of the intermediate zwitterion.[181a,b] Quantum-mechanically, the zwitterion and singlet diradical represent different reference structures of the same species, a somewhat surprising and usually ignored fact (see Bergman[152] for a good discussion of this point). Ring-opened zwitterions and diradicals are also of great relevance in understanding methylenecyclopropanes and cyclopropanones, and heterocyclic derivatives such as oxiranes and aziridines. The reader is addressed to Sections 5.D.2.C and 5.D.3, respectively, for further discussion.

## 2.C Cyclobutane

### 2.C.1 COMPARISONS WITH CYCLOPROPANE AND ACYCLIC SATURATED HYDROCARBONS

We earlier compared cyclopropane with cycloethane (Sections 2.A.3 and 2.A.5) and found that the latter was more strained. Intuitively one expects cyclobutane to be less strained than cyclopropane, because the C—C—C angle in the former should better approximate the unstrained situation of acyclic hydrocarbons, such as propane (see Section 2.C.2). Comparison of the C—H bonds shows these also to be intermediate for cyclobutane. From free-radical data of Ref. 9 we find the C—H bond strengths or dissociation energies or ethylene, cyclopropane, cyclobutane, and propane to be 109, 100, 97, and 95 kcal/mole, respectively. This trend in decreasing CH—H bond strengths is in accord with the general systematics for hydrocarbons: "highly strained rings have a proclivity commensurate with the degree of internal stresses present for acidity ... (the) resistance to H-atom abstraction, and large $J(^{13}C-H)$ coupling constants."[182] All the conceptual models we discussed for cyclopropane have counterparts for cyclobutane. Commencing with the role of d orbitals, it is shown by quantum-chemical calculations[24a,b] that they are more important for cyclopropanes than for cyclobutanes, and in turn that they are more important for cyclobutanes than acyclic hydrocarbons. In particular, in order to obtain the correct order of stability for the isomers methyl cyclopropane and cyclobutane, it is necessary to include d

[181a] D. J. Cram and A. Ratajczak, *J. Am. Chem. Soc.* **90**, 2198 (1968).
[181b] E. W. Yankee and D. J. Cram, *J. Am. Chem. Soc.* **92**, 6328, 3269, 6331 (1970).
[182] L. N. Ferguson (Ref. 8), p. 94.

orbitals on carbon.[24b] (We note at this time that the strain energies of cyclopropane and cyclobutane are so nearly comparable that the former isomer is
in fact more stable.) Maximum-overlap calculations[30,31a,b] and bent-bond
analyses[39a,b] both show cyclobutane to be more "normal" than cyclopropane.
(Strain energies may be directly computed from the former and give good
agreement with experiment for cyclobutane.[183]) A Walsh-orbital-based study
for cyclobutane and other cycloalkanes[74] showed an interesting dependence
on the number of atoms in the ring. Applications were made to hetero
analogues to explain experimentally observed ring-size-derived alternation in
numerous chemical reactions. We note analogous alternation in the chemistry
of carbocycles[184] and that a study of the thermodynamic vs. kinetic origins
of this alternation phenomenon is in progress.[185] Regrettably, little data were
presented for acyclic analogues in both Peterson[74] and Wilson and Goldhamer.[184] Walsh-orbital logic has also been successfully used to explain that,
while cyclobutane is considerably more "saturated" than cyclopropane,[186]
such "olefinic" features as conjugation with carbocationic and other unsaturated groups remain.[72,187-190] Perhaps the most incisive method is that of
group orbitals, applied to cyclobutane by Salem and Wright,[190] Jorgensen and
Salem.[191] These latter studies affirm the presence of transannular or intraring
(1,3 and 2,4) effects in cyclobutane and their large effect on the geometry of
this seemingly simple alicyclic hydrocarbon (see Section 2.C.2).[192]

### 2.C.2  GEOMETRY OF CYCLOBUTANE AND ITS DERIVATIVES

Intuitively, one would think cyclobutane should be planar and indeed have
a square (i.e., equiangular, equilateral) geometry. In this geometry, the
C—C—C angle would be 90° while any deviation from planarity would
further compress the already small C—C—C angle away from the idealized

[183] Z. B. Maksić, K. Kovacević, and M. Eckert-Maksić, *Tetrahedron Lett.* p. 101
(1975).

[184] A. Wilson and D. Goldhamer, *J. Chem. Educ.* **40**, 504 (1965).

[185] Steven D. Sprouse and Joel F. Liebman, unpublished investigations.

[186] K. B. Wiberg, K. C. Bishop, III, and R. B. Davidson, *Tetrahedron Lett.* p. 3169
(1973).

[187] A. Wilson and D. Goldhamer, *J. Chem. Educ.* **40**, 599 (1963).

[188] P. Bruckmann and M. Klessinger, *Angew. Chem. Int. Ed. Engl.* **11**, 524 (1972).

[189a] W. L. Jorgensen and W. T. Borden, *J. Am. Chem. Soc.* **95**, 6649 (1973).

[189b] W. L. Jorgensen and W. T. Borden, *Tetrahedron Lett.* p. 223 (1975).

[189c] W. L. Jorgensen, *J. Am. Chem. Soc.* **97**, 3082 (1975).

[189d] W. L. Jorgensen, *J. Am. Chem. Soc.* **98**, 6784 (1976).

[190] L. Salem and J. S. Wright, *J. Am. Chem. Soc.* **91**, 5947 (1969).

[191] W. L. Jorgensen and L. Salem (Ref. 71), pp. 26-27.

[192] J. D. Dunitz and V. Schomaker, *J. Chem. Phys.* **20**, 1703 (1952).

tetrahedral, 109.5°, or acyclic or propanelike,[1] 112° value. Or course, modifying the C—C—C angle affects both H—C—H angles and H–H interactions. While this is always true, cyclobutane is one of the few molecules in which the intricacies of the coupling of bond length and angles have been extensively unraveled.[45,193a,b] While cyclobutane has long been known to be nonplanar (24),[194] the barrier to planarity is so low[195,196a,b] that it comes as no surprise that numerous geometries have been chronicled for cyclobutane derivatives.[182,197–202] Nonetheless, Ottersen et al.[202] mention five alicyclic, unsubstituted species containing planar cyclobutanes: the diaza cycles 25 and 26,[202] bicyclo[2.1.0]pentane[203] (27), cubane[204] (28), and "basketene" dimer,[205] (29). Recent theoretical studies provide interesting insights into the nature and magnitude of these intraring interactions. In the first study,[10] it was noted that one could quantum-chemically mimic the strain energy of a distorted carbon in an alicyclic hydrocarbon with the analogously distorted methane. The strain energy for cyclobutane would then be four times that of the appropriate methane strain energy, and, indeed, good numerical agreement is achieved. However, these authors suggested that strain or deformation energy is to be related with internuclear repulsion or, in cyclobutane, the 1,3 and 2,4 carbon interactions. Since there are but two of these terms, the calculated strain energy would then be only twice the methane energy, i.e., no longer in numerical agreement. Inclusion of energies associated with eclipsing hydrogens (as in eclipsed ethane—see Chapter 1) then yields a total strain energy again in close agreement with experiment. We may thus conclude that

[193a] R. Pasternak and A. Y. Meyer, J. Mol. Struct. 13, 201 (1972).

[193b] R. Pasternak and A. Y. Meyer, J. Mol. Struct. 20, 351 (1974).

[194] A. Almenningen and O. Bastiansen, Acta Chem. Scand. 15, 711 (1961).

[195] J. Laane, in "Vibrational Spectra and Structure" (J. R. Durig, ed.), Vol. 1, pp. 25–50. Dekker, New York, 1971.

[196a] W. J. Lafferty, in "Critical Evaluation of Chemical and Physical Structural Information," pp. 386–409. Natl. Acad. Sci., Washington, D.C., 1974, D. R. Lide, Jr. and M. A. Paul (eds.).

[196b] T. B. Malloy, Jr. and W. J. Lafferty, J. Mol. Spectrosc. 54, 20 (1975).

[197] L. N. Ferguson (Ref. 8), p. 80.

[198] F. A. Cotton and B. A. Frenz, Tetrahedron 30, 1587 (1974).

[199] R. M. Moriarty, Top. Stereochem. 8, pp. 271–421 (1974).

[200] C. Cistaro, G. Fronza, R. Mondelli, S. Bradamente, and L. A. Pagini, J. Magn. Reson. 15, 367 (1974).

[201] J. R. Durig, Y. S. Li, B. A. Hudgens, and E. A. Cohen, J. Mol. Spectrosc. 459 (1976).

[202] T. Ottersen, C. Rommin, and J. P. Snyder, Acta Chem. Scand., Ser. B 30, 417 (1976).

[203] R. D. Suenram and M. D. Harmony, J. Chem. Phys. 56, 3827 (1972).

[204] E. B. Fleischer, J. Am. Chem. Soc. 86, 3889 (1964).

[205] W. J. Jones, W. D. Deadman, and E. Legoff, Tetrahedron Lett. p. 3827 (1970).

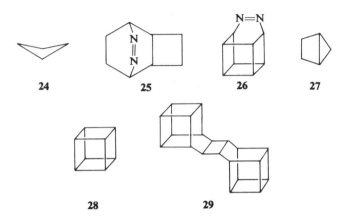

one-half the total strain energy is due to these intraring effects. The second study[206a] entailed semiempirical calculations in which these 1,3 and 2,4 interactions were "turned on and off," i.e., "the difference between the energy obtained in a normal calculation and that obtained in a corresponding calculation in which Fock matrix elements representing the 1,3 interactions were zeroed." These authors found that essentially all the strain energy of cyclobutane was due to these interactions. These authors argued against looking at a single energy component and more recently have recast most four-membered ring chemistry in their approach.[206b] In particular, the puckering of cyclobutane is a natural consequence of this analysis. Relief of steric strain is, in part, accomplished by tilting of the methylene groups.[206c]

### 2.C.3 CYCLOBUTANES, TETRAMETHYLENES, THE DIMERIZATION OF OLEFINS, AND THE CHEMISTRY OF RADICAL IONS

It has long been known that dimerization of olefins must be induced in order to form the corresponding cyclobutane.[207,208] Indeed, organic chemistry and biochemistry would differ markedly if this energetically favorable reaction were chemically spontaneous as well as thermodynamically spontaneous (i.e., $\Delta G^{\ddagger}$ was smaller though $\Delta H^{0}$ and $\Delta G^{0} < 0$). To emphasize one point, we cite the relative rarity of cyclobutanes as opposed to olefins as well as the comparative lack of reactivity of the former.[187] Additionally, we note that cyclobutanes do not react with boranes in a reaction analogous to hydro-

[206a] N. L. Bauld and J. Cessac, *J. Am. Chem. Soc.* **99**, 942 (1977).
[206b] N. L. Bauld, J. Cessac, and R. L. Holloway, *J. Am. Chem. Soc.* **99**, 8140 (1977).
[206c] D. Cremer, *J. Am. Chem. Soc.* **99**, 1307 (1977).
[207] J. D. Roberts and C. M. Sharts, *Org. React.* **12**, 1 (1962).
[208] P. D. Bartlett, *Chem. Soc. Rev.* **5**, 149 (1976).

**30**

boration.[138] We also recall the consequences of thymine (**30**) dimerization[209,210] on DNA structure and cellular function. Considerations based on "conservation of orbital symmetry"[169a,b] as well as other essentially quantum-chemical approaches (e.g., Zimmerman,[176a,211] Goddard,[176b] and Epiotis[212]) have been successfully used to explain why such dimerization reactions are generally thermally "forbidden" but photochemically "allowed." Likewise, the role of substituents in the stereospecificity and regiospecificity of this surprisingly unsimple class of reactions has been understood. For our purposes, it is only necessary to give, at this time, cross references to some other discussions of this class of reactions in our text: the transition-metal-catalyzed dimerization and dismutation of olefins (Section 5.C.2); the reaction of singlet oxygen with olefins to form 1,2-dioxetanes and subsequent "photochemistry without light" (Section 5.D.4.b); the interconversion of the valence isomers of benzene (Section 5.B.2); and the dimerization and other reactions of strained olefins (Section 3.E) and cyclobutadienes (Section 3.D.3).

In principle, olefins may dimerize to derivatives of the comparatively well-studied tetramethylene diradical,[152,153,213-216] but in general this is energetically disfavored. Cycloethanes are not so strained as to break two "two-membered rings" or $\pi$ bonds just to form one new $\sigma$ bond. A new term, "twixtyl," has been introduced[211] to describe such species as the above diradical and any other "molecule or a range of molecular conformations which is not a minimum in a potential-energy surface but which operationally behaves as a true intermediate." (For example, a species on a saddle point is a minimum on one potential curve but a maximum on another and is thus stable to some modes of vibration but not to all.) Since most studies of these diradicals started out with cyclobutanes, it is tempting to consider them as

[209] A. L. Lehninger, "Biochemistry," 2nd Ed., p. 881. Worth Publ., New York, 1975.
[210] J. D. Watson, "Molecular Biology of the Gene," 2nd Ed., pp. 293–294. Benjamin, New York, 1970.
[211] H. E. Zimmerman, *Acc. Chem. Res.* **4**, 272 (1971).
[212] N. D. Epiotis, *J. Am. Chem. Soc.* **94**, 1924, 1935, 1941 (1972).
[213] R. Hoffmann, S. Swaminathan, B. G. Odell, and R. Gleiter, *J. Am. Chem. Soc.* **92**, 7091 (1971).
[214] J. S. Wright and L. Salem, *J. Am. Chem. Soc.* **94**, 322 (1972).
[215] G. A. Segal, *J. Am. Chem. Soc.* **96**, 7892 (1974).
[216] J. J. Gajewski, *J. Am. Chem. Soc.* **98**, 5254 (1976).

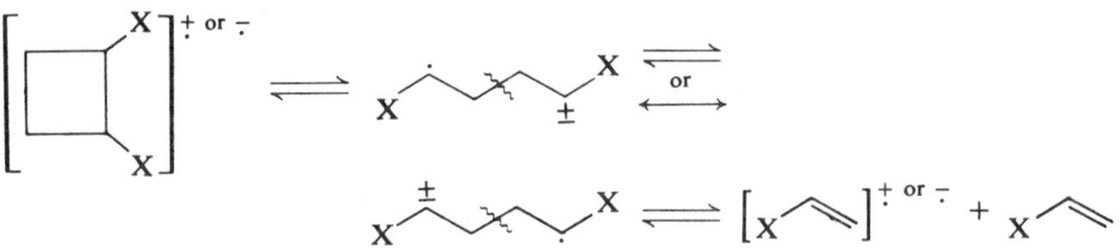

$$\left[\begin{array}{c} X \\ \square \\ X \end{array}\right]^{+\ or\ -} \rightleftharpoons \underset{X}{\overset{\cdot}{\diagup}}\!\!\diagdown\!\!\underset{\pm}{\diagup}\!\!\diagdown\ X \quad \overset{\rightleftharpoons}{\underset{or}{\longleftrightarrow}}$$

$$\underset{X}{\overset{\pm}{\diagup}}\!\!\diagdown\!\!\diagup\!\!\diagdown\ X \rightleftharpoons \left[X\diagdown\!\!\triangle\right]^{+\ or\ -} +\ X\diagdown\!\!\triangle$$

**Scheme 1**

"acyclic cyclobutanes." This allows us to consider together the radical ions of tetramethylene (or butane-1,4-diyl) and the radical ions of cyclobutanes. For example, the radical ions of 1,2-disubstituted cyclobutanes are in principle in equilibrium with those of 1,4-disubstituted butane-1,4-diyl and so with those of monosubstituted olefins (Scheme 1). Commencing with radical cations, it

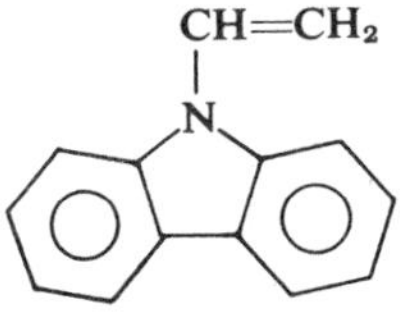

$$CH{=}CH_2$$

is found that *N*-vinyl carbazole, **31**, is catalytically dimerized to the cyclobutane[217a,b] (Scheme 2), although polymerization may also be achieved.[217a,b,218]

$$ZCH{=}CH_2 \longrightarrow ZCHCH_2{\overset{+}{\cdot}} + ZCHCH_2 \longrightarrow$$

$$(ZCHCH_2)_2{\overset{+}{\cdot}} + ZCHCH_2 \longrightarrow (ZCHCH_2)_2 + ZCHCH_2{\overset{+}{\cdot}}$$

**Scheme 2**

The reverse class of reaction, the fragmentation of cyclobutane radical cations to one molecule each of radical cationic and neutral olefin, is somewhat rare[219] but has been explained as part of the general phenomenon that mass spectrometric and photochemical reactions generally correspond.[220] Studies of the geometry and energetics of cyclobutane radical cations are relatively limited.[187,221,222a,b] In the parent species,[221] a large Jahn–Teller distortion to a diamondlike shape was calculated.[101,102] Even less is known about cyclobutane dications. The surprisingly stable bicyclo[2.2.2]octane-1,4-

---

[217a] L. P. Ellinger, *Polymer* **5**, 559 (1964).

[217b] L. P. Ellinger, *Polymer* **6**, 549 (1965).

[218] N. G. Gaylord, *Macromol. Rev.* **4**, 183 (1971).

[219] T. F. Thomas, P. J. Conn, and D. F. Swinehart, *J. Am. Chem. Soc.* **91**, 7611 (1969).

[220] R. G. Dougherty, *Top. Curr. Chem.* **45**, 93 (1974).

[221] P. Bischof, E. Haselbach, and E. Heilbroner, *Angew. Chem. Int. Ed. Engl.* **9**, 953 (1970).

[222a] R. Gleiter, E. Heilbronner, M. Heckman, and H. D. Martin, *Chem. Ber.* **106**, 28 (1973).

[222b] G. Bieri, E. Heilbronner, T. Kobayashi, A. Schmelzer, M. J. Goldstein, R. S. Leight, and M. D. Lipton, *Helv. Chim. Acta* **58**, 2657 (1976).

dication,[223a] **32**, has been investigated by low-temperature nmr. The literature explanation for its stability entails resonating ($2\pi$) butane-1,4-dications (cyclobutane dications); we present an alternate explanation below. Less stability[223b] is found for the more patently acyclic **33** (**33b** is seemingly a better description than **33a**). By analogy to our description of paracyclophanes

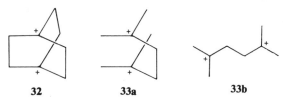

**32**                    **33a**                        **33b**

and related species (Section 3.H.3) and the "pyramidal carbonium ions" $C_5R_5^+$ and $C_6R_6^{2+}$ (Section 6.C), we suggest that the dication of interest, **32**, may also be written as $[C(CH_2)_3]_2^{2+}$. It may simply be shown that extra stability arises as the dication is the three-dimensional analogue of a "Y-aromatic" system (see Gund[224]).

We additionally interject, admittedly parenthetically, a formally iso-electronic equivalent of the reaction of an olefin dication with another olefin. From earlier comparison of carbocations and boron compounds,[134] we readily understand the facile reaction of the diboron compounds $B_2F_4$ and $B_2Cl_4$ with carbon–carbon double bonds to stereospecifically form *cis*-1,2-diboryl ethanes.[225]

The chemistry of cyclobutane radical anions and dianions is comparable to that of the radical cations and dications. Numerous cyclobutanes form radical anions, both $\sigma$ and $\pi$.[226,227] Perfluorocyclobutane,[227] as earlier predicted,[228] adds an electron in a low-lying $\sigma^*$ orbital. In contrast, benzo-cyclobutene adds an electron to a $\pi^*$ orbital as evidenced by its reduction[229] with $Li/NH_3/(CH_3)_2CHOH$ to the corresponding 1,4-cyclohexadiene (**34**).

**34**

[223a] G. A. Olah, G. Liang, P. von R. Schleyer, E. M. Engler, M. J. S. Dewar, and R. C. Bingham, *J. Am. Chem. Soc.* **95**, 6829 (1973).

[223b] G. A. Olah, J. L. Grant, R. J. Spear, J. M. Bollinger, A. Serianz, and G. Sipos, *J. Am. Chem. Soc.* **98**, 2150 (1976).

[224] P. Gund, *J. Chem. Educ.* **49**, 100 (1972).

[225] T. D. Coyle and J. J. Ritter, *Adv. Organomet. Chem.* **10**, 237 (1972).

[226] N. L. Bauld, F. R. Farr, and C. E. Hudson, *J. Am. Chem. Soc.* **96**, 5634 (1974).

[227] M. Shiatani and F. Williams, *J. Am. Chem. Soc.* **98**, 4006 (1976).

[228] J. F. Liebman, *J. Fluor. Chem.* **3**, 27 (1973–1974).

[229] P. Radlick and L. R. Brown, *J. Org. Chem.* **38**, 3412 (1973).

Both substituent and ring size changes effect marked changes on the addition of electrons. For example, 7,8-diphenyl benzocyclobutene is analogously reduced [230] to *ortho*-dibenzyl benzenes in a stereospecific, conrotatory manner. C—C bond cleavage is also observed in the related reduction of benzocyclopropene [231] to form toluene, 1-methylcyclohexa-1,4-diene, and 1,2-diphenylethane. The intermediate radical anion seems to be better expressed as **35** than as **36** (see Scheme 3). It would be interesting to know the relative electron

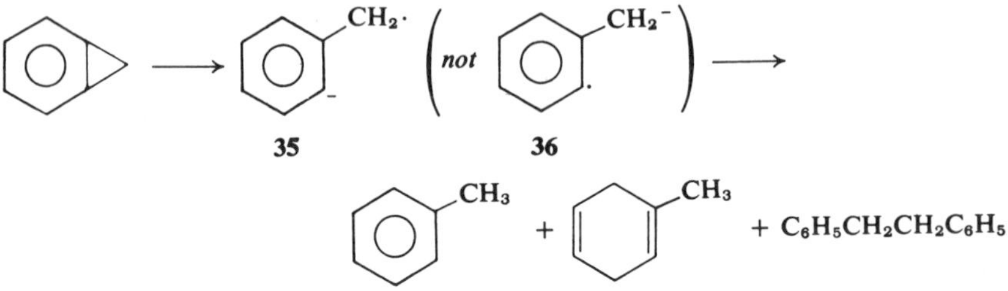

**Scheme 3**

affinities of these benzocycloalkenes and compare them with the related one-ring cycloalkenes and cycloalkanes. At the risk of mixing kinetics and thermodynamics, we may compare the capture rate of "thermal electrons" in the gas phase by perfluorocyclobutane and perfluorocyclopropane. It is found that the rate for the former exceeds that of the latter by $4 \times 10^5$.[232a] However, with the acyclic perfluorobutane and perfluoropropane, the former is but $10^4$ times faster.[232b]

With the goal of synthesizing propellanes (see Chapter 6.B.1), numerous dihalo polycyclic hydrocarbons have been reduced. Inconclusive evidence exists as to the role of cycloalkane radical anions in subsequent reduction steps. For example, reduction of dihalotryptycenes (**37**) yields tryptycene, **38**, and not tribenzo[2.2.2]propellane, **39**,[233,234] (Scheme 4). However, these radical anions are isoelectronic to the stable diamine radical cations[235,236] and the anionic species represent reasonable, if still conjectured, intermediates.

We turn now to the relationship of the radical anions and dianions of interest to the chemistry of olefins. We do not know of any catalytic dimerization of olefins to the cyclobutane via the radical anion in a way parallel to

[230] N. L. Bauld, C. S. Chang, and F. R. Farr, *J. Am. Chem. Soc.* **94**, 7164 (1972).
[231] P. Radlick and H. T. Crawford, *J. Chem. Soc., Chem. Commun.* p. 127 (1974).
[232a] K. M. Bansal and R. W. Fessenden, *J. Chem. Phys.* **59**, 1760 (1973).
[232b] K. M. Bansal and R. W. Fessenden, *J. Chem. Phys.* **53**, 3468 (1970).
[233] G. Markl and A. Mayl, *Tetrahedron Lett.* p. 1817 (1974).
[234] H. Bohm, J. Kalo, C. Varnitzky, and D. Ginsburg, *Tetrahedron* **30**, 217 (1974).
[235] R. Hoffmann, A. Imamura, and W. J. Hehre, *J. Am. Chem. Soc.* **90**, 1499 (1968).
[236] G. W. Eastland and M. C. R. Symons, *Chem. Phys. Lett.* **45**, 422 (1976).

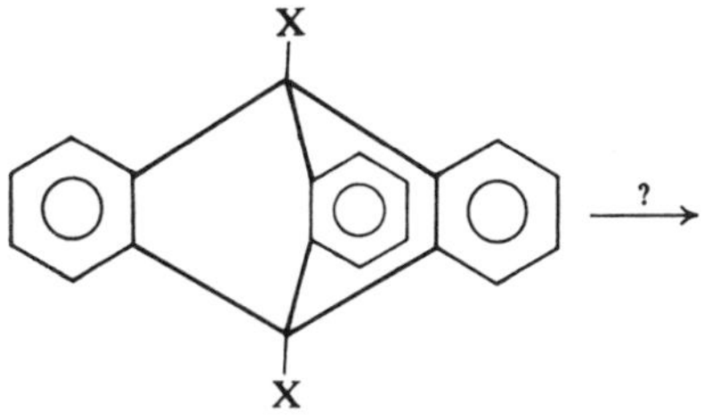

**Scheme 4**

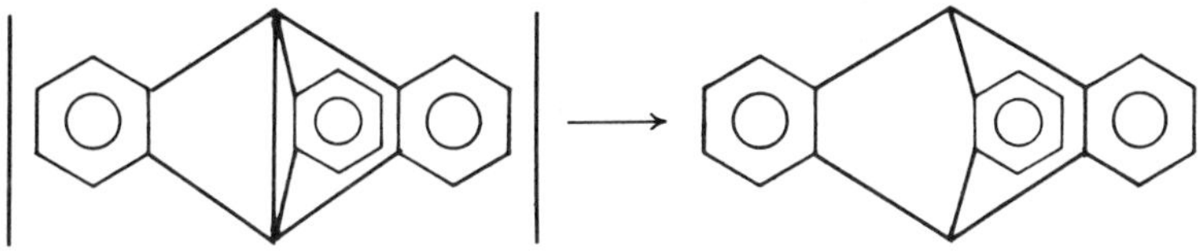

that of Scheme 1. However, depending on the olefin, $RR'C{=}CH_2$, and the precise reduction conditions, either dimeric $(RR'CHCH_2)_2$[237] or polymeric $-(RR'CCH_2)_n-(RR'CCH_2CH_2CRR')-(CH_2CRR')_n,-$[238] may be formed. It is seemingly inevitable that these reactions proceed through either the tetramethylene radical anion or dianion. With still other olefins, the tetramethylene dianion can be intercepted by reaction with triphenyl boron.[239] As this results in formation of the conjugated butadiene $RR'C{=}CH{-}CH{=}CRR'$, we have effectively oxidized the olefins by means of two species normally considered as reducing agents. Our final example parallels the reaction of olefins and olefin dianions. We recall the $\alpha$-effect,[240a,b] the anomalously high nucleophilicity of hydrazines, hydroxylamines, and hydroperoxides compared to their Brønsted basicity. A relatively recent interpretation[241] suggested a cyclic transition state for the reaction of these nucleophiles with $C{=}C$, $C{=}N$, and $C{=}O$ bonds. This transition state is isoelectronic to that formed by an olefin dianion reacting with an olefin.

[237] M. M. Baizer and J. P. Petrovich, *Prog. Phys. Org. Chem.* **7**, 189 (1970).

[238] N. G. Gaylord and S. S. Dixit, *Macromol. Rev.* **8**, 51 (1974).

[239] G. Wittig, G. Kicher, A. Ruckert, and P. Raff, *Justus Liebigs Ann. Chem.* **566**, 101 (1950).

[240a] J. O. Edwards and R. G. Pearson, *J. Am. Chem. Soc.* **84**, 16 (1962).

[240b] N. J. Fina and J. O. Edwards, *Int. J. Chem. Kinet.* **5**, 1 (1973).

[241] J. F. Liebman and R. M. Pollack, *J. Org. Chem.* **38**, 3444 (1973).

## CHAPTER 3

# Unique Strained Groupings or Building Blocks

In this chapter, we will examine groups which appear to manifest some unique aspects of chemical bonding. Thus, small monocyclic molecules such as cyclopropane and cyclobutane have unique "banana bonds" (Chapter 2), while bicyclo[1.1.0]butane and spiropentane exhibit thermochemical and chemical properties not simply explicable in terms of a sum of two cyclo-propane rings. The fact that tetrahedrane is nowhere near isolable argues that bicyclo[1.1.0]butane, which is isolable, is not adequate as a model. Various modes of distortion are examined in molecules having double bonds, triple bonds, cumulated linkages, and aromatic rings. The purpose is to provide a basic "vocabulary" of "conventional" modes of strain. Most of the polycyclic molecules discussed in Chapter 4 are well explained by the bonding present in their "building blocks" (e.g., three cyclohexane rings in adamantane; bent bonds in cubane similar to those of the six "cyclobutane faces" comprising it) modified by steric effects as well as special resonance and electrostatic effects arising from the relative orientations of these units. Chapter 4 will also briefly examine polycyclic molecules in a topological manner which ignores the seemingly artificial building-block approach. Chapter 6 will explore a basic vocabulary of "nonconventional" modes of distortion. The considerations in the present chapter are almost entirely confined to hydrocarbons; hetero-cyclic analogues are examined in Chapter 5.

It is impossible to document the reactivities of the molecules in this chapter without explicitly using conclusions and nomenclature associated with the theory of conservation of orbital symmetry or related approaches. While most readers of this book are probably well versed in these considerations, we will recommend References 10–19 in Chapter 5 in addition to our brief discussion in that chapter for purposes of introduction.

## 3.A Cycloalkanes

Table 3.1 provides a list of enthalpies of formation and strain energies for cycloalkanes up to $(CH_2)_{17}$. As discussed in the previous chapter, angle strain can, in a classical sense, be considered to be the greatest source of destabilization in cyclopropane and cyclobutane. Planar cyclopentane maintains an internal angle of 108°—rather close to tetrahedral as well as to the approximately 112° that acyclic molecules prefer. However, in order to relieve nonbonded repulsions, cyclopentane departs from planarity and exhibits a very facile pseudorotation conformational process.[1,2a,b] Cyclohexane exists in the chair conformation (1) in which bond angles and torsional angles are virtually ideal.[3] Appropriate substitution and bridging can produce cyclohexane rings locked in the boat (2) and twist-boat (3) conformations. For

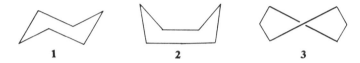

1         2         3

example, bicyclo[2.2.2]octane, introduced in Chapter 1, has three boat cyclohexane faces, while twistane (Chapter 4) has four twist-boat faces. Adamantane has four chair faces (Chapter 4), and the ubiquity of this and other diamondoid structures in hydrocarbon rearrangements as well as diamondoid conformations (see below) of certain cycloalkanes is largely attributable to minimization of angle and torsional strain. While cyclohexane has most frequently been assumed to be strainless [a strain-free —$CH_2$— increment can be calculated using this assumption by taking $\Delta H^0(C_6H_{12})$ and dividing by six to obtain $-4.93$ kcal/mole[4]], Schleyer et al. find a small destabilization (Table 3.1) attributable to the presence of nonbonded repulsions which an idealized acyclic molecule can avoid.[5] The enthalpy of the twist-boat conformer of cyclohexane is some 5.5 kcal/mole higher than that of the chair. However, reduced symmetry and especially the lack of rigidity in 3 give it a significantly higher entropy than 1. Thus, an ir spectrum of the twist boat has been obtained through heating cyclohexane to 800°K (at which the entropy term favoring the twist-boat approaches the enthalpy term in

[1] W. J. Adams, H. J. Geise, and L. S. Bartell, *J. Am. Chem. Soc.* **92**, 5013 (1970).

[2a] C. Altona, *in* "Conformational Analysis: Scope and Present Limitations" (G. Chiurdoglu, ed.), pp. 1–13. Academic Press, New York, 1970.

[2b] R. L. Lipnick, *J. Mol. Struct.* **21**, 411, 423 (1974).

[3] M. Davis and O. Hassel, *Acta Chem. Scand.* **17**, 1181 (1963).

[4] S. W. Benson, "Thermochemical Kinetics," 2nd Ed. Wiley, New York, 1976.

[5] E. M. Engler, J. D. Andose, and P. von R. Schleyer, *J. Am. Chem. Soc.* **95**, 8005 (1973).

**Table 3.1**

Experimental and Idealized ("Strainless") Enthalpies of Formation of Cycloalkanes and Strain Energies Based upon an Enthalpy of Formation of −5.15 kcal/mole for a Hypothetical Strainless —$CH_2$— Increment (see Ref. 5)[a]

| Ring size | $\Delta H_f$(g)(expt) | "Strainless" $\Delta H_f$ | Total strain | Strain/$CH_2$ |
|---|---|---|---|---|
| 2[b] | +12.45 | −10.30 | 22.8 | 11.4 |
| 3 | +12.73 | −15.45 | 28.3 | 9.4 |
| 4 | +6.78 | −20.60 | 27.4 | 6.85 |
| 5 | −18.44 | −25.75 | 7.3 | 1.45 |
| 6 | −29.50 | −30.90 | 1.4 | 0.2 |
| 7 | −28.21 | −36.05 | 7.8 | 1.1 |
| 8 | −29.73 | −41.20 | 11.5 | 1.4 |
| 9 | −31.73 | −46.35 | 14.6 | 1.6 |
| 10 | −36.88 | −51.50 | 14.6 | 1.45 |
| 11 | −42.87 | −56.65 | 13.8 | 1.25 |
| 12 | −55.03 | −61.80 | 6.8 | 0.6 |
| 13 | −58.88 | −66.95 | 8.1 | 0.6 |
| 14 | −57.13 | −72.10 | 15.0 | 1.1 |
| 15 | −72.04 | −77.25 | 5.2 | 0.35 |
| 16 | −76.88 | −82.40 | 5.5 | 0.35 |
| 17 | −87.07 | −87.55 | 0.5 | 0.33 |

[a] All values are in kcal/mole.
[b] Ethylene = "cycloethane."

magnitude), and immediately transferring it to a cell at 10°K (at which the conformational conversion to the chair is slow).[6a] (See the discussion in Chapter 1 concerning the relative importance of enthalpy and entropy.) The first example of a planar cyclohexane ring has been furnished by the molecule *syn*-benzene trioxide.[6b] Numerous studies have been made of the conformations of substituted cyclohexanes.[7a–j]

[6a] M. Aquillacte, R. S. Sheridan, O. L. Chapman, and F. A. L. Anet, *J. Am. Chem. Soc.* **97**, 3244 (1975).

[6b] W. Littke and U. Drück, *Angew. Chem., Int. Ed. Engl.* **13**, 539 (1974).

[7a] E. L. Eliel, "Stereochemistry of Carbon Compounds," Ch. 8. McGraw-Hill, New York, 1962.

[7b] E. L. Eliel, N. L. Allinger, S. J. Angyal, and G. A. Morrison, "Conformational Analysis," Ch. 2. Wiley (Interscience), New York, 1965.

[7c] E. W. Garbisch, Jr., B. L. Hawkins, and K. D. Mackay, *in* "Conformational Analysis: Scope and Present Limitations," G. Chiurdoglu (ed.), pp. 93–109. Academic Press, New York, 1970.

[7d] M. Hanack, "Conformation Theory," Ch. 2. Academic Press, New York, 1965.

[7e] F. R. Jensen and C. H. Bushweller, *Adv. Alicyclic Chem.* **3**, pp. 140–194. (1971).

The strain in medium and large cycloalkanes is almost entirely due to non-bonded or transannular repulsions.[8a-g] These molecules usually have a number of conformations of similar free energy and are therefore conformationally labile. The larger cycloalkanes are thus often described as "floppy," and experimental and theoretical investigations of the structures of these molecules are indeed challenging.

Cycloheptane's most stable conformation is the "twist chair" (TC) (4)

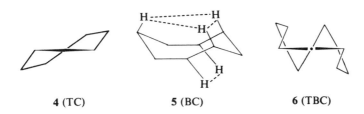

| 4 (TC) | 5 (BC) | 6 (TBC) |

having $C_2$ symmetry,[9,10] but other conformations are not much higher in energy and the molecule readily undergoes pseudorotation. Odd-membered cycloalkanes, including cycloheptane, are irregular in shape and especially labile conformationally. The strain in cyclooctane, which exists almost entirely in the boat-chair conformation (BC) (5), arises from virtual eclipsing

[7f] G. M. Kellie and F. G. Riddell, *Top. Stereochem.* **8**, pp. 225–269 (1974).

[7g] J. Reisse, *in* "Conformational Analysis: Scope and Present Limitations" (G. Chiurdoglu, ed.), pp. 219–228. Academic Press, 1971.

[7h] M. J. T. Robinson, *Tetrahedron* **30**, 1971 (1974).

[7i] R. D. Stolow, *in* "Conformational Analysis: Scope and Present Limitations" (G. Chiurdoglu, ed.), pp. 251–258. Academic Press, New York, 1970.

[7j] K. B. Wiberg and A. Shrake, *Spectrochim. Acta Part A* **29**, 583 (1973).

[8a] F. A. L. Anet, *in* "Conformational Analysis: Scope and Present Limitations" (G. Chiurdoglu, ed.), pp. 15–29. Academic Press, New York, 1970.

[8b] M. S. Baird, *MTP Int. Rev. Sci., Org. Chem., Ser. 1*, **5**, pp. 205–238 (1973).

[8c] J. D. Dunitz, *in* "Perspectives in Structural Chemistry" (J. D. Dunitz and J. A. Ibers, eds.), Vol. 2, pp. 1–70. New York, 1968.

[8d] E. L. Eliel, "Stereochemistry of Carbon," pp. 252–265. McGraw-Hill, New York, 1962.

[8e] E. L. Eliel, N. L. Allinger, S. J. Angyal, and G. A. Morrison, "Conformational Analysis," pp. 206–223. Wiley (Interscience), New York, 1965.

[8f] M. Hanack, "Conformation Theory," pp. 158–166. Academic Press, New York, 1965.

[8g] W. D. Ollis, J. F. Stoddart, and I. O. Sutherland, *Tetrahedron* **30**, 1903 (1974).

[9] J. B. Hendrickson, R. K. Bockman, Jr., J. D. Glickman, and E. Grunwald, *J. Am. Chem. Soc.* **95**, 494 (1973).

[10] E. S. Glazer, R. Knorr, C. Ganter, and J. D. Roberts, *J. Am. Chem. Soc.* **94**, 6026 (1972).

about two bonds as well as through nonbonded repulsions.[11] The most stable conformation of cyclononane is the twist-boat-chair (TBC) of $D_3$ symmetry (**6**).[12] Cyclodecane is most stable in the "diamond-lattice" boat–chair–boat (BCB) or [2323][13a–c] conformation (**7**).[14] Such "diamond-lattice" structures, which are possible for even-membered rings,[15] minimize angle and torsional strain. Any residual nonbonded interactions in an idealized "diamond-lattice" structure may be minimized further by slight distortion. The six repulsive interactions highlighted in **7** are the predominant source of its 14.6

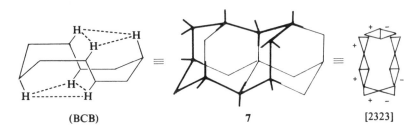

(BCB)            **7**            [2323]

kcal/mole of strain. While strain per $CH_2$ (Table 3.1) is a useful parameter for symmetric or virtually symmetric compounds such as cyclopropane, cyclobutane, cyclopentane, and cyclohexane, it is quite misleading in the analysis of cyclodecane, as only six methylene groups are contributing significantly to strain. While the larger rings are virtually strain free, because their large carbocyclic networks can apportion small destabilization energies among a large group of bonds, cyclotetradecane appears to be somewhat unique (15 kcal/mole of strain). This molecule now is known to exist in the diamondoid [3434] ($C_{2h}$) rectangular conformation, **8**.[16] The same kind of transannular repulsions present in **7** are also in this molecule, perhaps somewhat reinforced by buttressing. Cyclohexadecane exists predominantly in a diamondoid [4444] conformation and is virtually strain free.[16]

    Catenanes are molecules having interlocked rings. Molecular models

[11] F. A. L. Anet, *Fortschr. Chem. Forsch.* **45**, 169 (1974).

[12] F. A. L. Anet and J. J. Wagner, *J. Am. Chem. Soc.* **94**, 9250 (1972).

[13a] [2323] designates a four-sided structure of two, three, two, and three C—C bonds per side, in which corners are defined by groups having adjacent gauche bonds of identical (torsional angle) sign. Similarly, the [3434] conformation of cyclotetradecane is a four-sided structure of three, four, three, and four C—C bonds, respectively (see 13b and 13c below).

[13b] J. Dale, *Acta Chem. Scand.* **27**, 115 (1973).

[13c] J. Dale, *Top. Stereochem.* **9**, 199 (1976).

[14] E. A. Noe and J. D. Roberts, *J. Am. Chem. Soc.* **93**, 5266 (1971).

[15] M. Saunders, *Tetrahedron* **23**, 2105 (1967).

[16] F. A. L. Anet and A. K. Cheng, *J. Am. Chem. Soc.* **97** 2420 (1975).

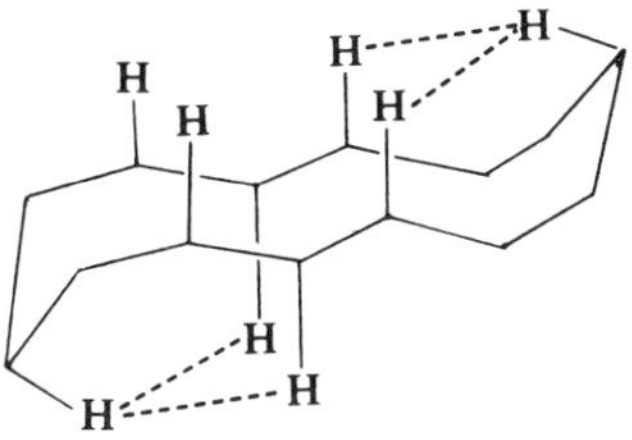

**8**

predict that (18,18)-catenane (**9**) should be the smallest isolable $(C_nH_{2n})_2$.[17] Here one should recall that space-filling models often overestimate steric repulsion and do not allow significant bond-angle and bond-length distortion.

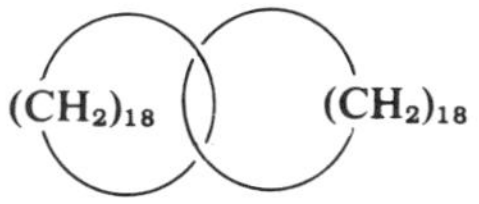

**9**

Molecular-mechanics calculations of catenanes are difficult because of the large size of even the smallest catenanes, the "floppiness" of the rings, and the highly sensitive and crucial dependence of strain energies on the proper choice of nonbonded potentials.[18] With somewhat softened potentials, perhaps the total strain in (16,16)-catenane will not quite total the C—C bond dissociation energy of ca. 85 kcal/mole, which should be reflected in the $E_{act}$ of a homolytic unlocking of this catenane. Furthermore, one might consider seemingly impossible small catenanes to see whether they are capable of survival. For example, a planar symmetric conformer of cyclo-dodecane has a cavity which might hold a methylene chain. An exceedingly crude estimate of the strain in planar cyclododecane (C—C—C angle of 150°, distortion per —CH$_2$— of 38°) can be obtained by using a strain energy of ca. 8.5 kcal per CH$_2$ through comparison of the strain per CH$_2$ of cyclo-propane (C—C—C angle distortion 52°, strain 9.4 kcal/mole) and cyclo-butane (—CH$_2$— angle distortion 22°, strain 6.9 kcal/mole). The total strain in planar cyclododecane would be about 100 kcal/mole, and that in (12,12)-catenane about 200 kcal/mole, ignoring steric repulsions. Homolytic scission of a C—C bond of one ring would have a transition state in which one ring is still intact. The release in strain upon cleavage of one C—C bond would not entirely relieve the strain in both rings. Although the unlocking of (12,12)-catenane would be enormously exothermic, it may exhibit an activation barrier large enough to allow its observation at low temperature. Synthesis

---

[17] G. Schill, "Catenanes, Rotaxanes, and Knots." Academic Press, New York, 1971.
[18] Profs. Paul von R. Schleyer and Eiji Osawa, personal communication to Arthur Greenberg.

of this and other very small catenanes would be an immense challenge, because the extrusion reactions most obvious as candidates might well cause premature unlocking of larger homologous catenanes. Perfluorinated rings might also be worthy of investigation, as the $-CF_2-CF_2-$ bond is relatively strong (see Benson[4]).

## 3.B  Bicyclic, Spirocyclic, and Bridged Hydrocarbons

### 3.B.1  cis-BICYCLIC MOLECULES[19a,b]

Cyclopropane and cyclobutane have a single angular constraint per carbon, while molecules such as bicyclo[1.1.0]butane (11), bicyclo[1.1.1]pentane (14), and spiropentane (45) contain one or more carbons having two independently constrained angles.[20] Strain in these molecules is explicable in the same manner as for cyclopropane.[20] Chapter 6 will consider molecules having more than two independently constrained angles per carbon for which some non-classical bonding schemes have been proposed.

Ethyl-1-bicyclo[1.1.0]butanecarboxylate, the first reported member of the bicyclobutane family, was obtained via an internal displacement reaction.[21] The next compound to be characterized was tricyclo[4.1.0.$^{2,7}$]heptane (10),

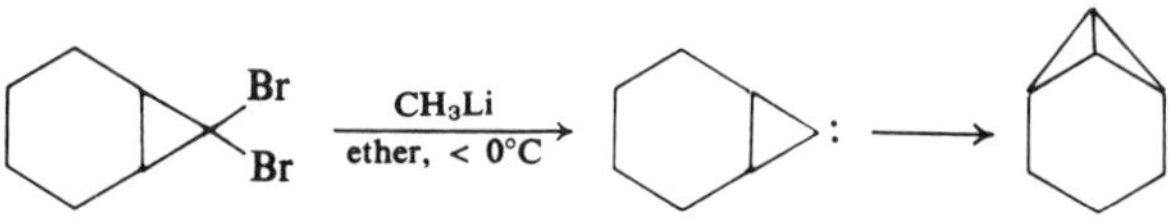

$$(1)$$

**10**

which, in contrast to the first derivative, was stable in the absence of acid.[22] Three different methods for synthesis of bicyclo[1.1.0]butane itself appeared during 1963. The first involved generation of allyl methylene and an internal cycloaddition which eventually produced 1,3-butadiene as the major product

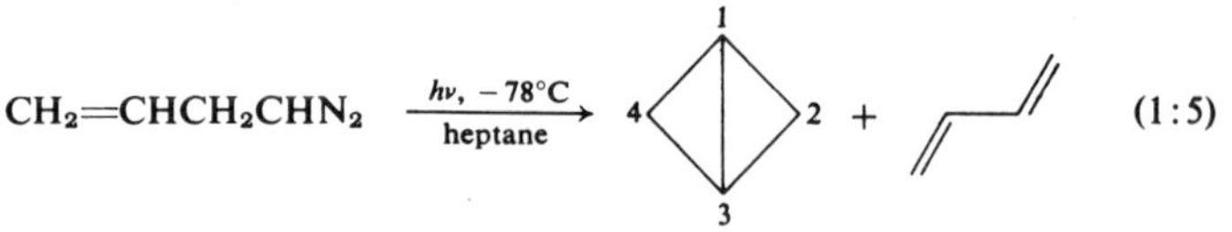

$$(1:5) \quad (2)$$

**11**

---

[19a] J. Meinwald and Y. C. Meinwald, *Adv. Alicyclic Chem.* **1**, pp. 1–51 (1966).

[19b] K. B. Wiberg, *Adv. Alicyclic Chem.* **2**, pp. 185–254 (1968).

[20] K. B. Wiberg and G. J. Burgmaier, *J. Am. Chem. Soc.* **94**, 7396 (1972).

[21] K. B. Wiberg and R. P. Ciula, *J. Am. Chem. Soc.* **81**, 526 (1959).

[22] W. R. Moore, H. R. Ward, and R. F. Merritt, *J. Am. Chem. Soc.* **83**, 2019 (1961).

[Eq. (2)].[23] A second synthesis, involving reduction of 1-bromo-3-chloro-cyclobutane, produced **11** in more than 90% yield [Eq. (3)].[24] A third method

$$Cl-\diamond\!\!-Br \xrightarrow[\text{dioxane}]{\text{Na}} \diamond + \square \quad (19:1) \tag{3}$$

**11**

involved photochemical synthesis from 1,3-butadiene.[25] Photochemical synthesis was not successful in the vapor phase because any bicyclobutane created was presumably in a "hot" state and could not be collisionally deactivated before reverting to starting material. Success was achieved in carbon tetrachloride solution, where such "hot" bicyclobutane could collide with inert solvent molecules and become deactivated. The early chemistry reported for bicyclo[1.1.0]butane included enormous sensitivity to acid, conversion in solution to 1,3-butadiene in about four minutes at 150°C, and very facile addition of iodine across the central bond, indicating its considerable olefinic character.[23-25] The chemical and physical properties of the bicyclo[1.1.0]butane system have been summarized.[26a,b] (Mechanisms for conversion of bicyclo[1.1.0]butane to 1,3-butadiene will be discussed briefly in Section 5.C.3.a.) In line with theoretical expectations, a high $J(^{13}C{-}H)$ coupling constant indicates that the bridgehead carbon orbital directed to hydrogen has about 40% s character.[27] This would lead one to expect relative acidic bridgehead protons. In fact, tert-butoxide/tert-butanol induces rapid isotopic exchange of bridgehead hydrogen in **10**.[27]

As Table 3.2 shows, the strain in bicyclobutane is more than 10 kcal/mole greater than the strain of two cyclopropanes (in contrast, the strain in bicyclo[2.1.0]pentane is equal to the strain of a cyclopropane and a cyclobutane). If it is assumed that bridgehead carbons in **11** are twice as strained as bridge carbons, one calculates a destabilization value of about 22 kcal/mole for each bridgehead carbon and 11 kcal/mole for each bridge carbon, which might be compared to 9.4 kcal/mole for each carbon in cyclopropane. Alternatively, if one arbitrarily assumes that the central bond is twice as strained as the other carbon–carbon bonds, then this linkage is destabilized by 22 kcal/mole. To what might the extra 10 kcal/mole strain increment be attributed? Qualitative Walsh-orbital descriptions of bicyclobutane predict a

[23] D. M. Lemal, F. Menger, and G. W. Clark, *J. Am. Chem. Soc.* **85**, 2529 (1963).

[24] K. B. Wiberg and G. M. Lampman, *Tetrahedron Lett.* p. 2173 (1963).

[25] R. Srinivasan, *J. Am. Chem. Soc.* **85**, 4045 (1963).

[26a] K. B. Wiberg, *Rec. Chem. Prog.* **26**, 143 (1965).

[26b] K. B. Wiberg, G. M. Lampman, R. P. Ciula, D. S. Connor, P. Schertler, and J. Lavanish, *Tetrahedron* **21**, 2749 (1965).

[27] G. L. Closs and L. E. Closs, *J. Am. Chem. Soc.* **85**, 2022 (1963).

**Table 3.2**

**Strain Energies of Some *cis*-Bicycloalkanes[a]**

| Compound | Strain energy (kcal/mole) |
|---|---|
| Bicyclo[1.1.0]butane | 66.5 |
| Bicyclo[2.1.0]pentane | 57.3 |
| Bicyclo[3.1.0]hexane | 33.9 |
| Bicyclo[4.1.0]heptane | 30.3 |
| *cis*-Bicyclo[2.2.0]hexane | 50.7 |
| *cis*-Bicyclo[3.2.0]heptane | 30.5 |
| *cis*-Bicyclo[3.3.0]octane | 12.5 |
| Bicyclo[1.1.1]pentane | 60–64[b] |
|  | 92.5[c] |
| Bicyclo[2.1.1]hexane | 41.2 |
| Bicyclo[3.1.1]heptane | 35.9 |
| Bicyclo[2.2.1]heptane | 17.0 |
| Bicyclo[2.2.2]octane | 13.0 |
| Bicyclo[3.2.1]octane | 12.1 |
| Bicyclo[3.3.1]nonane | 9.6 |
| Bicyclo[3.3.3]undecane | ca. 14[d] |

[a] For experimental enthalpies of formation and strain energies, see Refs. in this chapter.

[b] Estimated: M. D. Newton and J. M. Schulman, *J. Am. Chem. Soc.* **94**, 773 (1972).

[c] Molec. mechanics: N. L. Allinger, M. T. Tribble, M. A. Miller, and D. W. Wertz, *J. Am. Chem. Soc.* **93**, 1637 (1971).

[d] W. Parker, W. V. Steele, W. Stirling, and I. Watt, *J. Chem. Thermodyn*, **7**, 795 (1975).

significant difference in the lengths of the $C_1$—$C_3$ and $C_1$—$C_2$ bonds.[28] CNDO[29] and *ab initio* calculations[30] also have been performed on this molecule. A microwave study of bicyclo[1.1.0]butane disclosed some very striking features: a dipole moment of 0.7 D; $C_1$—$C_2$ and $C_1$—$C_3$ distances of 1.498 Å, and 1.497 Å respectively; a short $C_1$—H distance of only 1.071 Å (olefinic C—H is 1.070 Å); and unusually large $C_1$—$C_3$—H and $C_2$—$C_3$—H angles (128° and 130°, respectively).[31a,b] The *ab initio* calculation cited above reproduces this abnormally large dipole moment and attributes its magnitude and direction (colinear with $C_2$ axis, positive pole exocyclic) to endocyclic distortion of the $C_1$—$C_2$, $C_1$—$C_4$, $C_3$—$C_2$, $C_3$—$C_4$ bonds that exceeds in

[28] M. Pomerantz and E. W. Abrahamson, *J. Am. Chem. Soc.* **88**, 3970 (1966).

[29] K. B. Wiberg, *Tetrahedron* **24**, 1083 (1968).

[30] M. D. Newton and J. M. Schulman, *J. Am. Chem. Soc.* **94**, 767 (1972).

[31a] M. D. Harmony and K. W. Cox, *J. Am. Chem. Soc.* **88**, 5049 (1966).

[31b] K. W. Cox, M. D. Harmony, G. Nelson, and K. B. Wiberg, *J. Chem. Phys.* **50**, 1976 (1969).

polarization the exocyclic distortion of the $C_1$—$C_3$ bond.[30] Furthermore, the highest occupied orbital is calculated to be a canted system composed of virtually p orbitals.[30] Determination of a natural abundance $C_1$—$C_3$ $^1J(^{13}C$—$^{13}C)$ of 1-cyanobicyclo[1.1.0]butane, which might in principle indicate hybridization in the bridgehead–bridgehead bond, was exceedingly difficult owing to the reactivity of the molecule and low signal-to-noise ratio. Thus, the published disagreement with the predictions of theory[32] is presently disregarded.[33] An indirect determination of the hybridization of the central bond in bicyclobutane, through use of $^1J(^{13}C$—H) of the bridgehead carbons, $^1J(^{13}C$—$^{13}C)$ between the bridgehead ($C_1$ or $C_3$) and bridging carbons ($C_2$ or $C_4$), and assumption of unit or 100% s character of carbon, led to an estimate of 90% p character in this bond.[34] There is, however, no theoretical basis for this last assumption.[35] The magnitude and negative sign of the bridgehead–bridgehead $^1J(^{13}C$—$^{13}C)$ coupling constant of diethyl-1-methyl-3-phenyl-bicyclo[1.1.0]butane-2,4-dicarboxylate enriched in $^{13}C$ is in excellent agreement with Eq. (4)[32] when the central bond has a bit more than 91% p character.[36a,b]

$$^1J(^{13}C\text{—}^{13}C) = 0.0621(\% \, S_A)(\% \, S_B) - 10.2(\pm 2.4) \text{ Hz} \qquad (4)$$

Much of the chemistry of the central bonds in various bicyclo[1.1.0]butanes is reminiscent of the reactions of $\pi$ bonds.[19a,26a,b,37a–c] Reactive unsaturated molecules attack the bicyclobutane central bond from its electron-deficient endocyclic direction (thus, the aforementioned relative stability of 10 is due to steric hindrance opposing this favored mode of attack).[37a–c] The chemical, physical, and especially thermodynamic properties of this molecule endow it with unique properties not quantitatively predictable through reference to such model compounds as cyclopropane.

The strain energy of bicyclo[2.1.0]pentane (12) (see Table 3.2), a molecule first reported in 1957,[38] is equal to the strain energy of cyclopropane plus that of cyclobutane. A microwave study of this molecule indicates the following

[32] J. M. Schulman and M. D. Newton, *J. Am. Chem. Soc.* 96, 6295 (1974).

[33] M. Pomerantz and D. F. Hilderbrand, *J. Am. Chem. Soc.* 95, 5809 (1973); also Prof. Martin Pomerantz, personal communication to Arthur Greenberg.

[34] R. D. Bertrand, D. M. Grant, E. L. Allred, J. C. Hinshaw, and A. B. Strong, *J. Am. Chem. Soc.* 94, 997 (1972).

[35] M. Pomerantz and J. F. Liebman, *Tetrahedron Lett.* p. 2385 (1975).

[36a] M. Pomerantz and R. Fink, *J. Chem. Soc., Chem. Commun.* 430 (1975).

[36b] M. Pomerantz, R. Fink, and G. A. Gray, *J. Am. Chem. Soc.* 98, 291 (1976).

[37a] M. Pomerantz, *J. Am. Chem. Soc.* 88, 5349 (1966).

[37b] M. Pomerantz and R. N. Wilke, *Tetrahedron Lett.* p. 463 (1969).

[37c] M. Pomerantz, R. N. Wilke, G. W. Gruber, and U. Roy, *J. Am. Chem. Soc.* 94, 2752 (1972).

[38] R. Criegee and A. Rimmelin, *Chem. Ber.* 90, 414 (1957).

bond lengths: $C_1—C_4$ (1.536 Å), $C_2—C_3$ (1.565 Å), $C_1—C_2$ (1.528 Å), and $C_1—C_5$ (1.507 Å).[39] These results contradict an earlier electron-diffraction study,[40] now assumed to be in error, which found an abnormally long $C_2—C_3$ bond and a very short $C_1—C_4$ bond. Bicyclo[2.1.0]pentane is also attacked by reactive unsaturated molecules from the endocyclic direction.[41a-d] Bicyclo-[3.1.0]hexane (13) and its heterocyclic analogues are most stable in the boat conformation, which staggers vicinal (1,2 and 4,5) hydrogens better than the chair conformation.[42] Calculated strain energies of a number of *cis*-bicyclic

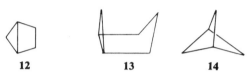

**12**                    **13**                    **14**

molecules are presented in Table 3.2. A somewhat novel feature of bicyclo-[2.2.0]hexane[43] is its relative ease of thermal decomposition (to 1,5-hexadiene, $E_{act} = 36.0$ kcal/mole[44]) compared to more strained molecules, including bicyclopentane (to cyclopentene, $E_{act} = 46.6$ kcal/mole[45]).

Bicyclo[1.1.1]pentane (14) was obtained in low yield through an intramolecular Wurtz reaction [analogous to that in Eq. (3)] involving 3-bromo-1-bromomethylcyclobutane and lithium amalgam in dioxane.[46a,b] Two gas-phase electron-diffraction investigations of the parent hydrocarbon[47,48] and microwave[49] and x-ray studies[50] of derivatives are in general agreement about the structure of the bicyclo[1.1.1]pentane framework. The molecule's most striking feature is a very short distance (1.85 Å) between nonbonded bridgehead carbons. Overlap of the back lobes of the carbon–hydrogen orbitals is invoked to explain the very high value for $^4J(H—H)$ (18 Hz).[46a,b,50] Addition-

[39] R. D. Suenram and M. D. Harmony, *J. Chem. Phys.* **56**, 3837 (1972).

[40] R. K. Bohn and Y H. Tai, *J. Am. Chem. Soc.* **92**, 6447 (1970).

[41a] P. G. Gassman, K. T. Mansfield, and T. J. Murphy, *J. Am. Chem. Soc.* **90**, 4746 (1968).

[41b] P. G. Gassman, K. T. Mansfield, and T. J. Murphy, *J. Am. Chem. Soc.* **91**, 1684 (1969).

[41c] P. G. Gassman and G. D. Richman, *J. Am. Chem. Soc.* **92**, 2090 (1970).

[41d] P. G. Gassman, *Acc. Chem. Res.* **4**, 128 (1971).

[42] R. L. Cook and T. B. Malloy, Jr., *J. Am. Chem. Soc.* **96**, 1703 (1974).

[43] S. Cremer and R. Srinivasan, *Tetrahedron Lett.* p. 24 (1960).

[44] C. Steel, R. Zand, P. Hurwitz, and S. G. Cohen, *J. Am. Chem. Soc.* **86**, 679 (1964).

[45] M. L. Halberstadt and J. P. Chesick, *J. Am. Chem. Soc.* **84**, 2688 (1962).

[46a] K. B. Wiberg, D. S. Connor, and G. M. Lampman, *Tetrahedron Lett.* p. 531 (1964).

[46b] K. B. Wiberg and D. S. Connor, *J. Am. Chem. Soc.* **88**, 4437 (1966).

[47] J. F. Chiang and S. H. Bauer, *J. Am. Chem. Soc.* **92** 1614 (1970).

[48] A. Almenningen, B. Andersen, and B. A. Nyhus, *Acta Chem. Scand.* **25**, 1217 (1971).

[49] K. W. Cox and M. D. Harmony, *J. Mol. Spectrosc.* **36**, 34 (1970).

[50] A. Padwa, E. Shefter, and E. Alexander, *J. Am. Chem. Soc.* **90**, 3717 (1968).

ally, the bridgehead reactivity of bicyclo[1.1.0]butane with *tert*-butyl hypo-chlorite,[46a,b] which contrasts with the inertness of norbornane, and the solvolytic activity of the 1-chloro derivative,[51] are manifestations of the mutual proximity of $C_1$ and $C_3$, as well as other factors. (See the next section for discussion of bridgehead carbonium ions.) The three cyclobutane faces of this symmetric ($D_{3h}$) molecule are puckered by 60°.[47,48] *Ab initio* calculations reproduce the experimental geometry of bicyclo[1.1.1]pentane, including its short bridgehead–bridgehead contact, quite well.[52] An estimated value of 60–64 kcal/mole for the strain energy in bicyclo[1.1.1]pentane,[52] which is considerably lower than a less reliable value obtained from molecular mechanics calculations,[53] allows the calculation of about 11 kcal/mole of strain per carbon–carbon bond in this molecule. Bicyclo[2.1.1]hexane[54,55] is also more active at the bridgehead than norbornane and maintains a non-bonded $C_1$—$C_4$ distance of only 2.17 Å.[56]

Larger bicyclo[*l.m.n*]alkanes are less strained than the smaller molecules (see Table 3.2), although there are no smooth trends because each molecule has its own structural idiosyncrasies. Even bicyclo[3.3.3]undecane (manx-ane),[57,58] which has bridgehead carbons almost coplanar with the methylenes, is only strained by about 14–15 kcal/mole,[59] a value 11–14 kcal/mole lower than calculated using molecular mechanics. (See Section 4.D.5.b for discussion of 1-azabicyclo bridgehead carbons.) The isomeric bicyclo[8.8.8]-hexacosanes **15** and **16** do not have this feature.[60,61] Conformational inter-conversion of **16** and the as yet unknown **17** (predicted to be less stable than **16**) has not been observed, and is said to be feasible only in higher homologues

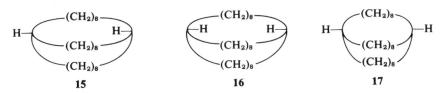

**15**          **16**          **17**

[51] K. B. Wiberg and V. Z. Williams, Jr., *J. Am. Chem. Soc.* **89**, 3373 (1967).

[52] M. D. Newton and J. M. Schulman, *J. Am. Chem. Soc.* **94**, 773 (1972).

[53] N. L. Allinger, M. T. Tribble, M. A. Miller, and D. W. Wertz, *J. Am. Chem. Soc.* **93**, 1637 (1971).

[54] R. Srinivasan, *J. Am. Chem. Soc.* **83**, 2590 (1961).

[55] K. B. Wiberg, B. R. Lowery, and T. H. Colby, *J. Am. Chem. Soc.* **83**, 3998 (1961).

[56] J. F. Chiang, *J. Am. Chem. Soc.* **93**, 5044 (1971).

[57] M. Doyle, W. Parker, P. A. Gunn, J. Martin, and D. D. MacNicol, *Tetrahedron Lett.* p. 3619 (1970).

[58] N. J. Leonard and J. C. Coll, *J. Am. Chem. Soc.* **92**, 6685 (1970).

[59] W. Parker, W. V. Steele, W. Stirling, and I. Watt, *J. Chem. Thermodyn.* **7**, 795 (1975).

[60] C. H. Park and H. E. Simmons, *J. Am. Chem. Soc.* **94**, 7184 (1972).

[61] P. G. Gassman and R. P. Thummel, *J. Am. Chem. Soc.* **94**, 7183 (1972).

containing bridges of ten or more methylene groups.[60] Analogous macro-bicyclic bridgehead diamines isomerize, but they do so via nitrogen inversions rather than by rotation about carbon–carbon bonds.[62]

While the tricyclohexanes **18**,[63] **19**,[64] and **20**[65] have been completely characterized, molecule **21** is the only representative of the tricyclo $[2.2.0.0^{2,5}]$

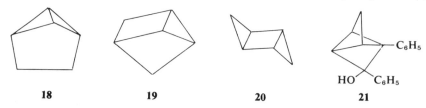

|  18  |  19  |  20  |  21  |

hexane system to be reported.[66] Only derivatives of tricyclo$[1.1.1.0^{4,5}]$pentane (**22**)[67,68] as well as its ketone (**23**)[69–71] are known. An interesting feature of **22** is the suggestion of a discrete "bond-stretch isomer," **22a**, separated by a tangible energy barrier arising from the predicted "symmetry-forbiddenness" (see Section 5.B.1) of this isomerization.[72] Novel tricycloheptanes include **10**,[22] **24**,[22,73,74] and **25**,[75] while only derivatives of **26**[76] and **27**[77,78] are known. This

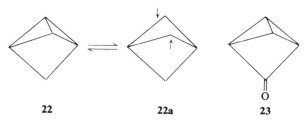

|  22  |  22a  |  23  |

[62] H. E. Simmons and G. H. Park, *J. Am. Chem. Soc.* **90**, 2428, 2429, 2431 (1968).

[63] Tricyclo[3.1.0.0$^{2,6}$]hexane: J. Meinwald, C. Swithenbank, and A. Lewis, *J. Am. Chem. Soc.* **85**, 1880 (1963).

[64] Tricyclo[2.1.1.0$^{2,5}$]hexane: D. M. Lemal and K. S. Shim, *J. Am. Chem. Soc.* **86**, 1550 (1964); Microwave Structure: R. D. Suenram, *J. Am. Chem. Soc.* **97**, 4869 (1975); Related ketone: C.-Y. Ho and F. T. Bond, *J. Am. Chem. Soc.* **96**, 7355 (1974).

[65] *anti*-Tricyclo[3.1.0.0$^{2,4}$]hexane: E. L. Allred and J. C. Hinshaw, *J. Am. Chem. Soc.* **90**, 6885 (1968).

[66] E. C. Alexander and J. Uliana, *J. Am. Chem. Soc.* **98**, 4324 (1976).

[67] G. L. Closs and R. B. Larrabee, *Tetrahedron Lett.* p. 287 (1965).

[68] Y. Hosokawa and I. Moritani, *Chem. Commun.* p. 905 (1970).

[69] S. Masamune, *J. Am. Chem. Soc.* **86**, 735 (1964).

[70] W. Von E. Doering and M. Pomerantz, *J. Am. Chem. Soc.* **86**, 961 (1964).

[71] H. Ona, H. Yamaguchi, and S. Masamune, *J. Am. Chem. Soc.* **92**, 7495 (1970).

[72] W. D. Stohrer and R. Hoffmann, *J. Am. Chem. Soc.* **94**, 1661 (1972).

[73] Tricyclo[4.1.0.0$^{3,7}$]heptane: H. C. Brown and H. M. Bell, *J. Am. Chem. Soc.* **85**, 2324 (1963).

[74] S. Winstein, A. H. Lewin, and K. C. Pande, *J. Am. Chem. Soc.* **85**, 2324 (1963).

[75] Tricyclo[4.1.0.0$^{2,5}$]heptane: H. Tanida, S. Terataka, Y. Hata, and M. Watanabe,

last molecule has a single carbon atom at the hub of three fused cyclobutane rings. Tricyclooctanes of note include tricyclo[4.2.0.0$^{2,5}$]octane (**28**) [79] and

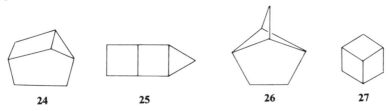

**24**          **25**          **26**          **27**

the symmetrical, highly stable tricyclo[3.3.0.0$^{2,6}$]octane (**29**).[80a,b] Tricyclo-[4.1.0.0$^{2,7}$]hept-4-en-3-one (**30**) is a potential precursor[81a] of the recently reported tricyclic carbonium ion, **31**.[81b] This ion should be stabilized by the

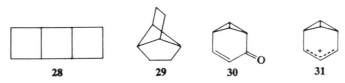

**28**          **29**          **30**          **31**

intramolecular interaction between "allyl cation" and "bicyclobutane" (see Jorgensen[81c]).

### 3.B.2 BRIDGEHEAD CARBONIUM IONS

Small bicyclic molecules such as norbornane (bicyclo[2.2.1]heptane) have played a critical role in the history of physical organic chemistry because they maintain well-defined spatial relationships between attached substituents while they also possess constrained bond lengths, bond angles, and torsional angles. Their role as "chemical tweezers"[82] may be emphasized by noting that the preferred planar structure of carbonium ions first was deduced through observations of enormously reduced solvolytic activity of compounds such as 1-bromonorbornane (**32**), in which prohibitive angular strain makes

*Tetrahedron Lett.* p. 5345 (1969). For the reader's amusement, compounds **24** and **25** both have three-, four-, five, six-, and seven-membered rings.

[76] A. Padwa and W. Eisenberg, *J. Am. Chem. Soc.* **92**, 2590 (1970).

[77] J. M. Harless and S. A. Monti, *J. Am. Chem. Soc.* **96**, 4714 (1974).

[78] J. Meinwald and J. Mioduski, *Tetrahedron Lett.* p. 4137 1974).

[79] M. Avram, I. G. Dinulescu, E. Marica, G. Mateescu, E. Sliam, and C. D. Nenitzescu, *Chem. Ber.* **97**, 382 (1964).

[80a] R. Srinivasan, *J. Am. Chem. Soc.* **83**, 4923 (1961).

[80b] R. Srinivasan, *J. Am. Chem. Soc.* **85**, 819 (1963).

[81a] H. Prinzbach, H. Babsch, and H. Fritz, *Tetrahedron Lett.* p. 2129 (1976).

[81b] M. Christl and G. Freitag, *Angew Chem., Int. Ed. Engl.*, **15**, 493 (1976).

[81c] W. L. Jorgensen, *J. Am. Chem. Soc.*, **97**, 3082 (1975).

[82] R. C. Fort, Jr. and P. von R. Schleyer, *Adv. Alicyclic Chem.* **1**, p. 283 (1966).

the bridgehead carbonium ion appreciably nonplanar.[82,83] Thus, the reduced solvolysis rates of other bridgehead bicyclic bromides such as **33** and **34**[82,83] follow this pattern, and there exists an excellent correlation between the logarithms of the relative rates of solvolysis of bridgehead-substituted compounds and the *differences* in the strain enthalpies of the corresponding hydrocarbons and bridgehead carbonium ions.[84,85a,b] Table 3.3 lists the strain energies of some bridgehead carbonium ions and differences in strain energies relative to the hydrocarbon, as well as references to their observation in acidic medium via nmr.[86–89] Compounds **32–34** clearly obey this relationship, which is valid because the structures of the activated complexes are very

|  | **32** | **33** | **34** |
|---|---|---|---|
| rel. $k_{solv}$(80% EtOH) = 1 | rel. $k_{solv} = 10^{-13}$ | rel. $k_{solv} = 10^{-6}$ | rel. $k_{solv} = 10^{-2}$ |

similar and differences in solvent effects negligible. Although the 1-bicyclo-[2.2.1]heptyl and 1-bicyclo[2.2.2]octyl cations are quite strained, they are formed during neopentyl-type rearrangements [Eq. (5)[90] and (6)[91]] which involve energetically favorable isomerization of a primary to a tertiary carbonium ion as well as some relief of strain. Because the bicyclo[3.3.3]-undecyl- (manxyl-) system approaches coplanarity at the bridgehead, the carbonium ion is actually less strained than the neutral hydrocarbon (Table 3.3). Thus, 1-chlorobicyclo[3.3.3]undecane solvolyzes $10^4$ times more rapidly than *tert*-butyl chloride,[92] and the bridgehead ion has itself been observed in

[83] R. C. Fort, Jr., *in* "Carbonium Ions" (G. A. Olah and P. von R. Schleyer, eds.), Vol. 4, pp 1783–1835. Wiley (Interscience), New York, 1973.

[84] P. von R. Schleyer, *in* "Conformational Analysis: Scope and Present Limitations" (G. Chiurdoglu, ed.), pp. 241–249. Academic Press, New York, 1970.

[85a] R. C. Bingham and P. von R. Schleyer, *J. Am. Chem. Soc.* **93**, 3189 (1971).

[85b] J. L. Fry, E. M. Engler, and P. von R. Schleyer, *J. Am. Chem. Soc.* **94**, 4628 (1972).

[86] G. A. Olah, D. P. Kelly, and R. G. Johanson, *J. Am. Chem. Soc.* **92**, 4137 (1970).

[87] P. von R. Schleyer, R. C. Fort, Jr., W. E. Watts, M. B. Comisarow, and G. A. Olah, *J. Am. Chem. Soc.* **86**, 4195 (1964).

[88] G. A. Olah and G. Liang, *J. Am. Chem. Soc.* **95**, 194 (1973).

[89] G. A. Olah, G. Liang, J. R. Wiseman, and J. A. Chong, *J. Am. Chem. Soc.* **94**, 4927 (1972).

[90] K. B. Wiberg and B. R. Lowry, *J. Am. Chem. Soc.* **85**, 3188 (1963).

[91] R. L. Bixler and C. Niemann, *J. Org. Chem.* **23**, 742 (1958).

[92] W. Parker, R. L. Tranter, C. I. F. Watt, L. W. K. Chang, and P. von R. Schleyer, *J. Am. Chem. Soc.* **96**, 7121 (1974).

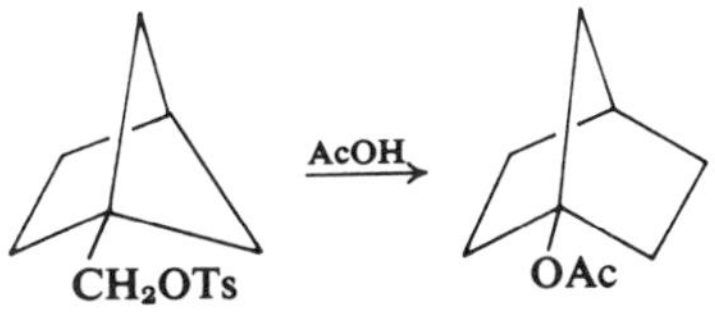

$$\text{CH}_2\text{OTs} \xrightarrow{\text{AcOH}} \text{OAc} \qquad (5)$$

$$\text{CH}_2\text{OTs} \xrightarrow{\text{AcOH}} \text{OAc} \qquad (6)$$

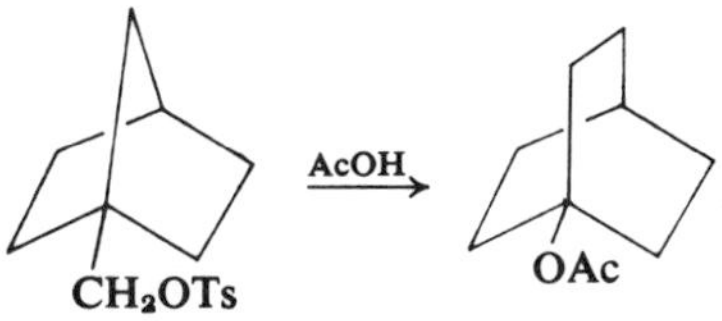

"superacid medium" and is stable to $-30°C$.[93b] The solvolyses of bridgehead-substituted bicyclo[1.1.1]pentanes and bicyclo[2.1.1]hexanes [e.g., Eq. (7)[51] and (8)[90]] are surprisingly rapid (e.g., 1-bromobicyclo[1.1.1]pentane solvolyzes

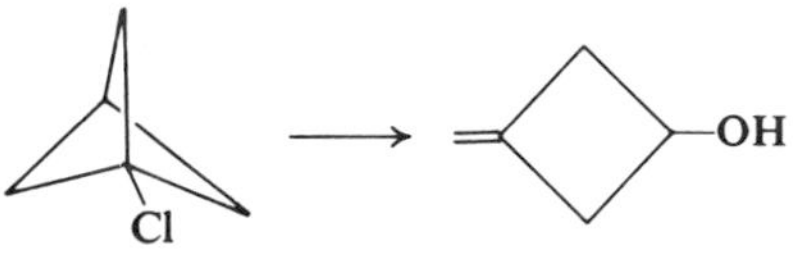

$$\underset{\text{Cl}}{\triangle} \longrightarrow =\square-\text{OH} \qquad (7)$$

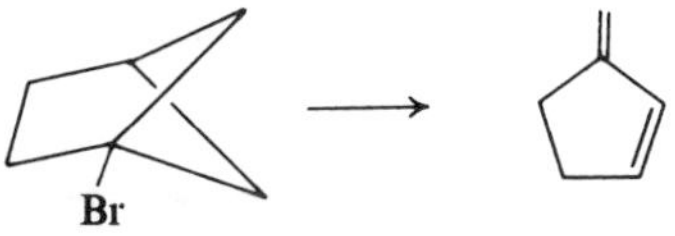

$$\underset{\text{Br}}{} \longrightarrow \qquad (8)$$

more rapidly than *tert*-butyl chloride). Part of this rate enhancement may be explained by postulated rear-lobe assistance from the other bridgehead carbons, which, as noted previously, are quite close. Ring opening in the rate-determining step might also be a factor in the enhanced rate, and would in any case prohibit the inclusion of these molecular systems in the above-mentioned strain–reactivity relativity relationship, since a different activated complex is postulated. The 10-perhydrotriquinacyl system (35) solvolyzes $10^{11}$ times more slowly than expected solely on the basis of strain, and this has been explained in a hyperconjugation context.[84,85a,b] Although the 1-bicyclo-[2.2.2]octyl cation has not yet been observed in solution because of its rapid rearrangement,[94] the strikingly stable (to $-60°C$) 1,4-bicyclo[2.2.2]octyl dication (36) has been characterized by nmr.[93a] The stability of this species is said to be a result of its resemblance to the aromatic cyclobutadiene dication

[93a] G. A. Olah, G. Liang, P. von R. Schleyer, E. M. Engler, M. J. S. Dewar, and R. C. *Chem. Soc.* **99**, 966 (1977).

[93b] G. A. Olah, G. Liang, P. von R. Schleyer, W. Parker, and C. I. F. Watt, *J. Am. Bingham, J. Am. Chem. Soc.* **95**, 6829 (1973).

[93c] A. de Meijere and O. Schallner, *Angew. Chem., Int. Ed. Engl.* **12**, 399 (1973).

[94] G. A. Olah and G. Liang, *J. Am. Chem. Soc.* **93**, 6873 (1971).

**Table 3.3**

**Calculated Strain Energies of Some Bridgehead Carbonium Ions**[a]

| Cation | Strain[b] (kcal/mole) | Strain difference,[c] (kcal/mole) |
|---|---|---|
| 1-Bicyclo[4.4.0]decyl[d] | 6.7 | 4.8 |
| 1-Bicyclo[3.3.2]decyl | 16.4 | −2.8 |
| 1-Bicyclo[3.3.3]undecyl (1-manxyl) | 18.7 | −6.8 |
| 1-Adamantyl[e] | 19.2 | 12.3 |
| 3-Homoadamantyl[f] | 21.1 | 6.5 |
| 1-Noradamantyl | 38.9 | 18.8 |
| 1-Bicyclo[3.3.1]nonyl | 17.9 | 8.3 |
| 1-Bicyclo[3.2.2]nonyl[g] | 22.0 | 6.6 |
| 1-Bicyclo[2.2.2]octyl | 29.3 | 16.3 |
| 1-Tricylo[4.4.0.0$^{4,9}$]decyl (1-twistyl-) | 43.7 | 17.6 |
| 1-Bicyclo[3.2.1]octyl | 30.8 | 18.7 |
| 1-Bicyclo[2.2.1]heptyl (1-norbornyl-) | 40.5 | 23.5 |
| 10-Perhydrotriquinacyl | 25.0 | 9.3 |
| 4-Tricyclo[2.2.1.0$^{2,6}$]heptyl (4-nortricyclyl) | 75.5 | 28.5 |

[a] A reference is included where a carbonium ion has been observed in solution.

[b] Strain energies are obtained from the calculated differences in strain between the carbonium ion and corresponding hydrocarbon as listed above and the strain energies of the hydrocarbons (cf. Hendrickson et al.[9]).

[c] (ion) − (hydrocarbon).

[d] G. A. Olah, D. P. Kelly, and R. G. Johanson, *J. Am. Chem. Soc.* **92**, 4137 (1970).

[e] P. von R. Schleyer, R. C. Fort, Jr., W. E. Watts, M. B. Comisarow, and G. A. Olah, *J. Am. Chem. Soc.* **86**, 4195 (1964).

[f] G. A. Olah and G. Liang, *J. Am. Chem. Soc.* **95**, 194 (1973).

[g] G. A. Olah, G. Liang, J. R. Wiseman, and J. A. Chong, *J. Am. Chem. Soc.* **94**, 4927 (1972).

(see Section 3.D.3), as well as to (calculated) delocalization of positive charge on the 12 hydrogen atoms which has the effect of minimizing electrostatic destabilization. $^{13}$C nmr shifts suggest close contact between the bridgehead carbons in **36**, and this is consistent with a $C_1$—$C_4$ distance of only 1.99 Å calculated by MINDO/3.[93a] A test of the homoaromaticity (or pseudo-aromaticity) postulated above might be furnished in the stability or instability and structure of the 1,5-manxyl dication (**37**). This dication has been generated and observed by nmr.[93a] Although flattening at the bridgeheads in the manxyl system would tend to make $C_1$ and $C_5$ close, apparently they are farther apart than in **36** and no special resonance interaction is indicated.

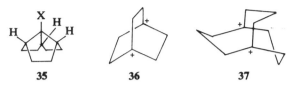

**35**       **36**       **37**

Moreover, while dication **37** decomposes above $-50°C$, 1-manxyl cation is stable to $-30°C$.[93a] A 1,5-trishomobarrelyl dication, stable to $0°C$, has also been postulated, although nmr analysis is not yet fully conclusive.[93c]

1-Bromobicyclo[2.2.1]heptane and similar molecules are very unreactive to nucleophilic substitution, as the $S_N2$ mechanism is sterically thwarted and the $S_N1$ mechanism is difficult owing to strain in the carbonium ion. However, 1-halo-1-silabicyclo[2.2.1]heptanes are extremely susceptible to bridgehead substitution by nucleophiles (see Section 5.D.5.c).

### 3.B.3 *trans*-FUSED BICYCLIC MOLECULES

When cyclopropane or cyclobutane is *trans*-fused to small rings (six or less vertices), the fusion bond is distorted in two planes and is termed "twist-bent"[95a,b] (see Fig. 3.1). The overlap in such a bond is particularly poor and molecules including this feature should be more reactive than their *cis*-fused isomers[95a,b] unless steric effects dictate otherwise. Derivatives of *trans*-bicyclo[6.1.0]nonane (**38**),[96–98] *trans*-bicyclo[5.1.0]octane (**39**),[95a,b,99–104] and

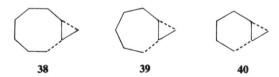

**38**       **39**       **40**

*trans*-bicyclo[4.1.0] heptane(**40**)[105] have been reported. Compound **38** is 2.9 kcal/mole higher in free energy than its *cis*-isomer,[104] and this difference may

[95a] P. G. Gassman, *Chem. Commun.* p. 793 (1967).

[95b] P. G. Gassman and F. J. Williams, *J. Am. Chem. Soc.* **93**, 2704 (1971).

[96] A. C. Cope and J. K. Hecht, *J. Am. Chem. Soc.* **85**, 1780 (1963).

[97] E. J. Corey and J. I. Schulman, *Tetrahedron Lett.* p. 3655 (1968).

[98] C. H. DePuy and J. L. Marshall, *J. Org. Chem.* **33**, 3326 (1968).

[99] P. G. Gassman, F. J. Williams, and J. Seter, *J. Am. Chem. Soc.* **90**, 6893 (1968).

[100] W. Kirmse and C. Hause, *Angew. Chem., Int. Ed. Engl.* **7**, 891 (1968).

[101a] K. B. Wiberg and A. de Meijere, *Tetrahedron Lett.* p. 519 (1969).

[101b] K. B. Wiberg, *Angew. Chem., Int. Ed. Engl.* **11**, 332 (1972).

[102] A. J. Ashe, III, *Tetrahedron Lett.* p. 523 (1969).

[103] P. G. Gassman, J. Seter, and F. J. Williams, *J. Am. Chem. Soc.* **93**, 1673 (1971).

[104] W. H. Pirkle and W. B. Lunsford, *J. Am. Chem. Soc.* **94**, 7201 (1972).

[105] J. V. Paukstelis and J. Kao, *J. Am. Chem. Soc.* **94**, 4783 (1972).

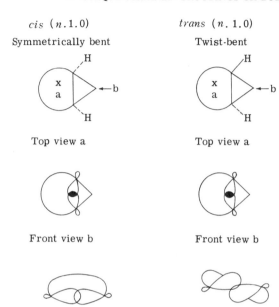

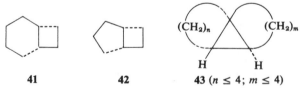

**Figure 3.1** Overlap of the strained central bond in *cis*- and *trans*-bicyclo[*n*.1.0]-alkanes. (See Gassman[95a] and Gassman and Williams.[95b])

be compared to a 9.2 kcal/mole enthalpy difference favoring *cis*-cyclooctene over *trans*-cylooctene (see Section 3.E.2).[101a] The liquid-phase enthalpy difference favoring *cis*-bicyclo[5.1.0]octane over **39** is about 9 kcal/mole,[104] which would allow one to anticipate a strain energy of about 40 kcal/mole in the latter. The strain in the central bond of **39** is manifested in its abnormally high sensitivity to acid.[98] *Trans*-fused bicyclo[6.1.0]nona-2,4,6-triene,[106] as well as many *trans*-bicyclo[*n*.2.0]compounds[107] (e.g., **41** and **42**), have also been characterized.

**41**          **42**          **43** ($n \leq 4$; $m \leq 4$)

Another means of producing "twist-bent" bonds is through fusion of rings in the manner of **43**.[95a,b] While "2,5-diphenylbisnorcaradiene" could be observed in solution and has a half-life of only 16 minutes at 25°C

[106] G. Moshuk, G. Petrowski, and S. Winstein, *J. Am. Chem. Soc.* **90**, 2179 (1968).
[107] See footnote 6 in Ref. 95b.

($\Delta G^{\ddagger} = 21.5$ kcal/mole),[108a](2,3:4,5) dibenzobisnorcaradiene (**44a**) is an isolable molecule (m.p. 108°) which is more stable than its tropylidene isomers (**44b,c**), to which it rearranges upon heating.[108b] The intermediacy of a highly

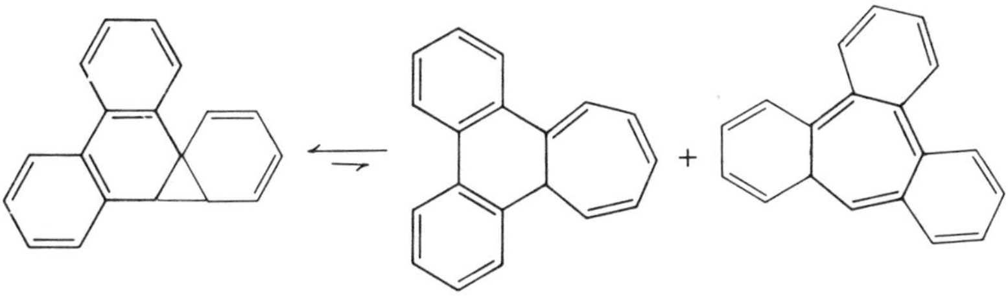

| **44a** | **44b** | **44c** |

substituted olefinic derivative of **43** (where $n = 4, m = 3$) has been indi-cated,[109a] and stable, synthetically useful molecules having $n = 4, m = 3$ have been reported and a derivative having $n = m = 3$ has recently been isolated.[109b,c]

### 3.B.4 SMALL-RING SPIRO HYDROCARBONS

Spiropentane (**45**), apparently first synthesized in 1896,[110] is destabilized by 65 kcal/mole and thus is about 10 kcal/mole more strained than two

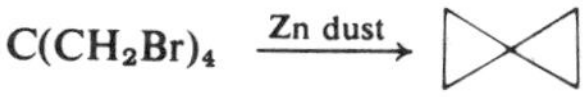

$$C(CH_2Br)_4 \xrightarrow{\text{Zn dust}} \text{\Large$\bowtie$} \qquad (9)$$
$$\mathbf{45}$$

cyclopropane rings. A Coulson–Moffitt model (Chapter 2) of bonding in this molecule would assume that the spiro junction carbon is $sp^3$ hybridized. Thus, the deviation of the interorbital angle (109.5°) from the internuclear angle ($62°$[111]) is some 5° greater than in cyclopropane. However, the sym-metry of spiropentane is $D_{2d}$, not $T_d$, and $sp^3$ hybridization of the central carbon is not required. A Walsh-type orbital picture of spiropentane may be assembled from an sp-hybridized central carbon and four $sp^2$-hybridized peripheral carbons.[112] The extra 10 kcal/mole strain increment in spiropentane

[108a] H. Dürr, M. Kausch, and H. Kober, *Angew. Chem., Int. Ed. Engl.* **13**, 670 (1974).

[108b] K.-H. Pauly and H. Dürr, *Tetrahedron Lett.* p. 3659 (1976).

[109a] H. Dürr and V. Fuchs, *Tetrahedron Lett.* p. 4049 (1976).

[109b] J. F. Ruppert and J. D. White, *J. Chem. Soc., Chem. Commun.* p. 976 (1976).

[109c] R. F. Heldeweg, H. Hogeveen, and L. Zwart, *Tetrahedron Lett.* p. 2535 (1977).

[110] See D. E. Applequist, G. F. Fanta, and B. W. Henrikson, *J. Org. Chem.* **23**, 1715 (1958).

[111] G. Dallinga, P. K. Vand der Draai, and L. H. Toneman, *Rec. Trav. Chim. Pays-Bas* **87**, 897 (1968).

[112] W. A. Bernett, *J. Chem. Educ.* **44**, 17 (1967).

is certainly a contributor to its relatively facile thermal epimerization and rearrangement to methylenecyclobutane.[113-115] Some other small spiro hydrocarbons include **46**,[116] **47**,[117] **48**,[118] **49**,[119] **50**,[119] **51**,[120] **52**,[121] and **53**.[122]

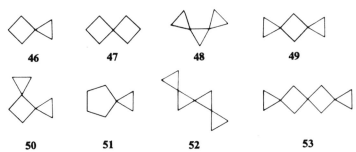

"Rotanes" such as **54**,[123a,b] **55**,[124a,b] and **56**[125a,b] are, in principle, capable of cyclic delocalization of electrons in the cyclopropane bent-bond orbitals. [5]-Rotane (**56**) displays a uv spectrum consistent with enhanced delocalization,[125a,b] while [3]-rotane (**54**), which has a greater separation between exocyclic three-membered rings, does not exhibit spectral properties attributable to such a feature.[123a,b] Most recently, [6]-rotane has also been reported.[126a,b] The molecule is a rather constrained chair whose ring

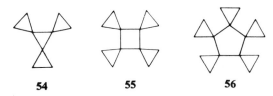

[113] M. C. Flowers and H. M. Frey, *J. Chem. Soc.* p. 5550 (1961).

[114] J. C. Gilbert, *Tetrahedron* **25**, 1459 (1969).

[115] J. J. Gajewski and L. T. Burka, *J. Am. Chem. Soc.* **94**, 8857, 8860, 8865 (1972).

[116] D. E. McGreer, *Can. J. Chem.* **38**, 1638 (1960).

[117] B. Winstein, A. H. Fenselav, and J. G. Thorpe, *J. Chem. Soc.* p. 2281 (1965).

[118] J. M. Denis, C. Girard, and J. M. Conia, *Synthesis* p. 549 (1972).

[119] P Binger, *Angew. Chem., Int. Ed. Engl.* **11**, 433 (1972).

[120] W. R. Roth and M. Martin, *Tetrahedron Lett.* p. 4695 (1967).

[121] L. Fitjer and J. M. Conia, *Angew. Chem., Int. Ed. Engl.* **12**, 761 (1973).

[122] F. Buchta and W. Merk, *Justus Liebigs Ann. Chem.* **716**, 106 (1968).

[123a] J. M. Denis and J. M. Conia, *Tetrahedron Lett.* p. 461 (1973).

[123b] L. Fitjer and J. M. Conia, *Angew. Chem., Int. Ed. Engl.* **12**, 334 (1973).

[124a] J. M. Conia and J. M. Denis, *Tetrahedron Lett.* p. 3545 (1969).

[124b] P. Le Perchec and J. M. Conia, *Tetrahedron Lett.* p. 587 (1970).

[125a] J. L. Ripoll and J. M. Conia, *Tetrahedron Lett.* p. 979 (1969).

[125b] J. L. Ripoll, J. C. Limassett, and J. M. Conia, *Tetrahedron* **27**, 2431 (1971).

[126a] E. Proksch and A. de Meijere, *Tetrahedron Lett.* p. 4851 (1976).

[126b] L. Fitjer, *Angew. Chem., Int. Ed. Engl.* **15**, 763 (1976).

inversion barrier ($\Delta G^{\ddagger} = 22$ kcal/mole) is the highest recorded for a cyclohexane.

The spiropentane system has been further destabilized by twisting as in molecules 57[127a,b] and 58.[128] Such distortion will be treated further in Section 3.F.

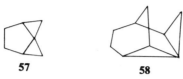

57                                    58

## 3.C  Tetrahedrane

Tetrahedrane (tricyclo[1.1.0.0²,⁴]butane, 59) and its derivatives remain enticing synthetic targets for organic chemists, who have succeeded during the past decade or so in making cyclobutadienes (see Section 3.D.3; cyclobutadiene is the valence isomer[129] of tetrahedrane). The development of new

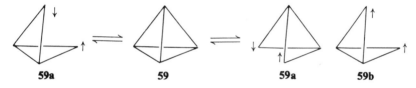

59a                    59                    59a            59b

high-energy syntheses, spectrometers capable of detecting low concentrations of short-lived intermediates, and especially low-temperature matrix isolation methods, coupled with new knowledge about bonding theory, may well make these molecules feasible subjects for current investigation. This is not to say that earlier investigators did not try to produce tetrahedranes.[129] An early description of the synthesis of a tetrahedrane[130] is now thought to be based upon experiments never in fact performed.[131] A more modern claim for the synthesis of diphenyltetrahedrane[132a] was quickly withdrawn.[132b,133]

[127a] L. Skattebøl, *Chem. Ind.* (*London*) p. 2146 (1962).

[127b] L. Skattebøl, *J. Org. Chem.* **31**, 2789 (1966).

[128] M. S. Baird, *J. Chem. Soc., Chem. Commun.* p. 197 (1974).

[129] L. T. Scott and M. Jones, Jr., *Chem. Rev.* **72**, 181 (1972).

[130] R. M. Beesley and J. F. Thorpe, *J. Chem. Soc.* **117**, 591 (1920).

[131] H. O. Larson and R. B. Woodward, *Chem. Ind.* (*London*) p. 193 (1959).

[132a] S. Masamune and M. Kato, *J. Am. Chem. Soc.* **87**, 4190 (1965).

[132b] S. Masamune and M. Kato, *J. Am. Chem. Soc.* **88**, 610 (1966).

[133] E. H. White, G. E. Maier, R. Graeve, V. Zirngibl, and E. W. Friend, *J. Am. Chem. Soc.* **88**, 611 (1966).

Schemes 1–3 depict the results of some recent studies[134–137] which either implicate tetrahedranes or "tetrahedrally symmetric" species (see below). Scheme 1 summarizes the results of a photoinduced reaction between labeled cyclopropene and carbon suboxide that produced a 24% yield of isotopic acetylenes in statistical abundance, consistent with tetrahedrane's intermediacy.[135] If we equate the carbene bicyclo[1.1.0]butanylidene (Scheme 1)

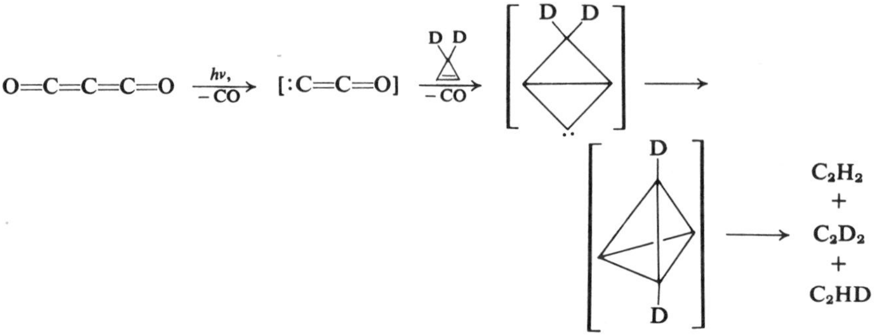

$$O=C=C=C=O \xrightarrow[-CO]{hv,} [:C=C=O] \xrightarrow{-CO} \quad \longrightarrow \quad C_2H_2 + C_2D_2 + C_2HD$$

**Scheme 1**

with the bent 1,2-cyclobutadiene ("homocyclopropylidene") of Pople's calculation, this singlet carbene is then calculated to be some 15.5 kcal/mole more stable than tetrahedrane (4–31G).[138a] On the other hand, if we use a calculated $\Delta H_f$ of 106 kcal/mole for cyclopropylidene,[138b] comparison of the heats of formation of cyclopropane and bicyclobutane would allow an estimate of 146 kcal/mole for bicyclobutanylidene, which would make it from 6 to 25 kcal/mole less stable than tetrahedrane (see below). (It is clear that simple bond-additivity considerations do not consider the resonance stabilization in "homocyclopropylidene" and are inadequate in the calculation of bicyclobutanylidene.) In Scheme 2, formation of acetylene, propyne, and 2-butyne in low yields upon photolysis of 1,5-dimethyltricyclo[2.1.0.0$^{2,5}$]-pentan-3-one (60) is consistent with the intermediacy of dimethyltetrahedrane.[136] Pyrolysis of the dilithium salt of deuterated *trans*-butenedial bistosylhydrazone (61) produces isotopic acetylenes which are consistent with some formation of tetrahedrane (Scheme 3).[137] One can obtain a very crude

[134] R. F. Peterson, Jr., R. T. K. Baker, and R. L. Wolfgang, *Tetrahedron Lett.* p. 4749 (1969).

[135] P. B. Shevlin and A. P. Wolf, *J. Am. Chem. Soc.* **92**, 406, 5291 (1970).

[136] H. Ona, H. Yamaguchi, and S. Masamune, *J. Am. Chem. Soc.* **92**, 7495 (1970).

[137] L. B. Rodewald and H.-K. Lee, *J. Am. Chem. Soc.* **95**, 623, 3084 (1973).

[138a] W. J. Hehre and J. A. Pople, *J. Am. Chem. Soc.* **97**, 6941 (1975).

[138b] W. A. Lathan, L. Radom, P. C. Hariharan, W. J. Hehre, and J. A. Pople, *Top. Curr. Chem.* **40**, 1 (1973).

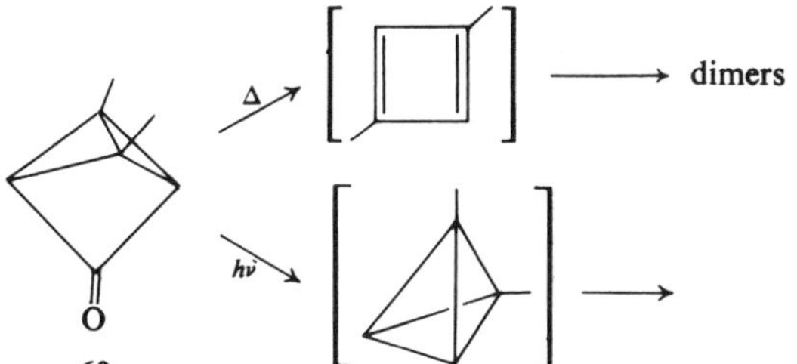

C$_2$H$_2$ + CH$_3$C≡CH + CH$_3$C≡CCH$_3$ + other products
(2%)        (6%)         (2%)

**Scheme 2**

estimate for the heat of formation of (3-cyclopropenyl)carbene (Scheme 3) by adding the enthalpies of formation of singlet methylene (101 kcal/mole[138c]), 3-methylcyclopropene (59 kcal/mole, estimated), and subtracting that of methane. The value calculated is some 35–60 kcal/mole higher than tetrahedrane, and even with a "fudge factor" of about −20 kcal/mole (see Section 3.D.2) for resonance stabilization, it is clear that the carbene is higher in enthalpy. However, in none of these cases can one say with any certainty

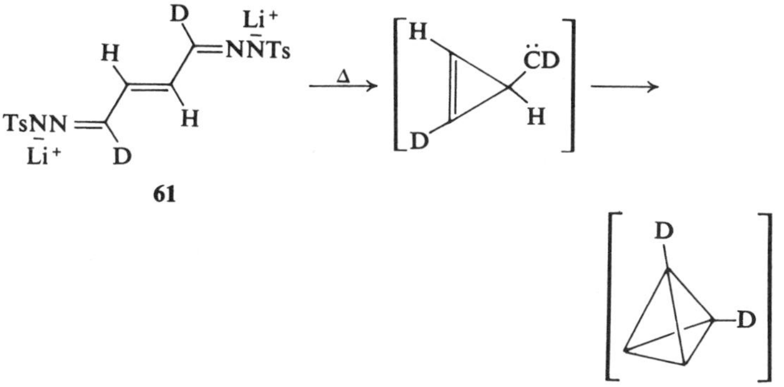

C$_2$H$_2$
+·
C$_2$D$_2$
+
C$_2$HD

**Scheme 3**

whether tetrahedrane has been formed or whether "tetrahedrally symmetric species" (i.e., rapidly equilibrating bicyclobutyl diradicals similar to **59a** having tetrahedrane as a "chemically insignificant" transition state) have been produced. Such a decision must await low-temperature spectroscopic studies of the species generated in the above reactions. More will be said about the results of low temperature matrix studies later in this section.

[138c] H. M. Frey, G. E. Jackson, R. A. Smith, and R. Walsh, *J. Chem. Soc. Faraday Trans. I*, **71**, 1991 (1975).

MINDO/2 calculations[139a-c] and two recent *ab initio* studies[138a,b,140] obtain enthalpies of formation for tetrahedrane of 120–140 kcal/mole, corresponding to strain energies in the range 130–150 kcal/mole. This would allow one to estimate a strain of 22–25 kcal per bond, not very different from the extra enthalpy estimated for the central bond of bicyclo[1.1.0]butane, or alternatively, 32–38 kcal/mole per carbon, which is an unprecedented value. [Chapter 4 discusses the utility, if any, of differentiating strain per atom (group) and strain per bond.] In any case, the thermal symmetry-forbidden[128] (see Section 5.B.1) concerted conversion to two acetylenes is calculated to be exothermic by 20–40 kcal/mole, while symmetry-forbidden thermal isomerization to cyclobutadiene (singlet) is calculated to be exothermic by up to 70 kcal/mole.[141] The most pessimistic expectations for the isolation of tetrahedrane arise from very simple calculations which predict that the loss in strain energy (63–84 kcal/mole) upon transformation of tetrahedrane to the bicyclobutyl diradical approaches the dissociation energy of a C—C bond. On the other hand, both recent *ab initio* calculations cited predict an appreciable barrier (at least 18 kcal/mole) separating tetrahedrane from bicyclobutyl diradical (this transformation is thermally forbidden[129]).[138a,b,140] Another interesting point is that the C—C bond length in tetrahedrane of 1.473 Å, calculated at the *ab initio* STO-3G level,[138a,b,140] is not that different from the bond lengths in bicyclobutane, and the first ionization energies of the two molecules are calculated to be similar.[140]

Stabilization of tetrahedrane relative to bicyclobutyl diradical is not easy to envision because no groups significantly destabilize radical centers. Perfluoroalkyl groups appear to stabilize strained rings (see Section 5.E.3), and tetrakis(perfluoromethyl)tetrahedrane might be amenable to study. Indeed, it is possible that this molecule (or the diradical) was generated in the photochemical decomposition of the stable compound hexakis(perfluoromethyl)-benzvalene ozonide (**62**) known to produce tetrakis(perfluoromethyl)cyclo-butadiene (Scheme 4[142]). This last intermediate may also be produced in the thermolysis represented by Scheme 5.[143] Matrix studies at 4°–10°K show the presence of another transient species in addition to tetrakis(perfluoromethyl)-cyclobutadiene generated from the benzvalene ozonide decomposition of

[139a] N. C. Baird and M. J. S. Dewar, *J. Am. Chem. Soc.* **89**, 3966 (1967).

[139b] N. C. Baird and M. J. S. Dewar, *J. Chem. Phys.* **50**, 1262 (1969).

[139c] N. C. Baird, *Tetrahedron* **26**, 2185 (1970).

[140] J. M. Schulman and T. J. Venanzi, *J. Am. Chem. Soc.* **96**, 4739 (1974).

[141] R. J. Buenker and S. D. Peyerimhoff, *J. Am. Chem. Soc.* **91**, 4342 (1969).

[142] T. Kobayashi, I. Kumaki, A. Ohsawa, Y. Hanzawa, M. Honda, and Y. Iataka, *Tetrahedron Lett.* p. 3001 (1975).

[143] T. Kobayashi, I. Kumaki, A. Ohsawa, Y. Hanzawa, and M. Honda, *Tetrahedron Lett.* p. 3819 (1975).

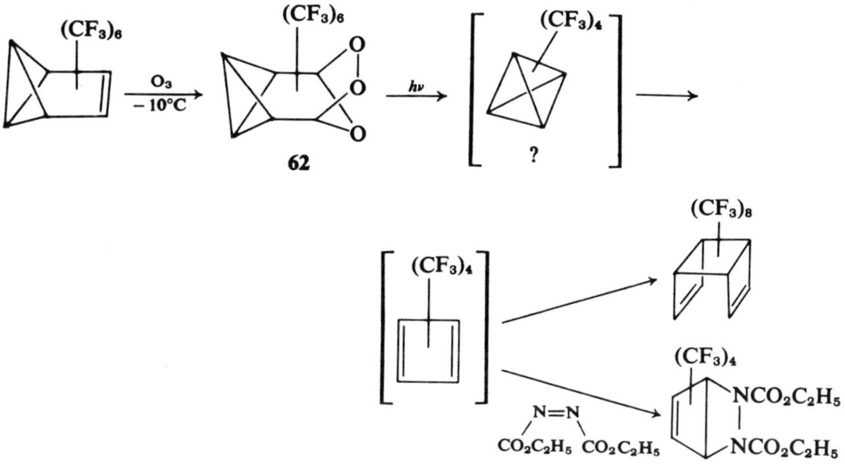

**Scheme 4**

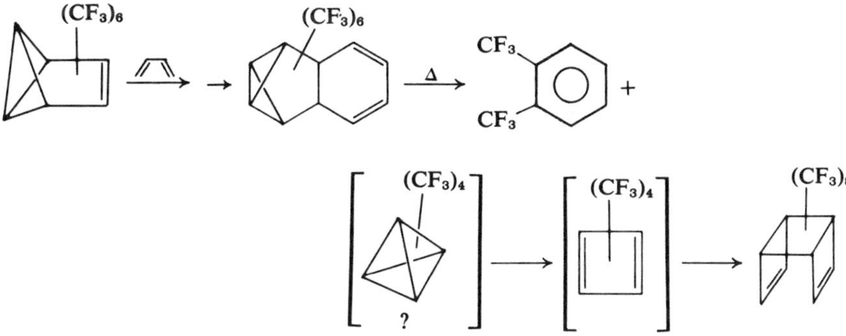

**Scheme 5**

Scheme 4.[144] While the cyclobutadiene comprises at least 90% of the primary photoproduct, the identity of the other molecule is of great interest.[144] Evidence for the trapping in a matrix at $-196°C$ of either tetramethyltetrahedrane or a diradical has also been offered.[145] Two unconventional ideas for obtaining tetrahedrane or at least its carbon skeleton involve formation of tetralithio-tetrahedrane (Section 5.E.4) or the LiH complex of tetrahedrane, which is isoelectronic with the relatively stable $C_2B_3H_5$. We also note another potential approach to a tetrahedrane in which a rigid molecule such as **63a** places two crowded triple bonds in a rigid and favorable orientation for photochemically allowed cycloaddition.[146] Another possibility might be a "push–pull" tetra-hedrane like **63b**.

[144] S. Masamune, T. Machiguchi, and M. Aratani, *J. Am. Chem. Soc.* **99**, 3524 (1977).
[145] G. Maier, *Angew. Chem., Int. Ed. Engl.* **13**, 425 (1974).
[146] H. A. Staab, E. Wehninger, and W. Thorwart, *Chem. Ber.* **105**, 2290 (1972).

$$>N \underset{\delta+}{\overset{}{=}} \cdots \cdots \underset{}{\overset{\delta-}{=}} CO_2R$$

| 63a | 63b |

We conclude this discussion by examining some heteroanalogues of tetra-hedrane. The hypothetical molecule tetrahedral $N_4$ is calculated to be unstable relative to two nitrogen molecules by more than 100 kcal/mole,[147,148] but tetrahedral $P_4$ molecules comprising white phosphorus are more stable than diatomic phosphorus. Comparison of single-bond strengths [$D$(N—N) 38 kcal/mole; $D$(P—P) 51 kcal/mole[149]] and triple-bond strengths [$D$(N≡N) 226 kcal/mole, $D$(P≡P) 116 kcal/mole[150]] along with an estimate of only 20 kcal/mole strain[149] in $P_4$ clarify this situation. Although tetrahedral $(SiH)_4$ is presently unknown, relatively strong silicon–silicon single bonds and weak multiple bonds[151] may allow this species to one day be observed in an inert atmosphere. The tetrahedral compound $(BCl)_4$ is difficult to synthesize in quantity, but isolable.[152]

### 3.D    Strained Alkenes: Molecules Lacking Torsional Strain

In an overly simplistic manner, one may identify four modes of distortion of olefinic linkages (64–67). Torsionally strained olefins (64) are discussed in Section 3.E. No organic examples of "*anti*-bent" organic molecules which include the strain feature 67[153a,b] are known. (However, molecule 68[154] has this feature, which has been rationalized in terms of weak double bonding between tin atoms in concert with a high inversion barrier for tricoordinate tin species[155a,b]). We will dwell briefly on the subject of "*syn*-bent" molecules,

[147] I. H. Hillier and V. R. Saunders, *J. Chem. Soc., Chem. Commun.* p. 1233 (1970).

[148] J. S. Wright, *J. Am. Chem. Soc.* 96, 4753 (1974).

[149] F. A. Cotton and G. Wilkinson, "Advanced Inorganic Chemistry," 3rd Ed. Wiley (Interscience), New York, 1972.

[150] G. G. Pimentel and R. D. Spratley, "Understanding Chemistry," Holden-Day, San Francisco, California, 1971.

[151] J. Simpson, *MTP Int. Rev. Sci., Inorg. Chem., Ser. 1* 1, 221 (1973).

[152] E. L. Muetterties, "The Chemistry of Boron and Its Compounds," pp. 364–366. Wiley, New York, 1967.

[153a] W. L. Mock, *Tetrahedron Lett.* p. 475 (1972).

[153b] L. Radom, J. A. Pople, and W. L. Mock, *Tetrahedron Lett.* p. 479 (1972).

[154] M. F. Lappert, *Adv. Chem.* 150, 256 (1976).

[155a] J. J. Zuckerman and J. F. Liebman, unpublished observations.

[155b] W. Cherry, N. Epiotis, and W. T. Borden, *Acc. Chem. Rev.*, 10, 167 (1977).

having strain feature **66**, but most of this section will concern itself with molecules such as cyclopropene, which has a normal $\pi$ system "superimposed" on a strained $\sigma$ framework (**65**). Of course, distortions of olefinic linkages are

**64**              **65**              **66**              **67**

usually mixtures of the modes shown, which are themselves simplifications, as rehybridization accompanies such geometry changes.

$$[\text{Si}(\text{CH}_3)_3]_2\text{CH}\diagdown\text{Sn}\!\!=\!\!=\!\!\text{Sn}\diagup\!\!\!\overset{\text{CH}[\text{Si}(\text{CH}_3)_3]_2}{\diagup}\!\!\!\overset{}{\diagdown}\!\!\text{CH}[\text{Si}(\text{CH}_3)_3]_2$$
$$[\text{Si}(\text{CH}_3)_3]_2\text{CH}\diagup$$

**68**

### 3.D.1  "*syn*-BENT" OLEFINIC LINKAGES

In the sense that few olefins having this molecular feature are known, the "*syn*-bent" carbon–carbon double bond is relatively rare. On the other hand, small cycloalkynes and arynes (Section 3.G) as well as small cycloallenes (Section 3.F) can be considered to possess such bonds. 9,9'-Didehydrodianthracene (**70**), which contains this feature, has a half-life in solution of about 30 minutes at 80°C.[156] Formation of a bis-*N*-aminotriazoline analogue of *N*-aminotriazoline **71** provided the precursor to 9,9',10,10'-tetradehydrodianthracene (**72**) (m.p. 388° dec.).[157] x-Ray studies of this molecule indicate normal olefinic bond length, but skeletal constraints as well as electron repulsion between the olefinic linkages force the symmetrical bending shown. The benzo-$\Delta^{3,7}$-tricyclo[3.3.2.0$^{3,7}$]decene **73** has also been generated recently and the dimer identified.[158]

### 3.D.2  CYCLOALKENES, BICYCLOALKENES, AND RELATED MOLECULES

The extra strain energy present in such molecules as cyclopropene, cyclobutene, methylene cyclopropane, and norbornene relative to the corresponding saturated molecules may be attributed to increases in the strain of the $\sigma$ framework. The strain energy of cyclopropene (**75**), a molecule first reported

[156] N. M. Weinshenker and F. D. Greene, *J. Am. Chem. Soc.* **90**, 506 (1968).

[157] R. L. Viavattene, F. D. Greene, L. D. Cheung, R. Majeste, and L. M. Trefonas, *J. Am. Chem. Soc.* **96**, 4342 (1974).

[158] R. Greenhouse, W. T. Borden, K. Hirotsu, and J. Clardy *J. Am. Chem. Soc.* **99**, 1664 (1977).

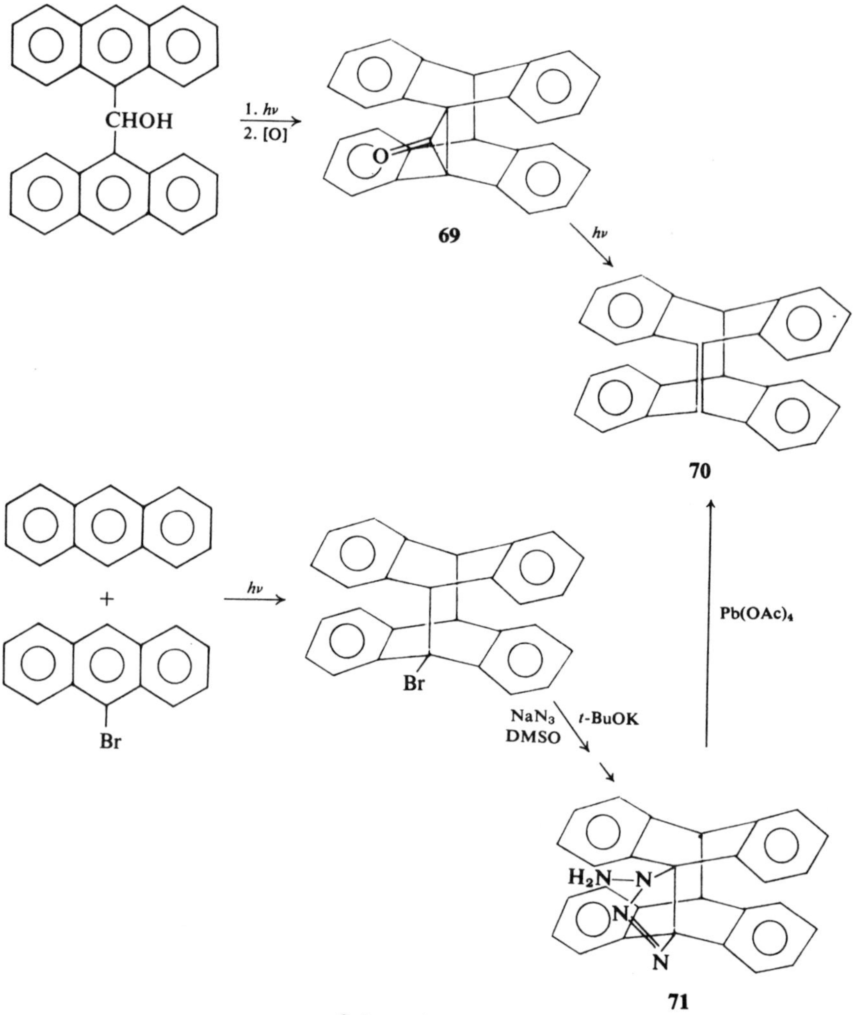

**Scheme 6**

in 1922,[159] is approximately 54.5 kcal/mole (Table 3.4), and it would be logical to assume that at least part of this might be attributable to a weak $\pi$ bond. The molecule is a highly reactive dienophile, in line with this view, but its short double bond (1.30 Å[159,160a,b]) and the high vibrational frequency

---

[159] G. L. Closs, *Alicyclic Chem.* **1**, pp. 53–127 (1966).

[160a] Most recent microwave data on cyclopropene: W. M. Stigliani, V. W. Laurie, and J. C. Li, *J. Chem. Phys.* **62**, 1890 (1975).

[160b] 1-methyl cyclopropene: M. K. Kemp and W. H. Flygare, *J. Am. Chem. Soc.* **91**, 3163 (1969).

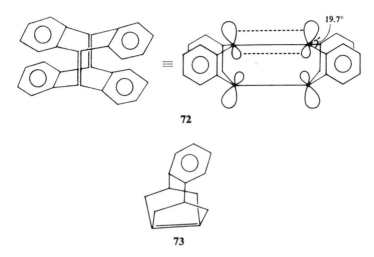

72

73

assigned to it[159] apparently contradict this assumption. Thus, the molecule's $\sigma$ framework would appear to be the source of destabilization, and if the double bond is abnormally strong, then strain in the $\sigma$ framework exceeds 54.5 kcal/mole. Significantly increased angular strain at all three carbon atoms is a major contributor to this high value for the strain energy [internal angles $C_1(51°)$, $C_2$ and $C_3(64.5°)$].[159,160a,b] Cycloaddition across cyclopropene's double bond reduces ring strain by 26 kcal/mole. Table 3.4 lists the strain energies of a number of cyclic and bicyclic olefins, including cyclopropene.

Suitably substituted cyclopropenes suffer the ene reaction (discussed in Section 5.A) and also readily undergo both cycloaddition (see above) and complexation with transition metals (Section 5.C.4.a) as a means of releasing strain. Here, we will briefly discuss their relatively facile ring opening to intermediates which have been described as vinylcarbenes. The energy of activation (32.6 kcal/mole) for racemization of optically active 1,3-diethylcyclopropene is equal, within experimental error, to the barrier for isomerization to acetylene and diene products, although a lower frequency factor causes the isomerization to occur more slowly.[161] A planar intermediate such as 74 (Scheme 7) has lost the asymmetry of the starting material, appears to be a viable intermediate, and would presumably be formed through concerted ring opening and bond rotation (see the discussion of thermal ring opening of cyclopropane in Chapter 2). Theoretical calculations[162] support

[161] E. J. York, W. Dittmar, J. R. Stevenson, and R. G. Bergman, *J. Am. Chem. Soc.* **94**, 2882 (1972); **95**, 5681 (1973).

[162] J. H. Davis, W. A. Goddard, III, and R. G. Bergman, *J. Am. Chem. Soc.* **98**, 4015 (1976).

Table 3.4

Calculated Strain Energies[a-c] of Some Cycloalkenes, Methylenecycloalkanes, and Bicycloalkenes

| Compound | Strain (kcal/mole) | Compound | Strain (kcal/mole) |
|---|---|---|---|
| Acetylene(cycloethene) | 9.1;[d] 58[e] | Bicyclo[2.1.1]hex-2-ene | Estd 50[f] |
| Cyclopropene | 54.5[a,c] | Bicyclo[2.2.1]hept-2-ene | 27.2;[a] 23.6[b] |
| Cyclobutene | 30.6;[a] 34.0[c] | Bicyclo[2.2.1]hepta-2,5-diene | 34.7;[a] 31.6[b] |
| Cyclopentene | 6.8;[a] 6.9[b] | Bicyclo[2.2.2]oct-2-ene | 16.0[b] |
| Cyclohexene | 2.5;[a] 2.6[b] | Bicyclo[2.2.2]octa-2,5,7-triene | 25.6[b] |
| cis-Cycloheptene | 6.7;[a] 7.25[b] | (barrelene) | |
| cis-Cyclooctene | 7.4;[a] 8.8[b] | 1,3-Cyclopentadiene | 2.9;[a] −0.9[c] |
| cis-Cyclononene | 11.5[a] | | |
| cis-Cyclodecene | 11.6[b] | (Bismethylene)cyclopropane | 39.6[c,g] |
| Methylenecyclopropane | 41.7;[a] 40.8[c] | (Trismethylene)cyclopropane | 28.8[c,g] |
| Methylenecyclobutane | 28.8[a] | Methylenecyclopropene | 41.5[c,g] |
| Methylenecyclopentane | 6.3;[a] 5.2[c] | | |

[a] See P. von R. Schleyer, J. E. Williams, Jr., and K. R. Blanchard, *J. Am. Chem. Soc.* **92**, 2377 (1970) and references cited therein.

[b] N. L. Allinger and J. T. Sprague, *J. Am. Chem. Soc.* **94**, 5734 (1972).

[c] Heats of formation calculated in N. C. Baird and M. J. S. Dewar, *J. Chem. Phys.* **50**, 1262 (1969) and strainless heats of formation from [a] above. For MINDO/3 calculations of some of these molecules, see R. C. Bingham, M. J. S. Dewar, and D. H. Lo, *J. Am. Chem. Soc.* **97**, 1294 (1975).

[d] Relative to the "strain" in ethylene: $2\Delta H_f$(ethylene) $-\Delta H_f$(ethane).

[e] Comparison of $\Delta H_f$(acetylene) with the sum of two strainless CH increments.

[f] Obtained by adding ca. 10 kcal/mole to the strain energy of bicyclo[2.1.1]hexane.

[g] The calculated *destabilization* energies appear to be too low and perhaps include an overestimate of resonance stabilization.

this general view of the opening of cyclopropene but indicate that the racemization intermediate is best described as the "1,3-diradical singlet" **76** "vinylcarbene." The calculated enthalpy difference of 36.6 kcal/mole between singlet vinylcarbene and cyclopropene[162] allows an estimate of 103 kcal/mole for the heat of formation of the former species. This is some 21 kcal/mole lower than a bond-additivity estimate [$\Delta H_f$(propene) $+ \Delta H_f$(singlet CH$_2$)[138c] $- \Delta H_f$(CH$_4$)]. This −21 kcal/mole discrepancy can be attributed to resonance stabilization of a singlet carbene and is the basis for our "fudge factor" employed in Sections 3.C and 3.G.3. At the risk of stating the obvious, we note that thermal generation of a carbene from a hydrocarbon is quite remarkable (see also Scheme 16 of this chapter). Although the reaction is not strictly comparable, one can use the decomposition of cyclopropane into ethylene and singlet methylene, estimated to be endothermic by about 100 kcal/mole, as a comparison. The great strain in cyclopropene, coupled

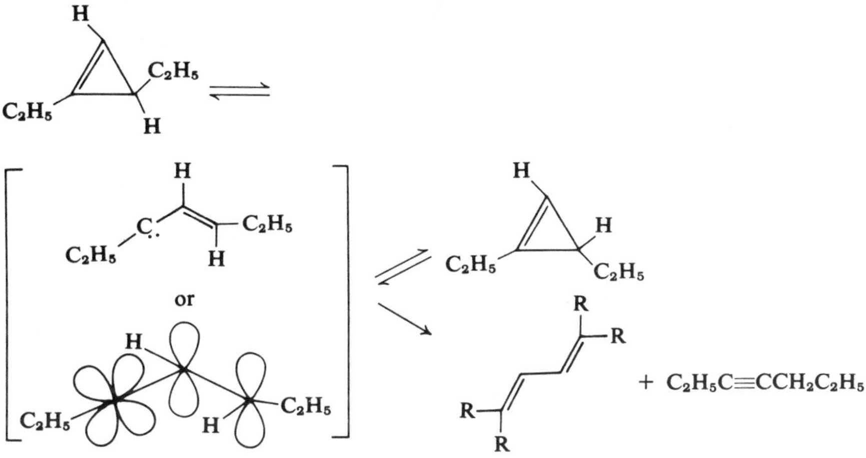

**Scheme 7**

with formation of a resonance-stabilized carbene, makes this reaction feasible. Triplet vinylcarbene has been calculated[162-164] as well as established

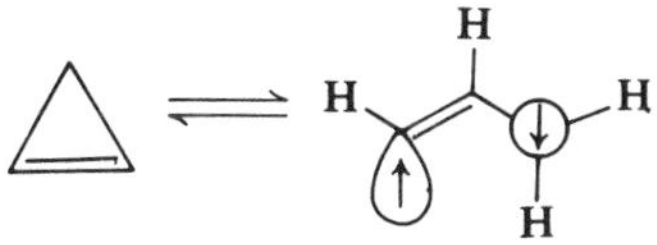

$$(10)$$

by esr studies[163,164] to be the ground state for the ring-opened isomer. Vinylcarbenes and related species have been generated via photochemical decomposition of diazoalkanes as well as through various thermal means, and have provided useful synthetic routes to cyclopropenes.[159] In Chapter 5 we will again examine the intermediacies of ring-opened isomers of cyclopropenes in transition-metal chemistry of the cyclic olefin as well as in metal-catalyzed rearrangements of 3,3′-bicyclopropenyls (benzene valence isomers).

Bicyclo[2.1.0]pent-2-ene ("homocyclobutadiene")[165] is destabilized by strain as well as, perhaps, some antiaromatic character. Another aspect of its instability is a calculated destabilizing interaction between the cyclopropane and ethylene fragments comprising the molecule.[166] (Any single-value

[163] R. S. Hutton, M. L. Marion, H. D. Roth, and E. Wasserman, *J. Am. Chem. Soc.* **96**, 4680 (1974).

[164] D. R. Arnold, R. W. Hunphreys, W. J. Leigh, and G. E. Palmer, *J. Am. Chem. Soc.* **98**, 6225 (1976).

[165] J. I. Brauman, L. E. Ellis, and E. E. van Tamelen, *J. Am. Chem. Soc.* **88**, 846 (1966).

[166] W. L. Jorgensen, *J. Am. Chem. Soc.*, **97**, 3082 (1975).

"destabilization energy" would include strain as well as electronic destabilization, and this is an example of the problems in apportioning approaches.) The molecule has a half-life of only four hours at 34°C and manifests its high energy by producing vibrationally excited states of 1,3-cyclopentadiene which are some 63 kcal/mole above the ground state.[167–169]

When a $\pi$ bond forms the bridge of a small fused bicyclic system, the underlying framework will be significantly strained. $\Delta^{1,4}$-Bicyclo[2.2.0]hexene (**77**)[170a–d] was produced by thermolysis of a tosylhydrazone in a manner which, unfortunately, formed 1,2-dimethylenecyclobutane as a coproduct. Although the molecule could be observed via nmr near $-60°C$, at $-20°C$ a Diels–Alder reaction with 1,2-dimethylenecyclobutane occurred. This problem was overcome by producing a platinum complex (see Section 5.C.4.a) of the strained olefin which could in turn release free **77**.[170b] Thus, pure $\Delta^{1,4}$-bicyclo[2.2.0]hexene could be observed in solution at ambient temperature. Its thermaldecomposition pathway is a bimolecular one rather than isomerization to 1,2-dimethylenecyclobutane.[171] Such dimerization is unique in that normal unstrained olefins and even cyclopropene do not react in this manner (torsion-

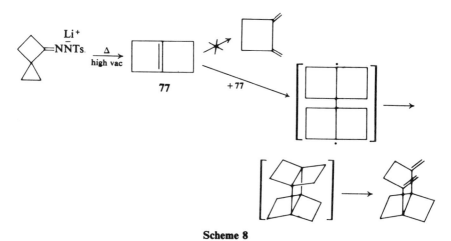

**Scheme 8**

[167] M. C. Flowers and J. M. Frey, *J. Am. Chem. Soc.* **94**, 8636 (1972).

[168] J. I. Brauman, W. E. Farneth, and M. B. D'Amore, *J. Am. Chem. Soc.* **95**, 5043 (1973).

[169] G. D. Andrews, M. Davalt, and J. E. Baldwin, *J. Am. Chem. Soc.* **95**, 5044 (1973).

[170a] K. B. Wiberg, G. J. Burgmaier, and P. Warner, *J. Am. Chem. Soc.* **93**, 246 (1971).

[170b] M. E. Jason, J. A. McGinnety, and K. B. Wiberg, *J. Am. Chem. Soc.* **96**, 6531 (1974).

[170c] C. J. Casanova and H. R. Rogers, *J. Org. Chem.* **39**, 3803 (1974).

[170d] K. B. Wiberg, W. F. Bailey, and M. E. Jason, *J. Org. Chem.* **39**, 3803 (1974).

[171] K. B. Wiberg and M. E. Jason, *J. Am. Chem. Soc.* **98**, 3393 (1976)

ally distorted olefins do react this way, see Section 3.E).[171] $\Delta^{2,5}$-Tricyclo-[4.2.1.0$^{2,5}$]nonene (78) was obtained through photochemical ring closure of 2,3-dimethylenenorbornane.[172] This same approach[172] was unsuccessful in obtaining 77 from 1,2-dimethylenecyclobutane, but produced a good yield of $\Delta^{1,5}$-bicyclo[3.2.0]heptene (79),[173] which had been earlier obtained in a manner analogous to that in Scheme 8. $\Delta^{1,5}$-Bicyclo[3.3.0]octene (80) has been

(11)

**78**

known for some time and is quite stable,[174] while 81[175] and 82[176a] have been reported more recently. *Ab initio* calculations of $\Delta^{1,3}$-bicyclo[1.1.0]butene (83)

**79**   **80**   **81**   **82**

suggest that this presently unknown and apparently nonplanar molecule is 80 kcal/mole (4–31G level) less stable than methylenecyclopropene and is also less stable than tetrahedrane.[138a,b] Recent work suggests the intermediacy of tricyclo[4.1.0.0$^{2,7}$]hept-1(7)-ene (83a), a species having this nonplanar structural feature, in apparent confirmation of the calculational study.[176b] There is some evidence for the intermediacy of $\Delta^{1,5}$-bicyclo[3.1.0]-hexene (84).[177] While 85[178,179] and 86[180a,b] are stable, isolable compounds,

[172] D. H. Aue and R. N. Reynolds, *J. Am. Chem. Soc.* **95**, 2027 (1973).

[173] W. Kirmse and K. H. Pook, *Angew. Chem., Int. Ed. Engl.* **5**, 594 (1966).

[174] E. Vogel, *Chem. Ber.* **85**, 25 (1952).

[175] P. E. Eaton, C. Giordano, and V. Vogel, *J. Org. Chem.* **41**, 2236 (1976).

[176a] J. Saver, B. Schroeder, and R. Wiemer, *Chem. Ber.* **100**, 306 (1967).

[176b] G. Szeimies, J. Harnisch, and O. Baumgärtel, *J. Am. Chem. Soc.* **99**, 5183 (1977).

[177] Prof. Philip Warner, unpublished data, personal communication to Joel F. Liebman.

[178] G. L. Closs and W. A. Böll, *J. Am. Chem. Soc.* **85**, 3904 (1963).

[179] G. L. Closs, W. A. Böll, H. Heyn, and V. Dev, *J. Am. Chem. Soc.* **90**, 173 (1968).

[180a] R. Breslow, J. Posner, and A. Krebs, *J. Am. Chem. Soc.* **85**, 234 (1963).

[180b] R. Breslow, L. J. Altman, A. Krebs, E. Mohacsi, I. Murata, R. A. Peterson, and J. Posner, *J. Am. Chem. Soc.* **87**, 1326 (1965).

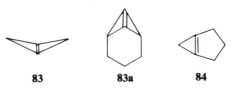

**83**          **83a**          **84**

the annelated cyclopropenone **87** has only been observed in solution at −60°C.[178,179] Attempts at isolating **88** have failed,[180a,b] but methyl substitution provides kinetic stability, and **89** is stable for five hours at 100°C.[181] (Cyclo-

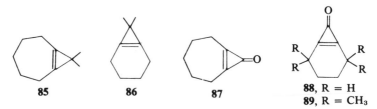

**85**          **86**          **87**          **88**, R = H
                                              **89**, R = CH₃

propenone is discussed in the next section). Additional constraints imposed upon methylenecyclopropane systems lead to instability, notably by making the corresponding trimethylenemethanes accessible. While 6-methylene-bicyclo[3.1.0]hexane (**90**) is quite stable,[182,183] attempts at isolating **91** have failed.[184]

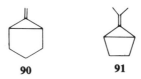

**90**          **91**

The rearrangement of phenylcarbenes to the corresponding cycloheptatri-enylidenes is now a well-known phenomenon.[185] One mechanism enjoying considerable support invokes a bicyclo[4.1.0]heptatriene as an intermediate. The first direct evidence for the intermediacy of such a species involved the generation and trapping of **92** (Scheme 9).[186] A later study produced dibenzo-bicycloheptatriene, which was trapped as a Diels–Alder adduct of 1,3-butadiene.[187] The existence of the bicyclohexadiene **93** has been deduced

[181] M. Suda and S. Masamune, *J. Chem. Soc., Chem. Commun.* p. 504 (1974).

[182] A. S. Kende and E. R. Riecke, *J. Chem. Soc., Chem. Commun.* p. 383 (1974).

[183] U. Langer and H. Musso, *Justus Liebigs Ann. Chem.* (7/8), 1180 (1976).

[184] J. A. Berson, R. J. Bushby, J. M. McBride, and M. Tremelling, *J. Am. Chem. Soc.* **93**, 1544 (1971).

[185] See footnote 1 in Ref. 186. See also R. L. Tyner, W. M. Jones, Y. Ohrn, and J. R. Sabin, *J. Am. Chem. Soc.* **96**, 3765 (1974) and references cited therein.

[186] W. E. Billups, L. P. Lin, and W. Y. Chow, *J. Am. Chem. Soc.* **96**, 4026 (1974).

[187] J. P. Mykytka and W. M. Jones, *J. Am. Chem. Soc.* **97**, 5933 (1975).

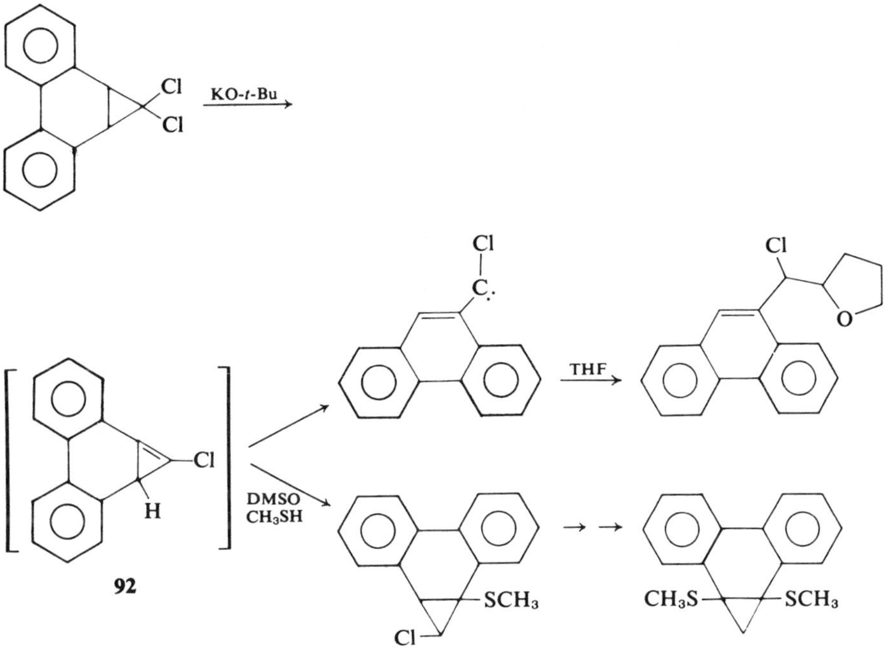

**Scheme 9**

from deuterium scrambling studies.[188] Flash photolysis of 1,4-benzenediazonium carboxylate produced a $C_6H_4$ species presumed to be 1,4-dehydrobenzene (*p*-benzyne) which was stable enough to be monitored for up to two

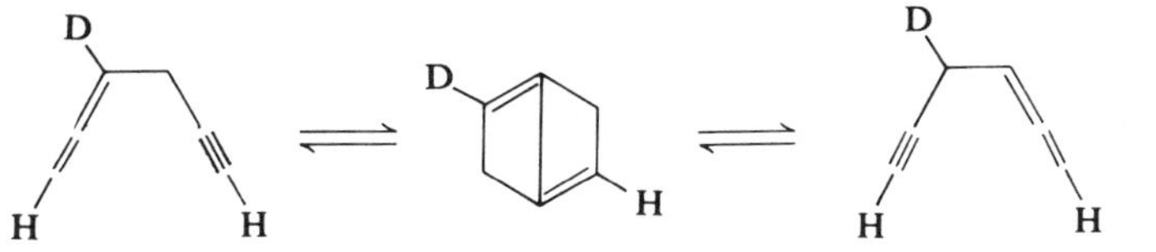

$$(12)$$

**93**

minutes under low-pressure conditions.[189] MINDO/3 calculations[190] point up the existence of two discrete singlet intermediates: 1,4-benzenediyl (e.g., **94**) and butalene (**95**); the latter is calculated to be considerably less stable than the former. However, no clear-cut choice is available between singlet or triplet 1,4-benzenediyl as the ground state for this species.[190] In line with these

[188] H. Hopf, *Chem. Ber.* **104**, 1499 (1971).

[189] R. S. Berry, J. Clardy, and M. E. Schafer, *Tetrahedron Lett.* p. 1003 (1965).

[190] M. J. S. Dewar and W.-K. Li, *J. Am. Chem. Soc.* **96**, 5569 (1974). For *ab initio* calculations of the three dehydrobenzenes which assume benzenoid structures, see D. L. Wilhite and J. L. Whitten, *J. Am. Chem. Soc.* **93**, 2858 (1971).

predictions, deuterium labeling studies coupled with kinetic and thermochemical considerations have implicated 1,4-benzenediyl as a discrete intermediate,[191,192] while butalene (95) has been generated and trapped.[193] Even more recently, 9,10-dehydroanthracene has been generated by photolysis of a

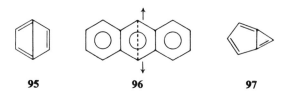

$$(13)$$

bisketene and monitored by uv spectroscopy in a 3-methylpentane glass at 77°K.[194] Because no esr signals attributable to the triplet (or quintet) are observed, the subject is assumed to be a singlet, and the general similarity of its uv spectrum to that of anthracene suggests 9,10-dehydroanthracenediyl (96) as the molecule formed. The above-cited MINDO/3 calculations also

**95**          **96**          **97**

find that singlet 1,3-dehydrobenzene ("*meta*-benzyne") is more stable than the 1,2- and 1,4-dehydrobenzene singlets, and also furnish a description of the molecule which indicates that it is perhaps better described as bicyclo[3.1.0]-hexa-1,3,5-triene (**97**). Convincing evidence for the intermediacy of this molecule,[195a,b] as well as a description of some of its chemistry,[195b] have appeared. One can note that **97** is to cyclopropenium as azulene is to tropylium.

### 3.D.3 STRAIN AND LATICYCLIC CONJUGATION

While a complete discussion of aromaticity, antiaromaticity, homoaromaticity, spiroaromaticity, and indeed the subject of "symmetry, topology,

[191] R. R. Jones and R. G. Bergman, *J. Am. Chem. Soc.* **94**, 660 (1972). See footnote 9 in this article for discussion of the diradical nature of the 1,4-dehydrobenzene generated at elevated temperature.

[192] R. G. Bergman, *Acc. Chem. Res.* **6**, 27 (1973).

[193] R. Breslow, J. Napierski, and T. C. Clarke, *J. Am. Chem. Soc.* **97**, 6275 (1975).

[194] O. L. Chapman, C.-C. Chang, and J. Kole, *J. Am. Chem. Soc.* **98**, 5703 (1976).

[195a] W. N. Washburn, *J. Am. Chem. Soc.* **97**, 1615 (1975).

[195b] W. N. Washburn and R. Zahler, *J. Am. Chem. Soc.* **98**, 3827, 7828 (1976).

and aromaticity"[196] is well beyond the scope and intention of this book,[197] we wish to briefly examine molecules and ions in which strain opposes aromatic stabilization or reinforces antiaromatic destabilization. We note that the increase in strain upon transformation of a cyclopropene derivative to an aromatic cyclopropenium ion (**97a**) (ca. 20 kcal/mole) is more than made up by the approximately 40 kcal/mole of resonance stabilization in the latter,[198] and that such ions (including the parent) have been observed as well as isolated as salts.[199a,b] In contrast, cyclopropenyl anion[198,200,201] combines strain energy of similar magnitude with appreciable antiaromatic destabilization, and has never been spectroscopically observed because of its instability.

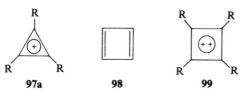

**97a**        **98**        **99**

A total destabilization energy of at least 55 kcal/mole (20 kcal/mole anti-aromatic destabilization[198] plus at least 35 kcal/mole of strain) has made cyclobutadiene (**98**) exceedingly elusive. Nonetheless, tri-*tert*-butylcyclobutadiene has been observed in solution via nmr.[202,203] "Push–pull" cyclobutadienes,[204] as well as cyclobutadienes which are relatively unperturbed electronically,[203,205] have been isolated and characterized. Cyclobutadiene has now been observed (in matrices at exceedingly low temperatures).[206-209]

[196] M. J. Goldstein and R. Hoffmann, *J. Am. Chem. Soc.* **93** 6193 (1971).

[197] These topics are well summarized by P. J. Garratt, "Aromaticity." McGraw-Hill, New York, 1971.

[198] R. Breslow, *Pure Appl. Chem.* **28**, 111 (1971).

[199a] R. Breslow, J. T. Groves and G. Ryan, *J. Am. Chem. Soc.* **89**, 5948 (1967).

[199b] R. Breslow and J. T. Groves, *J. Am. Chem. Soc.* **92**, 984 (1970).

[200] R. Breslow and M. Douek, *J. Am. Chem. Soc.* **90**, 2698 (1968).

[201] R. Breslow, *Angew. Chem., Int. Ed. Engl.* **8**, 565 (1969).

[202] G. Maier and A. Alzerreca, *Angew. Chem., Int. Ed. Engl.* **12**, 1015 (1973).

[203] S. Masamune, N. Nakamura, M. Suda, and H. Ono, *J. Am. Chem. Soc.* **95**, 8481 (1973).

[204] R. Gompper and G. Seybold, *Angew. Chem., Int. Ed. Engl.* **7**, 824 (1968).

[205] H. Kimling and A. Krebs, *Angew. Chem., Int. Ed. Engl.* **11**, 932 (1972).

[206a] C. Y. Lin and A. Krantz, *J. Chem. Soc., Chem. Commun.* p. 1111 (1972).

[206b] A. Krantz, C. Y. Lin, and M. D. Newton, *J. Am. Chem. Soc.* **95**, 2744 (1973).

[207] S. Masamune, M. Suda, H. Ono, and L. M. Leichter, *J. Chem. Soc., Chem. Commun.* p. 1268 (1972).

[208a] O. L. Chapman, C. L. McIntosh, and J. Pacansky, *J. Am. Chem. Soc.* **95**, 614 (1973).

[208b] O. L. Chapman, D. De La Cruz, R. Roth, and J. Pacansky, *J. Am. Chem. Soc.* **95**, 1337 (1973).

[209] G. Maier and B. Hoppe, *Tetrahedron Lett.* p. 861 (1973).

While theoretical (4–31G) calculations predict the cyclobutadiene ground state to be the square triplet,[206b] in line with tentative vibrational analyses,[208a,b] the fact that no esr signal has been detected, coupled with the direct observation via x-ray crystallography of rectangular geometry in methyl tri-*tert*-butylcyclobutadiene carboxylate,[210] would suggest a singlet as the ground state. Cyclobutadiene dications (**99**) have been monitored with nmr,[211a–c] while evidence implicating the intermediacy of cyclobutadiene dianion has also appeared.[212] While cyclooctatetraene dianion has been known since 1962,[213a] 1,3,5,7-tetramethylcyclooctatetraene dication has only recently been monitored by nmr.[213b] (Antiaromatic heterocycles including oxirenes will be discussed in Section 5.D.6.a).

The first known methylenecyclopropene, **100**,[214] had a great deal of dipolar character, and until very recently, all derivatives of this system were similar in nature. Tetramethylmethylenecyclopropene (**101**, $R_1 = R_2 = R_3 = R_4 = CH_3$), the first simple alkyl derivative to be observed, is stable in solution at

| 100 | 101 | 102 |

−20°C but polymerizes at room temperature,[215] while a more hindered derivative ($R_1 = R = t\text{-}C_4H_9$, $R_2 = H$, $R_3 = Br$) can be monitored at ambient temperature.[216] Dehydrohalogenation of 1,2-dichloro-1-methyl-cyclopropane produces methylenecyclopropene itself as a transient intermediate which can, however, be trapped.[217] Factors which contribute to the instability

[210] L. T. J. Delbaere, M. N. G. James, N. Nakamura, and S. Masamune, *J. Am. Chem. Soc.* **97**, 1973 (1975).

[211a] G. A. Olah, J. M. Bollinger, and A. M. White, *J. Am. Chem. Soc.* **91**, 3667 (1969).

[211b] G. A. Olah and G. D. Mateescu, *J. Am. Chem. Soc.* **92**, 1430 (1970).

[211c] G. A. Olah and J. S. Staral, *J. Am. Chem. Soc.* **98**, 6290 (1976).

[212] J. S. McKennis, L. Brener, J. R. Schweiger, and R. Pettit, *J. Chem. Soc., Chem. Commun.* p. 365 (1972).

[213a] T. J. Katz, W. H. Reinmuth, and D. E. Smith, *J. Am. Chem. Soc.* **84**, 802 (1962).

[213b] G. A. Olah, J. S. Staral, and L. A. Paquette, *J. Am. Chem. Soc.* **98**, 7817 (1976).

[214] A. S. Kende, *J. Am. Chem. Soc.* **85**, 1882 (1963).

[215] P. J. Stang and M. C. Mangum, *J. Am. Chem. Soc.* **97**, 3854 (1975).

[216] W. E. Billups and A. J. Blakeney, *J. Am. Chem. Soc.* **98**, 7817 (1976).

[217] W. E. Billups, A. J. Blakeney, and W. T. Chamberlain, *J. Org. Chem.* **41**, 3771 (1976).

of methylenecyclopropene include about 78 kcal/mole of strain (see below), a high index of free valency at the exomethylene carbon, and unhindered approach of reagents to this molecule. *Ab initio* calculations obtain a large dipole moment (2.08 D), consistent with considerable dipolar-type resonance.[138a,b] Cyclopropenone (**102**) is isolable,[218a-c] and its structure as determined by microwave spectroscopy has been reported.[219] The di-*tert*-butyl derivative is, in fact, quite stable.[220] The fact that cyclopropenones do not form hydrates in aqueous solution as cyclopropanones do,[218c] as well as similar observations of reduced reactivity along with theoretical calculations,[221] suggest that cyclopropenone is aromatic. However, it would appear that cyclopropenone is essentially no more aromatic in its ground-state properties than cyclopropene is,[219,222] although significant resonance stabilization is suggested and supported through photoelectron spectroscopic studies.[223] Aromatic properties thus seem to be manifested in cyclopropenone's reactivity. The standard enthalpy of formation of diphenylcyclopropenone, another stable derivative, has been used to estimate a lower limit of 78 kcal/mole of strain in cyclopropenone itself (this work is interesting to read because it illustrates the difficulties in apportioning "destabilization," "strain," and "delocalization energy" in a complex molecule).[224] Because the standard heat of formation of gaseous diphenylacetylene is unknown, one cannot calculate the energy for decomposition of diphenylcyclopropenone into this molecule and carbon monoxide. Dihydroxycyclopropenone (deltic acid) has been reported recently.[225]

Despite their potential for cyclic delocalization, radialenes* (**103**,[226-228] **104**,[229] and **105**[230]) are essentially normal olefins.[231,232] If one employs

[218a] R. Breslow and G. Ryan, *J. Am. Chem. Soc.* **89**, 3073 (1967).

[218b] R. Breslow, G. Ryan, and J. T. Groves, *J. Am. Chem. Soc.* **92**, 988 (1970).

[218c] R. Breslow and M. Oda, *J. Am. Chem. Soc.* **94**, 4787 (1972).

[219] R. C. Benson, W. H. Flygare, M. Oda, and R. Breslow, *J. Am. Chem. Soc.* **95**, 2772 (1973). For a general review, see K. T. Potts and J. S. Baum, *Chem. Rev.* **74**, 189 (1974).

[220] J. Ciabattoni and E. C. Nathan, III, *J. Am. Chem. Soc.* **91**, 4766 (1969).

[221] B. A. Hess, Jr., L. J. Schaad, and C. W. Holyoke, Jr., *Tetrahedron* **28**, 5299 (1972).

[222] We thank Prof. Ronald H. Levin for discussion of his [13]C nmr data on cyclopropenones.

[223] W. Schafer, A. Schweig, G. Maier, T. Sayrac, and J. K. Crandall, *Tetrahedron Lett.* p. 1213 (1974).

[224] H. P. Hopkins, Jr., D. Bostwick, and C. J. Alexander, *J. Am. Chem. Soc.* **98**, 1355 (1976).

[225] D. Eggerding and R. West, *J. Am. Chem. Soc.* **97**, (1975).

[226] G. Köbrich and H. Heinemann, *Angew. Chem., Int. Ed. Engl.* **4**, 594 (1905).

[227] E. A. Dorko, *J. Am. Chem. Soc.* **87**, 5519 (1965).

[228] P. A. Waitkus and L. I. Peterson, *J. Am. Chem. Soc.* **85**, 2268 (1963).

[229] G. W. Griffin and L. I. Peterson, *J. Am. Chem. Soc.* **85**, 2268 (1963).

[230] H. Hopff and A. K. Wick, *Helv. Chim. Acta* **44**, 380 (1961).

* See Addendum.

thermochemical data reported for biphenylene (106), it might appear possible to evaluate the strain in [4]radialene. A 60 kcal/mole value obtained this way[233] suffers from the considerable uncertainty in assigning a resonance

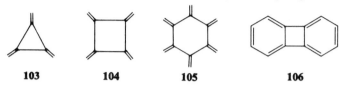

**103**          **104**          **105**          **106**

energy to this molecule and would appear to be unrealistically high. That there is little important interaction among the olefinic groups of these radialenes is supported by observations of significant puckering in perchloro-[4]radialene[234] and a chair conformation for 1,2,3,4,5,6-hexamethyl[6]-radialene.[235]

Small-ring, spiro-connected cycloalkenes can have, in principle, some interesting properties resulting from spiro conjugation.[236,237] Although the "(1,1)spirene" (spiropentadiene) 107 is unknown (calculated strain of 145 kcal/mole[238]), derivatives of 108 have been isolated,[239] and spiroheptadiene

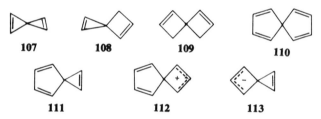

**107**          **108**          **109**          **110**

**111**          **112**          **113**

(109) has been characterized.[240] Spectral evidence appears to support the presence of significant spiroconjugation in 109,[240] 110,[241a,b] and 111.[242] Spiro[4.3]octa-1,3,5-triene, a potential precursor of the ($2\pi$, $4\pi$) spiroaromatic

[231] M. J. S. Dewar, *Chem. Soc., Spec. Publ.* No. 21, 177 (1967).

[232] V. H. Dietrich, *Acta Crystallogr.*, Sect. B. **26**, 44 (1970).

[233] R. C. Cass, H. D. Springall, and P. G. Quincey, *J. Chem. Soc.* p. 1188 (1955).

[234] F. P. van Remoortere and F. P. Boer, *J. Am. Chem. Soc.* **92**, 3355 (1970).

[235] W. Marsh and J. D. Dunitz, *Helv. Chim. Acta* **58**, 707 (1975).

[236] H. E. Simmons and T. Fukunaga, *J. Am. Chem. Soc.* **89**, 5208 (1967).

[237] R. Hoffmann, A. Imamura, and G. D. Zeiss, *J. Am. Chem. Soc.* **89**, 5219 (1967).

[238] M. J. S. Dewar, "The Molecular Orbital Theory of Organic Chemistry," p. 461. McGraw-Hill, New York, 1970.

[239] M. F. Semmelhack and R. J. DeFranco, *J. Am. Chem. Soc.* **94**, 8838 (1972).

[240] L. A. Hulshof and H. Wynberg, *J. Am. Chem. Soc.* **96**, 2191 (1974).

[241a] M. F. Semmelhack, J. S. Foos, and S. Katz, *J. Am. Chem. Soc.* **94**, 8638 (1972).

[241b] M. F. Semmelhack, J. S. Foos, and S. Katz, *J. Am. Chem. Soc.* **95**, 7325 (1973).

[242] H. Dürr, B. Ruge, and H. Schmitz, *Angew. Chem., Int. Ed. Engl.* **12**, 577 (1973).

ion **112** has been reported.[243] The hydrocarbon has a half-life in solution of only 90 minutes at $-4.5°C$, perhaps because heterolysis will produce a zwitterion having both allylic carbonium ion and cyclopentadienide character.[243] Ion **113** does not appear to exhibit the special stability which would be expected of a spiroaromatic ion.[244] Although uv,[245] nmr,[246] and photoelectron[247] studies of spiro[2.4]hepta-1,3-diene (**114**) indicate considerable cyclopropane conjugation with the $\pi$ system, structural studies (electron diffraction) do not support this view.[247] Ultraviolet studies also suggest some cyclobutane conjugation in **115**.[248]

**114**       **115**

**3.D.4** RELIEF OF STRAIN UPON FORMATION OF UNSTABLE
INTERMEDIATES

High-energy molecules often provide relatively low-energy routes to high-energy intermediates. For example, in Chapter 5 we will examine "Dewar benzene's" ability to thermally generate an electronically excited state of benzene, as well as the generation of electronically excited acetone from 1,2-dioxetane. Relief of strain through ring opening to diradical or other species is also sometimes observed.

Methylenecyclopropanes readily rearrange via a perpendicular singlet trimethylenemethane (**116**).[249–253] This perpendicular singlet is calculated to

**116**

[243] M. F. Semmelhack, R. J. DeFranco, Z. Margon, and J. Stock, *J. Am. Chem. Soc.* **95**, 426 (1973).

[244] C. F. Wilcox, Jr., and R. R. Craig, *J. Am. Chem. Soc.* **83**, 4258 (1961).

[245] R. A. Clark and R. A. Fiato, *J. Am. Chem. Soc.* **92**, 4736 (1970).

[246] R. Gleiter, E. Heilbronner, and A. de Meijere, *Helv. Chim. Acta* **54**, 1029 (1971).

[247] J. F. Chiang and C. F. Wilcox, Jr., *J. Am. Chem. Soc.* **95**, 2885 (1973).

[248] R. D. Miller, M. Schneider, and D. L. Dolee, *J. Am. Chem. Soc.* **95**, 8468 (1973).

[249] J. P. Chesick, *J. Am. Chem. Soc.* **85**, 2720 (1963).

[250] J. J. Gajewski, *J. Am. Chem. Soc.* **90**, 7178 (1968).

[251] J. C. Gilbert and J. R. Butler, *J. Am. Chem. Soc.* **92**, 2168 (1970).

[252] W. von E. Doering and H. D. Roth, *Tetrahedron* **26**, 2825 (1970).

[253] M. J. S. Dewar, *J. Am. Chem. Soc.* **93**, 3081 (1971).

be from 2 to 6 kcal/mole more stable than planar singlet trimethylene-methane,[254–256] and a stereochemical investigation of a bridged methylene-cyclopropane is also consistent with a small energy difference (ca. 2–3 kcal/mole) favoring the perpendicular singlet.[257] Photolysis of 4-methylene-$\Delta^1$-pyrazoline [258] or 3-methylenecyclobutanone [259] (see Scheme 10) provides the planar triplet (117) ($D_{3h}$), known to be the ground state of trimethylene-methane.[253–256,258–262] This species is stable for several months at $-196°C$.[261] Quantum-chemical calculations [254–256] predict that 117 is more than 20 kcal/mole lower in energy than the planar singlet. This is an interesting point,

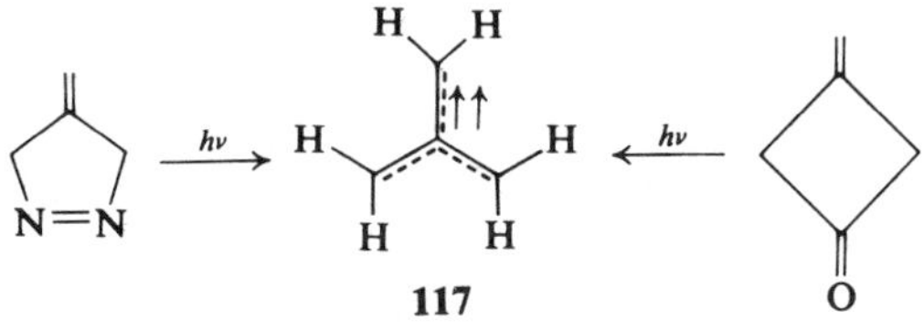

**Scheme 10**

because simple molecular calculations predict that the threefold symmetry of trimethylenemethane is responsible for its triplet character, and that decreased symmetry should break down the degeneracy of the highest occupied orbitals and presumably favor the singlet state. However, the triplet–singlet enthalpy difference is so large that substituted trimethylenemethanes of lower symmetry are also ground-state triplets.[263a,b] An added increment of strain, caused by cyclobutane fusion in 91, causes it to readily form a strained trimethylene-methane diradical which dimerizes before it can be observed.[184] Scrambling of deuterium labels in 1,2-dimethylenecyclobutane proceeds via tetramethylene-ethane (118, Scheme 11) and not through $\Delta^{1,4}$-bicyclo[2.2.0]hexene (77), which could also have been a plausible mechanism.[264a,b,265] This diradical

[254] D. R. Yarkony and H. F. Schaefer, III, *J. Am. Chem. Soc.* **96**, 3754 (1974).

[255] W. T. Borden, *J. Am. Chem. Soc.* **97**, 2906 (1975).

[256] J. H. Davis and W. A. Goddard, III, *J. Am. Chem. Soc.* **98**, 303 (1976).

[257] W. R. Roth and G. Wegener, *Angew. Chem., Int. Ed. Engl.* **14**, 758 (1975).

[258] P. Dowd, *J. Am. Chem. Soc.* **88**, 2587 (1966).

[259] P. Dowd and K. Sachdev, *J. Am. Chem. Soc.* **89**, 715 (1967).

[260] F. Weiss, *Quart. Rev., Chem. Soc.* **24**, 278 (1970).

[261] P. Dowd, *Acc. Chem. Res.* **5**, 242 (1972).

[262] R. J. Baseman, D. W. Pratt, M. Chow, and P. Dowd, *J. Am. Chem. Soc.* **98**, 5726 (1976).

[263a] B. K. Carpenter, R. D. Little, and J. A. Berson, *J. Am. Chem. Soc.* **98**, 5723 (1976).

[263b] M. S. Platz, J. M. McBride, R. D. Little, J. J. Harrison, A. Shaw, S. E. Potter, and J. A. Berson, *J. Am. Chem. Soc.* **98**, 5726 (1976).

[264a] J. J. Gajewski and C. N. Shih, *J. Am. Chem. Soc.* **89**, 4532 (1967).

[264b] P. A. Kelso, A. Yeshuron, C. N. Shih, and J. J. Gajewski, *J. Am. Chem. Soc.* **97**, 1513 (1975).

[265] W. von E. Doering and W. R. Dolbier, Jr., *J. Am. Chem. Soc.* **89**, 4534 (1967).

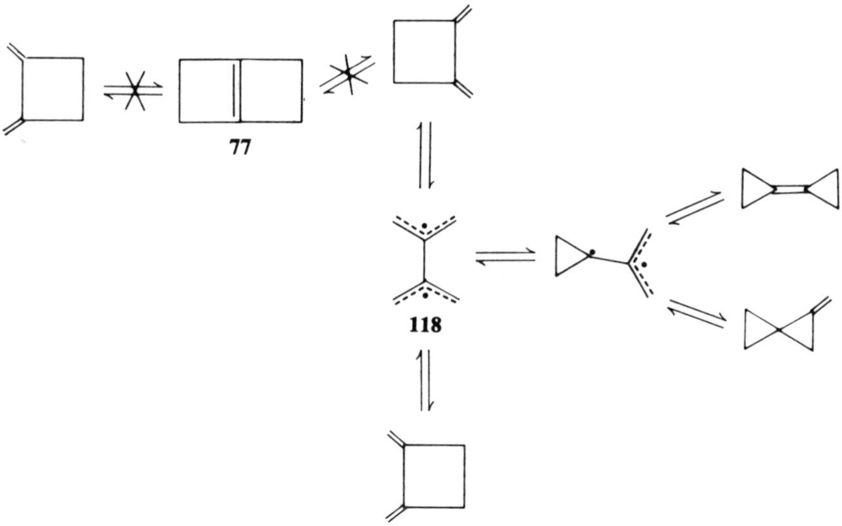

**Scheme 11**

may also be obtained through thermally induced isomerization of bis(cyclo-propylidene) or methylenespiropentane (Scheme 11).[266] Additional strain allows 2,3-dimethylenebicyclo[2.2.0]hexane (119) to form a tetramethylene-ethane even more readily.[267a,b]

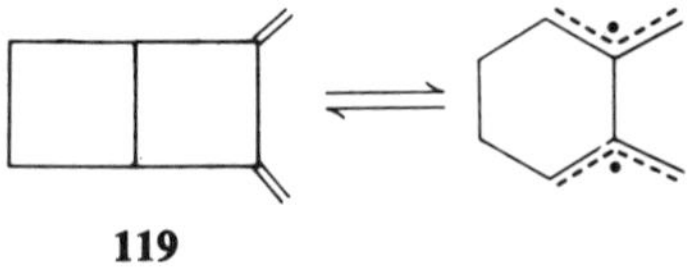

$$(14)$$

Substituted cyclopropanones (120)[268–270] rearrange and participate in cycloaddition reactions through intermediate oxyallyl isomers (121) (re-arrangement of cyclopropanone avoiding oxyallyl has been suggested,[271] although not yet supported). While an early theoretical study predicted that oxyallyl was more stable than cyclopropanone,[272] subsequent calculations as

[266] W. Dolbier, Jr. K. Akiba, J. M. Riemann, C. A. Harmon, M. Bertrand, A. Bezaquet, and M. Santelli, *J. Am. Chem. Soc.* **93**, 3933 (1971).

[267a] C.-S. Chang and N. C. Bauld, *J. Am. Chem. Soc.* **94**, 7593 (1972).

[267b] N. C. Bauld and C.-S. Chang, *J. Am. Chem. Soc.* **94**, 7595 (1972).

[268] N. J. Turro, *Acc. Chem. Res.* **2**, 25 (1969).

[269] N. J. Turro, R. B. Gagosian, S. S. Edelson, T. R. Darling, and W. B. Hammond, *Trans. N.Y. Acad. Sci.* **33**, 396 (1971).

[270] H. H. Wasserman, G. C. Clark, and P. C. Turley, *Top. Curr. Chem.* **47**, 73 (1974).

[271] M. E. Zandler, C. E. Choc, and C. K. Johnson, *J. Am. Chem. Soc.* **96**, 3317 (1974).

[272] R. Hoffmann, *J. Am. Chem. Soc.* **90**, 1475 (1968).

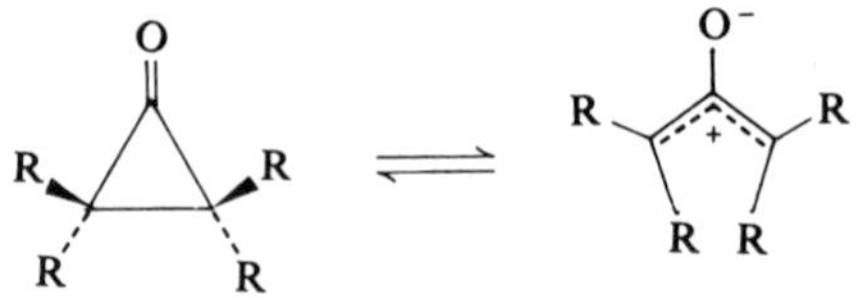

$$(15)$$

well as a microwave study of the parent molecule [273] support the greater stability of the cyclic isomer. Cyclopropanone itself has eluded the cycloaddition reactions characteristic of its derivatives,[268,269] and this is consistent with semiempirical [274] and *ab initio* [275] calculations suggesting enthalpy differences of 66 and 83 kcal/mole, respectively, favoring the closed form. (A value of 54 kcal/mole has been obtained through bond-additivity methods,[276] and this is also in good agreement with a 27 kcal/mole barrier for ring opening of *trans*-di-*tert*-butylcyclopropanone.[277]) The substituted cyclopropanone **69**[156] is a stable molecule because the four benzene rings attached to it are prevented from stabilizing its oxyallyl isomer. Allene oxides, which can also ring open to oxyallyls, as well as alpha-lactones, alpha-lactams, and other three-membered hetero rings capable of producing resonance stabilized open forms, will be considered in Section 5.D.2.C.

## 3.E   Torsionally Distorted π Bonds

### 3.E.1   STERICALLY CROWDED OLEFINS

In *cis*-2-butene, nonbonded repulsion between the methyl groups is, in large part, relieved through opening of the C=C—C angles to about 127°,[278,279a,b] and the strain in this molecule is about 1.2 kcal/mole. In *cis*-1,2-di-*tert*-butylethylene, these angles open to a calculated value of 135° and the strain energy due to angle distortion and residual nonbonded repulsions is considerable larger (calculated: 10.4–11.6 kcal/mole;[279a,b,280] experimental: 10.7 kcal/mole [281]). The calculated strain energy of 1,1-di-*tert*-butylethylene

---

[273] J. M. Pochan, J. E. Baldwin, and W. H. Flygare, *J. Am. Chem. Soc.* **91**, 1896 (1969).

[274] R. C. Bingham, M. J. S. Dewar, and D. H. Lo, *J. Am. Chem. Soc.* **97**, 1302 (1975).

[275] A. Liberles, A. Greenberg, and A. Lesk, *J. Am. Chem. Soc.* **94**, 8685 (1972).

[276] J. F. Liebman and A. Greenberg, *J. Org. Chem.* **39**, 123 (1974).

[277] R. L. Camp and F. D. Greene, *J. Am. Chem. Soc.* **90**, 7349 (1968).

[278] A. Almenningen, I. M. Anfinsen, and A. Haaland, *Acta Chem. Scand.* **24**, 43 (1970).

[279a] O. Ermer and S. Lifson, *J. Am. Chem. Soc.* **95**, 4121 (1973).

[279b] O. Ermer and S. Lifson, *Tetrahedron* **30**, 2425 (1974).

[280] N. L. Allinger and J. T. Sprague, *J. Am. Chem. Soc.* **94**, 5734 (1972).

[281] Cf. Ref. 280, explanation on p. 5737.

is 12.05 kcal/mole.[279a,b] Tri-*tert*-butylethylene (**122**)[282] also may, in principle, relieve nonbonded repulsions without disturbing π overlap. However, the molecule is calculated to have an olefinic torsional angle of 16° accompanied by a large C=C—C angle,[279a,b] and this is consistent with its ir[282] and uv[283] spectral properties. Despite its weakened double bond, tri-*tert*-butylethylene adds bromine slowly because of steric hindrance.[284] In-plane distortion cannot relieve the strain in tetra-*tert*-butylethylene (**123**), for which calculations

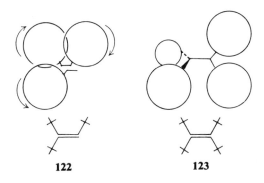

**122**          **123**

predict a 75° torsional angle about the double bond,[279a,b] which would appear to provide it with considerable diradical character. Although unsuccessful attempts at the synthesis of **123** have been discussed,[285a–c] if the molecule is ever produced it might be surprisingly stable, much as the sterically hindered tri-*tert*-butylmethyl radical[286] 1,1',3,3,3',3'-Hexamethyl-2,2'-binorbornylidene* (**125**, m.p. 125°–127°), synthesized through double extrusion of the selenodiazoline **124** (the corresponding thiodiazoline is also a successful precursor), is a "tied-back" tetra-*tert*-butylethylene whose chemical properties are apparently olefinic rather than diradical in nature.[285c] (Neither the seleno- nor the thiodiazoline precursors of **123** have yet been prepared.[285c]) An attempt at the preparation of **127** was thwarted because the intermediate **126** apparently fragmented to molecules which produced **124** and ultimately **125** (Scheme 12), which thus remains the most hindered olefin presently

[282] G. J. Abruscato and T. T. Tidwell, *J. Am. Chem. Soc.* **92**, 4125 (1970).

[283] G. J. Abruscato, R. G. Binder, and T. T. Tidwell, *J. Org. Chem.* **37**, 1787 (1972).

[284] G. J. Abruscato and T. T. Tidwell, *J. Org. Chem.* **37**, 4151 (1972).

[285a] D. H. R. Barton, F. S. Guziec, Jr., and I. Shahak, *J. Chem. Soc., Perkin Trans. 1* p. 1794 (1974).

[285b] T. G. Back, D. H. R. Barton, M. R. Britten-Kelly, and F. S. Guziec, Jr., *J. Chem. Soc., Chem. Commun.* p. 539 (1975).

[285c] T. G. Back, D. H. R. Barton, M. R. Britten-Kelly, and F. S. Guziec, Jr., *J. Chem. Soc., Perkin Trans. 1* p. 2079 (1976).

[286] G. D. Mendenhall and K. U. Ingold, *J. Am. Chem. Soc.* **95**, 3422 (1973).

* See Addendum.

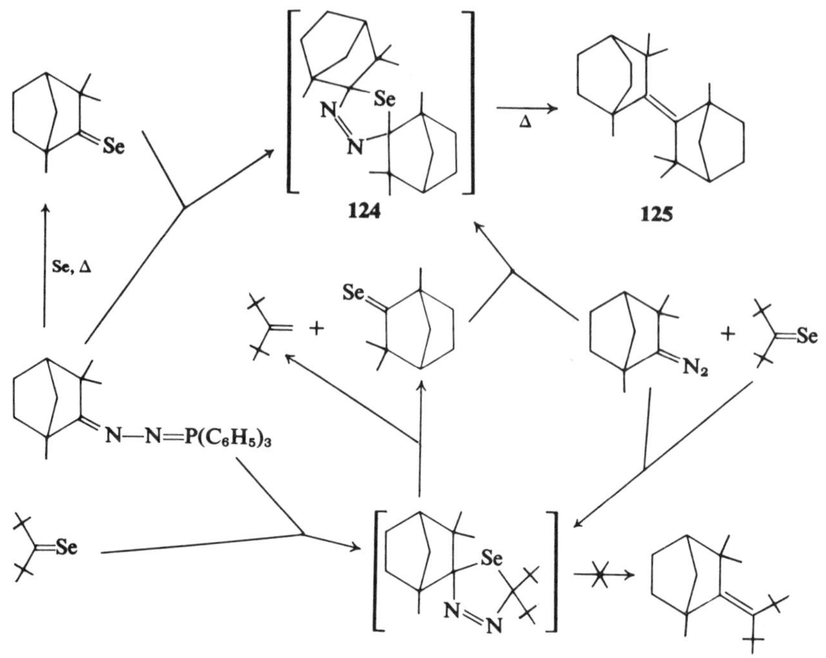

**Scheme 12**

known.[285c] Biadamantylidene (**128**)[287a,b] maintains an untwisted double bond in spite of the presence of significant nonbonded hydrogen repulsions. Tetracyclopropylethylene[288,289] and tetraisopropylethylene[290,291] have been reported, and the photoelectron spectrum of the latter indicates that it is an essentially normal olefin.[292] The torsional angle of 16° found in **129**[293] is in fair agreement with a twist of 22° calculated in di-*tert*-butyl-*trans*-2-butene.[279a,b] Steric hindrance in tetraneopentylethylene explains this molecule's inertness

[287a] J. H. Wieringa, J. Strating, H. Wynberg, and W. Adam, *Tetrahedron Lett.* p. 169 (1972).

[287b] S. C. Swen-Walstra and G. J. Visser, *Chem. Commun.* p. 82 (1971).

[288] A. Nierth, H. M. Ensslin, and M. Hanack, *Justus Liebigs Ann. Chem.* **733**, 187 (1970).

[289] T. Teraji, I. Moritani, E. Tsuda, and S. Nishida, *J. Chem. Soc. C.* p. 3252 (1971).

[290] R. F. Langler and T. T. Tidwell, *Tetrahedron Lett.* p. 777 (1975).

[291] D. S. Bomse and T. H. Morton, *Tetrahedron Lett.* p. 781 (1975).

[292] P. D. Mollere, K. N. Houk, D. S. Bomse, and T. H. Morton, *J. Am. Chem. Soc.* **98**, 4732 (1976).

[293] D. Mootz, *Acta Crystallogr., Sect. B* **24**, 839 (1968).

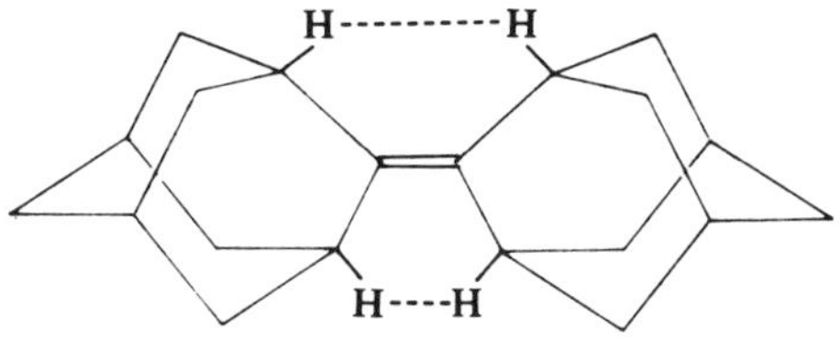

**128** 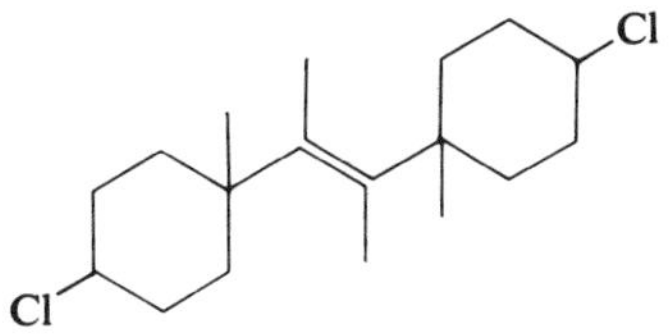                  **129**

to $Br_2/CCl_4$ and $FSO_3H/SO_2ClF(!)$.[294] A hindered olefin, camphenylidene-adamantane, has been employed to implicate perepoxides as intermediates in the reaction between olefins and singlet oxygen (see Section 5.D.2.d). The moral: Strained molecules often provide rewards far greater than those anticipated.

If bifluorenylidene (**130**, R = H) were coplanar, it would suffer exceedingly severe nonbonded hydrogen–hydrogen repulsions. These destabilizing interactions may be relieved by torsion about the double bond (**130a**) or by folding at the olefinic termini (**130b**). While early x-ray studies favored **130b**,[295] more recent work supports the twisted structure **130a** for a derivative

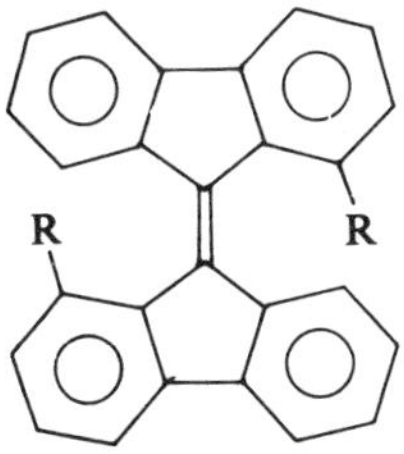

**130** 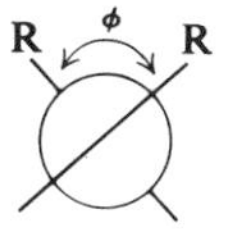        **130a** (*trans*) 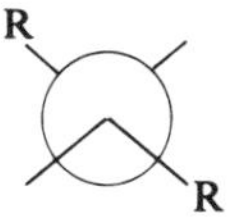        **130b** (*trans*)

[R = $CO_2CH(CH_3)_2$, $\phi = 40°$ [296]] as well as the parent molecule (R = H, $\phi = 43°$ [297]). The low rotational barriers in bifluorenylidenes [**130**, R = $CO_2CH(CH_3)_2$, 20–21 kcal/mole [298]; **130**, R = $CH_3$, 19 kcal/mole [299a]; **131**, 23.5 kcal/mole [299b]] are attributed to steric destabilization of the ground state

[294] G. A. Olah and G. K. S. Prakash, *J. Org. Chem.* **42**, 580 (1977).

[295] S. C. Nyburg, *Acta Crystallogr.*, Sect. B **7**, 779 (1954).

[296] N. A. Bailey and S. E. Hull, *J. Chem. Soc.*, *Chem. Commun.* p. 960 (1971).

[297] H. L. Ammon and G. L. Wheeler, *J. Am. Chem. Soc.* **97**, 2326 (1975). This finding was cited in their Table III as N. A. Bailey, personal communication.

[298] I. R. Gault, W. D. Ollis, and I. O. Sutherland, *J. Chem. Soc.*, *Chem. Commun.* p. 269 (1970).

[299a] I. Agranat, M. Rabinovitz, and A. Weitzen-Dagan, *J. Chem. Soc.*, *Chem. Commun.* p. 732 (1972).

[299b] I. Agranat, M. Rabinovitz, and A. Weitzen-Dagan, *Tetrahedron Lett.* p. 1241 (1974).

and resonance stabilization of the diradical transition state. Octachloropenta-fulvene (132)[300] is stable, in contrast to the hydrocarbon, and its blue-violet color is associated with a 41° torsional angle about its olefinic linkage.[301] Bianthrone (133) and its derivatives exhibit thermochromism (heat-dependent color changes). The A isomer (A, B, and C refer to distinct isomers of decreasing stability) maintains a folded conformation, similar to 130b.[302] The B isomer is photostable but thermally labile and is responsible for the thermo-chromic properties of 133.[303] This isomer exists in the twisted conformation, has a torsional angle equal to 57°, and is less stable than the A isomer by 5 kcal/mole.[303] (The C isomer of 133 has not yet been identified.[303]) The low rotational barriers (ca. 21 kcal/mole) of bianthrones are predominantly a result of destabilization of the ground state, and such conformational changes are definitely associated with color changes.[304]

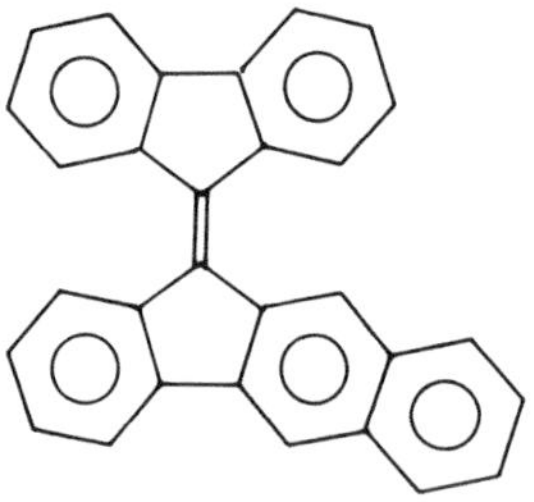

131 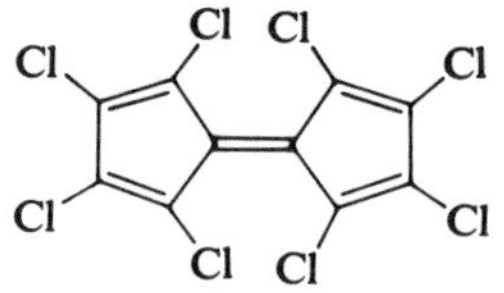                          132 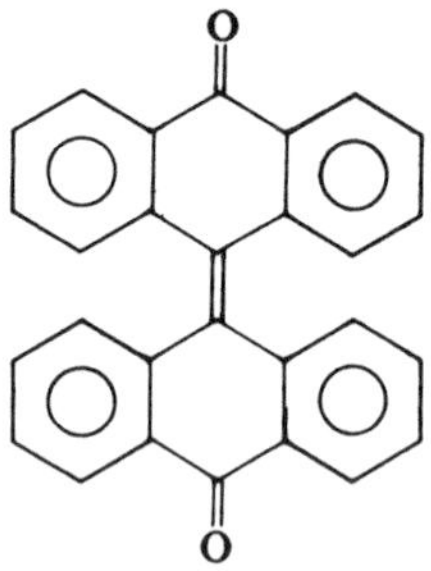                          133

### 3.E.2   *trans*-CYCLOALKENES

*trans*-Cyclooctene, first reported in 1953 as the product of thermal decom-position of *N,N,N*-trimethylcyclooctylammonium hydroxide,[305] remains the smallest isolable *trans*-cycloalkene. Its strain energy (16.7 kcal/mole, see Table 3.5) largely arises from its twisted olefinic linkage. *trans*-Cycloheptene (134), calculated to have about 27 kcal/mole of strain (Table 3.5), has been generated from *trans*-1,2-cycloheptene thionocarbonate and trapped by 2,5-diphenyl-3,4-isobenzofuran [Eq. 16],[306] but it has not been monitored spectroscopically. The only published experimental evidence for a *trans*-

[300] V. Mark, *Tetrahedron Lett.* p. 33 (1961).

[301] H. L. Ammon, G. L. Wheeler, and I. Agranat, *Tetrahedron* 29, 2695 (1973).

[302] E. Harnik and G. M. J. Schmidt, *J. Chem. Soc.* p. 3295 (1954).

[303] R. Korenstein, K. A. Muszkat, and S. Sharafy-Ozeri, *J. Am. Chem. Soc.* 95, 6177 (1973).

[304] I. Agranat and Y. Tapuhi, *J. Am. Chem. Soc.* 98, 615 (1976).

[305] A. C. Cope, R. A. Pike, and C. F. Spenser, *J. Am. Chem. Soc.* 75, 3212 (1953).

[306] E. J. Corey, F. A. Carey, and R. A. E. Winter, *J. Am. Chem. Soc.* 87, 934 (1965).

Table 3.5

**Experimental and Calculated Thermodynamic and Kinetic Parameters for Small and Medium *trans*-Cycloalkenes**

| Compound | $\Delta G^0$ (kcal/mole) (*trans* − *cis*) | $\Delta H^0$ (kcal/mole) (*trans* − *cis*) | Strain (kcal/mole) | $\Delta G^{\ddagger}$ rac (kcal/mole) |
|---|---|---|---|---|
| *trans*-Cycloheptene | | Calcd + 20.3[b] | + 27[b] | |
| *trans*-Cyclooctene | | + 9.2[c] | + 16.7 | 36[d] |
| *trans*-Cyclononene | + 4.04[a] | + 2.9[a,c] | + 14.4 | 20[d] |
| *trans*-Cyclodecene | + 1.86[a] | + 3.6[a] + 3.3[c] | | 10[d] |
| *trans*-Cycloundecene | − 0.67 | − 0.12[a] | | |
| *trans*-Cyclododecene | − 0.49 | + 0.41[a] | | |

[a] A. C. Cope, P. T. Moore, and W. R. Moore, *J. Am. Chem. Soc.* **82**, 1744 (1960).

[b] Add strain in *cis*-cycloheptene (6.7 kcal/mole) to calculated (*trans* − *cis*) enthalpy difference of + 20.3 kcal/mole. See N. L. Allinger and J. T. Sprague, *J. Am. Chem. Soc.* **94**, 5734 (1972).

[c] R. B. Turner and W. R. Meador, *J. Am. Chem. Soc.* **79**, 4133 (1957).

[d] A. C. Cope and B. A. Pawson, *J. Am. Chem. Soc.* **87**, 3649 (1965).

cyclohexene considers the formation of *trans*-1-phenylcyclohexene to result from laser photolysis of a methanolic solution of the *cis* isomer at ambient temperature.[307a] The transient intermediate formed and monitored ($\lambda_{max}$ = 380 nm) has a lifetime of nine microseconds, which is not affected by oxygen

$$(16)$$

concentration (and so it is probably not a triplet) or by solvent polarity (thus, it is probably not dipolar), and which corresponds to an activation barrier of 7 kcal/mole for isomerization to the *cis* isomer.[307a] *trans*-1-Phenylcyclo-heptene ($\lambda_{max}$ = 305 nm) is stable for many hours at low temperature and can be monitored at ambient temperature in solution.[307b] In Section 5.C.4.b we note the stability of transition-metal complexes of *trans*-cyclooctene and other twisted olefinic species.

The conformation and precise structure of *trans*-cyclooctene have been subjects of some controversy.[308] x-Ray studies of a platinum complex of

[307a] R. Bonneau, J. Joussot-Dubien, L. Salem and A. J. Yarwood, *J. Am. Chem. Soc.* **98**, 4329 (1976); see also J. A. Marshall, *Science* **170**, 137 (1970).

[307b] R. Bonneau, J. Joussot-Dubien, Y. Yarwood, and J. Pereryne, *Tetrahedron Lett.* p. 235 (1977).

[308] See O. Ermer, *Struct. Bonding (Berlin)* **27**, 196–198 (1976), for a more detailed discussion of this problem.

*trans*-cyclooctene indicated that the ring was in the "twist" (or "crown") conformation (**135a**).[309] It may be argued that transition-metal complexation so markedly perturbs the bonding in *trans*-cyclooctene as to give this study little relevance to the free molecule, especially in light of the flexibility[308] of the ring. A gas-phase electron-diffraction study then indicated a distorted chair conformation (**135b**) for free *trans*-cyclooctene.[310] However, an x-ray study[311] of *trans*-2-cyclooctenyl-3′,5′-dinitrobenzoate as well as more recent electron-diffraction results[312] on *trans*-cyclooctene find that this molecule has the twist (crown) conformation (**135a**), in line with results of two molecular-mechanics calculations which favor it over the distorted chair by 2.4–3.2

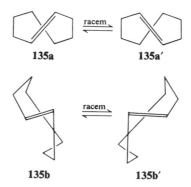

|  135a  |  135a′  |
| :---: | :---: |

|  135b  |  135b′  |
| :---: | :---: |

kcal/mole.[279a,b,280,308] The twisted double bond's torsional angle of 136°[312] is in excellent agreement with a value of 138° calculated by molecular mechanics,[308] and the olefinic bond length, which is essentially normal (1.33 Å), might be explained in terms of a weakened $\pi$ bond compensated for by strengthened $\sigma$ bonding. Coincidentally, the geometric features of the twist conformation of *trans*-cyclooctene indicated by these last two experimental studies are very similar to those of the metal complexes. The twist nature of *trans*-cyclooctene makes it dissymmetric, and it has been optically resolved.[313] Its very high specific rotation ($[\alpha]^{290} = \pm 411°$) is a property expected for an inherently dissymmetric chromophore.[314] The high barrier (36 kcal/mole, see Table 3.5) for racemization of *trans*-cyclooctene (**135a** $\rightleftharpoons$ **135a′**) allows an enantiomer to retain full optical activity after heating at 61°C for seven

[309] P. C. Manor, D. P. Shoemaker, and A. S. Parkes, *J. Am. Chem. Soc.* **92**, 5260 (1970).

[310] R. M. Gavin, Jr. and Z. F. Wang, *J. Am. Chem. Soc.* **95**, 1425 (1973).

[311] O. Ermer, *Angew. Chem., Int. Ed. Engl.* **13**, 604 (1974).

[312] M. Traetteberg, *Acta Chem. Scand., Ser. B* **29**, 29 (1975).

[313] A. C. Cope, G. R. Ganellin, and H. W. Johnson, Jr., *J. Am. Chem. Soc.* **84**, 3191 (1962).

[314] A. Moscowitz and K. Mislow, *J. Am. Chem. Soc.* **84**, 4605 (1962).

days.[315] This high barrier to racemization results from steric destabilization of the transition state caused by hydrogens forced into the interior of the ring en route to a 180° rotation about the double bond. *trans*-Cyclononene racemizes more readily because its increased ring size allows more ready passage of hydrogen through the interior.[315] This substance can only be optically resolved at temperatures of 0°C and lower.[315] It is apparent from Table 3.5 that *trans*-cycloundecene and *trans*-cyclododecene are more stable than their *cis* isomers.

*trans*-2-Cyclooctenone (**136**) has been obtained photochemically from its *cis* isomer.[316] This "synthesis," related to the production of a *trans*-cyclohexene noted above, relied upon the principle that olefin triplet excited states are more stable in the 90° twisted geometry. Thus, small *trans*-cycloalkene triplets are more stable than *cis* triplets. *trans*-2-Cycloheptenone (**137**) has also been generated photochemically, but it has not been isolated.[317,318] The cyclooctadienes **138**,[319] **139**,[320] and **140**[321a,b] (see also **141**[322]), which each

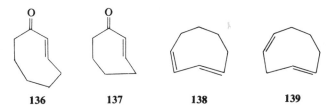

**136**            **137**            **138**            **139**

include a *trans* and a *cis* double bond, have been characterized, and *trans*, *trans*-1,5-cyclooctadiene has been monitored by nmr although no clear-cut decision between conformer **142a** or **142b** could be made on the basis of these data.[323] This molecule has been cited as a possible intermediate in the photochemical rearrangement of *cis,cis*-1,5-cyclooctadiene to **29**, much as the

[315] A. C. Cope, G. R. Ganellin, H. W. Johnson, J. V. Van Auken, and H. J. S. Winkler, *J. Am. Chem. Soc.* **85**, 3276 (1963).

[316] P. E. Eaton and K. Lin, *J. Am. Chem. Soc.* **86**, 2087 (1964).

[317] E. J. Corey, M. Tada, R. LaMahieu, and L. Libit, *J. Am. Chem. Soc.* **87**, 2051 (1965).

[318] P. E. Eaton and K. Lin, *J. Am. Chem. Soc.* **87**, 2052 (1965).

[319] A. C. Cope and C. L. Bumgardner, *J. Am. Chem. Soc.* **78**, 2812 (1956).

[320] E. Vedejs and P. L. Fuchs, *J. Am. Chem. Soc.* **93**, 4070 (1971). The first substituted *cis,trans*-1,4-cyclooctadienes were reported by M. S. Baird and C. B. Reese, *Chem. Commun.* p. 1644 (1970).

[321a] A. C. Cope, C. F. Howell, and A. Knowles, *J. Am. Chem. Soc.* **84**, 3190 (1962).

[321b] A. C. Cope, C. F. Howell, J. Bowers, R. C. Lord, and G. M. Whitesides, *J. Am. Chem. Soc.* **89**, 4024 (1967).

[322] K. Kraft and G. Koltzenberg, *Tetrahedron Lett.* p. 4357, 4723 (1967).

[323] G. M. Whitesides, G. L. Goe, and A. C. Cope, *J. Am. Chem. Soc.* **89**, 7136 (1967).

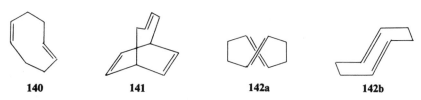

| | | | |
|---|---|---|---|
| **140** | **141** | **142a** | **142b** |

photochemical isomerization of **143**[324] to all-*cis*-cyclooctatetraene is thought to proceed through **144**,[325] which has not yet been observed spectroscopically.

$$\text{143} \xrightarrow{h\nu} \text{144} \longrightarrow \bigcirc \tag{17}$$

        **143**                    **144**

Dimers of *trans,cis*-2,4-cyclooctadienone (**145**) were obtained by irradiation of the *cis,cis* isomer.[326] The intermediacies of *cis,cis,cis,trans*-1,3,5,7-cyclooctatetraene[327] (whose tetraphenyl derivative **146** has a half-life of 18 hours at 25°C[328]), **147** (see Section 5.C.3.a),[329] **148**,[330,331] and **149**[332] have been suggested, while *cis,cis,trans,cis*-cyclononatetraene (**150**) has been trapped.[333]

| | | | |
|---|---|---|---|
| **145** | **146** | **147** | **148** |

| | | |
|---|---|---|
| **149** | **150** | **151** |

[324] I. Haller and R. Srinivasan, *J. Am. Chem. Soc.* **88**, 5084 (1966).

[325] J. Meinwald and H. Tsuruta, *J. Am. Chem. Soc.* **92**, 2579 (1970).

[326] G. L. Lange and E. Neidert, *Tetrahedron Lett.* p. 1349 (1972).

[327] H. E. Zimmerman and H. Iwamura, *J. Am. Chem. Soc.* **92**, 2015 (1970).

[328] E. H. White, E. W. Friend, Jr., R. L. Stern, and H. Maskill, *J. Am. Chem. Soc.* **91**, 523 (1969).

[329] K. B. Wiberg and G. Szeimies, *Tetrahedron Lett.* p. 1235 (1968).

[330] A. R. Brember, A. A. Gorman, and J. B. Sheridan, *Tetrahedron Lett.* p. 475 (1973).

[331] M. Christl and G. Bruntrup, *Angew. Chem., Int. Ed. Engl.* **13**, 208 (1974).

[332] S. W. Staley and T. J. Henry, *J. Am. Chem. Soc.* **91**, 7787 (1969).

[333] A. G. Anastassiou and R. C. Griffith, *J. Am. Chem. Soc.* **93**, 3083 (1971).

Two highly strained isomeric *trans,trans*-bicyclo[6.1.0]non-4-enes have been identified.[334a] It is worthwhile to recall here that the extra strain induced by *trans*-fusion (relative to *cis*-fusion) of an eight-membered ring to cyclopropane is about 6–7 kcal/mole less than the extra strain in *trans,trans*-bicyclo[6.1.0]-non-4-ene. Nevertheless, **151**, "parallel" isomer of *trans, trans*-bicyclo[6.1.0]-non-4-ene, exhibits an extraordinarily facile thermal conversion of an isolated *trans*-to-*cis* olefinic linkage.[334a] A *cis,trans*-homotropylidene (compare to **148**) has been suggested as a short-lived intermediate.[334b]

### 3.E.3 BRIDGEHEAD OLEFINIC LINKAGES

Based upon his investigations of camphor and pinane derivatives during the period between 1900 and 1924, Bredt formulated the rule presently bearing his name.[335] There is some question as to whether Bredt's rule [336a,b] was formulated as an *absolute* "prohibition" of bridgehead olefinic linkages[337] or "prohibition" of such bonding in small- and medium-sized bicyclic molecules only.[338] Systematic investigations of the limitations imposed by Bredt's rule, undertaken during the 1940's,[339] were included in a review of all relevant data published in 1950.[340] In this review, Fawcett defined a stability parameter $S$ (where $S = l + m + n$, see **152**; $l, m, n \neq 0$; in Section 3.F we will consider cases having a "zero" bridge), for which the lowest known value at the time was 9. Much later it was recognized that the size of the smallest

**152**          **153**          **154**

ring bearing a *trans* olefinic linkage in such bridgehead olefins was a much more accurate criterion for predicting stability.[338] Nevertheless, it has also been noted that neither criterion allows prediction of the relative stabilities

[334a] J. A. Deyrup and M. F. Betkouski, *J. Org. Chem.* **40**, 284 (1975).

[334b] R. T. Taylor and L. A. Paquette, *J. Am. Chem. Soc.*, **99**, 5824 (1977).

[335] J. Bredt, *Justus Liebigs Ann. Chem.* **437**, 1 (1924).

[336a] G. L. Buchanan, *Chem. Soc. Rev.* **3**, 41 (1974) includes references to earlier reviews of Bredt's rule.

[336b] R. Keese, *Angew. Chem., Int. Ed. Engl.* **14**, 528 (1975). This is a review of syntheses of bridgehead olefins.

[337] G. Köbrich, *Angew. Chem., Int. Ed. Engl.* **12**, 464 (1973).

[338] J. R. Wiseman, *J. Am. Chem. Soc.* **89**, 5966 (1967).

[339] V. Prelog, *J. Chem. Soc.* p. 420 (150) and references cited therein.

[340] F. S. Fawcett, *Chem. Rev.* **47**, 219 (1950).

of, for example, $\Delta^{1,2}$-bicyclo[4.3.1]decene (**153**) and $\Delta^{1,9}$-bicyclo[4.3.1]decene
(**154**), each of which has an $S$ value of 8 and a *trans*-cyclononene ring.[336a,b]
The chemical manifestations of Bredt's rule have been summarized.[82,336a,b]
An example is the exclusive presence of norcaradiene **155** rather than the
cycloheptatriene isomer, which would contain two bridgehead olefinic linkages[341] (conventional cycloheptatrienes are usually more stable[342]; however,
see Section 5.E.2). However, the homologous hydrocarbon is more stable as
a cycloheptatriene (**156**)[341] because the olefinic bonds are not quite as twisted
as those in the smaller ring.

**155**                    **156**

The smallest bridgehead olefins isolated to date all contain *trans*-cyclooctene
ring systems (all three have $S = 8$). This nicely fits Wiseman's stability
criterion noted above, because *trans*-cyclooctene is the smallest isolable
*trans*-cycloolefin. The first of these molecules to be characterized, $\Delta^{1,2}$-
bicyclo[3.3.1]nonene (**157**, see Scheme 13),[338,343] reacts with oxygen and

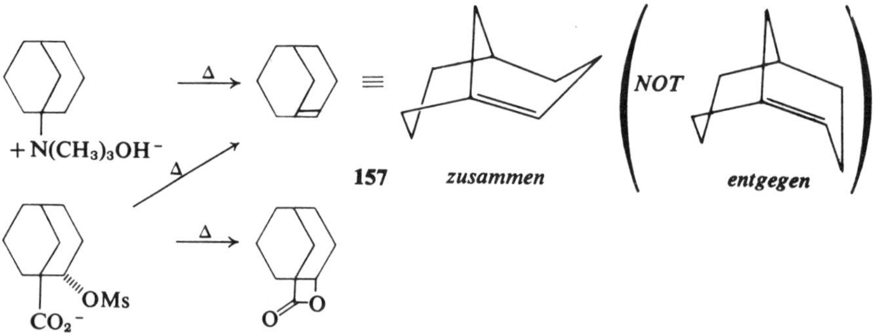

**Scheme 13**

polymerizes upon standing. The strain energy attributed to its bridgehead
olefinic linkage, 12 kcal/mole, is very similar to the strain in *trans*-cyclo-
octene.[344] Note that in order to maintain a *trans*-cyclooctene ring, **157** must
exist in the *zusammen* conformation, since the *entgegen* structure would

[341] E. Vogel, W. Wiedemann, H. D. Roth, J. Eimer, and H. Guenther, *Justus Liebigs Ann. Chem.* **759**, 1 (1972).
[342] G. Maier, *Angew. Chem. Int. Ed. Engl.* **6**, 402 (1967).
[343] J. A. Marshall and H. Faubl, *J. Am. Chem. Soc.* **89**, 5965 (1967).
[344] P. M. Lesko and R. B. Turner, *J. Am. Chem. Soc.* **90**, 6888 (1968).

entail the presence of a *trans*-cyclohexene ring[345] (however, see structure **251** in Chapter 5). $\Delta^{1,2}$-Bicyclo[4.2.1]nonene (**158**) and $\Delta^{1,8}$-bicyclo[4.2.1]nonene (**159**) were isolated following Hofmann elimination, which produced the latter,

$$ + \quad 1:5 \quad (18) $$

$^+N(CH_3)_3OH^-$          **158**               **159**

presumably less stable, molecule as the major product.[346a] Quite recently, $\Delta^{1,8}$-bicyclo[5.1.1]nonene (**152**, $l = 5$, $m = n = 1$) and $\Delta^{1,2}$-bicyclo[5.1.1]-nonene (**152**, $l = m = 1$, $n = 5$) have been isolated following Hofmann elimination which produced the former compound as the major product.[346b] $\Delta^{1,2}$-Bicyclo[3.2.2]nonene (**160**) and $\Delta^{1,7}$-bicyclo[3.2.2]nonene (**161**) are isomers of **157–159**, and both also have $S$ values of 7. However, these molecules, generated by Hofmann elimination, are only observable by nmr at $-80°C$ but dimerize rapidly at $0°C$, and it is apparent that the reason for their reactivities is the presence of *trans*-cycloheptene rings.[347] Once again, Hofmann elimination has produced more of the alkene (e.g., **161**) expected—and demonstrated—to be less stable. Hydrocarbon **160** as well as its [4.2.2] homologue are capable, in principle, of existing as a pair of enantiomers interconvertible by flipping of the largest carbon bridge. However, optically pure material has not yet been isolated for this interesting investigation.[346b] The facile dimerizations that **160** and **161** as well as many other bridgehead olefins undergo deserve some comment. Dimerization of untwisted ethylenes to form cyclobutanes is a rare phenomenon (see, however, compound **77**), but twisted olefinic linkages are expected to dimerize much more readily, as they offer an opportunity for concerted ($_\pi2_a + _\pi2_s$) cyclo addition (actually observed for molecule **141**).[348] $\Delta^{1,2}$-Bicyclo[3.2.1]octene (**162**) and $\Delta^{1,7}$-bicyclo[3.2.1]-octene (**163**) were generated via Hofmann elimination [thermolysis of a

**160**              **161**              **162**              **163**

[345] J. R. Wiseman and W. A. Pletcher, *J. Am. Chem. Soc.* **92**, 956 (1970).
[346a] J. R. Wiseman, H.-F. Chan, and C. J. Ahola, *J. Am. Chem. Soc.* **91**, 2812 (1969).
[346b] J. R. Wiseman, *Int. Symp. Chem. Strained Rings*, Binghamton, New York, 1977.
[347] J. R. Wiseman and J. A. Chong, *J. Am. Chem. Soc.* **91**, 7775 (1969).
[348] R. B. Woodward and R. Hoffmann, "The Conservation of Orbital Symmetry," Ch. 6, Sect. 1. Academic Press, New York, 1970.

thionocarbonate, see, for example Eq. (16), also provided **162**], and, while these two molecules containing *trans*-cycloheptene rings could not be observed spectroscopically, their diphenylisobenzofuran adducts were identified.[349] Another highly constrained bridgehead olefin, **165**, containing a *trans*-cycloheptene ring, has been implicated as an intermediate because a dimer as well as acetic-acid addition and furan cycloaddition products have been identified (Scheme 14).[350a,b] This is an interesting reaction because while a compound

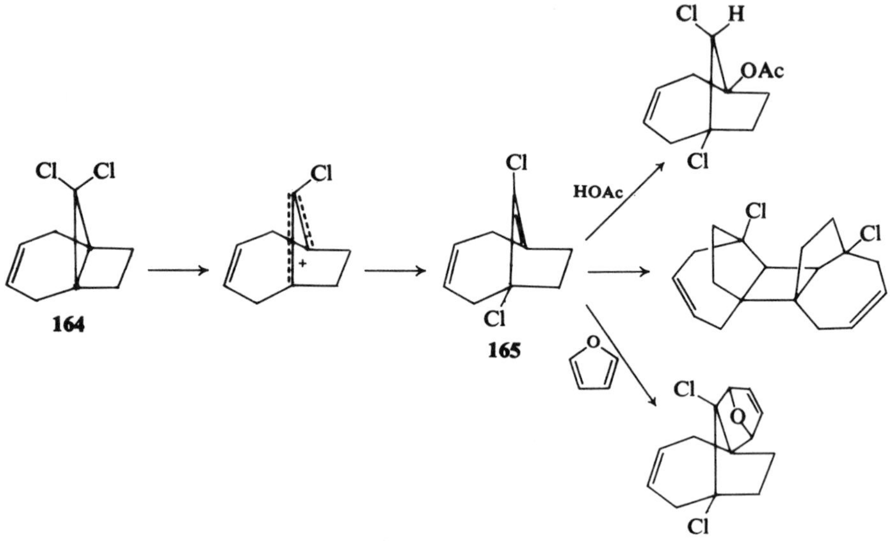

**Scheme 14**

such as 7,7-dichlorobicyclo[4.1.0]heptane will isomerize readily to 1,7-dichlorocycloheptene, a small propellane (see Chapter 6) should isomerize only with difficulty, as the product would have a bridgehead double bond. Apparently, the strain in the small propellane **164** is large enough to favor formation of **165**.[350a,b] Reports which implicate the intermediacies of other intermediates similar to **165** have also appeared.[351,352] $\Delta^{1,2}$-Adamantene (**166**) has been generated by the Hofmann elimination route,[353] as well as by a novel carbene-insertion reaction[354] (see Scheme 15).

[349] J. A. Chong and J. R. Wiseman, *J. Am. Chem. Soc.* **94**, 8627 (1972).

[350a] P. Warner, R. LaRose, C. Lee, and J. C. Clardy, *J. Am. Chem. Soc.* **94**, 7607 (1972).

[350b] P. Warner, R. C. LaRose, R. F. Palmer, C. Lee, D. O. Ross, and J. C. Clardy, *J. Am. Chem. Soc.* **97**, 5507 (1975).

[351] C. B. Reese and M. R. D. Stebles, *J. Chem. Soc., Chem. Commun.* p. 1231 (1972).

[352] K. Taguchi and F. H. Westheimer, *J. Am. Chem. Soc.* **95**, 7413 (1973).

[353] B. L. Adams and P. Kovacic, *J. Am. Chem. Soc.* **95**, 8206 (1973).

[354] M. Farcasiu, D. Farcasiu, R. T. Conlin, M. Jones, Jr., and P. von R. Schleyer, *J. Am. Chem. Soc.* **95**, 8207 (1973).

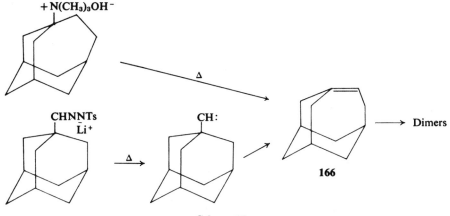

**Scheme 15**

In the early discussion presented in this section, we noted the paucity of data for even the transient existence of simple *trans*-cyclohexenes. There is, however, more support for *trans*-cyclohexene rings included in bicyclic systems which (a) cannot isomerize to the *cis* isomer, and (b) are afforded some kinetic stability as the result of steric hindrance. For example, $\Delta^{1,2}$-adamantene (**167**) has been generated and its intermediacy ascertained through the identification of products of its dimerization, cycloaddition with 1,3-butadiene (furan is unsuitable because the expected cycloadduct should be highly strained), and addition to the double bond.[355-357] (Note that the additional bridge in **167**, when compared to **157**, imposes the "*entgegen*-type" structure shown; see also compound **251** in Chapter 5.) Bicyclo[2.2.2]oct-1-ene (**168**) was obtained via the spontaneous decomposition at room temperature of 1-ethoxybicyclo[2.2.2]oct-2-yl-lithium[358] as well as rearrangement of

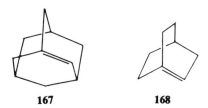

**167**          **168**

[355a] D. Grant, M. A. McKervey, J. J. Rooney, N. G. Samman, and G. Step, *J. Chem. Soc., Chem. Commun.* p. 1186 (1972).

[355b] W. Burns and M. A. McKervey, *J. Chem. Soc., Chem. Commun.* p. 858 (1974).

[355c] W. Burns, D. Grant, M. A. McKervey, and G. Step, *J. Chem. Soc., Perkin Trans. 1* p. 234 (1976).

[356] D. Lenoir, *Tetrahedron Lett.* p. 4049 (1972).

[357] J. E. Gano and L. Eizenberg, *J. Am. Chem. Soc.* **95**, 972 (1973).

[358] H. H. Grootveld, C. Blomberg, and F. Bickelhaupt, *J. Chem. Soc., Chem. Commun.* p. 542 (1973).

(1-norbornyl) carbene (analogous to the rearrangement in Scheme 15).[359a] The transient intermediacy of this bridgehead olefin was deduced from isolation of the product of *t*-BuLi addition in the first case and through deuterium labeling studies of the rearrangement product, 3-methylene-1,6-heptadiene, formed in the latter investigation. Scheme 16 illustrates the generation and

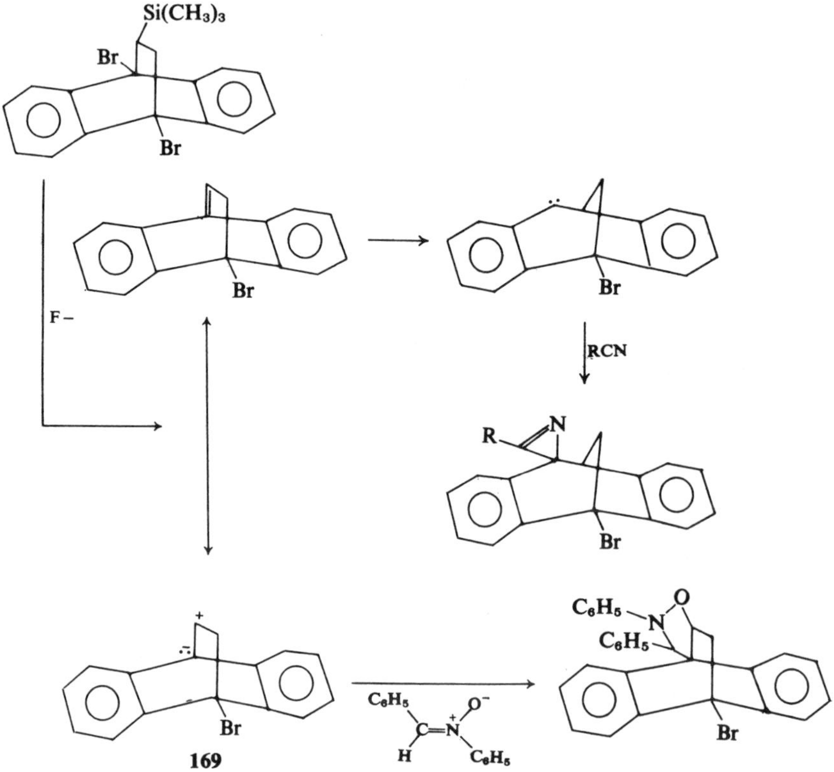

**Scheme 16**

some chemistry of **169**, a dibenzo analogue of $\Delta^{1,2}$-bicyclo[2.2.2]octene.[359b] One striking aspect is the thermal rearrangement of the bridgehead olefin into a carbene (recall that carbenes have been used to obtain bridgehead olefins, e.g., Scheme 15). This would appear to be the only case of a hydrocarbon other than a cyclopropene (Section 3.D.2) forming a carbene upon heating. Another interesting feature is the postulated zwitterionic ("carbonylide") nature of the bridgehead olefin consistent with regiospecific cycloaddition of a nitrone, as well as the seeming 1,2-alkyl shift which generates the carbene in Scheme 16.[359b] 1-Norbornene (**170**) was generated through reductive elimi-

---

[359a] A. D. Wolf and M. Jones, Jr., *J. Am. Chem. Soc.* **95**, 8209 (1973).
[359b] T. H. Chan and D. Massuda, *J. Am. Chem. Soc.* **99**, 936 (1977).

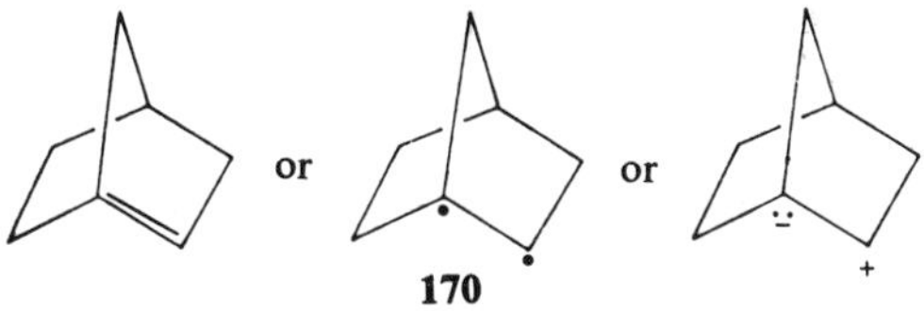

nation of various 1,2-dihalonorbornanes and its intermediacy established through interception by furan.[360a,b] Both this approach and elimination of ethoxylithium from 1-ethoxy-7-bromonorbornane could not produce $\Delta^{1,7}$-norbornene.[336b] As discussed in Section 5.E.3, fluorine substitution at the other bridgehead appears to stabilize derivatives of **170**. Silver-assisted rearrangement of 9,9-dibromotricyclo[3.3.1.0$^{1,5}$]nonane [Eq. (19)] is thought

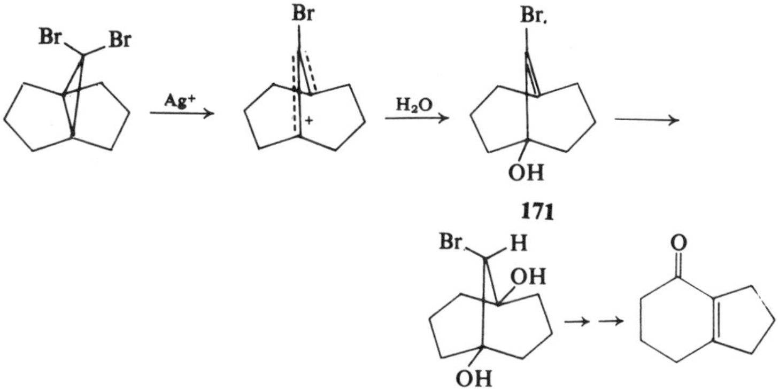

to produce the intermediate **171**.[361] Although this reaction is analogous to that depicted in Scheme 14, the apparent formation of **171** is a bit surprising, because the precursor propellane is not as strained as the one in Scheme 14, while the bridgehead olefin should be more strained.

Inclusion of an extra double bond in a bridgehead olefin introduces additional constraints. The diene **172**, an isolable liquid, was obtained through a wonderful Michael-type intramolecular Wittig reaction that takes place following addition of 3-cycloheptenone and ylide [Eq. (20)].[362] The acidity of the bridgehead hydrogen of **172** might conceivably reflect the stability of a homocyclopentadienide anion, but this remains to be tested (see **175** and **176**). The lower homologues **173** and **174** have been generated in an analogous manner and are not isolable, but have been trapped as their diphenylisobenzofuran and furan adducts, respectively.[362] Bridgehead olefin **175** is quite stable at $-79°C$, but readily reacts with phenyl azide at $0°C$ (Scheme 17).[363]

[360a] R. Keese and E. P. Krebs, *Angew. Chem., Int. Ed. Engl.* **10**, 262 (1971).
[360b] R. Keese and E. P. Krebs, *Angew. Chem., Int. Ed. Engl.* **11**, 518 (1972).
[361] P. Warner and S.-L. Lu, *J. Am. Chem. Soc.* **98**, 6752 (1976).
[362] W. G. Dauben and J. Ipaktschi, *J. Am. Chem. Soc.* **95**, 5088 (1973).
[363] T. V. Rajan Babu and H. Schechter, *J. Am. Chem. Soc.* **98**, 8261 (1976).

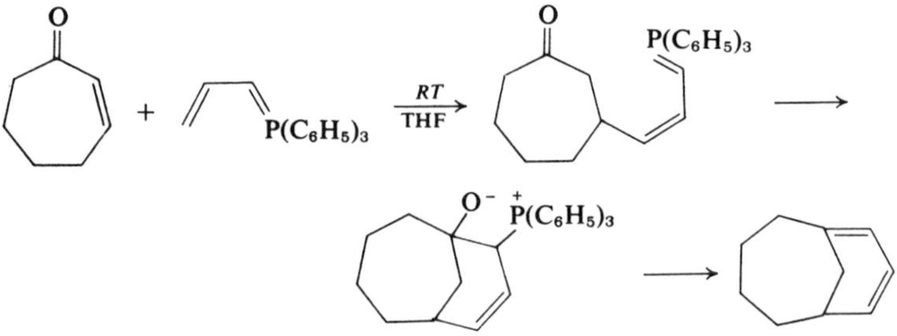

(20)

**172**

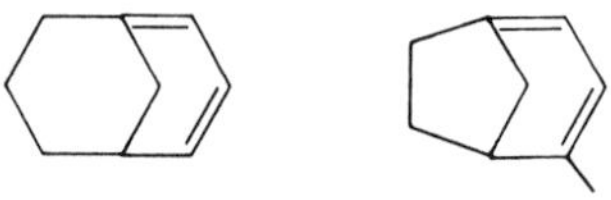

**173**          **174**

One should also note that the aromatic molecule 1,6-methano[10]annulene (**177**) [364] obviously maintains twisted $\pi$ overlap. The standard enthalpy of formation of this compound has been recently measured, and the difficulties

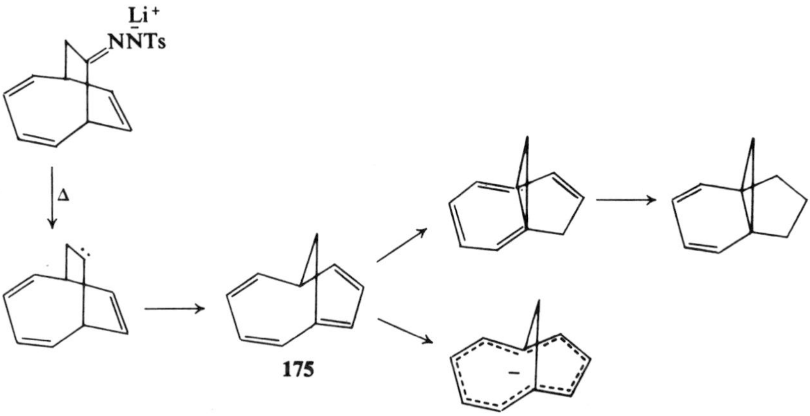

**175**

**176**

**Scheme 17**

in properly apportioning strain and aromatic stabilization have been reviewed.[365] Relatively slight perturbation of this system, as in **178**, is enough to favor the norcaradiene isomer.[366] 1,5-Methanocyclononatetraenyl anion

---

[364] E. Vogel, *Chem. Soc., Spec. Publ.* No. 21, 113 (1967).

[365] W. Bremser, R. Hagen, E. Heilbronner, and E. Vogel, *Helv. Chim. Acta* **52**, 418 (1969).

[366] R. Bianchi, A. Mugnoli, and M. Simonetta, *J. Chem. Soc., Chem. Commun.* p 1073. (1972).

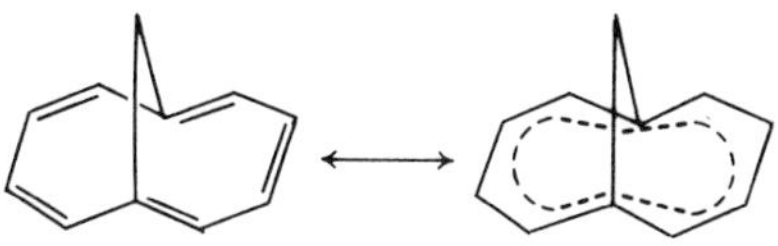

|     |     |     |
| :-: | :-: | :-: |
| **177** | **178** | **179** |

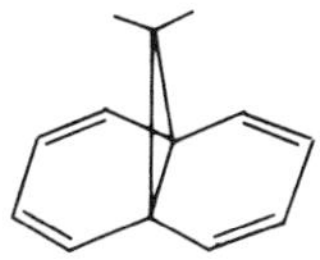

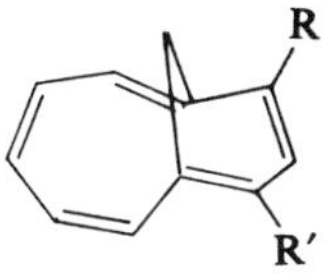

(**176**) [363,367] is a somewhat more twisted species which is, nonetheless, observable by nmr. Molecular-mechanics calculations predict that 1,5-methano[10]-annulene (**179**, R = R' = H) should have complete bond alternation and lack aromatic properties as the result of twisted olefinic linkages with torsional angles as large as 54°.[368] However, a number of derivatives of **179** as well as the parent hydrocarbon have now been synthesized [in a manner similar to that in Eq. (20)], and the nmr and uv spectra clearly indicate aromatic systems.[369] We note, in concluding this section, that [n]paracyclophanes and [n]metacyclophanes (see Section 3.H.3) may also be regarded as bridgehead olefins and are also related to such compounds as **177** and **179**. In spite of severe nonbonded repulsions, syn-1,6 : 8,13-bismethano[14]annulene is aromatic.[370a]

Finally, we note the synthesis of [10.10]betweenanene (**179a**), the first representative of a new class of molecules in which the same olefinic linkage is present in two different *trans* cycloalkene rings.[370b] The structure of the compound is reminiscent of **181** (see below), and its "buried" olefinic linkage can be likened to the triple bond in structure **93** of Chapter 6. An [8.10] betweenanene has also been isolated.[370c]

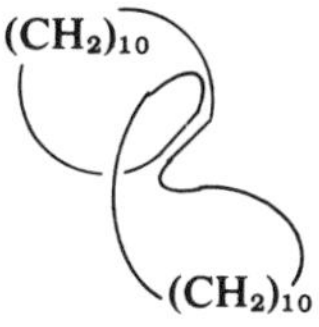

**179a**

In Chapter 5 a number of heterocyclic molecules containing twisted double bonds will be considered. This will include a brief discussion of bridgehead amides.

[367] P. Radlick and W. Rosen, *J. Am. Chem. Soc.* **88**, 3461 (1966).

[368] N. L. Allinger, and J. T. Sprague, *J. Am. Chem. Soc.* **95**, 3893 (1973).

[369] S. Masamune, D. W. Brooks, K. Morio, and R. L. Sobczak, *J. Am. Chem. Soc.* **98**, 8277 (1976).

[370a] E. Vogel, J. Sombroek, and W. Wageman, *Angew. Chem., Int. Ed. Engl.* **14**, 564 (1975).

[370b] J. A. Marshall and M. Lewellyn, *J. Am. Chem. Soc.* **99**, 3508 (1977).

[370c] M. Nakazaki, K. Yamamoto, and J. Yanagi, *J. Chem. Soc., Chem. Commun.* p. 346 (1977).

### 3.F "Bredt Compounds" and Cyclic Allenes

Köbrich has coined the term "Bredt compound," defined as ". . . bicyclic (and polycyclic) systems (alicycles and heterocycles) that, in addition to a strained σ-bond skeleton, have a twisted π bond at a bridgehead, and *purely* because of this ring strain, in contrast with compounds having the same structure but without this π bond, are unstable at room temperature." [337] According to this definition, bicyclo[3.3.1]non-1-ene (157) is not a "Bredt compound" because it is isolable at room temperature, while, for example, 160 is, because it cannot be so isolated. Twistene (180) [371] has a torsionally distorted double bond but is isolable at room temperature, and thus is not a Bredt compound, nor are 70, 72, and 73, whose strained bridgehead double bonds are not torsionally distorted. Tricyclo[3.3.3.0$^{2,6}$]undec-2(6)-ene (181) [372]

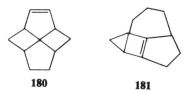

**180**          **181**

produced through desulfuration of a thionocarbonate in the presence of triethyl phosphite, has a twisted olefinic linkage, is not isolable (its dimer as well as diphenylisobenzofuran adduct have been characterized), and is not a "Bredt compound" since it is not a bridgehead π molecule. We will point out here that this definition does not consider the possible kinetic stabilization of such compounds by substituents such as alkyl groups. Köbrich also notes similarities in torsion, strain, and stability in the series 182–185, which should

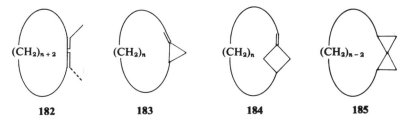

**182**          **183**          **184**          **185**

be of roughly comparable stability. [337] Thus, knowledge of the physical and chemical properties of one may allow reasonable expectations for the properties of the others. An idealized (C=C=C) angle of 180° and torsional

[371] M. Tichy and J. Sicher, *Tetrahedron Lett.* p. 4609 (1969). For synthesis of twista-4,9-diene, see H. G. Capraro and C. Ganter, *Helv. Chim. Acta* 59, 97 (1976).

[372] R. Greenhouse, T. Ravindranathan, and W. T. Borden, *J. Am. Chem. Soc.* 98, 6738 (1976).

angle of 90° dictates the requirement for relatively large rings in the cyclo-allene (182) series. Geometrical constraints opposing ring incorporation are less severe for 183 and 184, and even less so for 185.[337] Thus, 1,2-cyclo-nonadiene (182, $n = 4$) is the smallest stable cycloallene,[373,374] tricyclo-[4.1.0.0$^{1,3}$]heptane (57, i.e., 185, $n = 4$)[127a,b] and $\Delta^{1,2}$-bicyclo[5.1.0]octene (183, $n = 4$) are isolable, and indeed so is $\Delta^{1,2}$-bicyclo[5.1.1]nonene (184, $n = 4$).[346b] Diene 186 is a somewhat more strained member of this series which has also been reported.[375] In the lower homologous series, compound 187 (a derivative of 183, $n = 3$) has a half-life of about 70 hours at room temperature,[376] although 1,2-cyclooctadiene[377,378] can only be observed spectroscopically at low temperature[379,380] (perhaps a sterically hindered derivative might be isolable). Therefore, one might predict reasonable lifetimes for substituted derivatives of 184 and 185 if Köbrich's postulate holds. 1,2-Cycloheptadiene (182, $n = 2$) has been generated through dehydrohalo-genations of 1-halocycloheptenes[381,382a] and a platinum complex has been characterized (although the free molecule has thus far eluded observation;[380]

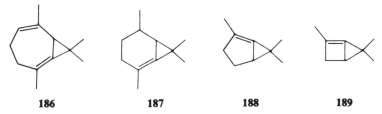

**186**          **187**          **188**          **189**

see Section 5.C.4.c. for discussion of such transition-metal complexes). Only dimers of 188 (derivative of 183, $n = 2$) have been isolated, and at tempera-tures higher than $-40°C$ this molecule forms a strained, highly reactive trimethylenemethane.[337] The same reaction conditions which produced 188 failed to provide any evidence for 189.[337] While 1,2-cyclohexadiene, a formal model for 189 in the Köbrich sense, has been produced and trapped;[382b,c]

[373] W. R. Moore and R. C. Bertelson, *J. Org. Chem.* 27, 4182 (1962).

[374] L. Skattebøl, *Acta Chem. Scand.* 17, 1683 (1963).

[375] See footnote 44c in Ref. 337.

[376] G. Köbrich and M. Baumann, *Angew. Chem., Int. Ed. Engl.* 11, 52 (1972).

[377] E. T. Marquis and P. D. Gardner, *Tetrahedron Lett.* p. 2793 (1966).

[378] H. Wittig, H. L. Dorsch, and J. Meske-Schuller, *Justus Liebigs Ann. Chem.* 711, 55 (1968).

[379] A. T. Bottini, F. P. Corson, R. Fitzgerald, and K. A. Frost, Jr., *Tetrahedron Lett.* p. 4757 (1970).

[380] J. P. Visser and J. E. Ramakers, *J. Chem. Soc., Chem. Commun.* p. 178 (1972).

[381] W. J. Ball and S. R. Landor, *J. Chem. Soc.* p. 2298 (1962).

[382a] G. Wittig and J. Meske-Schuller, *Justus Liebigs Ann. Chem.* 711, 76 (1968).

[382b] G. Wittig and P. Fritze, *Justus Liebigs Ann. Chem.* 711, 82 (1968).

[382c] G. Wittig and P. Fritze, *Angew. Chem., Int. Ed. Engl.* 5, 846 (1966).

we defer discussion of this species until later in this section, and note that as the allenic linkage becomes increasingly distorted, the relation used in the foregoing discussion must surely break down.

Observations related to the above discussion include the greater stability of $\Delta^{1,2}$-bicyclo[6.1.0]nonene (**183**, $n = 5$) compared to 8-methylenebicyclo-[5.1.0]octane (**190**, $n = 5$), which contrasts, as expected, with the findings that 7-methylenebicyclo[4.1.0]heptane (**190**, $n = 4$) is more stable than $\Delta^{1,2}$-bicyclo[5.1.0]octene (**183**, $n = 4$)[182] and 6-methylenebicyclo[3.1.0]hexane (**190**, $n = 3$) is more stable than $\Delta^{1,2}$-bicyclo[4.1.0]heptene (**183**, $n = 3$).[383]

$(CH_2)_n$
190          191a          191b          191c          191d

1,2-Cyclohexadiene (**191**) has indeed been generated as a transient species and trapped.[382b,c,384a,b] However, there is some debate about its structure and electronic properties.[385a] For example, while the reaction of a *gem*-dihalocyclopropane has been used successfully to obtain a variety of allenes, incorporation of the allenic group into a suitable small ring might prevent this reaction. Thus, as we have already noted in Eq. (1), the carbene or carbenoid species produced in the alpha elimination of 7,7-dibromobicyclo-[4.1.0]heptane suffers an insertion reaction in preference to rearrangement to 1,2-cycloheptadiene. It therefore came as something of a surprise to find that 6,6-dibromobicyclo[3.1.0]hexane reacted under similar conditions to provide products explicable in terms of the generation of 1,2-cyclohexadiene (Scheme 18).[384a,b] (We note that cyclopropylidene itself is calculated to be some 61 kcal/mole less stable than allene.[138b]) This distorted allenic linkage in this molecule probably precludes its description as **191a**, and makes either diradicals **191b** or **191c**, or allylic dipolar species **191d**, a better representation. The finding that 1,2-cyclohexadiene produced a cycloaddend, rather than a polymer, in the presence of styrene (Scheme 18) is one piece of evidence used to favor singlets **191b** or **191d**.[384a,b] However, semiempirical calculations suggest that triplet **191c** is the most stable form of 1,2-cyclohexadiene, and the authors have provided rationalization for the lack of polystyrene mentioned above.[385a] The same calculations also predict that 1,2-cyclopenta-diene should not be much more strained than 1,2-cyclohexadiene, and the

[383] U. Langer and H. Musso, *Justus Liebigs Ann. Chem.* 1180 (1976).

[384a] W. R. Moore and W. R. Moser, *J. Org. Chem.* **35**, 908 (1970).

[384b] W. R. Moore and W. R. Moser, *J. Am. Chem. Soc.* **92**, 5469 (1970).

[385a] P. W. Dillon and G. R. Underwood, *J. Am. Chem. Soc.* **96**, 779 (1974).

**Scheme 18**                                        tetramers

authors noted that no serious investigation of this molecule has yet appeared.[385a] Di-*tert*-butylnitroxide has been used to trap a dimethyl derivative of the 1, 2-cyclohexadiene dimer (diradical) of Scheme 18 but not 1,2-cyclohexadiene itself.[385b]

1,2-Cyclobutadiene has been calculated to be a bent species best described as a "homocyclopropenylidene" (**192**).[138a] As we have seen in Scheme 1, apparent production of 2-bicyclo[1.1.0]butanylidene (equivalent to **192**?) produces acetylenes (possibly via the tetrahedrane). However, "homocyclopropenylidene" has been calculated to be some 15 kcal/mole (4–31G) more stable than tetrahedrane and is thought to be a "tempting target" for matrix isolation.[138a] One cannot really consider "1,2-cyclopropadiene" without immediately realizing that its only realistic structure is cyclopropenylidene,

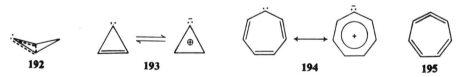

**192**                **193**                        **194**                        **195**

for which experimental precedent exists,[386] and whose calculated properties are consistent with the dipolar aromatic structure (**193**).[138a,b,386]

Toward the end of Section 3.D.2, we briefly considered the mechanism of the rearrangement of phenylcarbene to cycloheptatrienylidene, which is usually assumed to be a planar dipolar aromatic species (**194**).[386] However,

[385b] A. T. Bottini, L. J. Cabral, and V. Dev, *Tetrahedron Lett.* p. 615 (1977).
[386] H. Dürr, *Top. Curr. Chem.* **40**, 103 (1973).

INDO calculations have found nonplanar 1,2,4,6-cycloheptatraene (**195**) to be almost 14 kcal/mole more stable than cycloheptatrienylidene, although no chemical or physical evidence for **195** has yet appeared.[387] However, model calculations on benzo-annelated analogues[387] are fully consistent with the intermediacy of 3,8-methano[11]annulenylidene (**196**).[388] The intermediacy of homoaromatic carbene **197**, related in a formal sense to the

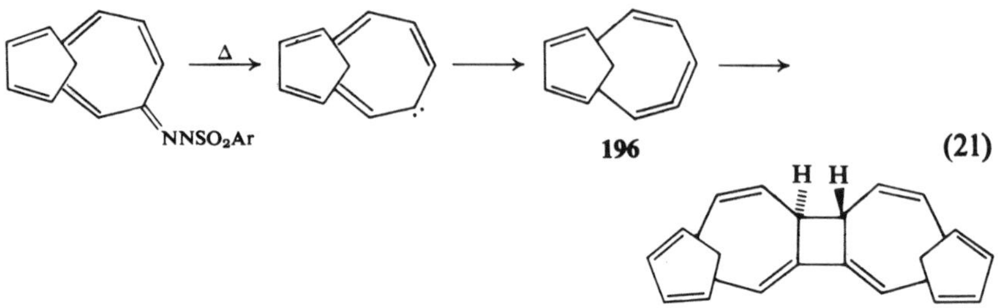

$$(21)$$

bridged 1,2-cyclohexadiene **198**, has been postulated.[389,390] 1,6-Methano[10]-

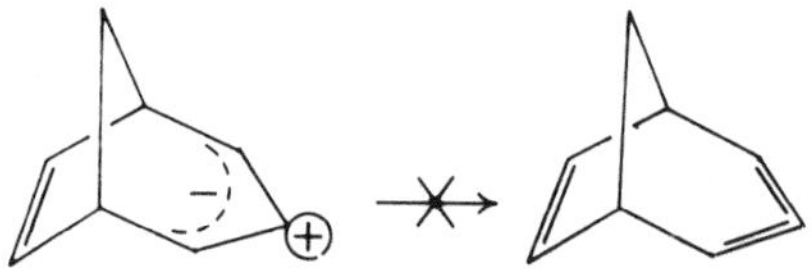

annulen-11-ylidene (**199**), which might be regarded as a fused di-1,2,4,6-cycloheptatetraene (see **195**), forms an unstable dimer which could not be characterized, but also reacts to produce a diphenylisobenzofuran adduct

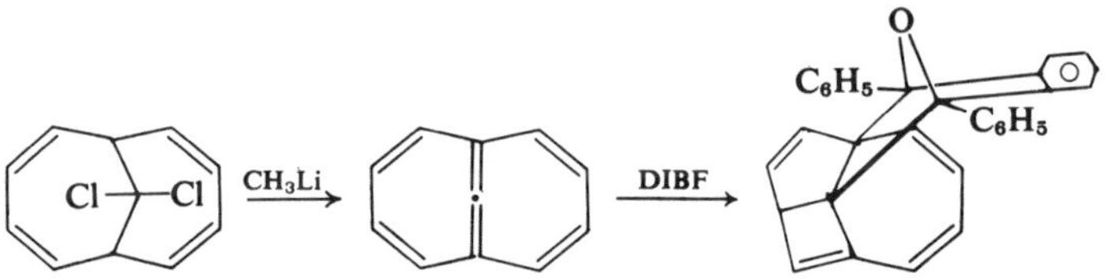

$$(22)$$

characterized by x-ray crystallography (labeling studies helped confirm the intermediacy of **199**).[391]

[387] R. L. Tyner, W. M. Jones, Y. Öhrn, and J. R. Sabin, *J. Am. Chem. Soc.* **96**, 3765 (1974).

[388] R. A. LaBar and W. M. Jones, *J. Am. Chem. Soc.* **96**, 3645 (1974).

[389] R. G. Bergman and V. J. Rajadhyaksha, *J. Am. Chem. Soc.* **92**, 2163 (1970).

[390] G. W. Klumpp and P. M. van Dijk, *Rec. Trav. Chim. Pays-Bas* **90**, 381 (1971).

[391] J. B. Carlton, R. H. Levin, and J. Clardy, *J. Am. Chem. Soc.* **98**, 6068 (1976).

Although 1,2,6-cyclononatriene[392] can be obtained through alpha elimination of 9,9-dibromobicyclo[6.1.0]non-4-ene, 1,2,4-cyclonoratriene (**200**) could not be obtained by the same route, and carbene insertion products were found instead [Eq. (23)].[393] However, 1,2,5-cyclononatriene (**201**) was

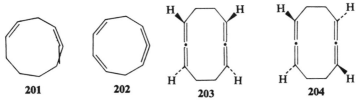

(23)

produced this way and exhibited reasonable thermal stability.[394] While an attempt to generate 1,2,4,8-cyclononatetraene was thwarted by an insertion reaction, 1,2,5,7-cyclononatetraene (**202**) is isolable when low-temperature work-up conditions are employed, and it manifests a dimerization half-life

**201**        **202**        **203**        **204**

at 0°C of 10–20 minutes in solution.[394] While *meso*-1,2,6,7-cyclodecatetraene (**203**) is known, the *dl* isomer **204** has yet to be characterized.[394,395] Attempts at generating 1,2,4,5-cyclodecatetraene under the same conditions failed, but 1,2,5,6-cyclodecatetraene (**205**) (apparently both diastereomers) could be so obtained.[394] The possible intermediacy of **206** has also been considered.[396a,b] The intermediacies of 1,2,5-cycloheptatriene[397] (**207**) and 1,2,4,6-cyclooctatetraene[398] (**208**) have been ascertained from the formation of their dimers.

[392] L. Skattebøl, *Tetrahedron Lett.* p. 167 (1961). The conformation of the molecule is now known to be the twist-boat-chair in solution; see F. A. L. Anet and I. Yavari, *J. Chem. Soc., Chem. Commun.* p. 927 (1975).

[393] C. G. Cardenas, B. A. Shoulders, and P. D. Gardner, *J. Org. Chem.* **32**, 1220 (1967).

[394] M. S. Baird and C. B. Reese, *Tetrahedron* **32**, 2153 (1976).

[395] E. Dehmlow and G. C. Ezimora, *Tetrahedron Lett.* p. 4047 (1970). For crystal structure of **203**, see H. Irngartinger and H.-U. Jagar, *Tetrahedron Lett.* p. 3595 (1976).

[396a] M. D'Amore and R. G. Bergman, *J. Am. Chem. Soc.* **91**, 5694 (1969).

[396b] M. D'Amore, R. G. Bergman, M. Kent, and E. Hedaya, *J. Chem. Soc. Chem. Commun.* p. 49 (1972).

[397] W. R. Dolbier, Jr., O. T. Garza, and B. H. Al-Sader, *J. Am. Chem. Soc.* **97**, 5038 (1975).

[398] M. Oda, Y. Ito, and Y. Kitahara, *Tetrahedron Lett.* p. 2587 (1975).

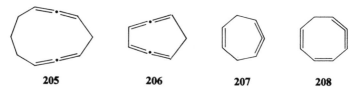

**205**        **206**        **207**        **208**

Allene **209** was trapped with styrene,[399] and it is apparent that no polystyrene was formed (see the earlier discussion of the ground state of 1,2-cyclohexadiene). However, the homologous reaction (25) did not apparently produce the

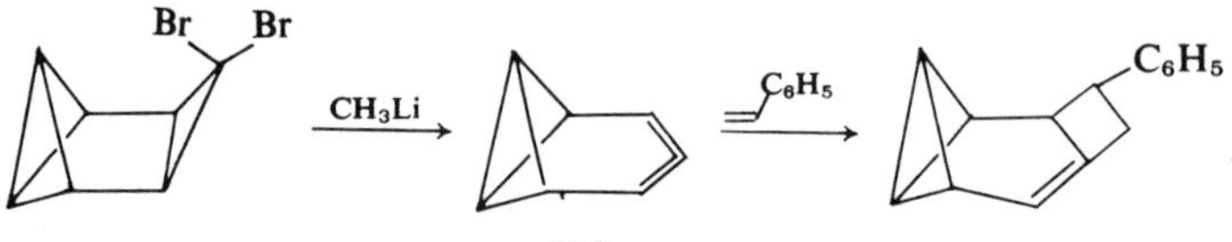

$$\text{(24)}$$

**209**

presumably less strained allene, but rather the high-energy intermediate **210**.[397] A plausible explanation of the ease of formation of **209** relative to its

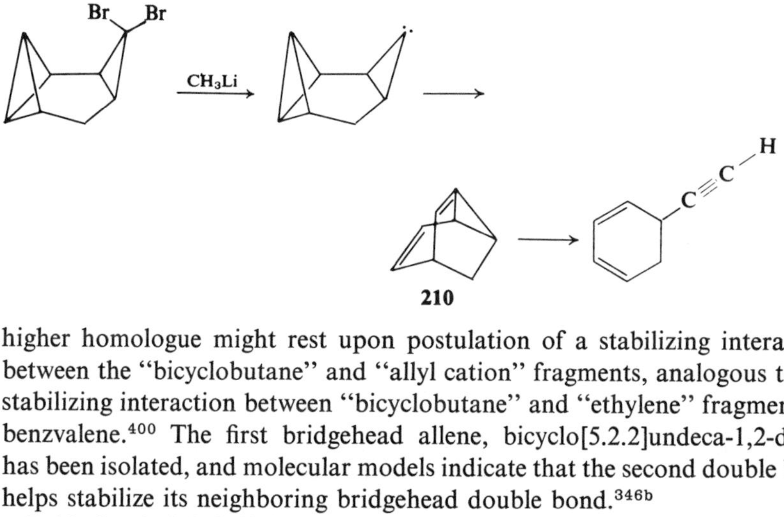

$$\text{(25)}$$

**210**

higher homologue might rest upon postulation of a stabilizing interaction between the "bicyclobutane" and "allyl cation" fragments, analogous to the stabilizing interaction between "bicyclobutane" and "ethylene" fragments in benzvalene.[400] The first bridgehead allene, bicyclo[5.2.2]undeca-1,2-diene, has been isolated, and molecular models indicate that the second double bond helps stabilize its neighboring bridgehead double bond.[346b]

Bridgehead olefinic linkages in suitably small bicyclo[m.n.o]alkenes possess considerable torsional character, and members of this series which are not isolable at room temperature are also regarded as "Bredt compounds." Thus, while $\Delta^{1,2}$-bicyclo[4.2.0]octene (**211**),[401] $\Delta^{1,2}$-bicyclo[3.3.0]octene (**212**),[402]

[399] M. Christl and M. Lecher, *Angew. Chem., Int. Ed. Engl.* **14**, 765 (1975).
[400] Ethylene and allyl cation are $\pi$-isoelectronic; see ref. 81C.
[401] L. Skattebøl and S. Solomon, *J. Am. Chem. Soc.* **87**, 4506 (1965).
[402] A. D. Ketley, *Tetrahedron Lett.* p. 1687 (1964).

and $\Delta^{1,2}$-bicyclo[3.2.0]heptene **(213)**[403a,b] are isolable at room temperature, bicyclo[3.2.0]hepta-1,3,6-triene **(214)** is not[404,405] (half-life of three hours in dilute solution at 25°C), nor is the benzologue, **215**.[406] Indications are that bicyclo[2.2.0]hex-1-ene **(216)** is probably not isolable.[337] The bridged Dewar benzene **217** has a half-life of 58 minutes in solution at room temperature,[407,408] while **218** has been generated but could not be observed.[409]

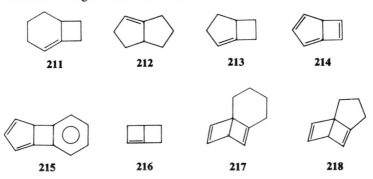

| 211 | 212 | 213 | 214 |

| 215 | 216 | 217 | 218 |

## 3.G Cycloalkynes, Benzyne, and Cyclocumulenes

The deformations present in cyclooctyne **(219)**, benzyne **(220)**, and 1,2,3-cyclodecatriene **(221)** (and other cyclocumulenes having an odd number of cumulated double bonds) are somewhat related. (Cyclocumulenes having an even number of cumulated bonds are related to cycloallenes.) In each instance, a normal $\pi$ bond and a relatively weak $\pi$ bond, perpendicular to the first one, are present. Comprehensive reviews of the chemistries of these types of molecules have appeared.[410-412] For the present, we will briefly examine the occurrence and properties of cycloalkynes, postpone consideration of benzyne until Section 3.H.1, and conclude this section with a very brief look at cyclocumulenes and related molecules.

[403a] R. A. Moss and J. R. Whittle, *Chem. Commun.* p. 341 (1969).

[403b] R. A. Moss, U.-H. Dolling, and J. R. Whittle, *Tetrahedron Lett.* p. 931 (1971).

[404] R. Breslow, W. Washburn, and R. G. Bergman, *J. Am. Chem. Soc.* **91**, 196 (1969).

[405] N. L. Bauld, C. E. Dahl, and Y. S. Rim, *J. Am. Chem. Soc.* **91**, 2787 (1969).

[406] M. P. Cava, K. Narasimham, W. Zieger, L. J. Radonovich, and M. D. Glick, *J. Am. Chem. Soc.* **91**, 2379 (1969).

[407] K. Weinges and K. Klessing, *Chem. Ber.* **107**, 1915 (1974).

[408] I. J. Landheer, W. H. de Wolf, and F. Bickelhaupt, *Tetrahedron Lett.* p. 2813 (1974).

[409] I. J. Landheer, W. H. de Wolf, and F. Bickelhaupt, *Tetrahedron Lett.* p. 349 (1975).

[410] G. L. Gilchrist and C. W. Rees, "Carbenes, Nitrenes, and Arynes." Appleton, New York, 1969.

[411] R. W. Hoffmann, "Dehydrobenzene and Cycloalkynes," Academic Press, New York, 1967.

[412] H. G. Viehe, "Chemistry of Acetylenes," Dekker, New York, 1969.

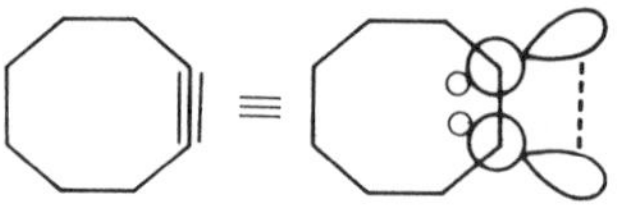

**219** 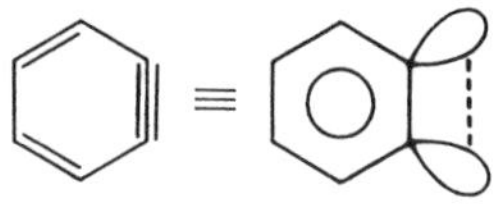                              **220**

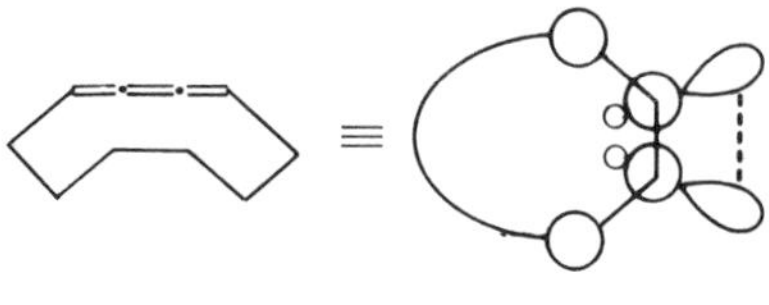

**221**

Cyclooctyne, the smallest isolable simple cycloalkyne, was first obtained through decomposition of 1,2-cyclooctanedionebishydrazone.[413] The strain in this molecule's triple bond, relative to that in 4-octyne (i.e., the difference in monohydrogenation enthalpies), is about 10 kcal/mole,[414,415] and is reflected in cyclooctyne's high susceptibility to oxygen as well as its explosive reaction with phenyl azide.[413] Part of the strain is certainly attributable to the weak $\pi$ bond in this molecule (see **219**). As the groups attached to an acetylenic linkage are deformed from their preferred colinear arrangement, the initial degeneracy of the two perpendicular orbitals is lost. While one $\pi$ bond remains essentially normal, the other one becomes higher in energy. Photoelectron studies confirm this expectation by finding 0.32- and 0.31-eV splittings between the $\pi$ orbitals in **222** and **223**, respectively, although there is negligible splitting in cyclooctyne.[416] As we shall see shortly, the high-energy, weak $\pi$ bond is capable of initiating some unique chemistry. Before discussing more highly strained cycloalkynes, we note that the strain in cyclononyne (relative to 4-octyne) is only 2.9 kcal/mole, while cyclodecyne is essentially strain-free.[414] Cycloheptyne, generated by magnesium-induced elimination of 1,2-dihalocycloheptenes as well as through decomposition of 1,2-cycloheptanedionebishydrazone, has been trapped with diphenylisobenzofuran.[417a,b] Molecular-mechanics calculations obtain a value of about 31 kcal/mole for the total strain in cycloheptyne (compare to *total* strain enthalpies of 20.8 kcal/mole for cyclooctyne and 16.4 kcal/mole for cyclononyne, respectively; the calculated thermochemical data are in very good agreement with experi-

---

[413] A. T. Blomquist and L. H. Liu, *J. Am. Chem. Soc.* **95**, 790 (1973).

[414] R. B. Turner, A. D. Jarratt, P. Goebel, and B. J. Mallon, *J. Am. Chem. Soc.* **95**, 790 (1973).

[415] N. L. Allinger and A. Y. Meyers, *Tetrahedron* **31**, 1807 (1975).

[416] H. Schmidt, A. Schweig, and A. Krebs, *Tetrahedron Lett.* p. 1471 (1974).

[417a] G. Wittig and A. Krebs, *Angew. Chem.* **72**, 324 (1960).

[417b] G. Wittig and A. Krebs, *Chem. Ber.* **94**, 3268 (1961).

mental hydrogenation enthalpies for these two molecules).[415] The substituents in 2,2,6,6-tetramethylcycloheptyne (222) provide this molecule with some kinetic stability, and it is indeed isolable.[418] Tetramethylthiacycloheptyne 223 is even more stable because its long carbon–sulfur bonds decrease the angular distortion of the acetylenic linkage.[419a,b] Nevertheless, it is more reactive to cycloaddition reactions than cyclooctyne.[419a,b] This thiacycloheptyne participates in a remarkable reaction with molecular oxygen. Its high-energy

222          223

distorted $\pi$ bond apparently forms a complex with (triplet) oxygen which is thought to spin-invert to the singlet oxygen complex that ultimately provides a dione in an electronically excited state [see Eq. (96) in Chapter 5].[420] Apparently, the very nature of the strained bond and its interaction with oxygen facilitates spin inversion of triplet to singlet oxygen.[420] Cyclohexyne (224) and cyclopentyne have also been generated via 1,2-dihalocycloalkenes and 1,2-cycloalkanedionebishydrazones and trapped with diphenylisobenzofuran.[417a,b] These cycloalkynes are also produced in reactions between 1-halocycloalkenes and phenyllithium and their intermediacies detected by $^{14}C$ label scrambling [e.g., Eq. (26)].[421a-c] There is presently no evidence for even the transient existence of cyclobutyne,[417a,b,421a,b] which is calculated to be more than

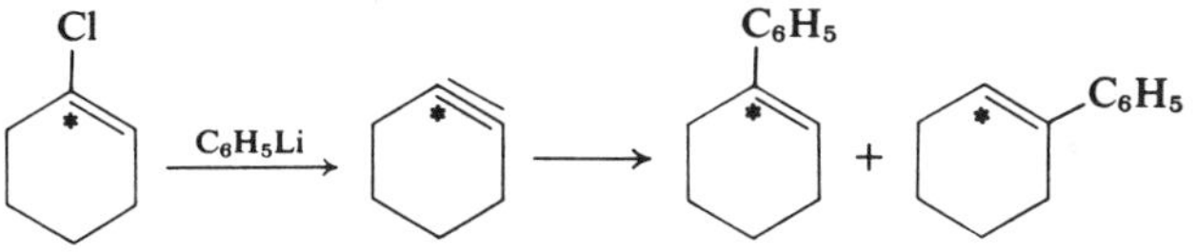

(26)

224

20 kcal/mole less stable than its isomer, tetrahedrane.[138a,b] Norbornyne (225) [a "$4\frac{1}{2}$" or "(5-)cycloalkyne" if cyclopentyne is a "5-cycloalkyne"] has been generated from 2-chlorobicyclo[2.2.1]hept-2-ene through use of n-butyllithium

[418] A. Krebs and H. Kimling, *Angew. Chem., Int. Ed. Engl.* 10, 509 (1971).

[419a] A. Krebs and H. Kimling, *Tetrahedron Lett.* p. 761 (1970).

[419b] J. Haase and A. Krebs, *Z. Naturforsch. A* 27, 624 (1972).

[420] N. J. Turro, V. Ramamurthy, K.-C. Liu, A. Krebs, and R. Kemper, *J. Am. Chem. Soc.* 98, 6758 (1976).

[421a] F. Scardiglia and J. D. Roberts, *Tetrahedron* 13, 343 (1957).

[421b] L. K. Montgomery and J. D. Roberts, *J. Am. Chem. Soc.* 82, 4750 (1960).

[421c] L. K. Montgomery, F. Scardiglia, and J. D. Roberts, *J. Am. Chem. Soc.* 87, 1917 (1965).

but not with methyllithium.[422a,b] Isotopic labeling studies and the finding that the reaction proceeds with almost complete racemization [Eq. (27)] strongly implicate norbornyne's discrete existence.[422a,b]

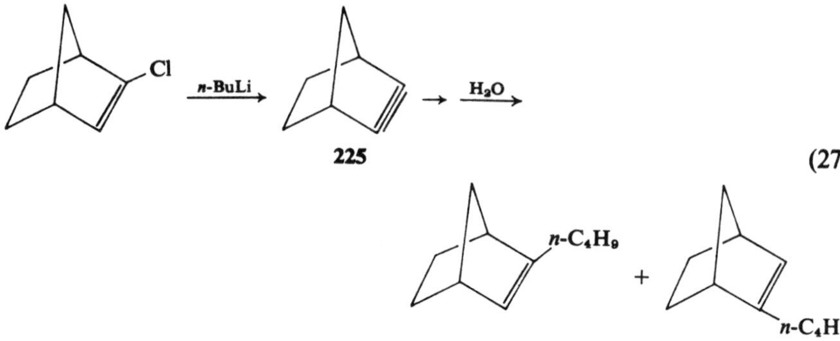

**225**                                                                    (27)

1,5-Cyclooctadiyne (**226**), obtained in low yield from the highly exothermic dimerization of butatriene, is a crystalline material that is stable at 0°C under an inert atmosphere.[423a,b] Such stability is noteworthy when one considers that the (C≡C—C) angles are less than 160° and the two bonds are closer (2.60 Å) than in any known cyclophane, although not quite so close as in compound **72**.[423b] The heat of hydrogenation of 1,6-cyclodecadiyne indicates the presence of essentially normal triple bonds in this molecule (although the complexities in estimating strain in this molecule have been discussed).[414]

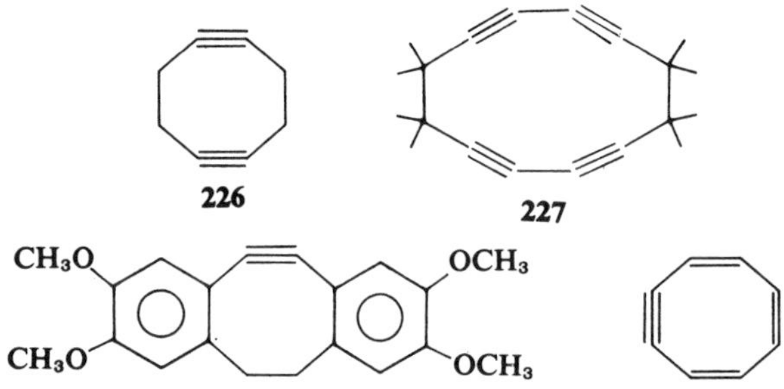

**226**                 **227**                                            (28)

**228**                 **229**

In contrast to the behavior of 1,3,7,9-cyclododecatetrayne, the tetramethyl derivative **227** can be isolated and purified (it is stable at 150°C) despite the angle distortions (C≡C—C angle of 166°) and nonbonded π interactions in

[422a] P. G. Gassman and J. J. Valcho, *J. Am. Chem. Soc.* **97**, 4768 (1975).
[422b] P. G. Gassman and T. J. Atkins, *Tetrahedron Lett.* p. 3035 (1975).
[423a] E. Kloster-Jensen and J. Wirz, *Angew. Chem., Int. Ed. Engl.* **12**, 671 (1973).
[423b] E. Kloster-Jensen and J. Wirz, *Helv. Chim. Acta* **58**, 162 (1975).

this molecule.[424] Compound **228** has been characterized,[425] while 1,3,5-cyclo-octatrien-7-yne (**229**) has eluded isolation or even spectroscopic observation.[426] Compounds **230**,[427a,b] **231**,[427a] and **232**[428] are among the very few compounds

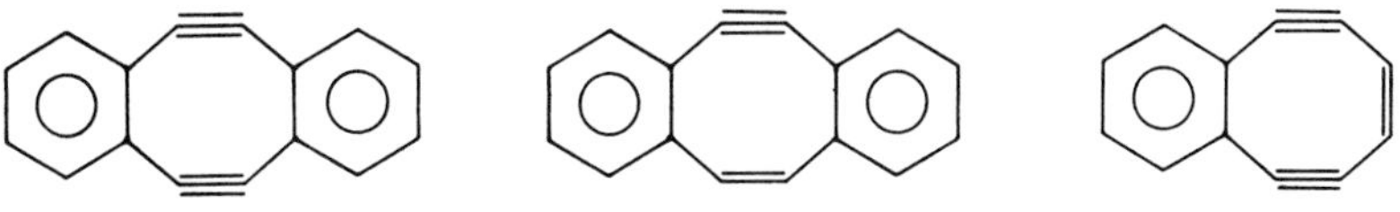

|  |  |  |
|---|---|---|
| **230** | **231** | **232** |

containing planar, neutral, fully conjugated eight-membered rings. The last of these compounds decomposes within minutes at 0°C and exhibits the paratropic ring current expected of a planar $4n$ $\pi$ monocycle.[428] The inter-mediates **233**,[429] **234**,[425] and **235**[430] have been trapped. Relative rates of

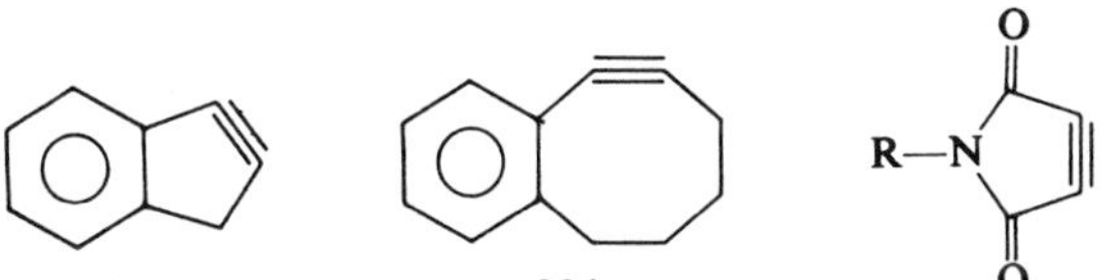

|  |  |  |
|---|---|---|
| **233** | **234** | **235** |

elimination to form **236** again reflect reduced strain in molecules having long carbon–sulfur bonds.[431]

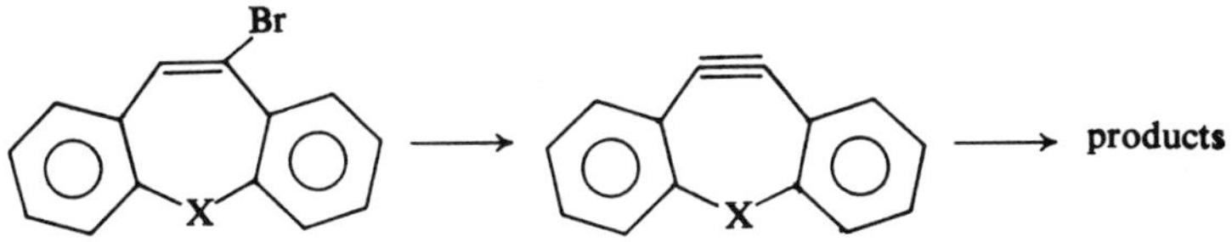

**236**

relative rate = 1 (X = CR₂)
relative rate = $10^5$ (X = SO₂)

1,5,9-Cyclododecatriyne (**237**) offers some intriguing structural features because it might well be in equilibrium with **238** (it could actually be a

[424] L. T. Scott and G. J. DeCicco, *Tetrahedron Lett.* p. 2663 (1976).

[425] G. Seitz, L. Pohl, and R. Pohlke, *Angew. Chem., Int. Ed. Engl.* **8**, 447 (1969).

[426] A. Krebs, *Angew. Chem., Int. Ed. Engl.* **4**, 953 (1965).

[427a] H. N. C. Wong, P. J. Garratt, and F. Sondheimer, *J. Am. Chem. Soc.* **96**, 5604 (1974).

[427b] H. M. C. Wong, P. J. Garratt, and F. Sondheimer, *J. Am. Chem. Soc.* **97**, 658 (1975).

[428] H. N. C. Wong, and F. Sondheimer, *Angew. Chem., Int. Ed. Engl.* **15**, 117 (1976)

[429] G. Wittig and H. Heyn, *Chem. Ber.* **97**, 1609 (1964).

[430] W. Draber, *Angew. Chem., Int. Ed. Engl.* **6**, 75 (1967).

[431] W. Tochtermann, K. Oppenlander, and M. N.-D. Hoang, *Justus Liebigs Ann. Chem.* **701**, 117 (1967).

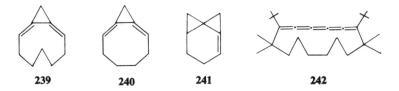

$$\text{(28)}$$

resonance hybrid of **237** and **238**) or even with [6]radialene* (**105**).[432a] While the molecule must maintain significant $\pi$ overlap of ethylenic p orbitals, its uv, ir, and [13]C nmr properties clearly indicate the presence of acetylenic functionality. However, its relatively facile addition reaction, depicted in Eq. (28), suggests a low-energy isomerization to **238**, which is estimated to be some 25 kcal/mole more stable than **237**.[432a] The dodecafluoro derivative of **238** has been identified.[432b–d]

The smallest isolable cyclic cumulene is 1,2,3-cyclodecatriene (**221**).[433] Recalling the analogies suggested in the previous section, one might anticipate some stability for **239–241**, the last of which has been observed spectroscopically.[434,435] The higher cyclocumulene **242** has been characterized.[436]

| 239 | 240 | 241 | 242 |

[432a] A. J. Barkovich and K. P. C. Vollardt, *J. Am. Chem. Soc.* **98**, 2667 (1976).

[432b] G. Camaggi, *J. Chem. Soc. C* p. 2382 (1971).

[432c] R. L. Soulen, S. K. Choi, and J. D. Park, *J. Fluor. Chem.* **3**, 141 (1973–1974).

[432d] R. P. Thummel, J. D. Korp, I. Bernal, R. L. Harlow, and R. L. Soulen, *J. Am. Chem. Soc.* **99**, 6916 (1977).

[433] W. R. Moore and T. M. Ozretich, *Tetrahedron Lett.* 3205 (1967).

[434] W. R. Roth and G. Erker, *Angew. Chem., Int. Ed. Engl.* **12**, 505 (1973).

[435] W. Grimme and H.-J. Rother, *Angew. Chem., Int. Ed. Engl.* **12**, 505 (1973).

[436] T. Negi, T. Kaneda, Y. Sakata, and S. Misumi, *Chem. Lett.* p. 703 (1972).
* See Addendum.

## 3.H  Distorted Aromatic Rings

In this section we will examine benzene rings which are distorted by virtue of their constrained annelation or bridging as well as from the presence of repulsive steric interactions. The normally flat, hexagonal aromatic ring may suffer in-plane distortion as well as ring deformation to structures reminiscent of cyclohexane chairs, boats, and twist boats. It is well to remember that any apparent destabilization energy for a strained aromatic system not only arises from distortions in bond angles, bond lengths, torsional angles, and residual nonbonded repulsions, but is a measure of reduction in aromatic stabilization caused by poor $\pi$ overlap and/or bond fixation.

### 3.H.1  ortho-BRIDGED BENZENES AND peri-BRIDGED NAPHTHALENES

Benzyne (**219**), benzocyclopropene (**243**), and benzocyclobutene (**244**) comprise a series of 1,2-bridged benzenes of increasing stability. Although benzyne has been generated and is a transient intermediate,[411,412,437a-c] it has recently been spectroscopically (ir) monitored in an argon matrix at 8°K following its photochemical generation from phthaloyl peroxide.[438] When furan is added to the matrix, warming of the solution to 50°K allows one to record the disappearance of 1,2-benzyne and the simultaneous appearance of the Diels–Alder adduct. Preliminary[438] and more detailed analyses[439] of the ir spectrum of 1,2-benzyne indicate that it is well represented by a cycloalkyne structure (i.e., **219**, rather than cycloallenic structure) and that there is significant bond alternation ($C_1$—$C_2$ = 1.344 Å; $C_2$—$C_3$ = 1.410 Å; $C_3$—$C_4$ = 1.387 Å; $C_4$—$C_5$ = 1.405 Å[439]). This is also consistent with its chemistry, which is best explained in terms of a ground-state singlet.[437a-c] The most recent experimental value for the heat of formation of benzyne is 100 kcal/mole.[440] One may calculate a "lower limit" for the enthalpy of formation of hypothetical "unstrained benzyne" through "conceptual combination" of benzyne and a normal triple bond [Eq. (29)]. The calculated value (+56 kcal/mole) would indicate the presence of about 44 kcal/mole of destabilization

$$\Delta H_f(\text{benzyne}) = \Delta H_f(\text{benzene}) + \Delta H_f(\text{2-butyne}) - \Delta H_f(\text{cis-2-butene}) \qquad (29)$$

[437a] R. W. Atkins and C. W. Rees, Chem. Commun. p. 152 (1969).

[437b] M. Jones, Jr. and R. H. Levin, J. Am. Chem. Soc. **91**, 6411 (1969).

[437c] A. T. Bowne, T. A. Christopher, and R. H. Levin, Tetrahedron Lett. p. 4111 (1976).

[438] O. L. Chapman, K. Mattes, C. L. McIntosh, J. Pacansky, G. V. Calder, and G. Orr, J. Am. Chem. Soc. **95**, 6134 (1973).

[439] J. W. Laing and R. S. Berry, J. Am. Chem. Soc. **98**, 660 (1976).

[440] H. M. Rosenstock, J. T. Larkins, and J. A. Walker, Int. J. Mass. Spectrom. Ion Phys. **11**, 309 (1973).

energy. (If an ortho diradical were chosen as an "upper-limit" model, benzyne would appear to be resonance stabilized.) The destabilization energy of benzyne is largely attributable to poor overlap of the reactive bond, but distortion of the aromatic nucleus, bond alternation, and the reduced aromatic stabilization associated with it also contribute to the strain (see above discussion). [See Section 3.D.2 for discussion of 1,3-dehydrobenzene and 1,4-dehydrobenzene(s).] Before returning to other strained aromatic molecules, we shall very briefly discuss the so-called Mills–Nixon effect, which relates to bond fixation.

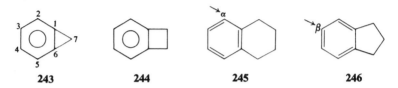

In 1930, Mills and Nixon explained predominant $\beta$ substitution in indane (**246**) and $\alpha$ substitution in tetralin (**245**) in terms of the bond-fixated structures shown.[441] Although the experimental data upon which these investigators based their conclusions have been shown to be incorrect, there remains ample evidence for some bond fixation in certain molecules. For example, benzyne's structural parameters, as noted above, very neatly satisfy a bond-localized structure. Bond alternation continues to be used as an important criterion for determining whether or not a molecule is aromatic. However, as we shall see below, many molecules appear to be bond fixated, but often their structures cannot be neatly represented in a Kekulé manner. The Mills–Nixon effect is sometimes used today as a term connoting nonequivalence of bonds in a substituted benzene ring arising from the distortion of bond angles in *ortho*-annellated aromatic derivatives. The reduced reactivity of the $\alpha$ hydrogen in **246** has been explained both in terms of strain[442a,b] and hybridization.[443]

The first known benzocyclopropene[444] (**248**) was generated in the photochemical decomposition of a $3H$-indazole,[445] a method useful only for generating 7,7-disubstituted benzocyclopropenes. The reaction is known to involve the intermediacy of a triplet species (**247**) related to vinylcarbene (see Section 3.D.2).[446] The parent molecule was reported during the following

[441] W. H. Mills and I. G. Nixon, *J. Chem. Soc.* p. 2510 (1930).

[442a] J. Vaughan, G. J. Welch, and G. J. Wright, *Tetrahedron* **21**, 1665 (1965).

[442b] R. Taylor, G. J. Wright, and A. J. Homes, *J. Chem. Soc. B* p. 780 (1967).

[443] A. Streitwieser, Jr., G. R. Ziegler, P. J. Mowery, A. Lewis, and R. G. Lawler, *J. Am. Chem. Soc.* **90**, 1357 (1968).

[444] B. Halton, *Chem. Rev.* **73**, 113 (1973).

[445] R. Anet and F. A. L. Anet, *J. Am. Chem. Soc.* **86**, 525 (1964).

[446] G. L. Closs, L. R. Kaplan, and V. I. Bendall, *J. Am. Chem. Soc.* **89**, 3376 (1967).

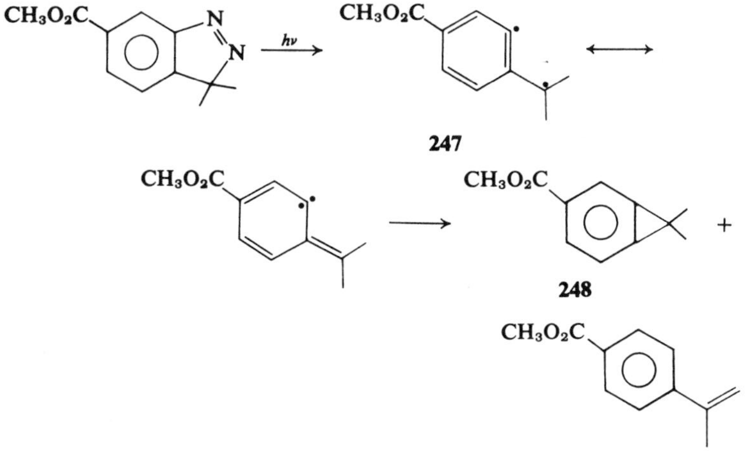

(30)

year as the product of a clever Diels–Alder-*retro*-Diels–Alder sequence [Eq. (31)].[447] The strain[448] in benzocyclopropene (ca. 68 kcal/mole[449a,b]) is manifested in its facile ring opening under electrophilic substitution reaction

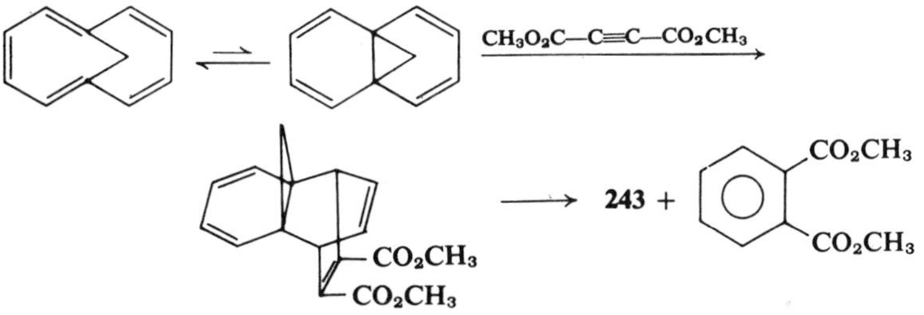

(31)

conditions.[444] Certainly a portion of this destabilization energy is attributable to some loss of aromatic stabilization which accompanies partial bond localization. However, the x-ray structure of dimethyl-3,6-diphenylbenzocyclopropene-7,7-dicarboxylate (**249**) indicates considerable bond localization but not in a manner explicable in terms of a Kekulé-type structure.[450] 1*H*-Cyclopropa-[*b*]naphthalene (**250**) also cannot be represented so simply.[449b] In both these molecules, the three-membered ring appears to have very similar dimensions to cyclopropene, a result supported in the analysis of the unusually large

[447] E. Vogel, W. Grimme, and S. Korte, *Tetrahedron Lett.* p. 3625 (1965).

[448] E. F. Ullman and E. Buncel, *J. Am. Chem. Soc.* **85**, 2106 (1963).

[449a] W. E. Billups and W. Y. Chow, *J. Am. Chem. Soc.* **95**, 4099 (1973).

[449b] W. E. Billups, W. Y. Chow, K. H. Levell, E. S. Lewis, J. L. Margrave, R. L. Sass, J. J. Shieh, P. G. Werness, and J. L. Wood, *J. Am. Chem. Soc.* **95**, 7878 (1963).

[450] E. Carstensen-Oeser, B. Muller, and H. Dürr, *Angew. Chem., Int. Ed. Engl.* **11**, 422 1972).

C₆H₅ 1.333 Å → structure **249** with CO₂CH₃ groups; bond lengths 1.392 Å, 1.419 Å, 1.387 Å, C₆H₅; and structure **250** with bond lengths 1.362 Å, 1.439 Å, 1.337 Å, 1.368 Å, 1.403 Å, 1.408 Å, 1.437 Å.

$^{1}J(^{13}C—^{13}C)$ between carbons 1 and 6 of benzocyclopropene itself.[451] Although semiempirical calculations also predict bond localization in benzo-cyclopropene (somewhat different from that manifested by **249**), the dia-magnetic ring current in this molecule indicates that it is most certainly aromatic.[452a,b] However, facile addition of iodine across the central (1,6) $\pi$ bond of benzocyclopropene could be taken as evidence of reduced aromati-city,[447] but this may be a reflection upon the transition state of the reaction rather than the ground-state properties of this molecule. 1,4-Dihydrodicyclo-propa[b,g]naphthalene (**251**) is isolable (m.p. 132°–133°), but it decomposes explosively upon melting and is very sensitive to shock.[453] These findings have caused Vogel and co-workers to be somewhat pessimistic about the chances for isolation of **252** at room temperature.[453] To the extent that the afore-mentioned structural studies indicate cyclopropene character in **249** and **250**, molecules **251** and **252** might be somewhat destabilized, because their Kekulé-type structures allow only one cyclopropenelike ring. 1H-Cyclopropa[a]-naphthalene (**253**) has been isolated and also exhibits the spectral properties of a naphthalene derivative.[454] An anomalously high dipole moment in 7,7-difluorobenzocyclopropene has been attributed to a large resonance

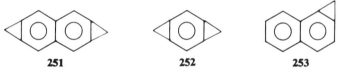

**251**          **252**          **253**

contribution of the ionic structure 7-fluorobenzocyclopropenium fluoride (analogous to **255**).[455] Benzocyclopropenone (**254**) has been monitored at 8°K in an argon matrix,[438] and the carbonium ion **255** has been observed in solution and isolated as a chloride salt.[456] Although these two are more

[451] H. Gunther and W. Herrig, *J. Am. Chem. Soc.* **97**, 5594 (1975).

[452a] C. S. Cheung, M. A. Cooper, and S. L. Manatt, *Tetrahedron* **27**, 101 (1971).

[452b] M. A. Cooper and S. L. Manatt, *J. Am. Chem. Soc.* **92**, 1605 (1970).

[453] J. Ippen and E. Vogel, *Angew. Chem., Int. Ed. Engl.* **13**, 736 (1974).

[454] S. Tanimoto, R. Schafer, J. Ippen, and E. Vogel, *Angew. Chem., Int. Ed. Engl.* **15**, 613 (1976); see also K. Grohmann, unpublished results cited here. (ref. 6).

[455] R. Pozzi, K. R. Ramaprasad, and E. A. C. Lucken, *J. Mol. Struct.* **28**, 111 (1975).

[456] B. Halton, A. D. Woodhouse, H. M. Hugel, and D. P. Kelly, *J. Chem. Soc., Chem. Commun.* p. 247 (1974).

strained than benzocyclopropene, they benefit from added resonance stabilization. Compound **256** is one of the first examples of a spirobenzocyclopropene.[457]

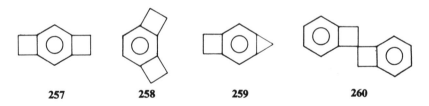

|  | 254 | | 255 | | 256 |

Although benzocyclobutene (**244**)[458a,b] is considerably less reactive than benzocyclopropene, it still has a tendency to ring-open under conditions of electrophilic substitution. Structural studies[459,460] of *cis*-1,2-dichlorobenzocyclobutene and benzo[1,2:4,5]dicyclobutene (**257**)[461] indicate little tendency for bond localization in these molecules. Although x-ray[462] and electron-diffraction studies[463] show the presence of considerable bond localization in biphenylene (**106**), it is clear that the driving force is minimization of anti-aromatic character in the central cyclobutadienoid ring. While the uv spectrum of **257** includes a bathochromic shift relative to 1,2,4,5-tetramethylbenzene, $\lambda_{max}$ of benzo[1,2,:3,4]dicyclobutene (**258**) is actually very slightly

|  | 257 | | 258 | | 259 | | 260 |

lower than that of 1,2,3,4-tetramethylbenzene.[464] Benzotricyclobutene **238** has only been inferred as an intermediate[432a], but its dodecafluoro derivative[432b,c]

[457] H. Dürr and H. Schmitz, *Angew. Chem., Int. Ed. Engl.* **14**, 647 (1975).

[458a] M. P. Cava and D. R. Napier, *J. Am. Chem. Soc.* **78**, 500 (1956).

[458b] M. P. Cava and A. A. Deana, *J. Am. Chem. Soc.* **81**, 4266 (1959).

[459] G. L. Hardgrove, L. K. Templeton, and D. H. Templeton, *J. Phys. Chem.* **72**, 668 (1968).

[460] J. L. Lawrence and S. G. G. MacDonald, *Acta Crystallogr., Section B*, **25**, 978 (1969).

[461] M. P. Cara, A. A. Deana, and K. Muth, *J. Am. Chem. Soc.* **82**, 2524 (1960).

[462] J. K. Fawcett and J. Trotter, *Acta Crystallogr.* **20**, 87 (1966).

[463] A. Yokozeki, C. F. Wilcox, Jr., and S. H. Bauer, *J. Am. Chem. Soc.* **96**, 1026 (1974).

[464] R. P. Thummel, *J. Am. Chem. Soc.* **98**, 628 (1976).

has been characterized by x-ray.[432d] Predictably, cyclopropa[4.5]benzocyclo-butene (259) [465,466] exhibits a greater bathochromic shift than does benzo-cyclopropene. The strain in compound 260 is sufficiently high to make its corresponding spiro-conjugated diradical relatively accessible.[467] The inter-mediacy of cyclobuta[1,2-d]benzyne (261) has been inferred through the isolation of its dimer (262) as well as a furan cycloadduct (Scheme 19).[468a] The fused cyclobutane ring is thought to furnish stabilization relative to higher homologues by augmenting the bond-fixated structure 219 known to be favored for o-benzyne.[468a] Dimer 262 is also a highly interesting molecule, because the central ring should be more antiaromatic than that in biphenylene. Although 262 is thermally stable, it is very easily hydrogenated, and this may reflect some form of resonance destabilization.[468a] Cyclobuta[1,2-c]benzyne has also been generated and trapped as a furan adduct.[468b]

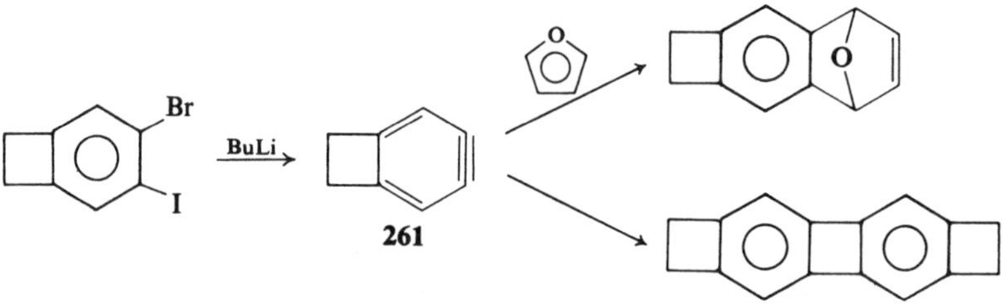

**Scheme 19**

The instability of benzocyclobutadiene (263) and its derivatives reflects the antiaromaticity of cyclobutadiene, albeit somewhat reduced, and the added strain of benzene fusion. The first isolable simple benzocyclobutadiene, 264, possessed two phenyl groups that served to mitigate antiaromatic destabili-zation.[469] The second such species to be reported, 265,[470a,b] is actually quite

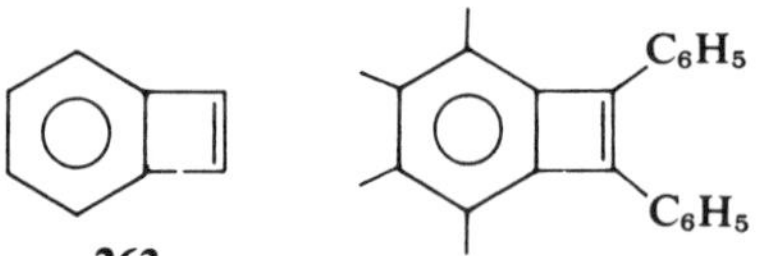

[465] D. Davalian and P. J. Garratt, *J. Am. Chem. Soc.* **97**, 6883 (1975).

[466] C. J. Saward and K. P. C. Vollhardt, *Tetrahedron Lett.* p. 4539 (1975).

[467] M. P. Cava and J. A. Kucskowski, *J. Am. Chem. Soc.* **92**, 5800 (1970).

[468a] R. L. Hillard, III and K. P. C. Vollhardt, *J. Am. Chem. Soc.* **98**, 3579 (1976).

[468b] R. L. Hillard, III and K. P. C. Vollhardt, *Angew. Chem., Int. Ed. Engl.*, **16**, 399 (1977).

[469] H. Straub, *Angew. Chem., Int. Ed. Engl.* **13**, 405 (1974).

[470a] F. Toda and M. Ohi, *J. Chem. Soc., Chem. Commun.* p. 506 (1975).

[470b] F. Toda and K. Tanaka, *J. Chem. Soc., Chem. Commun.* p. 177 (1976).

stable thermally (m.p. under $N_2 = 296°-297°C$), but in addition to its extreme sensitivity to oxygen, it very readily engages its aromatic ring in a Diels–Alder reaction which simultaneously destroys two cyclobutadienoid rings (Scheme 20).[470b] Benzocyclobutadiene (263) itself has been generated

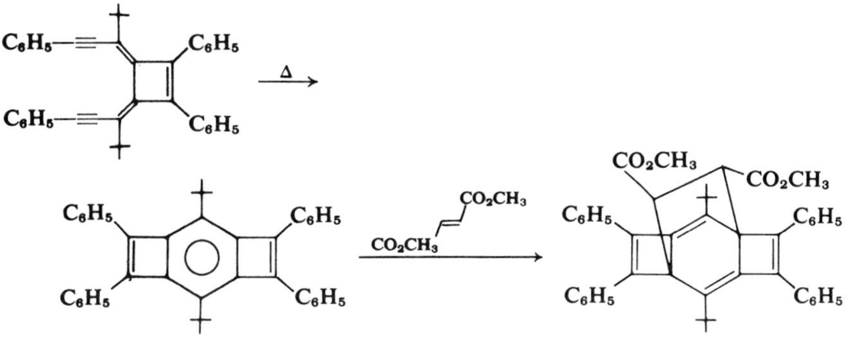

**Scheme 20**

and its ir and uv spectra recorded in an argon matrix.[471] This molecule dimerizes rapidly at temperatures exceeding 75°K. Naphthocyclobutadienes are somewhat more stable, the first isolable derivative having been reported in 1963,[472] and the parent molecule having been generated and trapped but not isolated.[473]

Although 1,8-methanonaphthalene (266)[474] is *peri*-bridged rather than *ortho*-bridged, the distortions introduced on the aromatic nucleus are similar to those in compounds discussed earlier. (Heterocyclic analogues of 266 are discussed at the end of Section 5.D.7) Acenaphthene (267) experiences angle pinching from the presence of its five-membered ring, accompanied by corresponding opening of the opposite angle.[475] The heat of formation of acenaphthene indicates a strain energy of about 9 kcal/mole relative to 1,5-dimethylnaphthalene (or a *hypothetically* unstrained 1,8-dimethylnaphthalene) and 1,2-diphenylethane,[476] which is only about 2 kcal/mole higher than the strain in 1,8-dimethylnaphthalene.[477] The symmetry of pyracene (268) precludes the cooperative angular distortion observed in acenaphthene, and the

[471] O. L. Chapman, C. C. Chang, and N. R. Rosenquist, *J. Am. Chem. Soc.* 98, 261 (1976).

[472] M. P. Cava, B. Y. Hwang, and J. P. Van Meter, *J. Am. Chem. Soc.* 85, 4032 (1963).

[473] M. P. Cava and A.-F. C. Hsu, *J. Am. Chem. Soc.* 94, 6441 (1972).

[474] R. J. Bailey and H. Schechter, *J. Am. Chem. Soc.* 96, 8116 (1974).

[475] V. Balasubramaniyan, *Chem. Rev.* 66, 567 (1966).

[476] A strain energy of 6.7 kcal/mole for 267 has also been calculated; H. H. Boyd, R. L. Christensen, and R. Pua, *J. Am. Chem. Soc.* 87, 3554 (1965).

[477] W. D. Good, *J. Chem. Thermodyn.* 5, 715 (1973).

corresponding bond angle is 115°.[478] The "ultimate" in *peri*-bridged naphthalenes is 1,8-dehydronaphthalene or 1,8-naphthalenediyl (**269**). This species
was postulated in the oxidation of the *N*-aminotriazene **270**[479a,b] as well as
in the thermolysis of a 1,8-naphthalene-based anhydride (**271**)[480] (Scheme
21). Among the most interesting chemical reactions of 1,8-naphthalenediyl

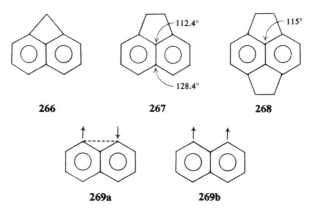

| 266 | 267 | 268 |

| 269a | 269b |

are its stereospecific 1,2-addition to olefins, and a marked propensity for
1,2-addition instead of 1,4-addition to 1,3-dienes.[479a,b,481a,b] In contrast,
benzyne prefers 1,4-addition to conjugated dienes and adds nonstereo-
specifically to olefins but stereospecifically to conjugated dienes.[437a,c] The
stereospecificity of 1,8-naphthalenediyl's addition to olefins would support
the presence of a singlet state (**269a**), and this intermediate's tendency to
undergo 1,2-addition in preference to 1,4-addition has been nicely rationalized
in terms of a species having an antisymmetric "ethylenelike" HOMO in
contrast to the symmetric "ethylenelike" HOMO of benzyne.[482] However,
the ability of 1,8-naphthalenediyl to form radical abstraction products such
as naphthalene and 1-phenylnaphthalene,[479a,b,480] as well as initially unanticipated 1,3-addition products with 1,3-dienes,[481a,b] might be best
explained in terms of the triplet **269b**. Extended Hückel calculations show a
HOMO–LUMO splitting of only 0.44 eV for **269**, in contrast to 1.52 eV for
1,2-benzyne.[482] Thus, while benzyne is clearly a singlet in the ground state,[437a–c]
the singlet and triplet states of 1,8-naphthalenediyl (**269a** and **269b**) may be

[478] G. L. Simmons and E. C. Lingafester, *Acta Crystallogr.* **14**, 872 (1961).

[479a] C. W. Rees and R. C. Storr, *Chem. Commun.* p. 193 (1965).

[479b] C. W. Rees and R. C. Storr, *J. Chem. Soc. C* p. 760 (1969).

[480] E. K. Fields and S. Meyerson, *J. Org. Chem.* **31**, 3307 (1966).

[481a] J. Meinwald and G. W. Gruber, *J. Am. Chem. Soc.* **93**, 3802 (1971).

[481b] J. Meinwald, L. V. Dunderton, and G. W. Gruber, *J. Am. Chem. Soc.* **97**, 681 (1975).

[482] R. Hoffmann, A. Imamura, and W. J. Hehre, *J. Am. Chem. Soc.* **90**, 1499 (1968).

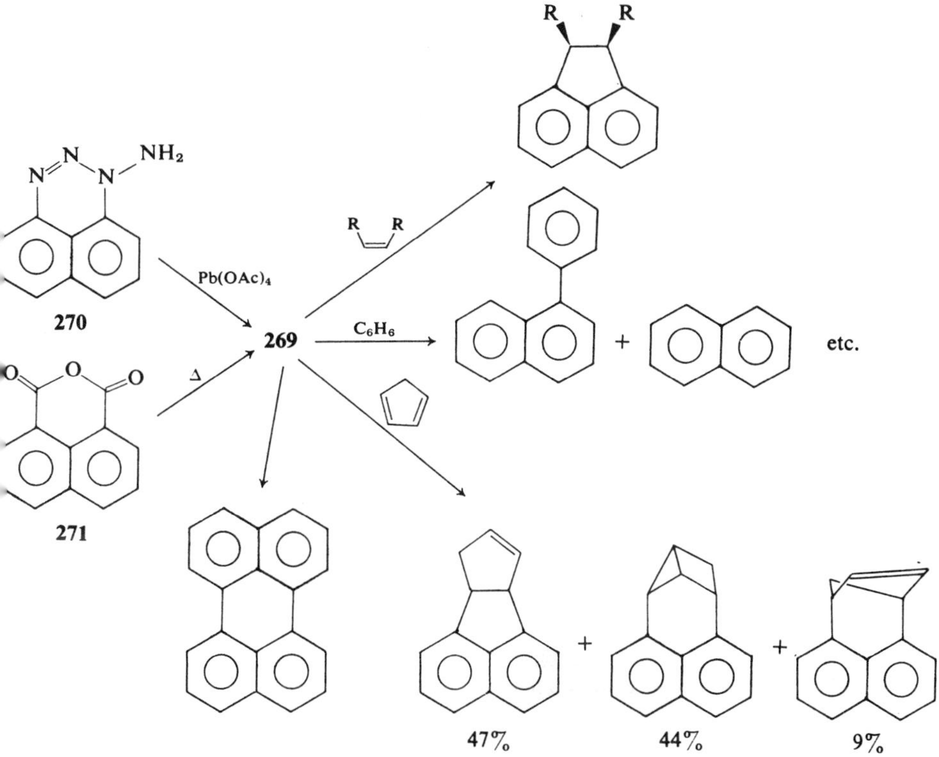

**Scheme 21**

somewhat comparable in energy. Efforts at forming a Pt(0) complex of **269** (analogous to the stable benzyne complex, Section 5.C.4.c) failed,[479b] but 1,8-naphthyne has been monitored in the mass spectrometer.[483] It may be that lead [Pb(OAc)$_4$] also plays a role.[437a–c,481b]

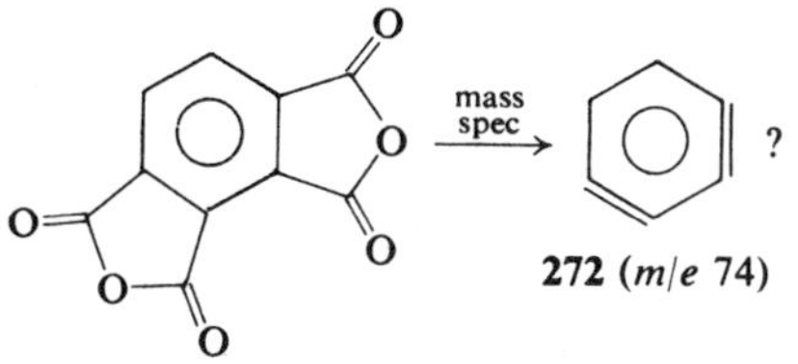

$$ (32) $$

1,2,3,4-Benzdiyne (**272**) (as well as the 1,2,4,5 isomer) have been postulated as intermediates.[484] The best evidence for their intermediacies appears to be the rather abundant (ca. 60% of base peak) ion at $m/e$ 74 formed during

[483] H. F. Gruetzmacher and W. R. Lehmann, *Justus Liebigs Ann. Chem.* 2116 (1975).
[484] E. K. Fields and S. Meyerson, *Adv. Phys. Org. Chem.* 6, 1 (1968).

mass-spectrometric analysis of the corresponding dianhydrides. In light of the extreme instability of benzyne noted above, we remain somewhat apprehensive about postulating 272 as a discrete species. Comparison of the heats of formation of benzene (20 kcal/mole) and benzyne (100 kcal/mole) suggests a value of at least 180 kcal/mole for the enthalpy of formation of 272. Conceptual addition of two propynes and a molecule of 2-butyne and subtraction of two ethanes suggests an upper limit for the enthalpy of formation of the isomeric 1,3,5-hexatriyne of only 64 kcal/mole, and perhaps this is the $C_6H_2$ species monitored.

### 3.H.2 AROMATIC MOLECULES DESTABILIZED BY STERIC STRAIN

Although benzene is flat, calculational studies suggests that it is surprisingly flexible, perhaps departing from planarity by up to 10° (angle $\alpha$, see Table 3.6) in vibrational modes operative at ambient temperatures.[485,486] While certain bridged benzenes must force "bending and battering"[487] of their aromatic rings (see next section), there had been until fairly recently some disagreement over the extent that nonbonded repulsions could induce such distortion. 1,2,4-Tri-*tert*-butylbenzene (273), the first compound to have *o-tert*-butyl

$$t\text{-Bu}—C{\equiv}CH \xrightarrow{\text{Co}_2(\text{CO})_8} \text{Co}_2(\text{CO})_4\,(HC{\equiv}C—t\text{-Bu})_3 \xrightarrow{\Delta} \quad \quad (33)$$

273

groups, was reported in 1961 as the product of a trimerization reaction involving *tert*-butylacetylene and $CO_2(CO)_8$.[488] A flurry of activity at that time subsequently led to production of *o*-di-*tert*-butylbenzene (274)[489–491] and 1,2,4,5-tetra-*tert*-butylbenzene (275)[492,493] via trimerization of alkynes as well as by "more conventional" methods (Scheme 22). The strain energies

[485] H. Wynberg, W. C. Nieuwpoort, and H. T. Jonkman, *Tetrahedron Lett.* p. 4623 (1973).

[486] N. L. Allinger, J. T. Sprague, and T. Liljefors, *J. Am. Chem. Soc.* 96, 5100 (1974).

[487] D. J. Cram and J. M. Cram, *Acc. Chem. Res.* 4, 204 (1971).

[488] U. Kuerke, C. Hoogzand, and W. Hubel, *Chem. Ber.* 94, 2817 (1961).

[489] C. Hoogzand and W. Hubel, *Angew. Chem.* 73, 680 (1961).

[490] E. M. Arnett and M. E. Strem, *Chem. Ind. (London)* p. 2008 (1961).

[491] A. W. Burgstahler and M. O. Abdel-Rahman, *J. Am. Chem. Soc.* 85, 173 (1963); see also L.R.C. Barclay, N. D. Hall, and J. W. MacClean, *Tetrahedron Lett.* p. 243 (1961).

[492] C. Hoogzand and W. Hubel, *Tetrahedron Lett.* p. 637 (1961).

[493] E. M. Arnett, M. E. Strem, and R. A. Friedel, *Tetrahedron Lett.* p. 658 (1961).

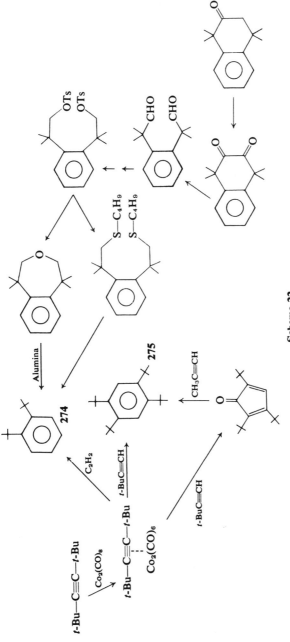

**Scheme 22**

of **273** and **274** (relative to *meta* or *para* isomers) are both about 22 kcal/mole.[494] With the exception of relatively facile acid-catalyzed di-*tert*-butylation, the reactivity and spectra of these two compounds indicate that they are essentially normal planar aromatics. This view is further confirmed by the planarity, determined by x-ray methods, of 2,3-di-*tert*-butylquinoxaline (**276**).[495] Despite about 30 kcal/mole of strain energy, 1,2,4,5-tetra-*tert*-butylbenzene (**275**) is planar.[496] That 1,2,4,5-tetra-*tert*-butylbenzene is considerably less than twice as strained as two *o*-*tert*-butylbenzenes is somewhat curious. Perhaps the first two groups cause major ring deformations which are essentially the same as in the tetra-*tert*-butyl case. The inter-*tert*-butyl angles are widened to about 130°, and long benzene–*tert*-butyl bonds (ca. 1.567 Å) further decrease steric repulsions.[496] Thus, while benzenes can undergo out-of-plane distortion without a significant enthalpy increase, they still appear to prefer in-plane distortion, a result confirmed by molecular-mechanics calculations.[368] Only two 1,2,3-tri-*tert*-butylbenzenes, **277**[497] and **278**,[498] have been reported, and while large bathochromic shifts are observed in their uv spectra, no conclusions have been drawn about ring geometry.[494] Some other interesting manifestations of crowding in related compounds include blocked rotation in hexaisopropylbenzene[498] as well as inhibition of styrenelike resonance in **279** coupled with the unreactivity of olefinic linkages in this compound.[499]

**276**          **277**          **278**          **279**

*Peri*(1,8) interactions in naphthalene derivatives are greater in magnitude than *ortho* interactions.[475] Thus, the strain in 1,8-dimethylnaphthalene (7 kcal/mole[477]) is considerably higher than that in *o*-xylene (1 kcal/mole). The

[494] E. M. Arnett, J. C. Sanda, J. M. Bollinger, and M. Barber, *J. Am. Chem. Soc.* **89**, 5389 (1967).

[495] G. J. Visser, A. Vos, A. de Groot, and H. Wynberg, *J. Am. Chem. Soc.* **90**, 3253 (1968).

[496] A. van Bruijnsvoort, C. Eilermann, H. van der Meer, and C. H. Stam, *Tetrahedron Lett.* p. 2527 (1968).

[497] H. G. Viehe, R. Merenyi, J. F. M. Oth, and P. Valange, *Angew. Chem., Int. Ed. Engl.* **3**, 746 (1964).

[498] E. M. Arnett and J. M. Bollinger, *J. Am. Chem. Soc.* **86**, 4729 (1964).

[499] E. M. Arnett, J. M. Bollinger, and J. C. Sanda, *J. Am. Chem. Soc.* **87**, 2050 (1965).

stress in this molecule is evident in its structure, wherein opening of angles $C_1$—$C_{8a}$—$C_8$ (125.5°) and $C_{8a}$—$C_1$—$CH_3$ (124.8°) lessens but does not eliminate nonbonded repulsions between methyls, which are only 2.93 Å apart.[500] The heavy atoms in this molecule are entirely coplanar, a finding which contrasts with an earlier study which found the ring system in 3-bromo-1,8-dimethylnaphthalene to be somewhat buckled.[501] Thus, the structure of 1,8-dimethylnaphthalene, at least, again would appear to support the view that aromatic compounds prefer in-plane distortion (naphthalenes should warp even more easily than benzenes[485]). The substituents in 1,4,5,8-tetra-chloronaphthalene (**280**)[502,503] and octamethylnaphthalene (**281**)[504,505] are

**280**                    **281**                    **282**                    **283**

forced above and below the average (but hypothetical) ring plane. A number of *peri-tert*-butylnaphthalenes, including **282**, have been synthesized,[506] and these clearly stagger their bulky substituents above and below the ring.[506,507] The strains in **282** as well as in **273** have been utilized as a driving force in the photochemical syntheses of their Dewar isomers (see Section 5.B.2). Two isomers of **283** are observable via nmr at low temperatures.[508] Recent examinations of the hyperfine splittings of the radical anions of **282** have indicated an outer edge distortion of the *peri-tert*-butyl-bearing carbons to be about 20°.[509] Analogous studies of the sterically hindered aromatic species would be of interest: Do the geometries of the neutral hydrocarbons and their

[500] D. Bright, I. E. Maxwell, and J. de Boer, *J. Chem. Soc., Perkin Trans. 2* p. 2101 (1973).

[501] M. D. Jameson and B. R. Penfold, *J. Chem. Soc.* p. 528 (1965).

[502] M. A. Davydova and Y. T. Struchokov, *J. Struct. Chem. (USSR)* **2**, 63 (1961); **3**, 170, 202 (1962); **6**, 98 (1965).

[503] G. Gafner and F. H. Herbstein, *Acta Crystallogr.* **15**, 1081 (1962).

[504] G. Gafner and F. H. Herbstein, *Nature (London)* **200**, 130 (1963) and references cited therein.

[505] G. Ferguson and J. M. Robertson, *Adv. Phy. Org. Chem.* **1**, 203 (1963).

[506] R. W. Franck and E. G. Leser, *J. Am. Chem. Soc.* **91**, 1577 (1969).

[507] H. van Bekkum, T. J. Nieuwstad, and J. van Barneveld, *Rec. Trav. Chim. Pays-Bas* **88**, 1028 (1969).

[508] D. L. Fields and T. H. Regan, *J. Org. Chem.* **37**, 2986 (1971).

[509] I. B. Goldberg, H. R. Crane, and R. W. Frank, *J. Phys. Chem.* **79**, 1741 (1975).

radical anions correspond? (We note that one may place an aromatic solvent in contact with potassium-metal film, and addition of crown ether will generate the encapsulated cation and solvated electrons which form radical anions of the aromatic compound.[510])

Steric destabilization in 4,5-dimethylphenanthrene (**284**) is even more severe (12.6 kcal/mole[511]) than *peri*-dimethyl repulsion, but still smaller than that in **285** (15 kcal/mole, attributed in part to "buttressing" by the *peri* hydrogen shown[511]) or in 1,12-dimethylbenzo[c]phenanthrene (**286**), despite the presence of more bonds to share deformation, in which the rings are actually skewed and indeed enantiomers are resolvable.[512] The energy of activation for racemization of hexahelicene (**287**) is about 35–36 kcal/mole.[513,514a,b] Racemization of hexahelicene is a conformational process

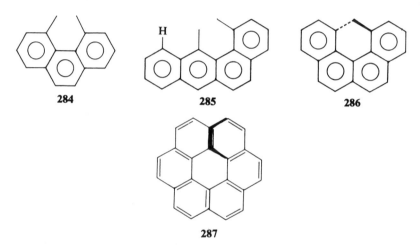

284    285    286

287

(i.e., no bond breaking and re-formation) despite the difficulties encountered in a molecular-model simulation. Apparently, this large molecule can distribute distortion among a large number of component groups and bonds so that strain at the transition state is not too severe.[514b] Pentahelicene's racemization barrier is not as high ($\Delta H^{\ddagger} = 23$ kcal/mole[515]), since overlap of its terminal rings is less pronounced than in hexahelicene. The six bonds on

[510] B. Kempf, S. Raynal, A. Collet, F. Schue, S. Boileau, and J. M. Lehn, *Angew. Chem., Int. Ed. Engl.* **13**, 611 (1974).

[511] H. A. Karnes, B. D. Kybett, M. H. Wilson, J. I. Margrave, and M. S. Newman *J. Am. Chem. Soc.* **87**, 5554 (1965).

[512] M. S. Newman and R. M. Wise, *J. Am. Chem. Soc.* **78**, 450 (1956).

[513] M. S. Newman and D. Lednicer, *J. Am. Chem. Soc.* **78**, 4795 (1956).

[514a] R. H. Martin and M. J. Marchant, *Tetrahedron* **30**, 347 (1974).

[514b] R. H. Martin, *Angew. Chem., Int. Ed. Engl.* **13**, 649 (1974).

[515] C. Goedicke and H. Stegemeyer, *Tetrahedron Lett.* p. 937 (1970).

the inner periphery of hexahelicene average 1.437 Å; those on the outer periphery average 1.334 Å.[514b] 1,16-Dehydrohexahelicene (288) is predicted, on the basis of molecular models, to be saddle-shaped and potentially resolvable. While the hydrocarbon has been synthesized,[516] no evidence concerning

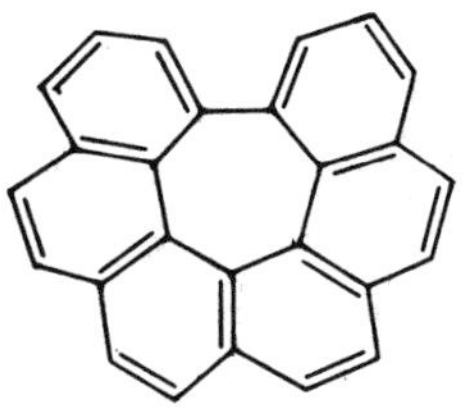

288 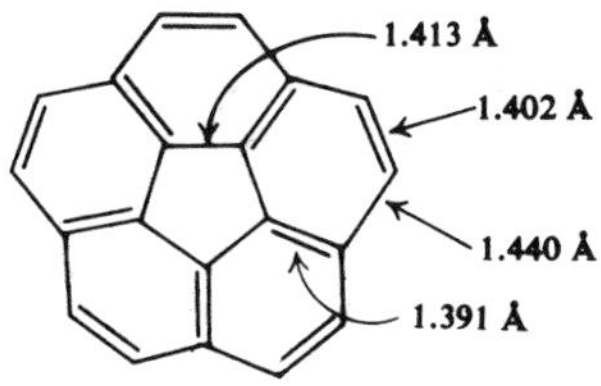          289

290, n = 12    293, n = 8
291, n = 10    294, n = 7
292, n = 9     295, n = 6

its chiral properties has yet appeared. Corannulene (289),[517a,b] which has been termed "[5]circulene,"[518] is a bowl-shaped molecule caught in a tenuous balance between strain and aromaticity.[519] A systematic study of the photoelectron spectra of the helicenes from benzene, naphthalene, and phenanthrene (i.e., [1]-, [2]-, and [3]helicene) through [14]helicene has been reported,[520] with explicit interest in $\pi$- and $\sigma$-orbital mixing and transannular effects.

### 3.H.3 PARACYCLOPHANES, METACYCLOPHANES, AND RELATED MOLECULES

Suitably short *para* or *meta* bridges induce significant departures from planarity in an aromatic ring. A most surprising finding during the past few years is that stability and even aromatic properties can coexist with rather severe distortion.[487,521a,b] This is consistent with the calculated flexibility of aromatic rings previously noted.[485,486] Thus, [12]paracyclophane (290),[522]

[516] P. J. Jessup and J. A. Reiss, *Tetrahedron Lett.* p. 1453 (1975).
[517a] W. E. Barth and R. G. Lawton, *J. Am. Chem. Soc.* 88, 380 (1966).
[517b] W. E. Barth and R. G. Lawton, *J. Am. Chem. Soc.* 93, 1730 (1971).
[518] J. H. Dopper and H. Wynberg, *Tetrahedron Lett.* p. 763 (1972).
[519] J. C. Hanson and C. E. Nordman, *Acta Crystallogr. Sect. B* 32, 1147 (1976).
[520] S. Obenland and W. Schmidt, *J. Am. Chem. Soc.* 97, 6633 (1975).
[521a] B. H. Smith, "Bridged Aromatic Compounds," Academic Press, New York, 1964.
[521b] F. Vogtle and P. Neumann, *Top. Curr. Chem.* 48, 67 (1974).
[522] D. J. Cram, N. L. Allinger, and H. Steinberg. *J. Am. Chem. Soc.* 76, 6132 (1954) and references cited therein.

[10]paracyclophane (**291**),[522] [9]paracyclophane (**292**),[522,523] and [8]para-cyclophane (**293**)[524a,b,525,526] have been known for some time and are all quite stable. [8]Paracyclophane, synthesized in the way of Scheme 23, is a stable liquid whose uv spectral properties are normal except for a slight (10 nm) bathochromic shift.[524a,b] The first simple derivative (**296**) (see, however,

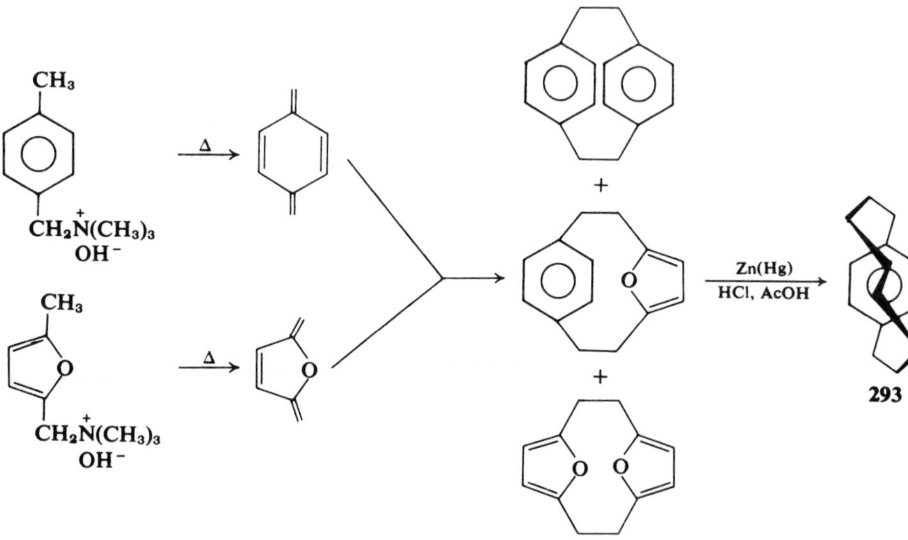

**Scheme 23**

**303**) of a [7]paracyclophane reported in 1972, some 11 years after publication of [8]paracyclophanes, was obtained in the manner depicted in Scheme 24.[527] The initial acyloin condensation is a time-honored method for generating macrocycles. The subsequent Wolff rearrangement (see Section 5.D.6.a), which involves elimination of nitrogen and formation of an α-ketocarbene, is exothermic enough to produce an appreciably strained molecule. The uv spectrum of this molecule exhibits a bathochromic shift of about 20 nm, and one of the bridge hydrogens has an nmr chemical shift of −1.4 ppm since it is forced into the aromatic "π cloud."[527] Corresponding experiments have not yet produced a derivative of [6]paracyclophane.[527] [7]Paracyclophane

[523] M. F. Bartlett, S. K. Figdor, and K. Wiesner, *Can. J. Chem.* **30**, 291 (1952).

[524a] D. J. Cram and G. R. Knox, *J. Am. Chem. Soc.* **83**, 2204 (1961).

[524b] D. J. Cram, C. S. Montgomery, and G. R. Knox, *J. Am. Chem. Soc.* **88**, 515 (1966).

[525] The 4-carboxylic acid: M. L. Allinger and L. A. Freiberg, *J. Org. Chem.* **27**, 1490 (1962).

[526] The 4-carboxylic acid: A. T. Blomquist and L. F. Chow, cited in A. T. Blomquist and F. W. Schaefer, *J. Am. Chem. Soc.* **83**, 4547 (1961).

[527] N. L. Allinger and T. J. Walter, *J. Am. Chem. Soc.* **94**, 9267 (1972).

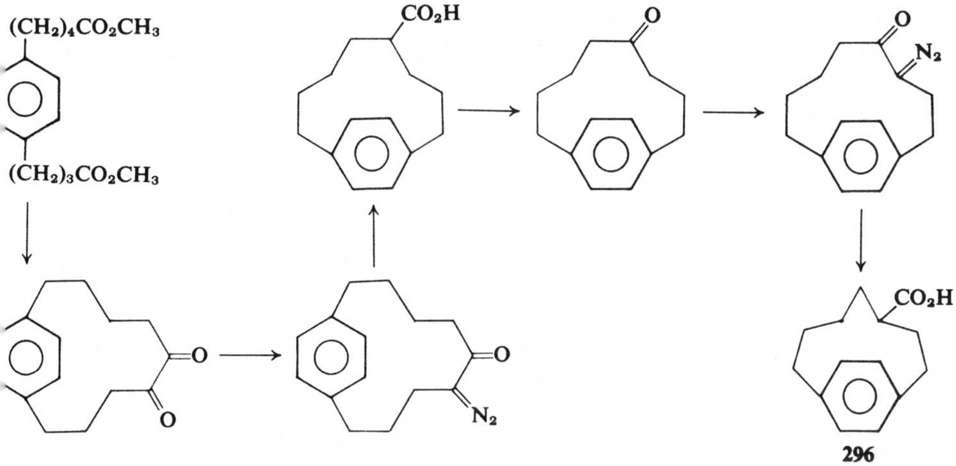

**Scheme 24**

(294) itself, as well as [6]paracyclophane (295), have been generated by analogous carbene routes (see Scheme 25 for [6]paracyclophane).[528,529a] Both of these paracyclophanes are isolable, and the uv and nmr spectral properties of [6]paracyclophane indicate that it is aromatic. Although not yet tested, these last two syntheses have been postulated to pass through bridged Dewar benzenes such as **298**.[528,529a] Dewar benzene **298** has been quantitatively formed through photolysis of [6]paracyclophane, the only case in which a

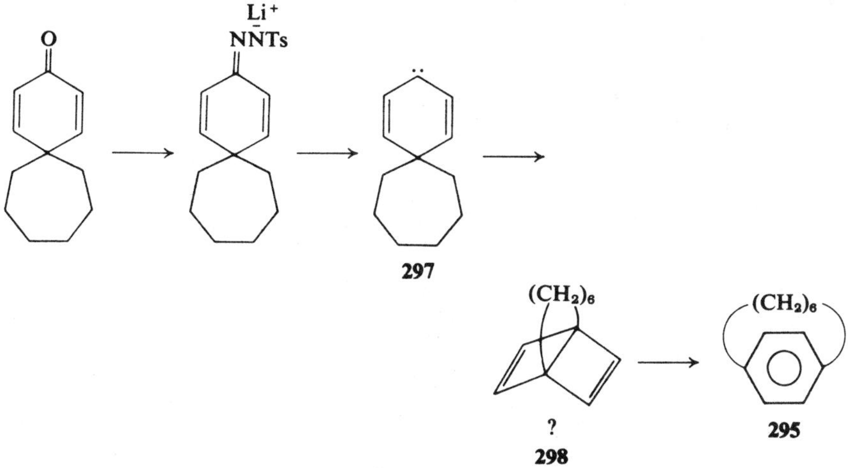

**Scheme 25**

[528] A. D. Wolf, V. V. Kane, R. H. Levin, and M. Jones, Jr., *J. Am. Chem. Soc.* **95**, 1680 (1973).

[529a] V. V. Kane, A. D. Wolf, and M. Jones, Jr., *J. Am. Chem. Soc.* **96**, 2643 (1974).

Dewar isomer has been the sole product of photolysis.[529b] Moreover, **298** aromatizes cleanly and relatively easily ($E_a = 19.9 \pm 0.9$ kcal/mole).[529b] It is possible to very crudely estimate the energetics of this process using bond-additivity methods. If one combines a recent experimental heat of formation of (singlet) $CH_2$ (101 kcal/mole[138c]) with group increments,[4] and assumes an arbitrary "fudge factor" of $-20$ kcal/mole (see Section 3.D.2) to correct for inductive and resonance stabilization as well as strain, a value of 100 kcal/mole is estimated for the heat of formation of **297**. Group increments[4] would allow one to predict a value of $-5$ kcal/mole for $\Delta H_f^0$ of *hypothetical* unstrained [6]paracyclophane. Recognizing that Dewar benzene is some 65 kcal/mole less stable than benzene (see Table 5.1), we obtain a lower limit of $+60$ kcal/mole for the heat of formation of **298**, some 40 kcal/mole lower than **297**. This merely serves to remind the reader how high in energy carbenes are. Furthermore, this calculation makes it apparent that an [*n*]paracyclophane should tolerate about 65 kcal/mole of strain before it becomes less stable than its Dewar isomer. Molecular-mechanics calculations find a strain of only about 29 kcal/mole (Table 3.6) in [6]paracyclophane, which seems surprisingly low.[530] Reactions analogous to those in Scheme 25 have been unsuccessful in obtaining [5]paracyclophane.[531] However, the intermediacy of [5]paracyclophane (**300**) has been postulated in the thermal rearrangement of 1,4-pentamethylene Dewar benzene (**299**, Scheme 26).[532,533] One strong piece of evidence supporting this view is the observed greater thermal stability of 1,4-trimethylene Dewar benzene, which, although more strained than **299**, would have to rearrange to an impossibly warped [3]paracyclophane. Molecular-mechanics computations predict a strain of about 40 kcal/mole for [5]paracyclophane (Table 3.6),[530] which should make it more than 25 kcal/mole more stable than **299** as well as more than 25 kcal/mole more stable than the benzylic diradical of Scheme 26. (See Section 5.B.2 for discussions of other Dewar isomers of small [*n*]paracyclophanes.) While we have dealt primarily with benzene distortion as the primary determinant of strain in these molecules, it is clear that distortion of polymethylene bridges also introduced a sizable enthalpy increment.[530]

The highly strained compound [2.2]paracyclophane (**301**) was first obtained in trace amounts in 1949 from the highly reactive intermediate *p*-xylylene[534a,b]

[529b] S. L. Kammula, L. D. Iroff, M. Jones, Jr., J. W. van Straten, W. H. de Wolf, and F. Bickelhaupt, *J. Am. Chem. Soc.*, **99**, 5815 (1977).

[530] N. L. Allinger, J. T. Sprague, and T. Liljefors, *J. Am. Chem. Soc.* **96**, 5100 (1974).

[531] Professor Maitland Jones, Jr., personal communication to Arthur Greenberg.

[532] J. W. van Straten, I. J. Landheer, W. H. de Wolf, and F. Bickelhaupt, *Tetrahedron Lett.* p. 4499 (1975).

[533] K. Weinges and K. Klessing, *Chem. Ber.* **109**, 793 (1976).

[534a] C. J. Brown and A. C. Farthing, *Nature (London)* **164**, 915 (1949).

[534b] C. J. Brown, *J. Chem. Soc.* p. 3265 (1953).

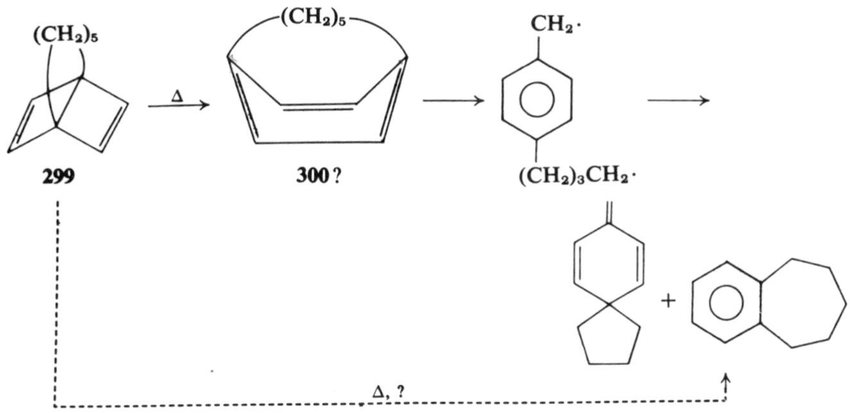

**Scheme 26**

(the reaction can be reversed photochemically;[535] see Scheme 23 for a mixed reaction involving p-xylylene). x-Ray data[534a,b,536,537] indicate that the aromatic rings in this molecule are boatlike in shape and deviate from co-planarity by 12°–13° (angle α in Table 3.6). Skewing of these rings relieves

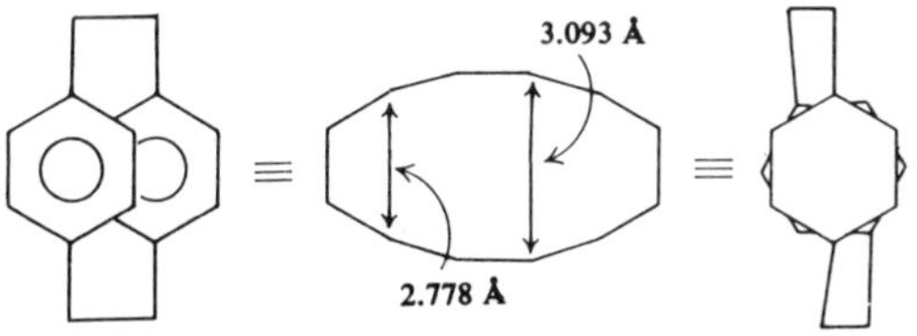

**301**

some of the nonbonded interactions which are incurred by their forced proximity. The molecule's strain energy, 31–33 kcal/mole,[538,539] is due to nonplanarity of the aromatic nucleus, deformation of the benzene-bridge bond angle (β in Table 3.6), van der Waals repulsion between the aromatic rings, and partial eclipsing of bridge methylene groups.[540] Although [2.2]-paracyclophane is a stable compound, distortion and proximity of the two

[535] G. Kaupp, *Angew. Chem., Int. Ed. Engl.* **15**, 442 (1976).

[536] K. Lonsdale, H. J. Milledge, and K. W. Krishna Rao, *Proc. R. Soc., Ser. A* **255**, 82 (1960).

[537] H. Hope, J. Bernstein, and K. N. Trueblood, *Acta Crystallogr., Sect. B* **28**, 1733 (1972).

[538] R. H. Boyd, *Tetrahedron* **22**, 119 (1966).

[539] D. L. Rodgers, E. F. Westrum, Jr., and J. T. S. Andrews, *J. Chem. Thermodyn.* **5**, 733 (1973).

[540] R. H. Boyd, *J. Chem. Phys.* **49**, 2574 (1968).

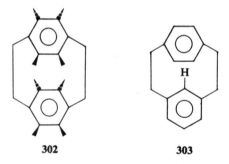

**301**                                                                    (34)

rings provide some novel departures from "normal" chemistry. Thus, one can imagine that bending of the aromatic rings in **301** forces electron density to the exterior of the molecule, and enhanced reactivity with electrophiles should be the result. Such behavior has in fact been observed in the relatively facile addition of methylene to [2.2]paracyclophane [Eq. (34); note the addition is to the bridgehead position, see Section 3.F.3].[541,542] The acid-catalyzed rearrangement of [2.2]paracyclophane to [2.2]metaparacyclophane (**303**)[543] is driven in part by loss of about 8–10 kcal/mole of strain.[544] This reduction in strain is mainly the result of decreased ring eclipsing in **303**, since the *para*-bridged ring in this molecule is even more distorted than in [2.2]paracyclo-phane (see Table 3.6). One should note that, effectively, **303** was the first [7]paracyclophane system to be characterized. The critical balance between strain and aromaticity in [2.2]paracyclophanes explains why increased steric

**302**               **303**

repulsion in octamethyl[2.2]paracyclophane **302** makes this molecule un-stable.[545] The "classically conjugated but orbitally unconjugated" compound [2.2]paracyclophane-1,9-diene (**304**)[546] features aromatic rings that are more distorted than those in [2.2]paracyclophane (see Table 3.6), but it also maintains a greater nonbonded distance between these rings.[547] A very

[541] K. Menke and H. Hopf, *Angew. Chem., Int. Ed. Engl.* **15**, 165 (1976).

[542] R. Nader and A. de Meijere, *Angew. Chem., Int. Ed. Engl.* **15**, 166 (1976).

[543] D. J. Cram, R. C. Helgeson, D. Lock, and L. A. Singer, *J. Am. Chem. Soc.* **88**, 1324 (1966).

[544] C. Shieh, D. C. McNally, and R. H. Boyd, *Tetrahedron* **25**, 3653 (1969).

[545] D. T. Longone and L. H. Simanyi, *J. Org. Chem.* **29**, 3245 (1964).

[546] K. C. Dewhirst and K. J. Cram, *J. Am. Chem. Soc.* **80**, 3115 (1958).

[547] R. K. Gantzel and K. N. Trueblood, *Acta Crystallogr.* **18**, 958 (1965).

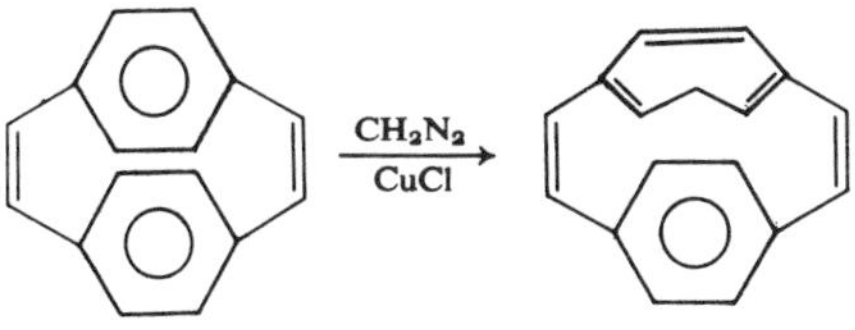

interesting aspect of the chemistry of this molecule is its preferred addition of methylene to the aromatic ring rather than across its olefinic linkage [Eq. (35)].[542] Outward "bending" of the aromatic ring's electron density and steric hindrance to attack of the carbon–carbon double bond combine to favor this unusual mode of addition.[542] Another curious feature is addition to the nonbridgehead linkage to form an "anti-Bredt" compound.

**Table 3.6**
**Strain Energies and Distortions of the Aromatic Rings of Some Paracyclophanes (or Substituted Derivatives)**

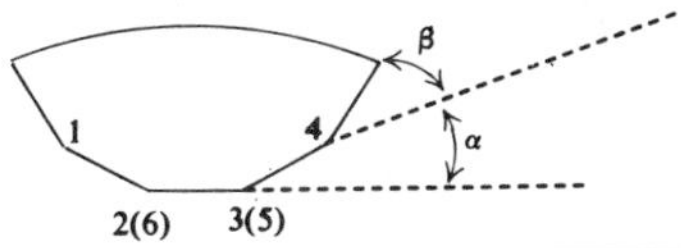

| Compound | Strain energy (kcal/mole) | Bending ($\alpha$) (deg) |
|---|---|---|
| [8]Paracyclophane | Calcd. 16.8[a] | 9;[b] calcd. 12.5[a] |
| [7]Paracyclophane | Calcd. 20.9[a] | 15–17;[c] calcd. 18.2 |
| [6]Paracyclophane | Calcd. 28.7[a] | Calcd. 22.4[a] |
| [5]Paracyclophane | Calcd. 39[a] | Calcd. 26.5[a] |
| [2.2]Paracyclophane | 31;[d] 33[b] | 12.6[e] |
| [3.3]Paracyclophane | 12[f] | 6.4[g] |
| [2.2]Metaparacyclophane | 24[f] | 14[h] (para ring) |
| [1.8]Paracyclophane | 2[f] | |
| [2.2]Paracyclophane-1,9-diene | Estd. 39[g] | 13.5[i] |
| [2.2]Metaparacyclophane-1,9-diene | | 18.4[j] (para ring) |

[a] N. L. Allinger, J. T. Sprague, and T. Liljefors, *J. Am. Chem. Soc.* **96**, 5100 (1974).
[b] D. L. Rodgers, E. F. Westrum, Jr., and J. T. S. Andrews, *J. Chem. Thermodyn.* **5**, 733 (1973).
[c] N. L. Allinger, T. J. Walter, and M. G. Newton, *J. Am. Chem. Soc.* **96**, 4588 (1974).
[d] R. H. Boyd, *Tetrahedron* **22**, 119 (1966).
[e] H. Hope, J. Bernstein, and K. N. Trueblood, *Acta Crystallogr.*, Sect. B **28**, 1733 (1972).
[f] C. Shieh, D. C. McNally, and R. H. Boyd, *Tetrahedron* **25**, 3653 (1969).
[g] R. K. Gantzel and K. N. Trueblood, *Acta Crystallogr.* **18**, 958 (1965).
[h] K. N. Trueblood and M. J. Crisp, footnote 17 in D. J. Cram and J. M. Cram, *Acc. Chem. Res.* **4**, 204 (1971).
[i] C. L. Coulter and K. N. Trueblood, *Acta Crystallogr.* **16**, 667 (1963).
[j] A. W. Hanson, *Acta Crystallogr.*, Sect. B **27**, 197 (1971).

Of the possible [n.m]paracyclophanes, we will mention only two examples. [1.7]Paracyclophane (305) remains the smallest derivative having a one-carbon bridge.[548] The compound [2.6]paracyclophane (306) was synthesized through a vinylogue of p-xylylene.[549]

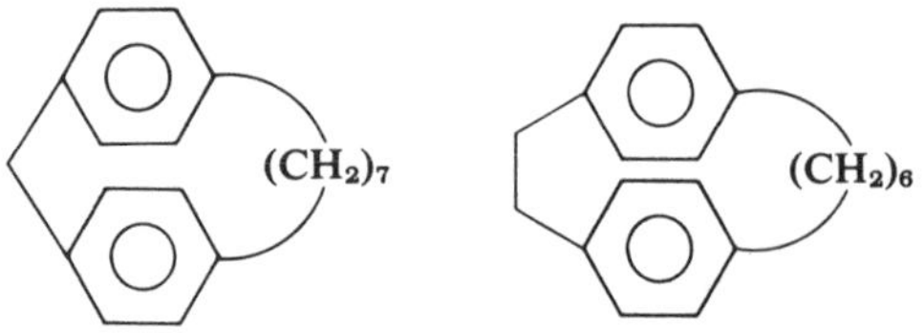

305                    306

Paracyclophanes have been studied extensively for their interesting stereo-chemical and conformational aspects, transannular-directing chemical effects (i.e., the ability of one ring to direct chemistry on the other ring), inter-ring interactions, basicity, spectra, and charge-transfer properties.[487,520] Rigid orientation and proximity of aromatic rings allow paracyclophanes, including derivatives of 301, to be used as "chemical tweezers." For example, see the published studies of the effect of *syn* and *anti* orientation on an intramolecular redox reaction.[550a,b]

The border between stability and instability in [n]metacyclophanes appears to be reached when n is 5. Thus, [7]metacyclophane and [6]metacyclophane (307) have been isolated, but their lower homologue* remains unknown.[551] These molecules are clearly aromatic, but their uv spectra show bathochromic shifts relative to unstrained benzene analogues. While the series of compounds 308 including n = 6 have been obtained, only tautomer 309 is known.[552a,b] Similarly, 310 and 311 have been isolated, while 312 has been generated only

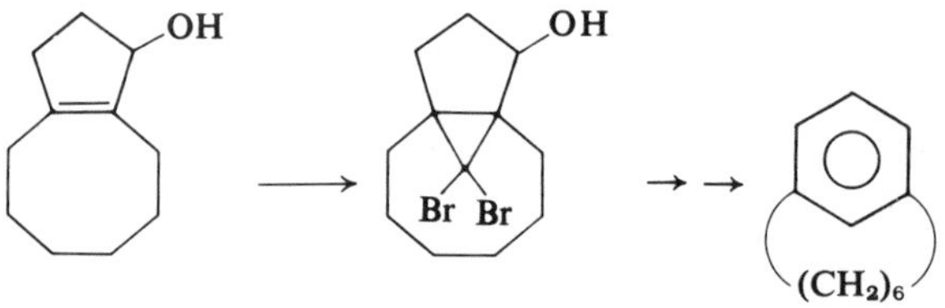

$$(36)$$

307

[548] D. J. Cram and M. F. Antar, *J. Am. Chem. Soc.* **80**, 3103 (1958).

[549] L. G. Kaufman and D. T. Longone, *Tetrahedron Lett.* p. 3229 (1974).

[550a] W. Rebafka and H. A. Staab, *Angew. Chem., Int. Ed. Engl.* **12**, 776 (1973).

[550b] W. Rebafka and H. A. Staab, *Angew. Chem., Int. Ed. Engl.* **13**, 203 (1974).

[551] S. Hirano, H. Hara, T. Hiyama, S. Fujita, and H. Nozaki, *Tetrahedron* **31**, 2219 (1975).

[552a] V. Prelog and K. Wiesner, *Helv. Chim. Acta* **30**, 1465 (1947).

[552b] V. Prelog, K. Wiesner, W. Ingold, and O. Hafliger, *Helv. Chim. Acta* **31**, 1325 (1948).

* See Addendum.

**308,** $n \geq 6$

**309**

**310,** $n = 8$; X = Br, Cl
**311,** $n = 6$
**312,** $n = 5$

**313**

as a transient intermediate.[553,a,b] The relatively low strain in [2.2]metacyclophane (**313**) (12 kcal/mole [544]) reflects slight distortion of aromatic nuclei and staggering of the aromatic rings.[487] While only the *anti* form is known for the parent (**313**) and most derivatives, a few *syn* isomers have now been made via the intermediacy of *syn*-[3.3]-dithiacyclophane precursors having electron-donating groups on one ring and electron-withdrawing groups on the other. The *syn*-metacyclophanes studied thus far have been isomerized to the *anti* forms upon heating above 215°C.[555d] Scheme 27 depicts the sequence of reactions employed in the synthesis of [2.2]metacyclophane-1,9-diene (**314**).[554a,b] This molecule was converted to the aromatic *trans*-15,16-dihydropyrene (**315**).[554a,b] The stability of **314** (indefinite lifetime as crystals under $N_2$) contrasts with the instability of various 8,16-dialkyl [2.2]metacyclophanes (**316**) which spontaneously isomerize to the corresponding *trans*-15,16-dihydropyrenes (**317**). (For example, one can observe in solution via nmr some *anti*-8,16-dimethyl[2.2]metacyclophane-1,9-diene **316**, R = $CH_3$) in addition to the dihydropyrene (**317**, R = $CH_3$), but work-up provides only the latter.[554a,b] [2.2]Metaparacyclophane-1,9-diene (**318**) was also obtained through the Stevens rearrangement–Hofmann elimination sequence (e.g., Scheme 27).[555a,b] Conformation rotation of the *meta* ring in this compound occurs quite readily ($\Delta G^{\ddagger}_{-96°} = 8.3$ kcal/mole). This is in very striking

[553a] W. E. Parham and J. K. Rinehart, *J. Am. Chem. Soc.* **89**, 5668 (1967).

[553b] W. E. Parham, D. R. Johnson, C. T. Hughes, M. K. Meilahn, and J. K. Rinehart, *J. Org. Chem.* **35**, 1048 (1970).

[554a] R. H. Mitchell and V. Boekelheide, *J. Am. Chem. Soc.* **92**, 3510 (1970).

[554b] R. H. Mitchell and V. Boekelheide, *J. Am. Chem. Soc.* **96**, 1547 (1974).

[555a] V. Boekelheide and P. H. Anderson, *Tetrahedron Lett.* p. 1207 (1970).

[555b] V. Boekelheide, P. H. Anderson, and T. A. Hylton, *J. Am. Chem. Soc.* **96**, 1558 (1974).

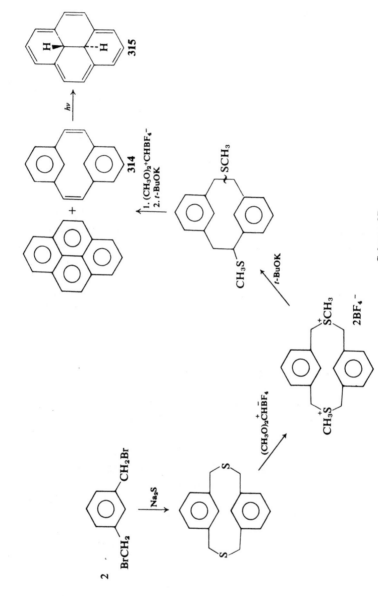

**Scheme 27**

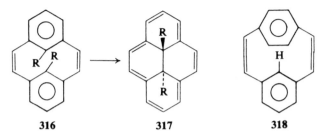

316            317            318

contrast to the high barrier ($\Delta^{\ddagger}_{140°}$ = 20.8 kcal/mole) for *meta* ring rotation in [2.2]metaparacyclophane (303). In line with x-ray structural work, it is clear that the $H_8$ atom shown in 303 must endure an appreciably deeper penetration into the $\pi$ cloud of the *para*-bridged ring than must the corresponding H atom in 318 in the corresponding conformational process.[555b] Shortening of the etheno linkage in 318 relative to the corresponding saturated linkage in 303 is more than compensated for by opening of the C=C—C angle of the former. The extreme sensitivity to steric effects of the conformational barrier in 303 and its various 8-substituted derivatives provides a subtle probe into the "size" of various groups. For example, a very large steric isotope effect ($k_D/k_H$ = 1.20) has been recorded for 303.[555c] (If we consider a lower zero-point vibrational energy for C—D than for C—H, then the former should be "shorter" and "sterically less demanding." However, one cannot separate out effects caused by differences in electron distribution in such an analysis.) Although fluorine and hydrogen are often assumed to be nearly isosteric, the *meta* ring in 8-fluoro[2.2]metaparacyclophane-1,9-diene (replace H shown in 318 by F) flips less than $10^{-11}$ times as rapidly as in the parent molecule.[555c] It is thus apparent that a very steep nonbonded potential has to be deduced. The *meta*-bridged ring in (9,10)anthraceno[2.2]-metacyclophane rotates approximately 45 times more rapidly than in meta-paracyclophane and this is, in part, attributed to relative ease of 9,10 bending in anthracene.[555c]

Cyclophane compounds having more than two bridges often exhibit novel properties. [2.2.2](1,3,5)Cyclophanetriene (319), synthesized from 1,3,5-tris-(bromomethyl)benzene using the Stevens rearrangement–Hofmann elimination sequence, manifests a unique inversion in the order of chemical shifts of its vinylic ($\tau$ 2.63) and aromatic ($\tau$ 3.76) protons as the combined result of strain and forced proximity of aromatic rings.[556a,b] The aromatic rings in [2.2.2](1,3,5)cyclophane (320) are even closer than those in [2.2]paracyclophane

[555c] S. A. Sherrod, R. L. da Costa, R. A. Barnes, and V. Boekelheide, *J. Am. Chem. Soc.* 96, 1565 (1974).

[555d] Doctoral dissertation of David Kamp, University of Oregon, June 1976; Prof. Virgil Boekelheide, personal communication to the authors.

[556a] V. Boekelheide and R. A. Hollins, *J. Am. Chem. Soc.* 92, 3512 (1970).

and also more distorted.[556a,b] The aromatic rings in **319** and **320** are constrained into chairlike conformations.[556b] Intuition as well as semiempirical (MINDO/2) studies indicate that aromatic rings prefer boatlike distortion (as in the [$n$]- and [$n.m$]paracyclophanes) to chairlike distortion.[557a] Thus, even cyclophane **321**, which could, in principle, relieve nonbonded repulsion if the central aromatic ring were to adopt a chairlike structure, maintains the all-boat geometry depicted.[557b] Photoelectron spectra of **319** and **320** are consistent

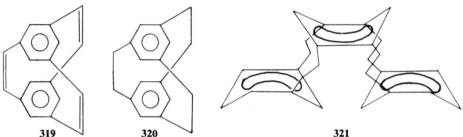

319          320                          321

with decreased nonbonded distances between aromatic rings relative to [2.2]paracyclophane.[558a] The molecule [2.2.2](1,2,4)(1,3,5)cyclophane, an isomer of **320**, is dissymmetrically twisted and has an aromatic proton exhibiting a chemical shift of 4.96τ.[558b] The distortion produced by a third ethano bridge in [2.2.2](1,2,4)cyclophane is manifested in some polyolefinic reactivity not observed for its homologue [3.2.2](1,2,4)cyclophane or in [2.2]paracyclophane.[559] The synthesis of [2.2.2.2](1,2,4,5)cyclophane (**322**, Scheme 28) very nicely illustrates the manner in which fixed geometry ("chemical tweezer" effect) in a cyclophane can introduce reaction specificity.[560a] Thus, ester groups direct chloromethylation to the pseudogeminal positions of the opposite decks, and proximity favors transannular carbene insertion reactions which form the last two ethano bridges. The compound [2.2.2](1,2,3,5)cyclophane, an isomer of **322**, is the first example of a cyclophane possessing three consecutive ethano bridges.[560b] Other examples of novel multibridged cyclophanes include **323** and **324**,[561] **325**,[562] and **326**,*

[556b] V. Boekelheide and R. A. Hollins, *J. Am. Chem. Soc.* **95**, 3201 (1973).

[557a] H. Iwamura, H. Kihara, S. Misumi, Y. Sakata, and T. Umemoto, *Tetrahedron Lett.* p. 615 (1976).

[557b] T. Umemoto, T. Otsubo, and S. Misumi, *Tetrahedron Lett.* p. 1573 (1974).

[558a] R. Boschi and W. Schmidt, *Angew. Chem., Int. Ed. Engl.* **12**, 405 (1973).

[558b] M. Nakazaki, K. Yamamoto, and Y. Miura, *J. Chem. Soc., Chem. Commun.* p. 206 (1977).

[559] E. A. Truesdale and D. J. Cram, *J. Am. Chem. Soc.* **95**, 5825 (1973).

[560a] R. Gray and V. Boekelheide, *Angew. Chem., Int. Ed. Engl.* **14**, 107 (1975).

[560b] W. Gilb, K. Menke, and H. Hopf, *Angew. Chem., Int. Ed. Engl.* **16**, 191 (1977).

[561] A. J. Hubert, *J. Chem. Soc. C* p. 13 (1967).

[562] A. J. Hubert and M. Hubert, *Tetrahedron Lett.* p. 5779 (1966).

* See Addendum.

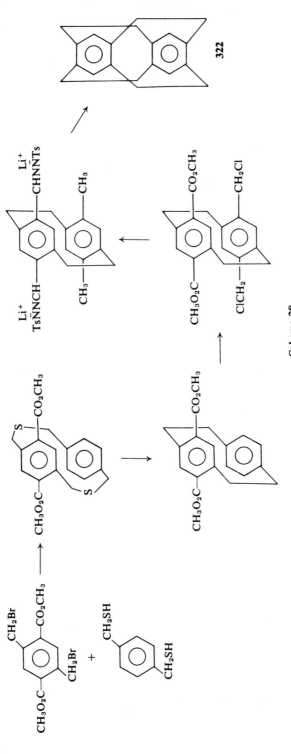

**Scheme 28**

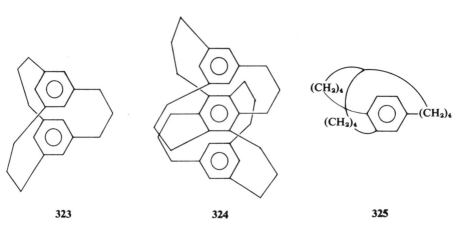

**323**                **324**                **325**

whose bridged methylene groups are forced by crowding to rotate in a concerted manner.[563] The molecule 5,6,7,8-tetrahydro-1,7-ethanonaphthalene (**327**), a [5]metacyclophane, has been claimed as the result of HF-induced (320°C) or $Pt-Al_2O_3$ catalyzed dehydrogenation of a mixture of isomeric tricyclo[5.3.1.1^{4,11}]dodecanes.[564a,b] The uv spectral properties of 1,4:2,5-bis(octamethylene)benzene (**328**; this has been referred to as "[8.8]paracyclophane" but the nomenclature is inconsistent with common usage) are consistent with the apparent twist-boat shape of its aromatic ring and the molecule's probable $D_2$ symmetry.[565a,b] A number of stacked paracyclophanes,[561,566–569] metacyclophanes,[570] and metaparacyclophanes[567a,b] have

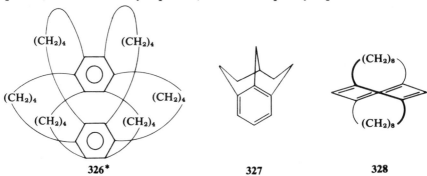

**326***                **327**                **328**

[563] R. D. Stephens, *J. Org. Chem.* **38**, 2260 (1973).

[564a] E. I. Bagrii, T. N. Dolgopolova, and P. I. Sanin, *Izv. Akad. Nauk SSSR, Ser. Khim.* p. 2648 (1973); *Chem. Abstr.* **80**, 59770f (1974).

[564b] See also *Chem. Abstr.* **82**, 97749 (1975).

[565a] M. Nakazaki, K. Yamamoto, and S. Tanaka, *Tetrahedron Lett.* p. 341 (1971).

[565b] M. Nakazaki, K. Yamamoto, and S. Tanaka, *J. Org. Chem.* **41**, 4081 (1976); see also M. Nakazaki, K. Yamamoto, and S. Tanaka, *Tetrahedron Lett.* p. 341 (1971).

[566a] D. T. Longone and H. S. Chow, *J. Am. Chem. Soc.* **86**, 3898 (1964).

[566b] D. T. Longone and H. S. Chow, *J. Am. Chem. Soc.* **92**, 994 (1970).

* See Addendum.

been reported, and their novel stereochemistry and electronic properties have been studied. A rather severe nonbonded interaction between two almost perpendicular benzene rings in **329** forces additional skewing of the two aromatic rings comprising the [2.2]paracyclophane part of this molecule.[571]

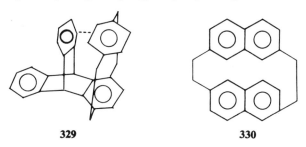

**329**                                **330**

The 2,7-bridged naphthalene ring in mixed naphthalenophane **330** has an energy barrier for rotation almost identical with that of **303**. Apparently, the destabilization resulting from simultaneous intrusion of two hydrogen atoms into the $\pi$ cloud of the other naphthalene ring is compensated for by the ability of the larger system to apportion strain among a larger number of bonds.[572] The corresponding diene flips too rapidly to be observed on the nmr time scale.[572] Another novel aspect of certain bridged polycyclic benzenoid compounds is their ability to form intramolecular excimers. While the mode of deactivation of excimer **331\*** is exclusively photochemical [Eq. (37)],[573a] [2.2](1,4)naphthalenophane excimer (**332\***) will fluoresce as well as produce **333**, a thermolabile ($t_{1/2}^{20°} = 76$ sec) molecule that reverts to starting material which in turn ultimately disappears to form stable dibenzoequinene

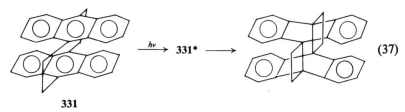

$$\xrightarrow{h\nu} \mathbf{331^*} \longrightarrow \tag{37}$$

**331**

[567a] T. Otsubo, S. Mizugami, Y. Sakata, and S. Misumi, *Tetrahedron Lett*. p. 4803 (1971).

[567b] T. Otsubo, Z. Tozuka, S. Mizogami, Y. Sakata, and S. Misumi, *Tetrahedron Lett*. p. 2927 (1972).

[568] H. Hopf, *Angew. Chem., Int. Ed. Engl*. **11**, 419 (1972).

[569] M. Nakazaki, K. Yamamoto, and J. Tanaka, *J. Chem. Soc., Chem. Commun*. p. 433 (1972).

[570] T. Umemoto, T. Otsubo, Y. Sakata, and S. Masamune, *Tetrahedron Lett*. p. 593 (1973).

[571] D. T. Longone and G. R. Chipman, *Chem. Commun*. p. 1358 (1969).

[572] V. Boekelheide and C.-H. Tsai, *Tetrahedron* **32**, 423 (1976).

[573a] H. H. Wasserman and P. M. Keehn, *J. Am. Chem. Soc*. **91**, 2374 (1969).

**334**
**Scheme 29**

(**334**).[573a,b] The radical anion of the diene of bis(2,7-naphthalenophane) closes to hydrocoronene derivatives.[574] The "super" cyclophanes [2.2](2,7)-pyrenophane (**335**) and its related diene have been reported.[575] The molecules formed by intramolecular cyclophane addition [products of Eq. (37) and product **333** in Scheme 29] have very long C—C bonds (see Section 4.E).

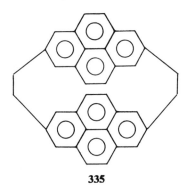

**335**

While even the most "routine" cyclophanes appear to have novel features, we nonetheless describe two of the most novel of novelties: (a) the triply clamped helical bis(triphenylmethane) system **336** (and its related triene) exhibits the characteristic propellerlike nature of each triphenylmethyl unit,

[573b] G. Kaupp and I. Zimmerman, *Angew. Chem., Int. Ed. Engl.* **15**, 441 (1976).
[574] C. Eischenbroich, F. Gerson, and J. A. Reiss, *J. Am. Chem. Soc.* **99**, 60 (1977).
[575] T. Umemoto, S. Satani, Y. Sakata, and S. Misumi, *Tetrahedron Lett.* p. 3159 (1975).

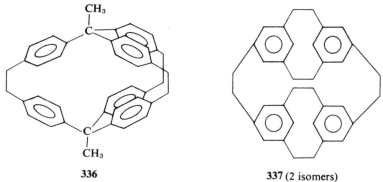

**336**                                              **337** (2 isomers)

which are in turn mutually displaced[576]; (b) the [2.2]paracyclo([2.2]meta-
cyclophane) (**337**) (two isomers) are the first examples of a "cyclophane
within a cyclophane," and these molecules can be transformed into **335** (note
analogy to chemistry in Scheme 27).[577] We also wish to note the generation
and internal trapping of **338**,[578] whose bridges are large enough to allow
aryne rotation unlike the bisethano-bridged analogue, which reacts inter-
molecularly.[571,579]

$$\text{(38)}$$

**338**

Severe distortion of the aromatic ring in **339**, which has a five-carbon *meta*
bridge, endows this compound with spectral and chemical properties strongly
suggestive of an olefin.[580] While model compound **340** seems to be aromatic[581]
(the aromatic ring of a carboxyl derivative is somewhat bent[582]), its benzologue
**341** appears be somewhat polyolefinic.[583] The middle ring in 7*bh*-indeno-

[576] F. Vogtle and G. Hohner, *Angew. Chem., Int. Ed. Engl.* **14**, 497 (1975).

[577] R. H. Mitchell, R. J. Carruthers, and J. C. M. Zwinkels, *Tetrahedron Lett.* p. 2585
(1976).

[578] D. T. Longone and J. A. Gladysz, *Tetrahedron Lett.* p. 4559 (1976).

[579] H. J. Reich and D. J. Cram, *J. Am. Chem. Soc.* **91**, 3527 (1969).

[580] H. Rapoport and G. Smolinsky, *J. Am. Chem. Soc.* **82**, 1171 (1960).

[581] H. Rapoport and J. Z. Pasky, *J. Am. Chem. Soc.* **78**, 3788 (1956).

[582] H. J. Lindner and P. Eilbracht, *Chem. Ber.* **106**, 2268 (1973).

[583] B. L. McDowell, H. Rapoport, and G. Smolinsky, *J. Am. Chem. Soc.* **84**, 3531
(1962).

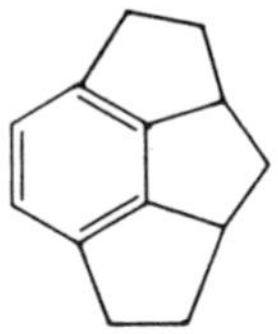

339 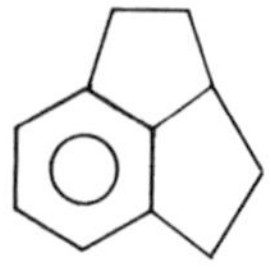    340 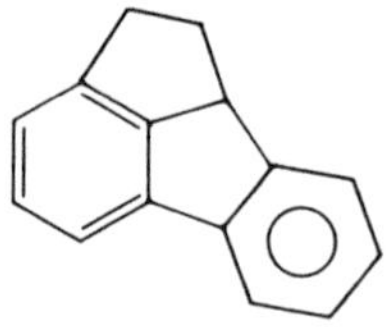    341

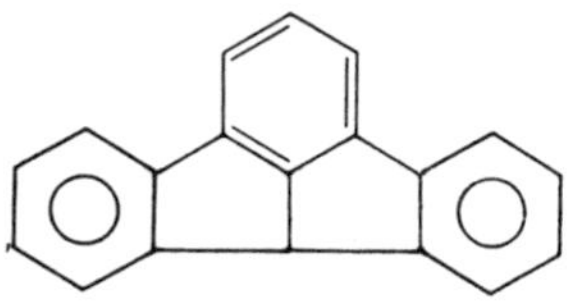

342 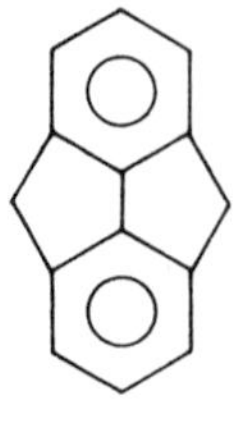    343

[1,2,3-*jk*]fluorene (**342**) is distorted and undergoes relatively facile Diels–Alder reactions.[584] The compound 4,8-dihydrodibenzo[*ck*,*gh*]pentalene (**343**) is, surprisingly, planar, and is estimated to have about 66 kcal/mole of strain.[585a–c]

### 3.H.4    CYCLOPHANES OF NONBENZENOID AROMATICS

During the last few years there has been a sudden surge of interest in nonbenzenoid aromatic cyclophanes, and this as well as the study of (4*n*)(4*n*)- and (4*n*)(4*n* + 2)cyclophanes appear to be major areas of future research interest. Thus, although 1,3-nonamethylenecyclopentadienide ("[9]cyclopentadienidophane") is stable[586] (it is unstrained), efforts to reduce (1,4)-pentamethylenecyclooctatetraene to the corresponding bridged dianion (**344**) failed,[587] as did attempts to make the bridged tropylium ion **345**.[588] While 2,7-nonamethylenetropone can be protonated by $CF_3COOH$ to form tropylium ion **346**, 2,7-hexamethylenetropone (**347**) could not be so protonated.[588] The yellow salt of [2.2](1,4)tropylioparacyclophane (**348**) has been

[584] H. Dietrich, D. Bladavski, M. Grosse, K. Roth, and D. Rewicki, *Chem. Ber.* **108**, 1807 (1975).

[585a] B. M. Trost and P. L. Kinson, *J. Am. Chem. Soc.* **92**, 2591 (1970).

[585b] B. M. Trost, P. L. Kinson, C. A. Maier, and I. C. Paul, *J. Am. Chem. Soc.* **93**, 7275 (1971).

[585c] B. M. Trost and P. L. Kinson, *J. Am. Chem. Soc.* **97**, 2438 (1975).

[586] S. Bradamante, A. Marchesini, and G. Pagani, *Tetrahedron Lett.* p. 4621 (1971).

[587] See footnote 188 in L. A. Paquette, *Tetrahedron* **31**, 2285 (1975).

[588] S. Hirano, T. Hiyama, and H. Nozaki, *Tetrahedron* **32**, 2381 (1976).

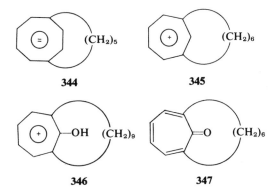

isolated,[589,590] as has the greenish-blue *anti*-[2,2](2,6)azulenophane (**349**),[591,592] a substance anticipated to be a particularly good electrical conductor. There is some evidence for the generation of [2.2](1,4)tropenidoparacyclophane (**350**)[593a] (see comments about its stability in the next section). We might also mention here capped porphyrins as examples of nonbenzenoid aromatic cyclophanes.[593b,c]

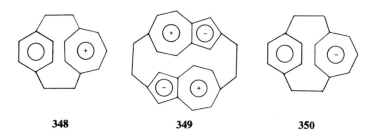

In the concluding section of this chapter we will examine the perturbations produced by bridging in the ordering of orbitals of simple aromatic species. In addition, "secondary orbital" effects between nonbonded rings will be examined with particular attention to nonbenzenoid aromatic species.

Suitable bridging and bending of a benzene ring will break orbital degeneracies which occur in the undistorted ring. In Fig. 3.2 are displayed π molecular

[589] H. Horita, T. Otsubo, Y. Sakata, and S. Misumi, *Tetrahedron Lett.* p. 3899 (1970).

[590] J. G. O'Connor and P. M. Keehn, *J. Am. Chem. Soc.* **98**, 8446 (1976).

[591] R. Luhowy and P. M. Keehn, *Tetrahedron Lett.* p. 1043 (1970).

[592] N. Kato, Y. Fukazawa, and S. Ito. *Tetrahedron Lett.* p. 2045 (1976).

[593a] Prof. Philip M. Keehn, personal communication to Joel F. Liebman.

[593b] J. Almog, J. E. Baldwin, R. L. Dyer, and M. Peters, *J. Am. Chem. Soc.* **97**, 226 (1975).

[593c] N. E. Kagan, D. Mauzerall, and R. B. Merrifield, *J. Am. Chem. Soc.*, **99**, 5484 (1977).

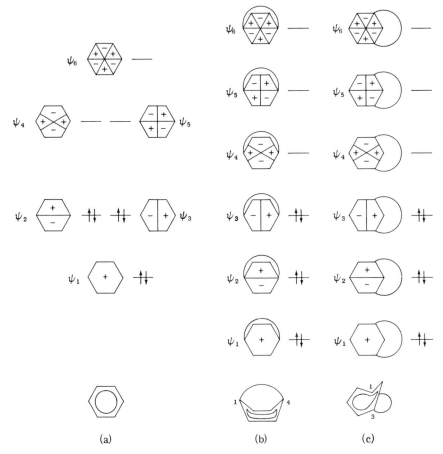

**Figure 3.2.** (a) The $\pi$ molecular orbitals of benzene. (b) The $\pi$ molecular orbitals of a small paracyclophane. (c) The $\pi$ molecular orbitals of a small metacyclophane.

orbitals of benzene, 1,4-bridged benzene,[594] and 1,3-bridged benzene (obviously, the mere fact of substitution breaks degeneracies also). Analysis of the esr spectrum of the metastable triplet of [7]paracyclophane fits the picture outlined in Fig. 3.2(b), because they both indicate greater separation of the unpaired electron (onto the 1,4 carbons) than in planar $p$-dialkyl-benzenes.[594] (As one of the two LUMO's for benzene is 1,4-bonding, we wonder if one-electron oxidation of the radical anion of [7]paracyclophane would form the Dewar benzene).

[594] E. Wasserman, R. S. Hutton, and F. B. Bramwell, *J. Am. Chem. Soc.* **98**, 7429 (1976).

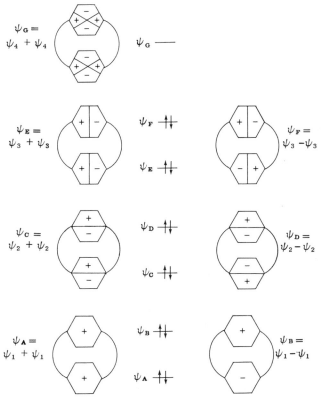

**Figure 3.3.** The occupied $\pi$ molecular orbitals and the lowest unoccupied $\pi$ molecular orbital of a generalized [m.n]paracyclophane.

We may consider orbital interactions between (planar) benzene rings in an [n.m]paracyclophane in the manner depicted in Fig. 3.3. Related but more rigorous descriptions of the bonding in [2.2]paracyclophane have appeared, with emphasis on ionization and substitution properties.[595] If distortion of the rings is taken into account, then orbital degeneracies are broken (perhaps $\psi_4$ and $\psi_5$ now become comparable in energy), but the conclusions to be drawn are unchanged. Six orbitals in Fig. 3.3 are occupied, and if we recall that antibonding interactions are usually more antibonding than corresponding bonding interactions are bonding, we conclude that there is a repulsive interaction between the two aromatic rings. This is consistent with recent evidence[596] that shows that gaseous benzene dimer is polar, and we note that

[595] E. Heilbronner and J. O. Maier, *Helv. Chim. Acta* **57**, 151 (1974).

[596] K. C. Janda, J. C. Hemminger, J. S. Winn, S. E. Novick, S. J. Harris, and W. Klemperer, *J. Chem. Phys.* **63**, 1419 (1975).

a structure having the rings perpendicular is consistent with a recent model for directionality of hydrogen bonds.[597] It also partially explains the skewing of benzenes previously noted for [2.2]paracyclophane (**301**), because this would decrease antibonding interactions between these rings. One should note that the rings are well under (0.3–0.4 Å), the van der Waals contact distance; in addition, skewing of the rings reduces nonbonded repulsions in the two ethano bridges.

With the orbital scheme of Fig. 3.3 in mind, we can anticipate some stabilization to be achieved upon removing an electron from the $\psi_3-\psi_3$ orbital (or even from the lower energy $\psi_2-\psi_2$ orbital). This may be accomplished either by ionization to the radical cation or by complexation with a strong acid. Not surprisingly, the ionization potentials for paracyclophanes are low,[558a] and strong complexes with tetracyanoethylene are easily formed.[597] Thus, there would appear to be some stabilization to be achieved in going from a $(4n + 2)\rho(4n + 2)$cyclophane to the $(4n + 2)\rho(4n)$ species. (In our nomenclature, $\rho$ connotes two rings, and $4n$, $4n + 1$, $4n + 2$ are the number of $\pi$ electrons in each of the rings.) Even greater stabilization might be anticipated for a $(4n + 2)\rho(4n)$ system. Electron-pair removal may be accomplished by protonation and if the ion is considered to have some homoaromatic character, a straightforward comparison can be drawn with other cyclophanes. Protonated [2.2]paracyclophane (**351**) is highly stable,[558a,b] and might be considered to include interaction between benzene and homocyclopentadienyl cation, as in **351a**. The extreme reactivity of [*n.m*]paracyclophanes to electrophilic substitution[599] may also be explained in this manner. (In writing the structures of protonated paracyclophanes, we note that Cram's initial investigations into this realm were the result of his interest in reactions of other classes of protonated aromatic species, in particular, proposed intermediates in the benzidine rearrangement.[600]) As we noted in the last section, there is some evidence for the generation of a $(4n)\rho(4n + 2)$ system,

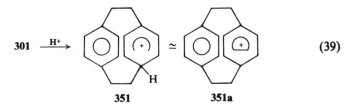

$$\mathbf{301} \xrightarrow{\ \mathrm{H}^+\ } \qquad \simeq \qquad\qquad (39)$$

**351**               **351a**

[597] P. Kollman, *J. Am. Chem. Soc.* **94**, 1837 (1972).

[598a] D. J. Cram, N. L. Allinger, and H. Steinberg, *J. Am. Chem. Soc.* **76**, 6132 (1954).

[598b] D. J. Cram and R. H. Bauer, *J. Am. Chem. Soc.* **81**, 5971 (1959).

[599] D. J. Cram, W. J. Wechter, and R. W. Kierstead, *J. Am. Chem. Soc.* **80**, 3126 (1958).

[600] Prof. Donald J. Cram, personal communication to Joel F. Liebman.

**350**,[593a] in which interaction with benzene should somewhat stabilize the antiaromatic tropylide anion.

All the results we have described are also compatible with the Goldstein–Hoffmann topological description of (three-dimensional) aromaticity.[196] That is, we consider interaction between the highest occupied molecular orbital of one ring, HOMO, and the lowest unoccupied molecular orbital of the other ring, LUMO. When the HOMO and LUMO are of the same symmetry, stabilization is expected; if they are of different symmetry, no stabilization may be expected. The HOMO of a benzene ring is $\psi_2$ and $\psi_3$ [Fig. 3.2(a), recall the degeneracy]. The LUMO of a homocyclopentadienyl cation is essentially $\psi_3$, by analogy to both the open-chain pentadienyl and (cyclic) cyclopentadienyl cations. As such, stabilization is expected. Odd-electron species, such as the above-mentioned paracyclophane radical cations, are harder to describe. We suspect that a radical will mimic the compound with one more electron but be neither as stabilized nor as destabilized as the corresponding electron-rich species. Therefore, the radical cation should be stabilized relative to the neutral $(4n + 2)\rho(4n + 2)$paracyclophane. This argument may be used to explain the facile formation of paracyclophane radical anions,[601–603b] a most surprising result when discussed solely in terms of $\pi$-electron repulsion effects. Indeed, [2.2]paracyclophane is electrochemically reduced more readily than benzene.[602b] (In the *vapor phase*, benzene radical anion is unbound.[602c]) We note that the thermal instability of these radical anions has been ascribed to the formation of the dianion,[603a,b] an even more surprising species that subsequently decomposes by simple $CH_2$—$CH_2$ bond cleavage. (The cyclophane dianion must be a singlet if it decomposes to the open-chain dianion by a spin-allowed process.) We note that [2.2]paracyclophane undergoes homolytic aromatic substitution more rapidly than benzene.[604]

We can reconcile an apparent paradox in [2.2]paracyclophane (PC) chemistry via the HOMO–LUMO aromaticity analysis. The solvolytic formation of PC—$CH_2{}^+$ ion is very facile[605a,b] and can be rationalized in terms of a $(4n + 2)\rho(4n)$cyclophane in the manner already described for **351**

[601] A. Ishitani and S. Nagakura, *Mol. Phys* **12**, 1 (1967).

[602a] F. Gerson and W. B. Martin, Jr., *J. Am. Chem. Soc.* **91**, 1883 (1969).

[602b] F. Gerson, H. Ohya-Nichiguchi, and C. Wydler, *Angew. Chem., Int. Ed. Engl.* **15**, 552 (1976).

[602c] K. D. Jordan, J. A. Michejda, and P. D. Burrow, *J. Am. Chem. Soc.* **98**, 1295 (1976).

[603a] J. M. Pearson, D. J. Williams, and M. Levy, *J. Am. Chem. Soc.* **93**, 5478 (1971).

[603b] D. J. Williams, J. M. Pearson, and M. Levy, *J. Am. Chem. Soc.* **93**, 5483 (1971).

[604] S. C. Dickerman and L. Milstein, *J. Org. Chem.* **32**, 852 (1967).

[605a] D. J. Cram and L. A. Singer, *J. Am. Chem. Soc.* **85**, 1075 (1963).

[605b] D. J. Cram and F. L. Harris, Jr., *J. Am. Chem. Soc.* **89**, 4642 (1967).

(or **351a**). However, formation of the PC—O$^-$ species by deprotonation of the phenol is seemingly inhibited, "primarily reflecting the loss of resonance stabilization of the anion due to the bent nature of the adjacent benzene ring."[606] There is no reason, *a priori*, to assume that resonance in PC—O$^-$ is any more "inhibited" than it is in PC—CH$_2$$^+$. A simple explanation would be that PC—O$^-$ may be described as a $(4n + 2)\rho(4n + 2)$cyclophane which does not enjoy the special stabilization found in PC—CH$_2$$^+$, a $(4n)\rho(4n + 2)$-cyclophane.

The HOMO–LUMO analysis allows prediction of a stabilizing interaction in a $(4n)\rho(4n + 2)$cyclophane and destabilization in the $(4n + 2)\rho(4n + 2)$ system when $n$ takes the same value on both sides of $\rho$. Thus, **350** is predicted to gain stability relative to **348**. That is not to say that **350** ($6\rho7^-$ system) will be intrinsically more stable than **348** ($6\rho7^+$ system). The tropylium ion ($7^+$) is aromatic and highly stable, while tropylide anion ($7^-$) is antiaromatic and very unstable. However, the reduction of **348** to **350** should be considerably easier than that of the corresponding monocyclic seven-membered compounds. (A parallel study of analogues of [$n$]paracyclophane would also be of interest because a corresponding effect is not predicted.) Analogous analyses can also be done for polyhedral boron hydride anions (e.g., B$_6$H$_6$$^{2-}$, B$_{10}$H$_{10}$$^{2-}$).[607] Another interesting result of the HOMO–LUMO analysis is our prediction that there should be greater stabilization in $(4n)\rho(4n)$ cyclophane than in $(4n)\rho(4n + 2)$ systems. Thus, we note the possibility of generating

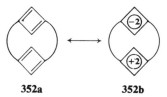

**352a**          **352b**

(as yet unknown) $(4n)\rho(4n)$cyclophanes such as **352** and **353**. Compound **322** is a logical precursor of **353**, but [2.2.2.2](1,3,5,7)cyclooctatetraenophane thus far eludes synthesis.[608] Lest the reader dismiss these compounds out of hand, one should recall that cyclobutadiene largely dimerizes to form the *syn*, i.e., face-to-face, product.[609] Even more striking is the fact that tri-*tert*-

[606] B. E. Norcross, D. Becker, R. I. Cukier, and R. M. Schultz, *J. Org. Chem.* **32**, 220 (1967).

[607] J. F. Liebman, unpublished results.

[608] Prof. Virgil Boekelheide, personal communication to the authors. Attempts at synthesis of this compound have been motivated by independent considerations analogous to the above HOMO–LUMO analysis.

[609] See a discussion of this specific type of "steric attraction" in R. B. Woodward and R. Hoffmann, "The Conservation of Orbital Symmetry," pp. 145–151. Academic Press, New York, 1970.

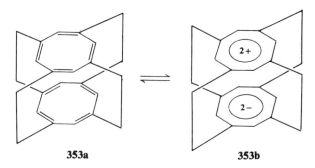

353a                                353b

butylcyclobutadiene dimerizes *syn*, with activation parameters ($\Delta S^{\ddagger} = -47 \pm 3$ eu; $\Delta H^{\ddagger} = 7.2 \pm 0.9$ kcal/mole) which suggest a highly ordered transition state in which stabilizing electronic factors more than compensate for steric repulsion.[610] (Recall also that substituted cyclobutadiene dication and dianion as well as cyclooctatetraene dication and dianion have all been observed at least spectroscopically, Section 3.E.3). It can be seen that the olefinic and zwitterionic two-cyclobutadiene aggregates in **352a** and **352b** are valence-bond equivalents of the molecular orbitals used in cyclobutadiene dimerization in a face-to-face orientation.[611]

In Chapter 5 we will discuss some heteroanalogues of distorted aromatic molecules. For the present we will note that heteroanalogues of [6]paracyclophane and [7]paracyclophane have been cited as possible structures for doubly charged molecular anions from $p$-$NO_2$-$C_6H_4$-$(CH_2)_n$-$CO_2R$ (where $n = 3$ or 4),[612] although other structures are possible, including a spiro nitronate ion analogous to the spiro compound of Scheme 26.[613]

[610] G. Maier and W. Saver, *Angew. Chem., Int. Ed. Engl.* **14**, 648 (1975).

[611] S. D. Frans, R. L. Kellert, and J. F. Liebman, *Int. Symp. Chem. Strained Rings*, Binghamton, New York, 1977.

[612] J. H. Bowie and B. J. Stapleton, *J. Am. Chem. Soc.* **98**, 6480 (1976).

[613] Steven D. Sprouse and Joel F. Liebman, unpublished results.

## CHAPTER 4

# Polycycles: Aesthetics, Rearrangements, and Topology

Polycyclic hydrocarbons may be viewed as being comprised of some of the "building blocks" (e.g., small cyclic and bicyclic molecular frameworks) described in Chapter 3. Thus, adamantane's strain is not too different from that attributed to its four "cyclohexane faces," and the strain in prismane and cubane are remarkably well reproduced as a sum of the strain of the "faces," respectively. However, one would not expect such a concept to hold universally. Numerous polycyclic hydrocarbons do indeed depart from the properties expected on the basis of the building block approach: tetrahedrane (Chapter 3) is a striking example, and *endo,endo*-tetracyclo[6.2.1.1.$^{3,6}0^{2,7}$]dodecane and lepidopterene, both to be discussed in Section 4.E, also exhibit properties not explicable in terms of a simple summation of "parts." Later in this chapter we will very briefly examine strikingly different topological methods of viewing such molecules in order to see what insights are to be gained from such approaches.

Polycyclic hydrocarbons are far too numerous for us to attempt complete coverage. Furthermore, a plethora of rearrangements between isomers is possible. We have selected only certain species and have discussed rearrangement pathways (most notably for adamantanes) in a manner aimed at illustrating the importance of molecular strain.

## 4.A   Adamantanes and Higher Homologues

### 4.A.1   ADAMANTANE

Many of the exotic molecules cited in this book have been created under markedly unearthly conditions with a view toward illustrating or testing the

178

limits of Nature's rules. Their syntheses almost seem to entail a matching of wits with Nature. In contrast, the discovery of acid-catalyzed rearrangements of various isomeric $C_{10}H_{16}$ molecules to adamantane (**1**), as well as rearrangements of a spectrum of hydrocarbons, including even nujol and cholesterol, to substituted adamantanes, have revealed a very slow and silent but irrevocable natural process: "everything rearranges to adamantane" (see, however, qualifications below). (Recent reviews of adamantane chemistry include Fort,[1a] McKervey,[1b] and Engler and Schleyer.[1c]) Indeed, the initial report[2] in 1933 of adamantane as a minor constituent of Czechoslovakian petroleum now may be rationalized as the result of rearrangements of decaying organic matter catalyzed by naturally occurring Lewis acids (e.g., aluminum silicates).[1a-c] The study of adamantane and related molecules also has modified some very basic concepts in strain, conformational analysis, carbonium-ion stabilities, and rearrangement pathways.

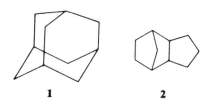

**1**                    **2**

The first synthetic adamantane was produced in 1941 in a very low-yield multistep sequence.[3] Despite subsequent improvements, "rational" syntheses continued to provide but minute quantities of adamantane. In 1957, Schleyer reported $AlCl_3$- or $AlBr_3$-catalyzed rearrangement of tetrahydrodicyclopentadiene (**2**, *exo* and *endo* isomers) to adamantane in yields subsequently maximized to about 20%.[4a,b] The driving force for rearrangement is relief of strain (note the norbornane framework having ca. 17 kcal/mole strain in **2**). While the rearrangement mechanism will be examined later in this section, the question of the stability of adamantane should be discussed at this point.

Originally, adamantane was assumed to be strain free, as its carbocyclic network appeared to impose "ideal" tetrahedral angles and bond lengths. In fact, the molecule does possess tetrahedral ($T_d$) symmetry, virtual tetrahedral

[1a] R. C. Fort, Jr., "Adamantane, the Chemistry of Diamond Molecules." Dekker, New York, 1976.

[1b] M. A. McKervey, *Chem. Soc. Rev.* **3**, 479 (1975).

[1c] E. M. Engler and P. von R. Schleyer, *MTP Int. Rev. Sci., Org. Chem. Ser. 1* **5**, 239 (1973).

[2] S. Landa and V. Machacek, *Collect. Czech. Chem. Commun.* **5**, 1 (1933).

[3] V. Prelog and R. Seiwerth, *Chem. Ber.* **74**, 1644, 1769 (1941).

[4a] P. von R. Schleyer, *J. Am. Chem. Soc.* **79**, 3292 (1957).

[4b] P. von R. Schleyer and M. M. Donaldson, *J. Am. Chem. Soc.* **82**, 4645 (1960).

angles, and bond lengths of 1.54 Å.[5-7] However, one would expect perfect tetrahedral angles only for $CX_4$ molecules such as methane and carbon tetrachloride, and almost perfect angles for diamond. Thus, the imposition of near tetrahedral angles is actually one source of adamantane's strain. Adamantane is composed of three chair cyclohexane rings (or equivalently, four chair cyclohexane faces; for schemes on ring count, see Section 4.F). If one assumes, as in the past, that cyclohexane is strainless, then the same might be true of adamantane. However, cyclohexane is no longer considered to be strainless, and it is calculated to be some 1.2–1.4 kcal/mole higher in enthalpy than the strain-free model compound.[8,9] These values are obtained through use of the Allinger and Schleyer force fields, respectively, which differ somewhat in that the former assumes harder H---H nonbonded repulsions while the latter a harder nonbonded C---C potential (see brief discussion of molecular mechanics in Chapter 1). Strain in cyclohexane is considered to result chiefly from nonbonded interactions which relatively unconstrained acylic molecules can relieve. Adamantane, being more rigid than cyclohexane and less capable of minimizing nonbonded repulsions, is more strained than the three cyclohexane rings (or four cyclohexane "faces") comprising it. For example, molecular-mechanics calculations based upon the Schleyer force field find that adamantane has 6.87 kcal/mole of strain relative to a "strain-free" model.[9] The most recent published determination of adamantane's heat of formation leads to the conclusion that the above force field overestimates this molecule's thermochemical stability by 0.74 kcal/mole (the Allinger force field leads to a value some 2.06 kcal/mole lower than that found experimentally).[10] Thus, combination of Schleyer's strain-free increments with McKervey's experimental enthalpy determination allows an estimate of about 7.6 kcal/ mole for the strain in adamantane. Another feature of adamantane's rigidity is the greater equatorial preferences exhibited by its substituents compared to those in the cyclohexane series. For example, the enthalpy difference of 2.5 kcal/mole between 2-methyladamantane (3, methyl group axial in one ring) and 1-methyladamantane (4, methyl group equatorial) is greater than the equatorial preference (1.7 kcal/mole) of a methyl group attached to cyclohexane.[1a]

[5] W. Norwacki and K. Hedberg, *J. Am. Chem. Soc.* **70**, 1497 (1948).

[6] J. Donohue and S. J. Goodman, *Acta Crystallogr.* **22**, 342 (1967).

[7] I. Hargittai and K. Hedberg, *J. Chem. Soc., Chem. Commun.* p. 1499 (1971).

[8] N. L. Allinger, M. T. Tribble, M. A. Miller, and D. W. Wertz, *J. Am. Chem. Soc.* **93**, 1637 (1971).

[9] E. M. Engler, J. D. Andose, and P. von R. Schleyer, *J. Am. Chem. Soc.* **95**, 8005 (1973).

[10] T. Clark, T. McO. Knox, H. Mackle, M. A. McKervey, and J. J. Rooney, *J. Am. Chem. Soc.* **97**, 3835 (1975).

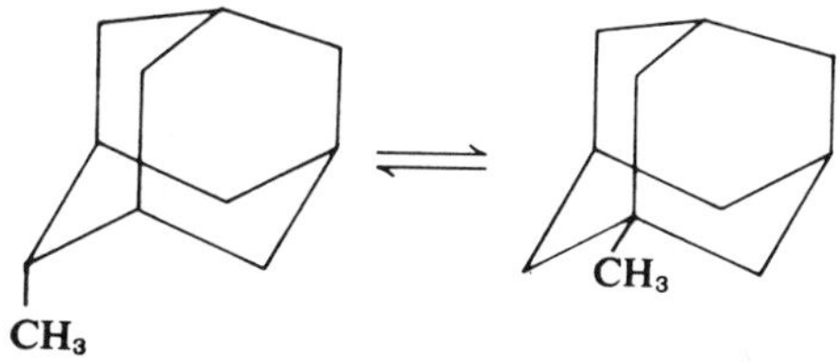

$$\text{CH}_3 \quad \mathbf{3} \qquad \rightleftharpoons \qquad \text{CH}_3 \quad \mathbf{4} \tag{1}$$

Although adamantane is strained, its highly symmetric carbocyclic framework rather evenly distributes deformation. Furthermore, it is certainly the most stable $C_{10}H_{16}$ isomer, hence the ubiquity of its occurrence in acid-catalyzed rearrangements. This is easy to demonstrate in a generalized manner. First, comparison of the enthalpies of formation of cyclohexane ($-29.4$ kcal/mole) and 2,3-dimethyl-2-butene ($-15.9$ kcal/mole) clearly indicates preference for an unstrained alicyclic ring over an olefinic linkage as an unbiased unit of unsaturation. Second, comparison of the enthalpies of formation of n-pentane ($-35.10$ kcal/mole), isopentane ($-36.85$ kcal/mole), and neopentane ($-40.27$ kcal/mole) illustrates the thermochemical stability of highly branched structures, assuming that strain does not enter the picture. Adamantane thus has three relatively unstrained cyclohexane rings and a high degree of branching (there are no 4° carbons in this molecule because their inclusion in a nonolefinic $C_{10}H_{16}$ would entail considerable strain). A nice illustration of the importance of branching is afforded by the acid-catalyzed rearrangement of *trans,syn,trans*-perhydroanthracene (5) to 1,3,5,7-tetra-methyladamantane (6).[1b,11] Strain in the two molecules is about equal, but

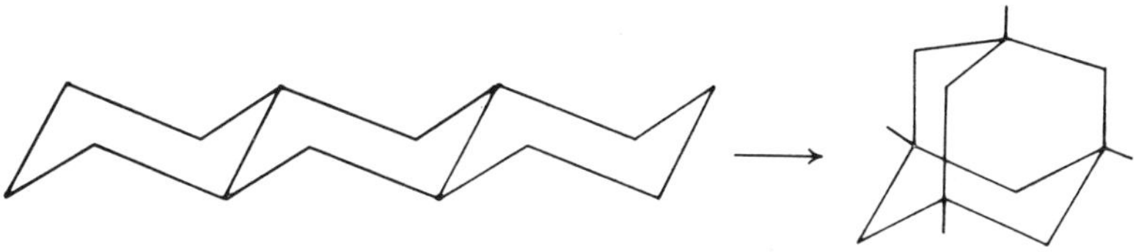

$$\mathbf{5} \qquad \longrightarrow \qquad \mathbf{6} \tag{2}$$

the reaction is exothermic by more than 10 kcal/mole because **6** is more highly branched and thus more stable.

Although rearrangements can be initiated in the presence of simple Lewis acids in a multiphase manner and even in the vapor phase in the presence of catalysts, the most effective means for their induction involve so-called "sludge catalysts." Typical sludge catalysts include cyclohexane, $AlCl_3$, and

[11] A. Schneider, R. W. Warren, and E. J. Janoski, *J. Org. Chem.* **31**, 1617 (1966).

HCl, or *tert*-butylbromide and AlBr$_3$.[12,13] Their chief virtues are twofold: first, they allow hydrocarbon to share the same phase as inorganic catalyst; second, they provide a steady-state concentration of carbonium ions necessary for initiation of the rearrangement as well as for the numerous intermolecular hydride transfers which are assumed to occur (see below). Use of sludge catalysts will thus often allow the products of a reaction to be thermodynamically controlled. However, the yield of adamantane is very dependent upon the choice of starting material, and the nature of the rearrangement pathway will determine whether intermediates may be observed and even whether conditions of thermodynamic control will in fact prevail. Thus, as noted previously, isomerization of the isomeric tetrahydrodicyclopentadienes (**2**) produces relatively low yields of adamantane at the relatively high temperatures required for rearrangement. On the other hand, twistane (**7**),[14] perhydrotriquinacene (**8**),[15] protoadamantane (**9**),[16,17] and hydrocarbons **10** and **11**[18] all rearrange rapidly and quantitatively to adamantane. The reasons for this dichotomy in behavior became apparent in a detailed discussion of the mechanism of rearrangement.

The rearrangement of tetrahydrocyclopentadiene (**2**) to adamantane was

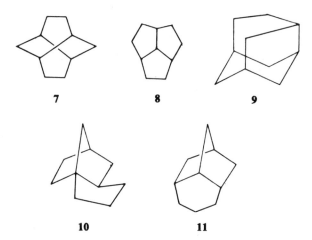

|   **7**   |   **8**   |   **9**   |
|:---:|:---:|:---:|

|   **10**   |   **11**   |
|:---:|:---:|

[12] A. Schneider, R. W. Warren, and E. J. Janoski, *J. Am. Chem. Soc.* **86**, 5365 (1964).

[13] V. Z. Williams, Jr., P. von R. Schleyer, G. J. Gleicher, and L. B. Rodewald, *J. Am. Chem. Soc.* **88**, 3862 (1966).

[14] H. W. Whitlock and M. W. Siefken, *J. Am. Chem. Soc.* **90**, 4929 (1968).

[15] L. A. Paquette, G. V. Meehean and S. J. Marshall, *J. Am. Chem. Soc.* **91**, 6779 (1969).

[16] B. R. Vogt, *Tetrahedron Lett.* p. 1575 (1968).

[17] J. E. Baldwin and W. D. Fogelsong, *J. Am. Chem. Soc.* **90**, 4303 (1968).

[18] E. M. Engler, A. Farcasiu, A. Sevin, J. M. Cense, and P. von R. Schleyer, *J. Am. Chem. Soc.* **95**, 5769 (1973).

initially thought to commence with a 1,3-alkyl shift[4] which still lacks precedent. A substantial advance in the understanding of this mechanism was achieved by Whitlock and Siefken.[14] These authors' analysis assumed the general nature of the reaction to be intermolecular hydride transfers and (intramolecular) 1,2-alkyl shifts, as depicted in Eq. (3). They generated a set

$$R_1 \longrightarrow R_1{}^+ \longrightarrow R_2{}^+ \longrightarrow R_2 \longrightarrow R_2{}^{+\prime} \longrightarrow$$
$$R_3{}^+ \longrightarrow R_3 \longrightarrow \text{etc.} \qquad (3)$$

of 16 tricyclodecanes interconnected by 1,2-alkyl shifts, omitting structures containing three- or four-membered rings as well as alkylated derivatives. The latter assumption implicitly omitted 1° carbonium ions, which should be formed more slowly than the rate of rearrangement. The level of approximation used did not consider the relative enthalpies of the tricyclodecanes and essentially considered all members of the manifold and interconversion pathways to be equal in energy and probability, respectively. The shortest pathway between twistane and adamantane was seen to involve only two steps, while the shortest pathway between tetrahydrodicyclopentadiene and adamantane involved five steps. However, proximity should play no quantifiable role, and in any case the authors assumed that many pathways might operate simultaneously. At the next level of sophistication, Whitlock and Siefken omitted all bridgehead carbonium ions (see Section 3.B.2), and came to a striking realization. The $C_{10}H_{16}$ manifold split into two subsets. The smaller one contained tetrahydrodicyclopentadiene, which was no longer connected via 1,2-alkyl shifts to adamantane, a member of the larger subset. Hydrocarbons 7–9 and 11 were also in the adamantane subset, while 10 was a member of the smaller subset. Here, apparently, was the reason for the difficulty in rearrangement of 2 relative to twistane (8). Relaxation of this restriction to include relatively stable bridgehead carbonium ions including 13 provided the rearrangement pathway depicted in Scheme 1, which could be one of a number of possible routes. Schleyer and co-workers[18] quantified these considerations by employing molecular mechanics to calculate enthalpies of formation of tricyclodecanes and the corresponding carbonium ions. They systematically examined sequential 1,2-alkyl shifts to discover the lowest-energy pathway. The calculations supported the sequence of Scheme 1. One difference that emerged was that the relatively slow rate of "adamantanization" of 2 was now seen to be the result of an initial endothermic and presumably rate-determining conversion to 12 (Scheme 1), rather than a consequence of the intermediacy of bridgehead carbonium ion 13. In fact, hydrocarbon 10 was seen to rearrange rapidly to adamantane despite the intermediacy of 13. At $-10°C$ this molecule initially isomerized to 11 and subsequently to protoadamantane (9), both of which could be monitored spectroscopically, prior to conversion to adamantane. No intermediates are

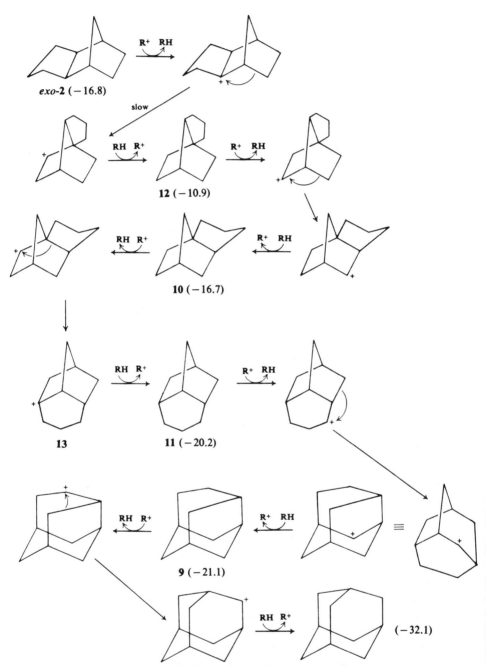

**Scheme 1.** Calculated enthalpies of formation are in parentheses.

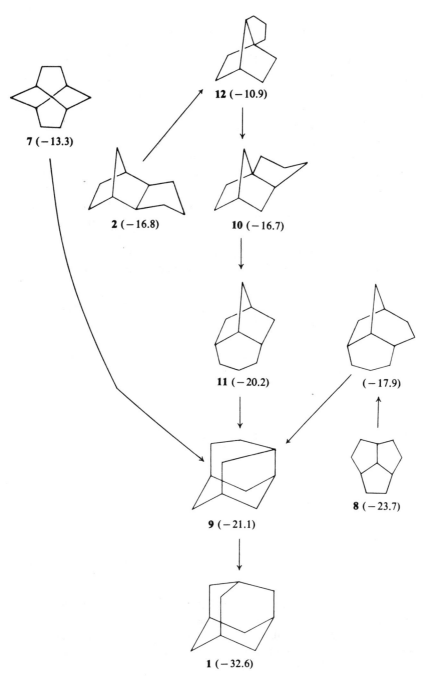

**Scheme 2.** $\Delta H_f$ in kcal/mole are in parentheses.

observed when tetrahydrodicyclopentadiene is the starting material. In Scheme 2, a much abbreviated version of "adamantane land" compares rearrangement pathways of tetrahydrodicyclopentadiene, twistane, and perhydrotriquinacene, along with calculated enthalpies of formation of the hydrocarbons. In all postulated pathways, including those not shown, protoadamantane is the final intermediate before adamantane.

In addition to the rearrangement factors already noted, it is well to point out that high-temperature catalyzed rearrangements produce significant amounts of fragmentation and disproportionation products. Thus, rearrangement of tetrahydrodicyclopentadiene produced many products besides adamantane. Another important factor is that intramolecular 1,2-alkyl shifts must satisfy certain stereoelectronic requirements in order to occur at a reasonable rate. That is, the migrating alkyl group and adjacent vacant p orbital should be eclipsed. This requirement is satisfied in the pathways depicted in Schemes 1 and 2. Because a vacant orbital lobe and adjacent alkyl groups are almost perpendicular, intramolecular 1,2-hydride and alkyl shifts in adamantanes have very high energy barriers and do not occur.

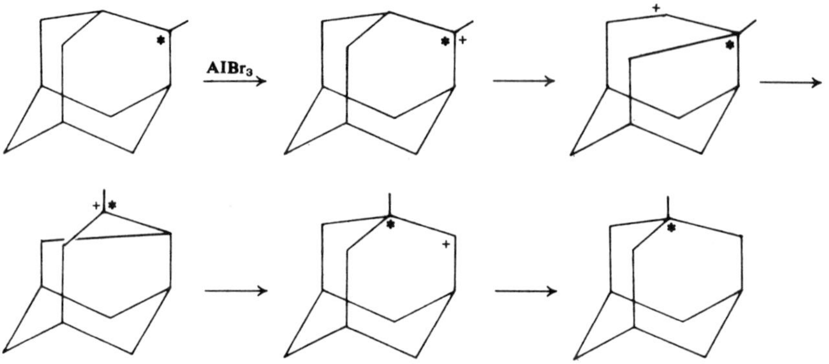

(4)

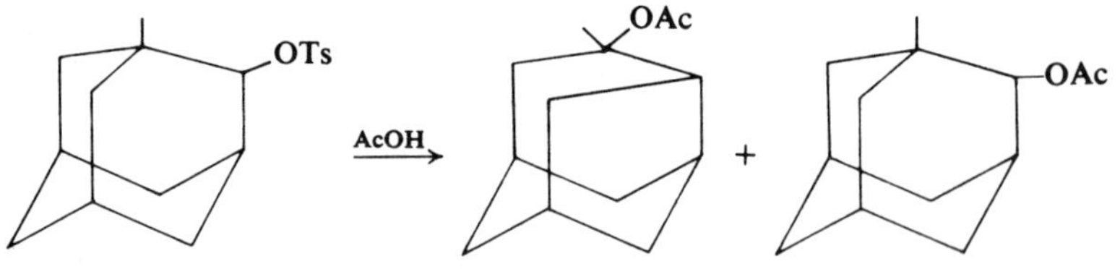

(5)

One last feature about the tricyclodecane manifold deserves some comment. Whitlock and Siefken described adamantane(s) "as a bottomless pit into which rearranging molecules may irreversibly fall."[14] We have earlier noted some side reactions, including fragmentation and disproportionation, which decrease the yield of adamantane. Furthermore, if a pathway which includes a local minimum, i.e., a stable intermediate other than adamantane, is traversed, for all practical purposes, "descent" might end at that point. It still may be argued that these by-products and stable intermediates will eventually, given time, heat, or the proper reaction conditions, rearrange to adamantane(s). However, it is now known that, once formed, adamantane can itself rearrange under some conditions. Two examples will suffice to illustrate this point. First, [14]C-labeling studies have shown that the *apparent* 1,2-alkyl shift which interconverts 2-methyl and 1-methyladamantane, expected not to occur on stereoelectronic grounds, is in fact a more deep-seated rearrangement, moving (C—CH₃) as a unit and passing through protoadamantane [Eq. (4)].[19] Equation (5) depicts conversion of an adamantane derivative into a protoadamantane upon solvolysis.[20a,b,21]

**4.A.2** DIAMANTANE, TRIAMANTANE, TETRAMANTANES, AND RELATED
COMPOUNDS

Adamantane may be viewed as the unsubstituted central unit of diamond (termed the "infinite adamantylogue of adamantane"[22a]). It is also the methane analogue in an homologous series in which diamantane ("congress-ane") (**14**) and triamantane (**15**) are analogues of ethane and propane, while *anti*-tetramantane (**16**) and skew-tetramantane (**17**) are related to *n*-butane conformers and isotetramantane (**18**) substitutes for isobutane.[22a–c] The Schleyer force field leads to a calculation of 10.69 kcal/mole of strain in diamantane.[9] The most recent experimental determination of the standard

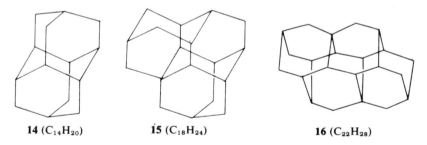

**14** ($C_{14}H_{20}$)        **15** ($C_{18}H_{24}$)        **16** ($C_{22}H_{28}$)

[19] Z. Majerski, P. von R. Schleyer, and A. P. Wolf, *J. Am. Chem. Soc.* **92**, 5731 (1970).

[20a] D. Lenoir and P. von R. Schleyer, *Chem. Commun.* p. 941 (1970).

[20b] D. Lenoir, D. J. Raber, and P. von R. Schleyer, *J. Am. Chem. Soc.* **96**, 2149 (1974).

[21] L. A. Spurlock and K. P. Clark, *J. Am. Chem. Soc.* **92**, 3829 (1970).

[22a] C. Cupas, P. von R. Schleyer, and D. J. Trecker, *J. Am. Chem. Soc.* **87**, 917 (1965).

[22b] T. M. Gund, V. Z. Williams, Jr., E. Osawa, and P. von R. Schleyer, *Tetrahedron Lett.* p. 3877 (1970).

[22c] T. M. Gund, E. Osawa, V. Z. Williams, Jr., and P. von R. Schleyer, *J. Org. Chem.* **39**, 2979 (1974).

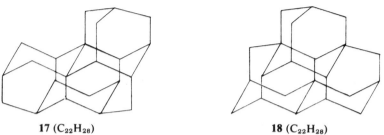

**17** ($C_{22}H_{28}$)                    **18** ($C_{22}H_{28}$)

enthalpy of formation leads to the conclusion that this force field has over-estimated diamantane's stability by some 4.77 kcal/mole (the discrepancy is 5.53 kcal/mole when the Allinger force field is employed).[10] Combination of this latest experimental determination with Schleyer's strain-free model diamantane allows the calculation of some 15.5 kcal/mole of strain in **14**, virtually double the value in adamantane.

Diamantane was originally obtained from hydrogenated norbornadiene dimer (**19**, mixture of stereoisomers) in yields of only 1–10% (Scheme 3).[22a–c] [2+4]Hydrogenated Katz dimers (7:1 mixture of **20** and **21** in Scheme 3) provide yields of up to 25% while tetrahydro-Binor-S (**22** or **23**) rapidly isomerizes to form diamantane in 70% yield.[22a–c,23a,b] Thus, just as in the

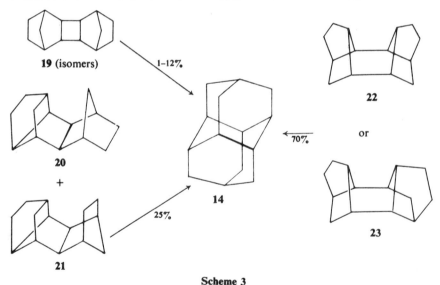

**19** (isomers)          1–12%

**20**
+

**21**                    25%

**14**

70%          or

**22**

**23**

**Scheme 3**

[23a] D. Faulkner, R. A. Glendinning, D. E. Johnston, and M. A. McKervey, *Tetrahedron Lett.* p. 1671 (1971).

[23b] T. Courtney, D. E. Johnson, M. A. McKervey, and J. J. Rooney, *J. Chem. Soc., Perkin Trans. 1* p. 2691 (1972).

case of adamantane, the choice of starting material and rearrangement pathway has a crucial influence upon yield. At 0°C, two abundant intermediates and five minor intermediates have been detected in the tetrahydro-Binor-S rearrangement. Schleyer and co-workers initially attempted to analyze the rearrangement mechanism using the Whitlock–Siefken approach, but the sheer number of isomers and the complexity of pathways made this a hopeless task (the reader is encouraged to peruse the original work cited because it wittily employs understatement to emphasize difficulties in such analyses).[24] Some simplification beyond Whitlock and Siefken's assumptions was achieved by ruling out pentacyclotetradecanes having quaternary carbons, because any small stabilization achieved by increased branching would be far outweighed by destabilization resulting from ring strain. Starting from 22, various 1,2-alkyl shifts were examined and the enthalpies of formation of the resulting hydrocarbons calculated. The isomer interconversion graph obtained did not allow passage of 22 to diamantane, so at least one isomer having quaternary carbon must be involved. However, three viable pathways were calculated which led to 24, one of the two major intermediates observed in the "adamantization" of 22 (Scheme 4). At this point, at least one isomer having quaternary carbon had to be traversed. The analysis then involved generation of isomers resulting from 1,2-alkyl shifts of 24 and comparison of their calculated enthalpies of formation. Hydrocarbon 25 was judged to be the lowest-energy isomer for escape from the subset. Various 1,2-alkyl shifts upon 25 produced hydrocarbons whose calculated enthalpies of formation were compared. The lowest-energy isomer on the road to diamantane was chosen and the process continued. Scheme 4 shows the proposed mechanism for isomerization of tetrahydro-Binor-S to diamantane (heats of formation in kcal/mole of the various pentacyclotetradecanes are in parentheses). The isomerization of 24 to 25 is the rate-determining step, thus explaining accumulation of 24 in significant amounts during rearrangement. This step is significantly endothermic and the 1,2-alkyl shift joining carbonium ions 24+ and 25+ is not as favorable in stereoelectronic terms as one in which the dihedral angle ($\phi$) between alkyl group and empty lobe is 0°.

A glance at Scheme 4 shows that the rearrangement of tetrahydro-Binor-S to diamantane must pass through two highly strained molecules (25 and 26) with quaternary carbons. Schleyer's analysis indicates that [2+4]hydrogenated Katz dimers (20 and 21) must pass through three intermediates containing quaternary carbons, while norbornene dimers (19) must pass through four such hydrocarbons en route to diamantane. Thus, a rather simple explanation of the relative yields from these starting materials is

[24] T. M. Gund, P. von R. Schleyer, P. H. Gund, and W. T. Wipke, *J. Am. Chem. Soc.* **97**, 743 (1975).

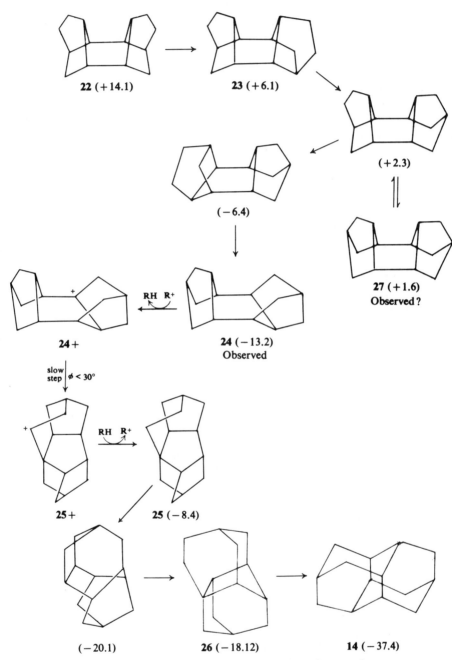

**22** (+14.1)    **23** (+6.1)

(+2.3)

(−6.4)

**27** (+1.6)
Observed?

**24**+    RH  R⁺    **24** (−13.2)
Observed

slow
step  φ < 30°

+    RH   R⁺    **25** (−8.4)

**25**+

(−20.1)    **26** (−18.12)    **14** (−37.4)

**Scheme 4.** $\Delta H_f$ values in kcal/mole are in parentheses.

apparent. In Scheme 4, hydrocarbon **27** exists in a kind of cul-de-sac in which it might build up to a reasonable degree. It is felt that this is the other major isomer, besides **24**, observed in the low-temperature rearrangement of tetra-hydro-Binor-S to diamantane.[24]

Triamantane has also been synthesized by the sludge-catalyzed rearrangement route.[25] The realization that one of the cyclooctatetraene dimers was heptacyclic and diolefinic suggested that upon suitable elaboration, a $C_{18}H_{24}$ heptacyclic hydrocarbon precursor of triamantane could be synthesized. The molecule in question, **28**, did in fact isomerize to triamantane, but maximum yield so obtained was very low (2.0–5.2%). The strain in triamantane has been calculated at 13.45 kcal/mole,[9] but might actually be closer to 22–23 kcal/mole if the experimental trend[10] established for adamantane and diamantane holds. The same cyclooctatetraene dimer was elaborated to the nonacyclodocosane level ($C_{22}H_{28}$) by its Diels–Alder reaction with cyclohexadiene and subsequent

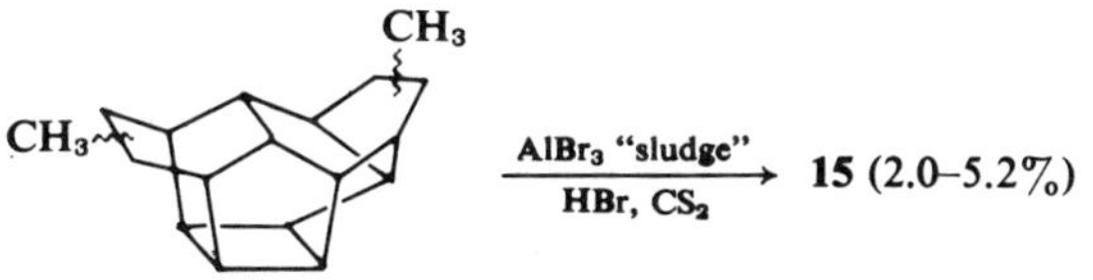

$$\text{CH}_3 \quad \text{CH}_3 \qquad \xrightarrow[\text{HBr, CS}_2]{\text{AlBr}_3 \text{ "sludge"}} \quad \textbf{15} \ (2.0\text{–}5.2\%) \qquad (6)$$

**28**

hydrogenation, which is thought to produce **29**.[26] A crystalline substance which could not be assigned any of the three tetramantane structures was obtained in 5–8% yield. The hydrocarbon obtained, **30**, was dubbed "bastardane." Prolonged heating of this molecule in the presence of catalysts produced

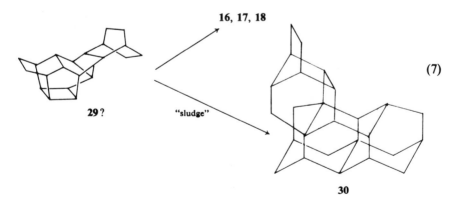

**16, 17, 18**

(7)

**29**?        "sludge"

**30**

[25] V. Z. Williams, Jr., P. von R. Schleyer, G. J. Gleicher, and L. B. Rodewald, *J. Am. Chem. Soc.* **88**, 3862 (1966).

[26] P. von R. Schleyer, E. Osawa, and M. G. B. Drew, *J. Am. Chem. Soc.* **90**, 5034 (1968).

decomposition but not rearrangement to any tetramantane. Although **30** is less stable than **16–18**, it is apparently in a fairly deep local minimum, and thus conditions of *true* thermodynamic control are apparently never achieved when **29** is the starting material. Obviously, some other nonacyclodocosanes must readily isomerize to tetramantane, but their location on the $C_{22}H_{28}$ manifold and the optimization of synthetic yield may well be problems awaiting the next generation of computers and chemists.

Recently, triamantane has been synthesized from diamantane in a relatively high-yield directed manner (Scheme 5).[27a] The final and key step involved multiple rearrangement of the precursor olefin in the vapor phase in the presence of hydrogen and silica-supported platinum catalyst. The vapor stream was passed over the catalyst at a temperature of 430°C. The detailed

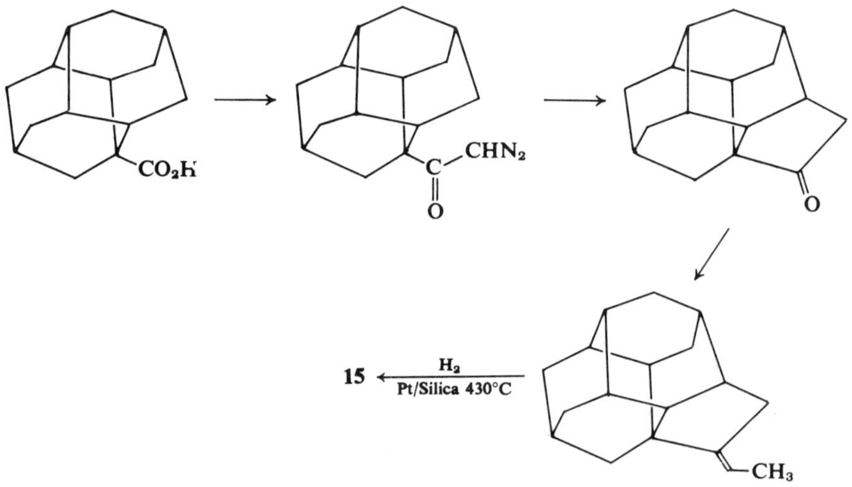

**Scheme 5**

mechanism of this transformation remains unknown;[27a] however, some aspects of gas-phase platinum-catalyzed rearrangements have been deduced.[27b] A partially related directed synthesis of diamantane from adamantane is shown in abbreviated form in Scheme 6.[28] The final rearrangement of proto-diamantane to diamantane is facile, as expected. Both these methods are, in principle, capable of continual stepwise homologation of adamantoids.

---

[27a] W. Burns, M. A. McKervey, and J. J. Rooney, *J. Chem. Soc., Chem. Commun.* p. 965 (1975).

[27b] W. Burns, M. A. McKervey, J. J. Rooney, N. G. Samman, J. Collins, P. von R. Schleyer, and E. Osawa, *J. Chem. Soc., Chem. Commun.* p. 95 (1977).

[28] D. Farcasiu, H. Bohm, and P. von R. Schleyer, *J. Org. Chem.* **42**, 96 (1977).

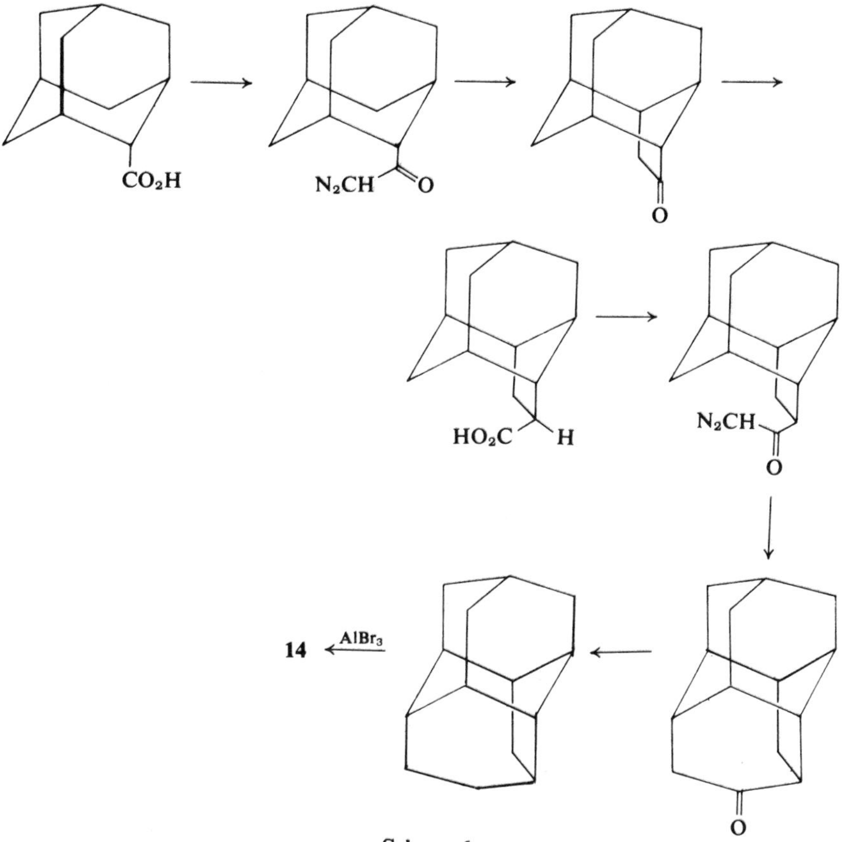

**Scheme 6**

Another even higher-yield synthesis of triamantane[29] involved rearrangement of Binor-S to form two olefins, **31** and possibly **32**, which are homologated with butadiene and reduced to produce hydrocarbons thought to be **33** and **34** (Scheme 7). This mixture of $C_{18}H_{24}$ hydrocarbons was converted in 60% yield to triamantane upon exposure to $AlCl_3$ in hot cyclohexane.

*Anti*-tetramantane (**16**) itself has finally ("legitimately") been prepared, using double homologation of diamantane and rearrangement of the resulting polycyclic diene in the presence of hydrogen and silica impregnated with platinum (Scheme 8), in the same manner that proved successful for triamantane (Scheme 5).[30] The *anti*-tetramantane structure was established by x-ray.

[29] R. Hamilton, M. A. McKervey, J. J. Rooney, and J. R. Malone, *J. Chem. Soc., Chem. Commun.* p. 1027 (1976).

[30] W. Burns, T. R. B. Mitchell, M. A. McKervey, J. J. Rooney, G. Ferguson, and P. Roberts, *J. Chem. Soc., Chem. Commun.* p. 893 (1976).

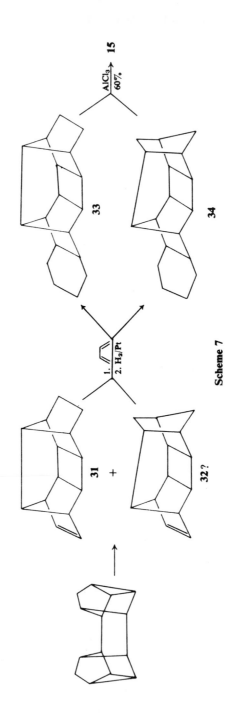

Scheme 7

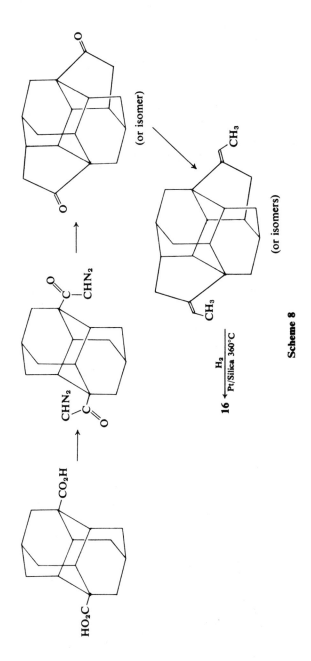

**Scheme 8**

Lest one become overly optimistic over instant wealth accruing from chemical synthesis, it is well to remember that about $2.5 \times 10^{21}$ homologations are required to produce one carat of "polyamantane." Diamond is, of course, slightly less stable than graphite ("diamonds are forever," but graphite is for longer); in fact, "gaseous diamond" can be said to be strained by ca. 2 kcal/mole carbon if one uses approximate methods for determining $\Delta H_f$ for a structure consisting entirely of quaternary C and if one also makes a comparison with $\Delta H_f$ obtained with unstrained $C(C)_4$ increments.[31]

The molecules [1]diadamantane (35)[32,33] and [2]diadamantane (36)[34] have been synthesized, but [3]diadamantane (37) remains unreported. The calculated strain energies of 35, 36, and 37 are 25.5, 22.3, and 23.7 kcal/mole,

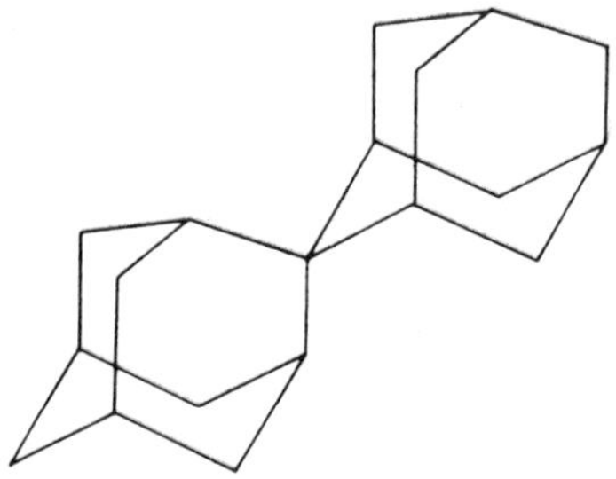

35 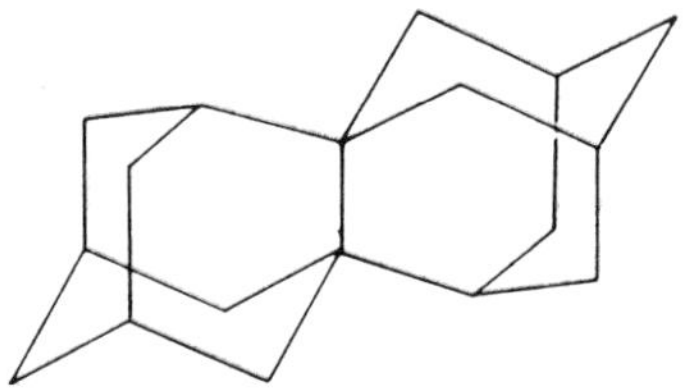                                                    36

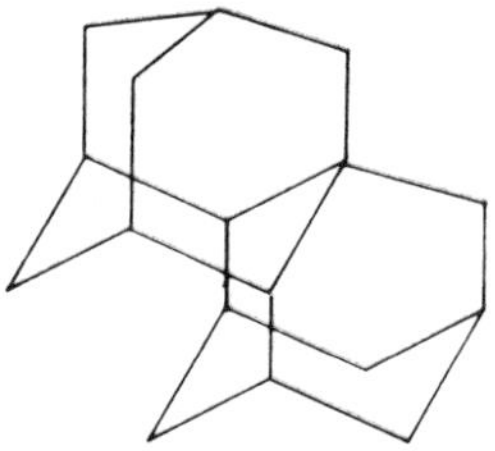

37

respectively.[9] To the extent that these calculational methods seem to have slightly overestimated stability in the adamantane series, the strain energies may be somewhat greater. Diamantane could be referred to as [6]diadamantane using the nomenclature employed above.

[31] David J. Schamp and Joel F. Liebman, unpublished observations.

[32] E. Boelema, J. Strating, and H. Wynberg, *Tetrahedron Lett.* p. 1175 (1972).

[33] W. D. Graham and P. von R. Schleyer, *Tetrahedron Lett.* p. 1179 (1972).

[34] W. D. Graham, P. von R. Schleyer, E. Hagaman, and E. Wenkert, *J. Am. Chem. Soc.* 95, 5785 (1973).

## 4.B  Iceane

In the preceding section we noted some of the difficulties in achieving conditions of thermodynamic control that must eventually produce the adamantoid member of a suitable isomer manifold. That is to say, all roads should eventually lead to formation of the appropriate adamantane compound. Nevertheless, the pitfalls encountered en route are reflected by the great variance in the yields of adamantane, diamantane, and triamantane, yields which depend crucially on the starting material (i.e., rearrangement pathway). Even more striking is the inability to obtain tetramantane from **29** because this rearrangement encounters "bastardane" (**30**), a fairly deep local minimum. The difficulties in winning iceane (**38**) (tetracyclo[5.3.1.1.$^{2,6}$0$^{4,9}$]dodecane, sometimes called "wurtzitane") from the tetracyclododecane manifold became obvious upon the realization that it is *not* the most stable tetracyclic $C_{12}H_{18}$. Iceane is the first member of a potential series of fused hydrocarbons having the general aspect of ice (recall that each oxygen is tetracoordinate in ice). There were two key considerations in obtaining the iceane skeleton from a rearrangement pathway: (a) start with a closely related polycyclic skeleton; (b), avoid conditions in which a significant steady-state concentration of carbonium ions (e.g., $AlBr_3$ sludge) would promote intermolecular hydride exchange and ultimately allow the most stable product to be formed.[35,36]

**38**

Iceane is a rigid symmetric molecule of $D_{3h}$ symmetry that features two chair cyclohexane faces and three boat cyclohexane faces. Its strain energy, calculated at about 25 kcal/mole,[36] is mostly the result of severe nonbonded interactions (i.e., three flagpole–flagpole interactions rigidly enforced as well as vicinal repulsions). Scheme 9 depicts the synthetic route (solid arrows) employed to obtain iceane. The initial carbonium ion, **39**+, is only two (stereoelectronically favorable) 1,2-alkyl shifts from the iceane cation **38**+. Intramolecular 1,2-alkyl shifts will not lead readily to the 2,4-ethanoadamantane cation **41**+. This is very important, because 2,4-ethanoadamantane is the most stable tetracyclic $C_{12}H_{18}$ isomer and is calculated to be almost

[35] C. A. Cupas and L. Hodakowski, *J. Am. Chem. Soc.* **96**, 4668 (1974).

[36] D. Farcasiu, E. Wiskott, E. Osawa, W. Thielecke, E. M. Engler, J. Slutsky, and P. von R. Schleyer, *J. Am. Chem. Soc.* **96**, 4669 (1974) and references cited therein.

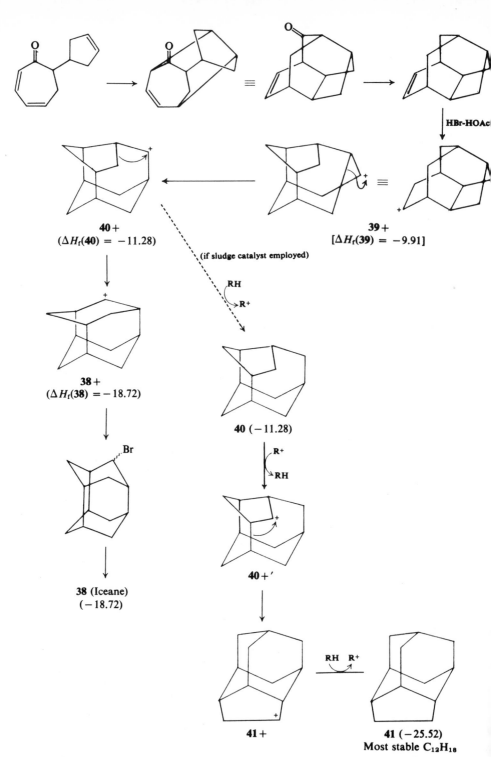

**40+**
$(\Delta H_f(\mathbf{40}) = -11.28)$

**39+**
$[\Delta H_f(\mathbf{39}) = -9.91]$

HBr-HOAc

(if sludge catalyst employed)

RH

R+

**38+**
$(\Delta H_f(\mathbf{38}) = -18.72)$

**40 (−11.28)**

R+

RH

Br

**40+′**

**38 (Iceane)**
**(−18.72)**

RH   R+

**41+**

**41 (−25.52)**
Most stable $C_{12}H_{18}$

**Scheme 9.** $\Delta H_f$ values in kcal/mole are in parentheses.

7 kcal/mole more stable than iceane, whose high symmetry also introduces an unfavorable entropy factor. If a sludge catalyst were utilized (dotted arrows in Scheme 9), one would expect intermolecular hydride shifts which would allow ultimate formation of 2,4-ethanoadamantane from **40+** under conditions of thermodynamic control. In fact, iceane rearranges to 2,4-ethanoadamantane when introduced to sludge-catalyst conditions.[35] Furthermore, **42** rearranges to 2,4-ethanoadamantane, while **43** fails to provide any rearrangement products. Hydrocarbon **44** did provide 2,4-ethanoadamantane following disproportionation.[36] Two additional syntheses of iceane that ultimately rely upon intramolecular rearrangements appeared shortly after the first report.[37,38]

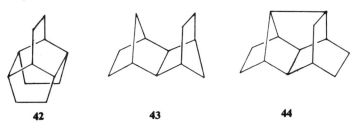

**42**          **43**          **44**

## 4.C  Miscellaneous Acid-Catalyzed Rearrangements

Hydrocarbon **45** rearranges to provide about an 84% yield of 2,4-ethanonoradamantane (**46**) and 2,8-ethanonoradamantane (**47**) as well as some methyladamantanes arising from disproportionation.[39] Although molecular mechanics predicts that **46** and **47** should be of equal stability, the equilibrium distribution shown suggests that **46** is actually more stable by about 2 kcal/mole and is, in fact, the energy minimum on the tetracycloundecane manifold.[39] Noriceane (**48**)[40] is a member of this manifold, but it could not be

**45**          →          **46**          +          **47**  +  methyladamantanes

**45**          **46**          **47**     97:3     (10%)

[37] D. P. G. Hamon and G. F. Taylor, *Tetrahedron Lett.* p. 155 (1975).

[38] H. Tobler, R. O. Klaus, and C. Ganter, *Helv. Chim. Acta* **58**, 1455 (1975).

[39] S. A. Godleski, P. von R. Schleyer, and E. Osawa, *J. Chem. Soc., Chem. Commun.* p. 38 (1976).

[40] T. Katsushima, R. Yamaguchi, and M. Kawanisi, *J. Chem. Soc., Chem. Commun.* p. 692 (1975).

detected in the rearrangement of **45** to **46** and **47**.[39] Noriceane is calculated to be some 3.4 kcal/mole more stable than **45** but 5.8 kcal/mole less stable than **46** or **47**.[39] While **46** is the most stable tricycloundecane and 1-methyladamantane the most stable tricycloundecane, pentacyclo[6.3.0.0$^{2,6}$0.$^{3,10}$0$^{5,9}$]-undecane (**49**),[41] is a trishomocubane of $D_3$ symmetry. This molecule has six equivalent cyclopentane faces (adamantane has four equivalent cyclohexane faces and may be thought of as hexakishomotetrahedrane[1c]). Although

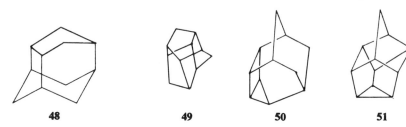

|   **48**   |   **49**   |   **50**   |   **51**   |

it has considerable strain (calculated at 42 kcal/mole), **49** is the only pentacycloundecane lacking a three- or four-membered ring, and its relative stability allows its formation via rearrangements of isomers.[42] A recently synthesized pentacycloundecane, **50**,[40] contains both a cyclopropane and a cyclobutane and is therefore less stable than **49**. Hexacyclo[5.4.0.0.$^{2,6}$0.$^{4,11}$0.-$^{5,9}$0$^{8,10}$]undecane (**51**)[43] is more highly branched and, with its smaller rings, even more strained.

The overwhelming majority of rearrangements discussed above employ Lewis-acid catalysts, and some striking differences are observed when such Brønsted acids as sulfuric or trifluoromethanesulfonic acids are employed. Thus, hydrocarbons **52–54** as well as homoadamantane (**55**) all rearrange rapidly under Lewis-acid catalysis to a mixture of 1- and 2-methyladamantane.[44a,b,45] 4-Homoisotwistane (**56**)[46] can be detected as an intermediate (see Scheme 10).[45] However, homoisotwistane suffers only slow rearrangement under Brønsted-acid catalysis, and this molecule, a local minimum on the $C_{11}H_{18}$ manifold, can be isolated in good yield.[44a,b,45] A somewhat similar situation occurs when 4-homoadamantene is rearranged via Lewis-acid catalysts to produce methyladamantanes, but this molecule

[41] G. R. Underwood and B. Ramamoorthy, *Tetrahedron Lett.* p. 4125 (1970).

[42] S. Godleski, P. von R. Schleyer, E. Osawa, and G. J. Kent, *J. Chem. Soc., Chem. Commun.*, p. 976 (1974).

[43] A. P. Marchand, T.-C. Chow, and M. Barfield, *Tetrahedron Lett.* p. 3359 (1975).

[44a] N. Takaishi, Y. Inamoto, and K. Aigami, *Chem. Lett.* p. 1185 (1973).

[44b] N. Takaishi, Y. Inamoto, and K. Aigami, *J. Org. Chem.* **40**, 276 (1975).

[45] M. Farcasiu, K. R. Blanchard, E. M. Engler, and P. von R. Schleyer, *Chem. Lett.* p. 1189 (1973).

[46] A. Krantz and C. Y. Lin, *Chem. Commun.* p. 1287 (1971).

forms 45–50% 4-homoisotwistene, 20–25% homoadamantane, and 20–25% 2-methyladamantane in the presence of $H_2SO_4$ and pentane.[47a,b] Homoadamantane (**55**) is calculated to be about 11.9 kcal/mole less stable than

**Scheme 10**

1-methyladamantane.[9] There are three isomeric bishomoadamantanes (in decreasing order of calculated stability): 1,1-bishomoadamantane (**57**)[48] (calculated strain: 23.46 kcal/mole[9]); 1,5-bishomoadamantane (**58**)[49–51] (calculated strain: 23.89 kcal/mole[9]); and 1,3-bishomoadamantane (**59**)[52,53] (calculated strain: 24.15 kcal/mole[9]). All three isomers are some 23–27

[47a] Z. Majerski and K. Mlinaric, *Chem. Commun.* p. 1030 (1972).
[47b] K. M. Majerski and Z. Majerski, *Tetrahedron Lett.* p. 4915 (1973).
[48] T. Sasaki, S. Eguchi, T. Toru, and K. Itoh, *J. Am. Chem. Soc.* **94**, 1357 (1972).
[49] F. N. Stepanov, M. I. Novikova, and A. G. Jurchenko, *Synthesis* p. 653 (1971).
[50] D. Skare and Z. Majerski, *Tetrahedron Lett.* p. 4887 (1972).
[51] H. Gerlach, *Helv. Chim. Acta* **55**, 2962 (1972).
[52] J. S. Polley and R. K. Murray, Jr., *J. Org. Chem.* **41**, 3294 (1976).
[53] T. Sasaki, S. Eguchi, and S. Hattori, *Tetrahedron Lett.* p. 97 (1977).

kcal/mole less stable than 1,3-dimethyladamantane. Lower homologues of adamantane include noradamantane (**60**)[54] (calculated strain: 20.07 kcal/mole[9]); the bisnoradamantane (**61**)[55,56] (calculated strain: 41.46 kcal/mole[9]) of $C_s$ symmetry analogous to **59**; and $D_{2d}$ bisnoradamantane (**62**)[57,58] (calculated strain: 47.15 kcal/mole[9]). This last molecule has four cyclopentane faces and is thus also an analogue of adamantane, which has four cyclohexane faces. Note that tetrahedrane can be termed hexanoradamantane.

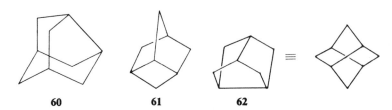

**60**                **61**                **62**

Aluminum tribromide-catalyzed equilibration establishes that 2,2'-biadamantane (**64**)[59] is less strained than either the 1,1' and 1,2' isomers (**63**[60] and **65**[61]) [Eq. (9)].[62]

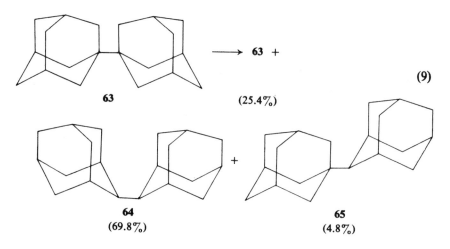

$$\text{(9)}$$

<sup>54</sup> P. von R. Schleyer and E. Wiskott, *Tetrahedron Lett.* p. 2845 (1967).

[54] P. von R. Schleyer and E. Wiskott, *Tetrahedron Lett.* p. 2845 (1967).
[55] P. K. Freeman, V. N. M. Rao, and G. E. Bigam, *Chem. Commun.* p. 5113 (1965).
[56] R. R. Sauers and R. A. Parent, *J. Org. Chem.* **28**, 605 (1963).
[57] O. W. Webster and L. H. Summer, *J. Org. Chem.* **29**, 3101 (1964).
[58] B. R. Vogt, S. R. Suter, and J. R. E. Hoover, *Tetrahedron Lett.* p. 1609 (1968).
[59] J. H. Wieringa, H. Wynberg, and J. Strating, *Synth. Commun.* **1**, 7 (1971).
[60] H. F. Reinhardt, *J. Org. Chem.* **27**, 3258 (1962).
[61] See footnotes in the following reference.
[62] J. Slutsky, E. M. Engler, and P. von R. Schleyer, *Chem. Commun.* p. 685 (1973).

## 4.D   Strained Molecules of Formula $(CH)_n$

A large number of $(CH)_n$ valence isomers have been reported and their thermal and photochemical interconversions mapped.[63] For the present we shall dwell upon only some of these species: first, the three $(CH)_n$ whose carbon frameworks have the topology of Platonic solids; and second, only those $(CH)_n$ that are especially strained or have appeared subsequent to the excellent review of Scott and Jones.[63]

### 4.D.1   "PLATONIC HYDROCARBONS"

Tetrahedrane (66), cubane (67), and dodecahedrane (68) are geometrically equivalent to the Platonic solids tetrahedron, hexahedron (cube or square prism), and (pentagonal) dodecahedron. The remaining Platonic solids, the octahedron and icosahedron, are not likely to have hydrocarbon equivalents (see, however, Chapter 6). To the extent that Platonic solids played a role in ancient theories of the structure of matter,[64] it almost seems that we have come full circle. Tetrahedrane has already been discussed in Section 3.C, and cyclobutadiene, its valence isomer, has been discussed in Section 3.D.3.

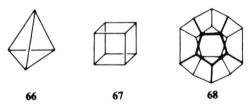

**66**      **67**      **68**

The first derivatives of cubane, **69** and **70** (Scheme 11), were reported in early 1964,[65a] and the parent hydrocarbon (**60**) later the same year.[65b] The reaction sequences (Scheme 11) employed for their syntheses relied crucially upon the expected *endo* orientation in the Diels–Alder reaction shown and also utilized double Favorskii contraction. Two subsequent syntheses[66,67] relied upon cyclobutadiene as an important intermediate (e.g., Scheme 12). The experimental strain energy of cubane (157 kcal/mole) may be considered to represent the sum of the strain energies of six cyclobutane faces.[68] This

[63] L. T. Scott and M. Jones, Jr., *Chem. Rev.* **72**, 181 (1972).

[64] A. J. Ihde, "The Development of Modern Chemistry," p. 5. Harper, New York, 1964.

[65a] P. E. Eaton and T. W. Cole, Jr., *J. Am. Chem. Soc.* **86**, 962 (1964).

[65b] P. E. Eaton and T. W. Cole, Jr., *J. Am. Chem. Soc.* **86**, 3157 (1964).

[66] J. C. Barborak, L. Watts, and R. Pettit, *J. Am. Chem. Soc.* **88**, 1328 (1966).

[67] C. G. Chin, H. W. Cutts, and S. Masamune, *Chem. Commun.* p. 880 (1966).

[68] B. D. Kybett, S. Carroll, P. Natalis, B. W. Bonnell, J. L. Margrave, and J. L. Franklin, *J. Am. Chem. Soc.* **88**, 626 (1966).

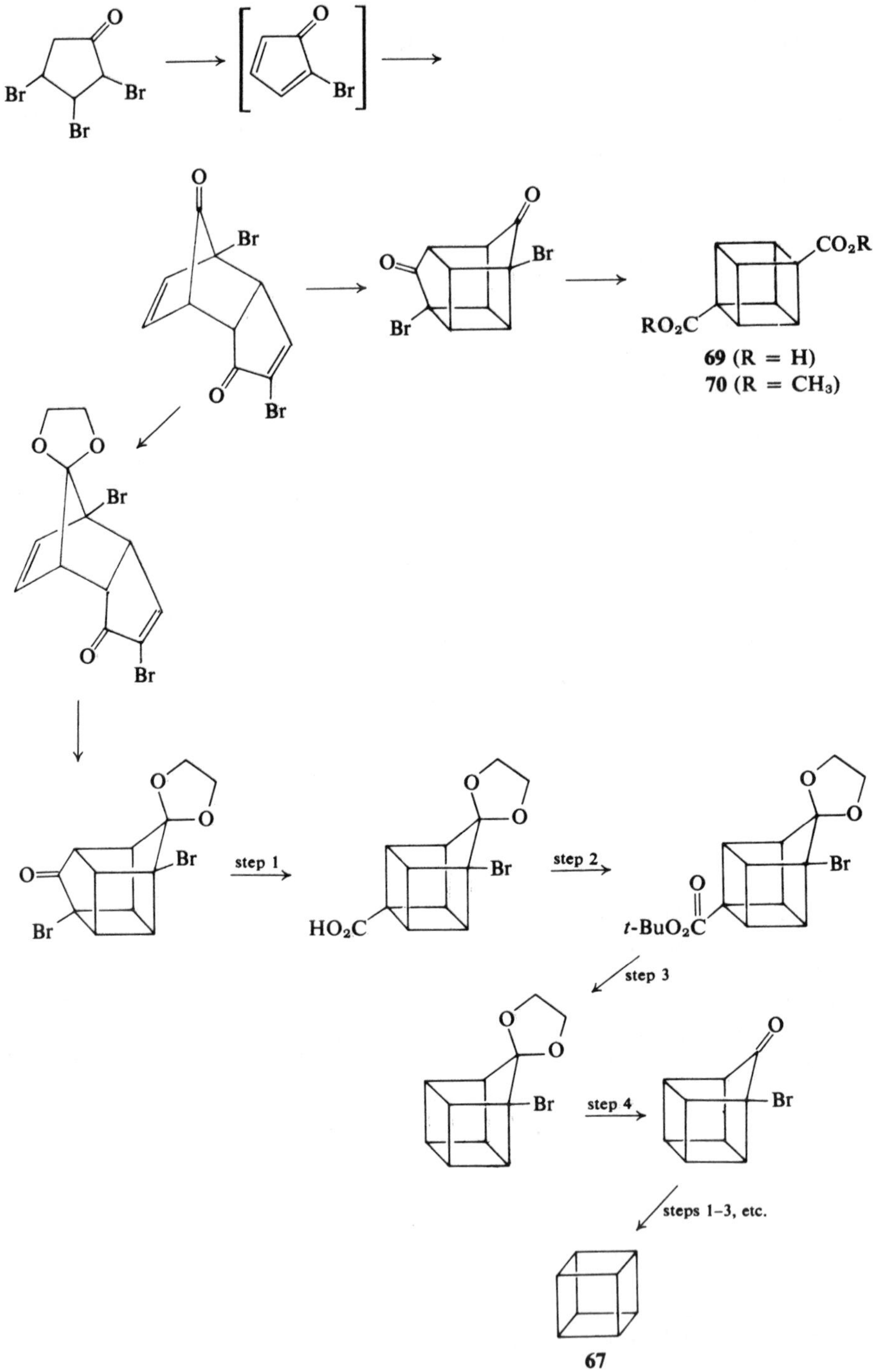

**69** (R = H)
**70** (R = CH₃)

step 1

step 2

step 3

step 4

steps 1–3, etc.

**67**

**Scheme 11**

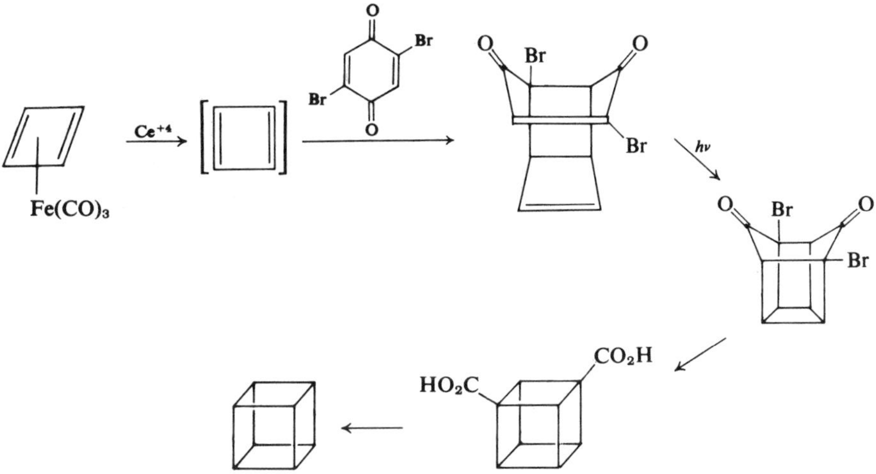

**Scheme 12**

figure translates as 20 kcal/mole per carbon or about 13 kcal/mole per framework bond, making these bonds more strained than those of cyclopropane but less so than the central bond of bicyclobutane (see Section 4.F). Discussion of the transition-metal chemistry of cubane and related species, including secocubane, homocubane, and bishomocubane, is deferred until Chapter 5.

Dodecahedrane (**68**) has eluded synthesis thus far. It is obtainable, at least plausibly, through ("photochemically allowed") dimerization of triquinacene [Eq. (10)],[69] but no success has been achieved using this approach (see below).

$$2 \qquad \xrightarrow{\quad ? \quad} \quad 68 \qquad (10)$$

One may approximate a standard heat of formation for triquinacene (assuming no homoaromatic stabilization) of about $+54$ kcal/mole [$\Delta H_f$(perhydroquinacene)[70] $+ 3 \Delta H_f$(cyclopentene) $- 3 \Delta H_f$(cyclopentane)] and compare this with calculated heats of formation for dodecahedrane. Unfortunately, there is an enormous discrepancy between the calculated values*: $-0.22$ kcal/mole (Schleyer force field) and $+45.28$ kcal/mole (Allinger force field).[9] (The source of this disagreement is discussed in Engler et al.[9]). In any case, it

[69] R. B. Woodward, T. Fukunaga, and R. C. Kelly, *J. Am. Chem. Soc.* **86**, 3162 (1964).

[70] T. Clark, T. M. Knox, H. Mackle, and M. A. McKervey, *J. Chem. Soc., Chem. Commun.* p. 666 (1975).

* See Addendum.

would appear that dimerization of triquinacene to form dodecahedrane should be exothermic by at least 50 kcal/mole. The major problem appears to be one of orientation (i.e., high negative entropy of activation), which allows competitive reactions to occur more rapidly. The extreme requirement for precise orientation is a reflection of the enormous increase in symmetry upon transformation of triquinacene ($C_{3v}$, symmetry number = 3) to dodecahedrane ($I_h$, symmetry number = 60),[71] not to mention the loss in degrees of freedom that occurs during dimerization. Perhaps the answer to the orientation problem is photochemical dimerization of triquinacene (perhaps with an ionic substituent at $C_{10}$) under enzyme-like conditions (e.g., in a micelle or in the cavity of a suitable cyclodextrin under aqueous conditions whereby the two rings might be constrained in a face-to-face orientation while occupying a small hydrophobic volume).

Dodecahedrane has been discussed topologically, in terms of its strain components, and with a view toward its molecular-orbital description.[72a] The molecule suffers almost negligible strain resulting from angle bending or bond stretching, but it maintains 30 pairs of eclipsing CH interactions that account for its calculated total strain, which in turn is evenly distributed among 20 CH centers or 30 framework bonds. The $I_h$ structure of dodeca-hedrane is calculated to be more stable than less symmetric structures since relief of eclipsing interactions is apparently insufficient to compensate for increased angle strain in the latter.[72b] Another interesting feature is the calculation of a small electron density in the center of this molecule, which is expected to "resemble a large rare gas molecule in its physical properties."[72a]

At present, at least three "rational" approaches to dodecahedrane are in progress. Eaton and Mueller[73] have taken the approach of making the peristylane derivative **71** (Scheme 13), which requires eventual "capping" by a cyclopentane ring with the formation of five C—C bonds in order to complete dodecahedrane. Key features include the cis-fusion in the starting material that helps direct formation of a closed molecule; this is further helped by Pd-catalyzed addition of hydrogen to the relatively unencumbered convex surface of the intermediate having four fused rings. Schleyer's force field[9] predicts that the strain in dodecahedrane is only about 4 kcal/mole higher than in peristylane, a value that would give additional reason for optimism for the eventual success of this approach.

Paquette's research group is pursuing at least two approaches to dodeca-hedrane. The first is based upon the recognition that two 2-substituted

[71] See any book on group theory, e.g., F. A. Cotton, "Chemical Applications of Group Theory," 2nd Ed. Wiley (Interscience), New York, 1971.

[72a] J. M. Schulman, T. Venanzi, and R. L. Disch, *J. Am. Chem. Soc.* **97**, 5335 (1975).

[72b] O. Ermer, *Angew. Chem., Int. Ed. Engl.*, **16**, 411 (1977).

[73] P. E. Eaton and R. H. Mueller, *J. Am. Chem. Soc.* **94**, 1014 (1972).

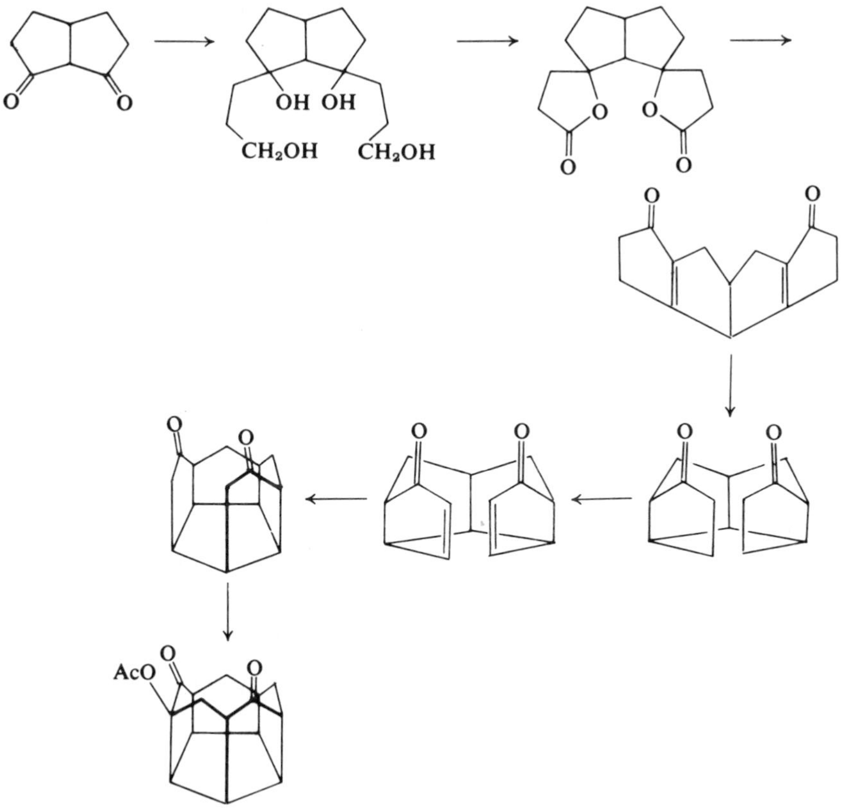

**Scheme 13**

triquinacenes of opposite configuration *cannot* generate a reasonable precursor for dodecahedrane if the two rings are conjoined in an $\alpha,\alpha'$ or $\beta,\beta'$ manner (Scheme 14*A*).[74a–c] However, linking of enantiomerically pure 2-substituted triquinacenes might provide a suitable precursor having all carbon atoms and lacking only five bonds (Scheme 14*B*). The first steps in this approach were the syntheses of (−)-triquinacene-2-carboxylic acid[74a] and (+)-2,3-dihydrotriquinacen-2-one (**72**).[74b,c] Reductive ($\alpha,\alpha'$) coupling of (+)-**72** led exclusively to diol **73** (Scheme 14*C*). In principle at least, coupling of racemic **72** might also lead to the desired products, because the transition state for formation of [(+)-**72**:(+)-**72**] (or the (−)(−) product, that is, the

[74a] L. A. Paquette, S. V. Ley, and W. B. Farnham, *J. Am. Chem. Soc.* **96**, 312 (1974).
[74b] L. A. Paquette, W. B. Farnham, and S. V. Ley, *J. Am. Chem. Soc.* **97**, 7273 (1975).
[74c] L. A. Paquette, I. Itoh, and W. B. Farnham, *J. Am. Chem. Soc.* **97**, 7280 (1975).

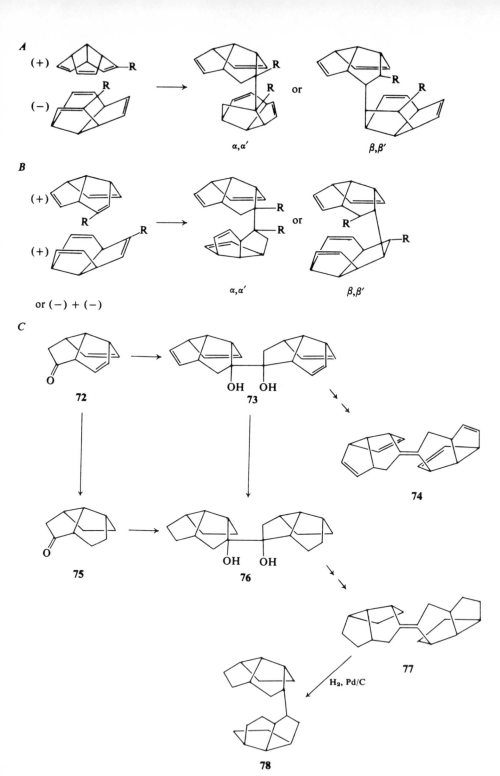

**Scheme 14**

*dl* compound) is diastereomeric with the one for coupling of (+)-**72** and (−)-**72** to form the *meso* adduct. In practice, however, racemic **72** provided an almost equal mixture of the *dl* and *meso* products.[74c] Catalytic hydrogenation of olefin **77**, obtained from ketone **75**, on its less encumbered side, produces *dl*-bivalvane (**78**), a pentasecododecahedrane (i.e., five C—C bonds

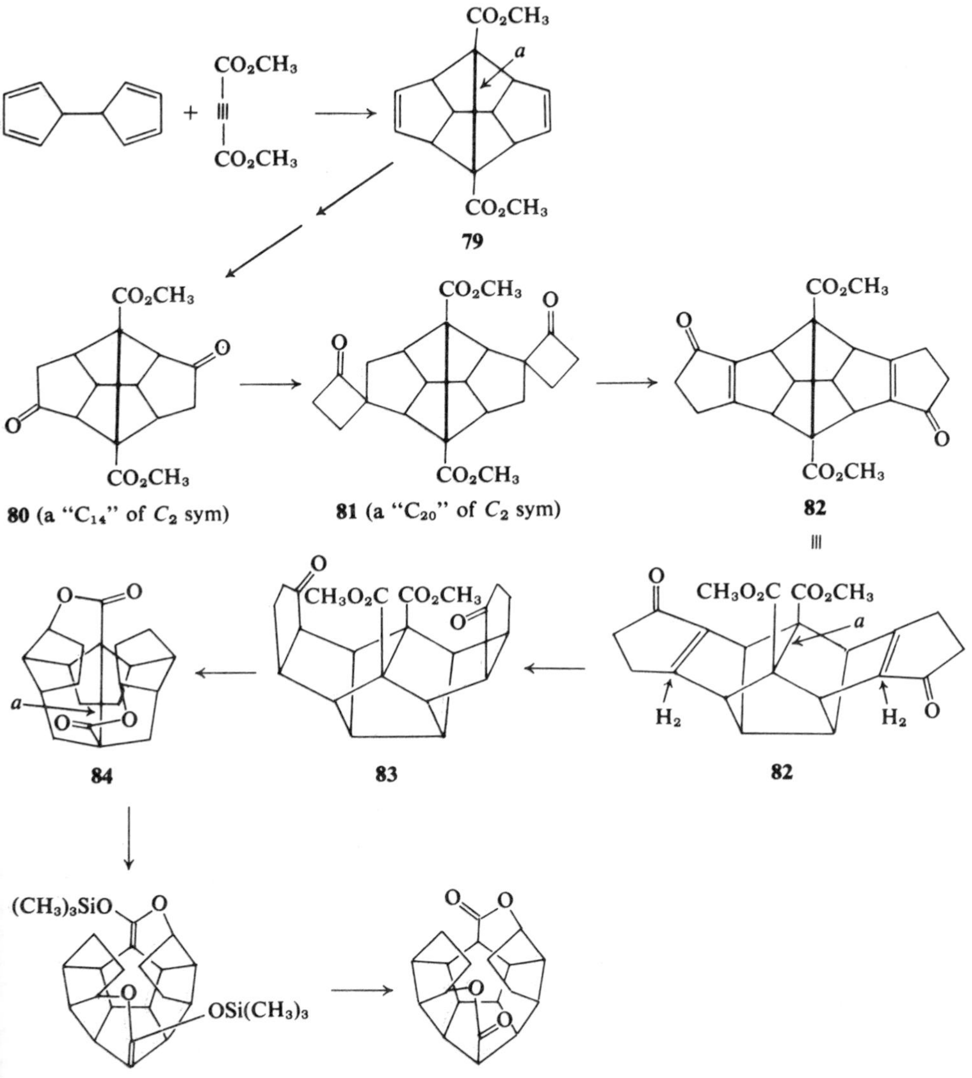


**79**

**80** (a "$C_{14}$" of $C_2$ sym) **81** (a "$C_{20}$" of $C_2$ sym) **82**

**84** **83** **82**

**85** **86**

**Scheme 15**

short of dodecahedrane)[74c] (see Scheme 14*C*). *Meso*-bivalvane has also been produced starting with racemic **72**. The very slow rate of hydrogenation of **77** as well as the nmr spectral properties of **78** are taken in tentative support of a conformation that interlocks the two component rings in *dl*-bivalvane in a manner well suited to its eventual transformation to dodecahedrane.[74c]

The second general approach, depicted in Scheme 15, has succeeded in generating trisecododecahedrane **86**, the most highly fused precursor to date.[75a,b] (The same general approach is also being explored by Hedaya's research group.[76]) Key features of this ingenious synthesis (molecular models must be utilized for best appreciation) include (a) recognition that **79** contains four of the eventual 12 fused rings of dodecahedrane (bond *a* in **79** helps enforce the stereochemistry of the synthesis and is eventually cleaved); (b) clever transformation of this molecule into **80** having $C_2$ symmetry; (c) stereospecific homologation to the $C_{20}$ level at intermediate **81**; (d) hydrogenation of **82** from its less hindered side to produce **83**, which is "closed" and nicely set up for conversion to **84**; (e) **84** can only react on its surface and essentially excludes even solvent molecules from its interior; and (f) the product **86** can be, in principle, elaborated by "a dehydrative *retro*-Baeyer–Villiger sequence" into a compound set up to form the remaining three framework bonds.[75b]

**4.D.2**  MISCELLANEOUS HIGHLY STRAINED $(CH)_n$

At this point we will catalogue some of the most strained, symmetric, and newest $(CH)_n$. A much more complete description of these classes of compounds, including their thermal and photochemical interconversions, has already been cited.[63] The benzene valence isomers, $(CH)_6$, will be treated in detail in Chapter 5. For the present we note Balaban's vast enumeration of all possible $(CH)_n$.[77] An excellent review of $(CH)_n{}^+$ carbonium ions has also appeared.[78]

In addition to cubane, some of the most novel and highly strained known $(CH)_8$ include: barrelene (**87**);[79] *syn-* and *anti*-tricyclo[4.2.0.0$^{2,5}$]octa-3,7-

[75a] L. A. Paquette and M. J. Wyvratt, *J. Am. Chem. Soc.* **96**, 4671 (1974).

[75b] L. A. Paquette, M. J. Wyvratt, O. Schallner, D. F. Schneider, W. J. Begley, and R. M. Blankenship, *J. Am. Chem. Soc.* **98**, 6744 (1976).

[76] D. McNeil, B. R. Vogt, J. J. Sudal, S. Theodoropulos, and E. Hedaya, *J. Am. Chem. Soc.* **96**, 4673 (1974).

[77] A. T. Balaban, *Rev. Roum. Chim.* **11**, 1097 (1966).

[78] R. E. Leone, J. C. Barborak, and P. von R. Schleyer, in "Carbonium Ions" (G. A. Olah and P. von R. Schleyer, eds.), Vol. 4, p. 1837. Wiley (Interscience), New York, 1973.

[79] H. E. Zimmerman and R. M. Paufler, *J. Am. Chem. Soc.* **82**, 1514 (1966).

diene (**88**)[80] (calculated by MINDO/1[81a] to be some 54 kcal/mole higher in energy than all-*cis*-cyclooctatetraene, to which they rapidly isomerize); tricyclo[3.3.0.0$^{2,6}$]octa-3,8-diene (**89**)[82a,b,83] (calculated to be about equal in thermodynamic stability to cyclooctatetraene[81a]); tetracyclo[3.2.1.0.$^{2,4}$0$^{3,8}$]-oct-6-ene(**90**)[84a,b](calculated by MINDO/1 to be almost 90 kcal/mole less stable

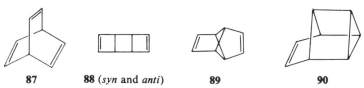

**87**       **88** (*syn* and *anti*)       **89**       **90**

than cyclooctatetraene[81a]); cuneane (**91**);[85] and tetracyclo[4.2.0.0.$^{2,4}$0$^{3,5}$]-oct-7-ene (**92**).[86] Semibullvalene (**93**) undergoes a very rapid degenerate Cope rearrangement.[87] Cuneane was obtained via silver-catalyzed rearrangement of cubane[85] (calculated to be 10.9 kcal/mole less stable than **91**[81a]); the mechanism of this fascinating transformation will be discussed in some depth in Chapter 5. It was hoped that thermolysis of **92** would produce the as yet unknown molecule "octvalene" (**94**), but cyclooctatetraene was produced

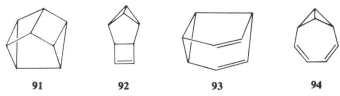

**91**       **92**       **93**       **94**

instead.[86] Another attempt at octvalene via cycloheptatrienylmethylidene (**95**) also failed.[88] Octvalene would be a fascinating hydrocarbon from a number of viewpoints, including the expectation that the interaction between bicyclobutane and butadiene fragments should be destabilizing.[89] Another $(CH)_8$ of related interest would be (**96**). MINDO/3 calculations[81b] of a few

[80] M. Avram, I. G. Dinelescu, E. Marica, G. Mateescu, E. Sliam, and G. Nenitzescu, *Chem. Ber.* **97**, 382 (1964).

[81a] H. Iwamura, K. Morio, and T. L. Kunii, *Chem. Commun.* p. 1408 (1971).

[81b] R. C. Bingham, M. J. S. Dewar, and D. J. Lo, *J. Am. Chem. Soc.* **97**, 1294 (1975).

[82a] J. Meinwald and D. Schmidt, *J. Am. Chem. Soc.* **91**, 5877 (1969).

[82b] J. Meinwald and H. Tsuruta, *J. Am. Chem. Soc.* **91**, 5877 (1969).

[83] H. E. Zimmerman, J. D. Robbins, and J. Schantl, *J. Am. Chem. Soc.* **91**, 5878 (1969).

[84a] G. W. Klumpp and J. Stapersma, *Tetrahedron Lett.* p. 747 (1977).

[84b] G. W. Klumpp, W. G. J. Rietman, and J. J. Vrielink, *J. Am. Chem. Soc.* **92**, 5266 (1970).

[85] L. Cassar, P. E. Eaton, and J. Halpern, *J. Am. Chem. Soc.* **92**, 6366 (1970).

[86] G. E. Gream, L. R. Smith, and J. Mainwald, *J. Org. Chem.* **39**, 3461 (1974).

[87] H. E. Zimmerman and G. L. Grunewald, *J. Amer. Chem. Soc.* **88**, 183 (1966).

[88] H. E. Zimmerman and L. R. Sousa, *J. Am. Chem. Soc.* **94**, 834 (1972).

[89] W. L. Jorgensen, *J. Am. Chem. Soc.* **97**, 3082 (1975) and references cited therein.

$(CH)_8$ isomers conflict seriously with some of the earlier MINDO/1 results. for example, cyclooctatetraene is calculated to be some 33 kcal/mole more

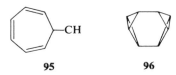

95                96

stable than semibullvalene and 74 kcal/mole more stable than **89**, in stark contrast with the predictions of MINDO/1. Unfortunately, experimental data are largely lacking.

The "star" of the $(CH)_{10}$ energy surface is certainly bullvalene (**97**), which undergoes rapid degenerate isomerization via $1.2 \times 10^6$ independent Cope pathways.[90a,b] Although rearrangement of hypostrophene (**98**), which also interconverts all carbons, is too slow to be detected on the nmr time (rearrangement to **99** occurs at 80°C), labeling studies establish the existence of

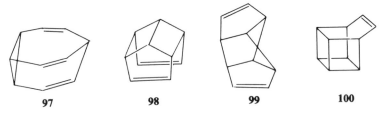

97            98          99              100

this process, in which deuterium is completely scrambled.[91] Some other novel $(CH)_{10}$ include basketene (**100**)[92,93], and hydrocarbons **101**,[94] **102**,[95] and **103**

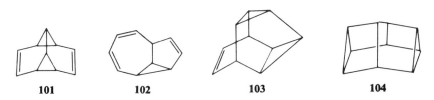

101         102           103            104

[90a] W. von E. Doering and W. R. Roth, *Angew. Chem., Int. Ed. Engl.* **2**, 115 (1963).
[90b] G. Schröder, *Angew. Chem., Int. Ed. Engl.* **2**, 481 (1963).
[91] J. S. McKennis, L. Brener, J. S. Ward, and R. Pettit, *J. Am. Chem. Soc.* **93**, 4957 (1971).
[92] S. Masamune, H. Cuts, and M. G. Hogben, *Tetrahedron Lett.* p. 1017 (1966).
[93] W. G. Dauben and D. L. Whalen, *Tetrahedron Lett.* p. 3743 (1966).
[94] S. Masamune, R. T. Seidner, H. Zenda, M. Wiesel, N. Nakatsuka, and G. Bigam, *J. Am. Chem. Soc.* **90**, 5286 (1968).
[95] E. Vedejs, R. A. Shepher, and R. P. Steiner, *J. Am. Chem. Soc.* **92**, 2158 (1970).

(snoutene).[96,97] The last molecule is formed upon silver-catalyzed rearrangement of basketene in a manner analogous to the cubane–cuneane interconversion (see Chapter 5). More recently, bicyclo[4.2.2]deca-2,4,7,9-tetraene,[98] barretane (**104**),[99] and the symmetric hydrocarbon **105**[100] have all been reported. Diademane (**106**),[101] obtained from snoutene (**103**), rearranges to

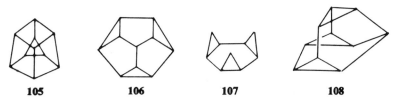

| **105** | **106** | **107** | **108** |

triquinacene, and includes an all-*syn*-trishomobenzene nucleus [note that all-*syn*-trishomobenzene (**107**) is not yet known, although heteroanalogues are]. Compound **108** is another highly strained (CH)$_{10}$.[102] At present, pentaprismane (**109**) (calculated strain: 135.7 kcal/mole[9]) remains unsynthesized despite the synthesis of a seemingly well-suited precursor.[103,104] (Homopenta-

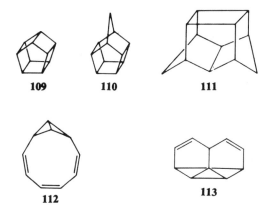

| **109** | **110** | **111** |

| **112** | **113** |

[96] W. G. Dauben, M. G. Bussolini, C. H. Schallhorn, D. J. Whalen, and K. J. Palmer, *Tetrahedron Lett.* p. 787 (1970).

[97] L. A. Paquette and J. C. Stowell, *J. Am. Chem. Soc.* **92**, 2584 (1970).

[98] H.-P. Loffler, *Tetrahedron Lett.* p, 787 (1974).

[99] D. Bosse and A. de Meijere, *Angew. Chem., Int. Ed. Engl.* **13**, 663 (1974).

[100] G. E. Gream, L. R. Smith, and J. Meinwald, *J. Org. Chem.* **39**, 3461 (1974).

[101] A. de Meijere, D. Kaufman, and O. Schallner, *Angew. Chem., Int. Ed. Engl.* **10**, 417 (1971); see also W. Spielmann, H.-H. Fisk, L.-U. Meyer, and A. de Meijere, *Tetrahedron Lett.* p. 4057 (1976) for 1,6-homodiademane.

[102] H. Prinzbach and D. Stusche, *Helv. Chim. Acta* **54**, 755 (1971).

[103] K.-W. Shen, *J. Am. Chem. Soc.* **93**, 3064 (1971).

[104] J. A. Berson and R. F. Davis, *J. Am. Chem. Soc.* **94**, 3658 (1972).

prismane **110** has been synthesized.[105-107] The initial report[108a] of its synthesis is now known to be incorrect; the compound obtained was actually **51**.[108b]) Bishomopentaprismane, or "bird-cage hydrocarbon" **111**, is also known.[109] Hydrocarbon **112** ("decvalene"?) is another of the many unknown $(CH)_{10}$ which remain tempting synthetic targets. Only a benzo derivative of **113** has been prepared at this time.[110]

The $(CH)_{12}$ isomers are of particular interest as "benzene dimers" (in formula at least, and sometimes in structure). For example, benzene dimer **114**[111a,b] (the *syn* dimer is unknown) can be calculated by group incremental schemes to be some 50 kcal/mole less stable than two benzene molecules. The free energy of activation for this process is about 25 kcal/mole,[111b]

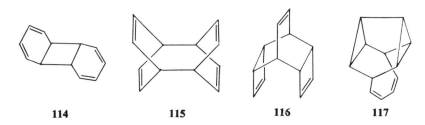

| **114** | **115** | **116** | **117** |

significantly lower than expected for a symmetry-forbidden process (see mechanism discussed in Scott and Jones[63]). The total energy "excess" of 50 + 25 kcal/mole is not quite enough to provide a molecule of benzene triplet (see Chapter 5). There is also substantial interest in benzene dimer **115**, whose conversion to two benzene molecules is also symmetry forbidden. A recent attempt at making this molecule was unsuccessful.[112] The reaction generating two benzene molecules is exothermic by roughly 40–45 kcal/mole, and is unlikely to generate triplet benzene unless a very high activation barrier is involved. Among the newest known $(CH)_{12}$ are **116**[113] (the *exo, endo* and

[105] P. E. Eaton, L. Cassar, R. A. Hudson, and D. R. Hwang, *J. Org. Chem.* **41**, 1445 (1976).

[106] E. C. Smith and J. Barborak, *J. Org. Chem.* **41**, 1433 (1976).

[107] A. P. Marchand, T.-C. Chou, J. D. Ekstrand, and D. Van der Helm, *J. Org. Chem.* **41**, 1438 (1976).

[108a] G. R. Underwood and B. Ramamoorthy, *Chem. Commun.* p. 12 (1970).

[108b] Prof. Alan P. Marchand, personal communication to the authors.

[109] L. De Vries and S. Winstein, *J. Am. Chem. Soc.* **82**, 5363 (1960).

[110] E. Vedejs, R. P. Steiner, and E. S. C. Wu, *J. Am. Chem. Soc.* **96**, 4040 (1974).

[111a] H. Rottele, W. Martin, J. F. M. Oth, and G. Schröder, *Chem. Ber.* **102**, 3985 (1969).

[111b] J. F. M. Oth, H. Rottele, and G. Schröder, *Tetrahedron Lett.* p. 61 (1970).

[112] N. C. Yang, C. V. Neywick, and K. Strinivasachar, *Tetrahedron Lett.* p. 4313 (1975).

[113] U. Erhardt and J. Daub, *J. Chem. Soc., Chem. Commun.* p. 83 (1974).

*exo,exo* isomers were known previously[114,115]), **117**,[116] **118**,[117] and **119**.[118a] The electronic and geometric requirements for degenerate Cope rearrangement are met in **118**, but this process could not be observed via nmr at 140°C. This molecule lacks the cyclopropane ring, present in bullvalene or semibullvalene, which apparently serves to increase the rate of Cope rearrangement; this, however, might be monitored upon the isolation of an enantiomer of

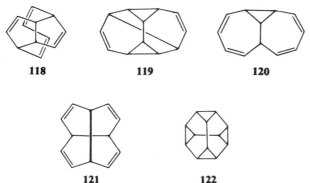

**118**          **119**          **120**

**121**          **122**

**118** because the conversion involves racemization.[117] Molecule **119** was synthesized as a potential precursor to molecules **120** and **121**, the latter, in principle, convertible upon photolysis to the truncated tetrahedron **122**.[118a,b] However, thermolysis of **119** generated **123**, which proceeded to ultimately provide benzene.[118b] Note that an interconversion analogous to **121** ⇌ **120** has been observed in the thermal isomerization of **124** to **125**.[119]

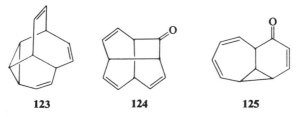

**123**          **124**          **125**

### 4.E   Additional Polycyclic Molecules

The number of miscellaneous strained hydrocarbons is far too large for us to attempt a complete accounting. Nonetheless, we will display some examples

[114] G. Schröder, and W. Martin, unpublished results, cited in G. Schröder and J. F. M. Oth, *Angew. Chem., Int. Ed. Engl.* **6**, 414 (1967).

[115] L. A. Paquette and J. C. Stowell, *Tetrahedron Lett.* p. 4159 (1969).

[116] L. A. Paquette and M. J. Kukla, *J. Chem. Soc., Chem. Commun.* p. 409 (1973).

[117] D. G. Farnum and A. A. Hagedorn, III, *Tetrahedron Lett.* p. 3987 (1975).

[118a] E. Vedejs and R. A. Shepher, *J. Org. Chem.* **41**, 742 (1976).

[118b] E. Vedejs, W. R. Wilber, and R. Twieg, *J. Org. Chem.* **42**, 401 (1977).

[119] R. Aumann, *Angew. Chem., Int. Ed. Engl.* **15**, 376 (1976).

of rather unique and often symmetric species. Propellanes and paddlanes are rather specialized classes of molecules that will be treated in Chapter 6.

Quadricyclane (126),[120,121] a homoprismane, has an experimental strain energy of 95 kcal/mole. [122] Continuing this series are 127[123] and 128, the latter a trishomoprismane dubbed "triasterane" having three boat cyclohexane

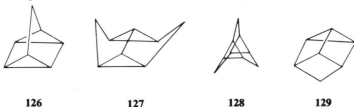

| 126 | 127 | 128 | 129 |

faces.[124] One can conceptually relate prismane to cubane through homoprismane 129[125] and the bishomoprismane 130 also known as secocubane.[126]

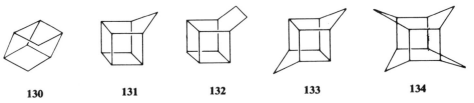

| 130 | 131 | 132 | 133 | 134 |

Homocubane (131),[127] basketane (132),[92,94] bishomocubane (133),[128] and tetraasterane (134)[129] are also known. Tetraasterane is unchanged after one hour of heating (in vacuo) at 500°C! Hydrocarbon 135 has bicyclo[2.2.2]-octane as its building unit and maintains eclipsed interactions between all pairs of nonbonded hydrogens.[130] The strain energy of perhydro[2.2]para-cyclophane (136),[131] a highly crowded molecule, is 26 kcal/mole.[132] Some

[120] W. G. Dauben and R. L. Cargill, Tetrahedron 15, 197 (1961).

[121] G. S. Hammond, N. J. Turro, and A. Fischer, J. Am. Chem. Soc. 83, 4674 (1961).

[122] R. B. Turner, P. Goebel, B. J. Mallon, W. von E. Doering, J. Coburn, Jr., and M. Pomerantz, J. Am. Chem. Soc. 90, 4315 (1968).

[123] P. K. Freeman, D. G. Kuper, and V. N. Mallikarjuna Rao, Tetrahedron Lett. p. 3301 (1965).

[124] V. Biethan, U. von Gizycki, and H. Musso, Tetrahedron Lett. p. 1477 (1965).

[125] A. R. Brember, A. A. Gorman, and J. B. Sheridan, Tetrahedron Lett. p. 481 (1973).

[126] W. G. Dauben, C. H. Schallhorn, and D. L. Whalen, J. Am. Chem. Soc. 93, 1446 (1971).

[127] See footnote 1 in Ref. 125.

[128] E. T. McBee, W. L. Dilling, and H. P. Braendlin, J. Org. Chem. 27, 2704 (1962).

[129] H.-M. Hutmacher, H.-G. Fritz, and H. Musso, Angew. Chem., Int. Ed. Engl. 14, 180 (1975).

[130] V. Boekelheide and R. A. Hollins, J. Am. Chem. Soc. 95, 3201 (1973).

[131] D. J. Cram and N. L. Allinger, J. Am. Chem. Soc. 77, 6289 (1955).

[132] C. Shieh, D. C. McNally, and R. H. Boyd, Tetrahedron 25, 3653 (1969).

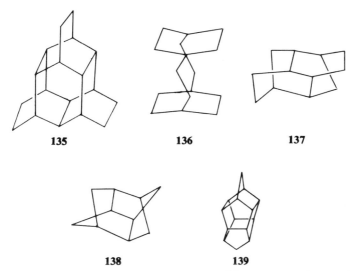

**135**                    **136**                    **137**

**138**                    **139**

other novel species include [8]twistane (**137**),[133] bisnortwistane (**138**),[133] and the dimer of norbornadiene (**139**).[134a,b,135] A series of dehydro compounds related to adamantane, noradamantane, and homoadamantane includes **140**,[136] **141**,[137] **142**,[138,139] and **143**.[140] While the parent molecule *cis*-bishomobenzene is unknown, two derivatives, **144**[141] and **145**,[142] have been identified,

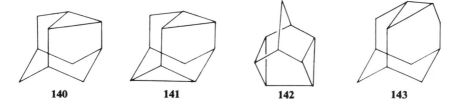

**140**            **141**            **142**            **143**

[133] K. I. Harao, T. Iwakuma, M. Taniguchi, E. Abe, O. Yonemitsu, T. Date, and K. Kotera, *J. Chem. Soc., Chem. Commun.* p. 691 (1974).

[134a] T. J. Katz and N. Acton, *Tetrahedron Lett.* p. 2601 (1967).

[134b] H.-D. Scharf, G. Weisgerber, and H. Höber, *Tetrahedron Lett.* p. 4227 (1967).

[135] S. C. Neely, D. Van der Helm, A. P. Marchand, and B. R. Hayes, *Acta Crystallogr., Sect. B* 32, 561 (1976).

[136] A. C. Udding, J. Strating, H. Wynberg, and J. L. M. A. Schlatmann, *Chem. Commun.* p. 657 (1966); see also D. Farcasiu and P. von R. Schleyer, *Tetrahedron Lett.* p. 3835 (1973) for a dehydrodiamantane.

[137] H. W. Geluk and T. J. De Boer, *J. Chem. Soc. Chem. Commun.* p. 3 (1972).

[138] H. Prinzbach and D. Hunkler, *Angew. Chem., Int. Ed. Engl.* 6, 247 (1967).

[139] P. K. Freeman and D. M. Balls, *J. Org. Chem.* 32, 2354 (1967).

[140] Z. Majerski, S. H. Liggero, and P. von R. Schleyer, *Chem. Commun.* p. 949 (1970).

[141] G. Kaupp and K. Rosch, *Angew. Chem., Int. Ed. Engl.* 15, 163 (1976).

[142] K. Menke and H. Hopf, *Angew. Chem., Int. Ed. Engl.* 15, 165 (1976).

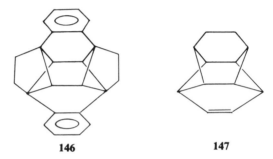

**144**                    **145**

the latter molecule obtained through the clever application of conformational control. Two novel hydrocarbons, whose highly stylized names invoke animal traits, are **146** and **147**. Dibenzoequinene (**146**),[143] a derivative of equinene that in principle is obtainable from a twofold intramolecular cycloaddition of [2.2]paracyclophane, has two highly puckered cyclobutane

**146**                    **147**

rings.[144]. Molecule **147** is a derivative of the presently unknown felicene system. Although it was not isolated, deuterium labeling studies establish its intermediacy in the thermal decomposition of an isomeric substance.[145]

This section is concluded with some examples of hydrocarbons in which the "total differs considerably from the sum of the parts." Although tetrahedrane is such a molecule, prismane and cubane are not very different in energy from the sum of their component rings. However, *exo,exo*-tetracyclo-[6.2.1.1.$^{3,6}$0$^{2,7}$]dodecane (**148**)[146] is calculated[9] to be about 16 kcal/mole more strained than two norbornanes, and the calculated strain in the *endo, endo* isomer **149**[146] is 78 kcal/mole higher than that of two norbornanes. Nonbonded repulsion is the problem, as demonstrated by the calculation of "normal" strain in "bird-cage hydrocarbon" **111**. A more startling effect[147] is

[143] H. H. Wasserman and P. M. Keehn, *J. Am. Chem. Soc.* **89**, 2770 (1967).

[144] A. V. Fratini, *J. Am. Chem. Soc.* **90**, 1688 (1968).

[145] A. Gilbert and R. Walsh, *J. Am. Chem. Soc.* **98**, 1606 (1976).

[146] S. Winstein and R. L. Hanson, *Tetrahedron Lett.* p. 25, 4 (1960).

[147] D. A. Dougherty, W. D. Hounshell, H. B. Schlegel, R. A. Bell, and K. Mislow, *Tetrahedron Lett.* p. 3479 (1976).

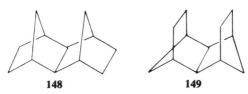

**148**          **149**

seen in the structures and energies of naphthalene dimer **150**,[148,149] anthracene dimers[150] **151**[151a] and **152**,[151b] tetracene dimer **153**,[152] as well as lepidopterene (**154**),[153,154] **155**,[155] and (presumably) **156**.[156] The 1,2-bonds are ca. 0.02 Å

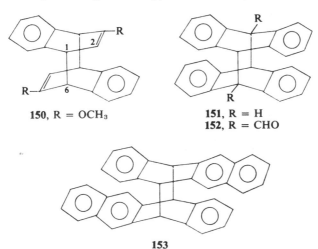

**150**, R = OCH₃          **151**, R = H
                          **152**, R = CHO

**153**

shorter and the 1,6-bonds ca. 0.04 Å longer in these molecules than one would predict on the basis of molecular-mechanics calculations, which are usually very reliable for related molecules.[147] The structure calculated for lepidopterene (**154**) is in fact ca. 6 kcal/mole lower in energy than the x-ray-determined structure. This highly significant departure has been attributed to

[148] J. S. Bradshaw and G. S. Hammond, *J. Am. Chem. Soc.* **85**, 3953 (1963).

[149] B. K. Selinger and M. Sters, *Chem. Commun.* p. 978 (1969).

[150] F. D. Greene, S. L. Misrock, and J. R. Wolfe, Jr., *J. Am. Chem. Soc.* **77**, 3852 (1955).

[151a] M. Ehrenberg, *Acta Crystallogr.* **20**, 177 (1966).

[151b] M. Ehrenberg, *Acta Crystallogr., Sect. B* **24**, 1123 (1968).

[152] J. Gaultier, C. Hauw, J. P. Desvergne, and R. Lapouyade, *Cryst. Struct. Commun.* **4**, 497 (1975).

[153] G. Felix, R. Lapouyade, A. Castellan, and H. Bouas-Laurent, *Tetrahedron Lett.* p. 409 (1975).

[154] J. Gaultier, C. Hauw, and H. Bouas-Laurent, *Acta Crystallogr., Sect. B* **32**, 1220 (1976).

[155] M. Ehrenberg, *Acta Crystallogr.* **20**, 183 (1966).

[156] G. Kaup and I. Zimmerman, *Angew. Chem., Int. Ed. Engl.* **15**, 441 (1976).

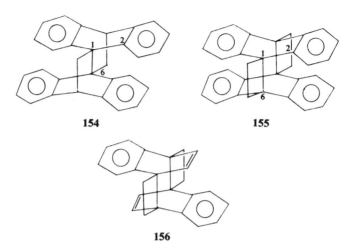

**154**          **155**

**156**

a new type of overlap interaction involving the 1,6-C—C $\sigma$ orbital and the four p orbitals which flank it.[147] (See also discussion in Chapter 1.) The 1,6-bond in **155** is reported at 1.77 Å,[155] almost 0.2 Å greater than the value calculated by means of molecular mechanics![147]

The photodimerization of anthracene to form **151** and related dimerizations of aromatics have been well investigated.[157] In principle, some of these molecules might be capable of generating a monomer in an electronically excited state (see Chapter 5). Although the 40–45 kcal/mole of excess energy in **115** (relative to two benzenes) should not allow the generation of a triplet benzene upon thermolysis ($E_{T_1} - E_{S_0} \approx 85$ kcal/mole), the corresponding triplet–ground-state differences in naphthalene and anthracene are only 62 and 42 kcal/mole, respectively.[158]

## 4.F Molecular Topology

In this book we have presented numerous compounds with unusual structures and/or bonding. By now, the reader may feel like "a stranger in a [strained] land."[159] We have tried to provide organizing principles derived from thermochemistry and our sense of structural chemical taxonomy. We have also tried to give the strain energy, the single number which denotes the instability relative to a well-defined reference state. In this section we wish to present some other aspects of molecular strain energy and of molecular structure. The usefulness and validity of these approaches will not be evalu-

[157] E. J. Bowen, *Adv. Photochem.* **1**, 23 (1963).
[158] J. G. Calvert and J. N. Pitts, Jr., "Photochemistry," Wiley, New York, 1966.
[159] Paraphrase from the Book of Exodus, II. 22.

ated here. Instead, it is left to further study by both the reader and the authors. However, as this section is highly speculative and mathematical and has no definitive applications, the reader may omit it with no loss of continuity or intelligibility for the rest of our book.

The reader should recall the method of group increments, in which the molecule is fragmented into the $CH_3$—, —$CH_2$—, —$CH\langle$, $\rangle C \langle$, —$CH=$ $CH$— groups, etc. To keep our discussion in this section brief, we will limit our attention only to saturated, unsubstituted alicyclic hydrocarbons. For this restricted but extremely large and important class of molecules, the only groups that need to be considered are the —$CH_2$—, —$CH\langle$, and $\rangle C \langle$. That is, any compound of interest may be written $(C)_\alpha (CH)_\beta (CH_2)_\gamma$.

Several immediate benefits arise from this formulation. First, comparison of molecules may naturally be limited to compounds with the same values of $\alpha$, $\beta$, and $\gamma$, which are alternately called valence isomers,[160] "isologous" compounds,"[161] or "informational homologues."[162] As such, while adamantane and twistane remain interrelatable, as both have $\alpha = 0$, $\beta = 4$, and $\gamma = 6$, we are exempt from having to consider 1,2-cyclodecadiene, 3,3,6,6-tetramethylcyclohexyne, monoterpenes, and most any other randomly chosen $C_{10}H_{16}$ compound.

A second benefit is that the strain energy takes on a simpler form[163,164] [see Eq. (11)].

$$\text{SE(cmpd)} = \Delta H_f^0(g, cmpd) - \alpha \Delta H_f^0(g, C) - \beta \Delta H_f^0(g, CH) - \gamma \Delta H_f^0(g, CH_2) \quad (11)$$

Generalizing,

$$\text{SE(cmpd)} = \Delta H_f^0(g, cmpd) - \sum_k \nu_k \Delta H_f^0(g, k) = \Delta H_f^0(g, cmpd) - \sum_k \nu_k \eta_k \quad (12)$$

Customarily, one chooses reference compounds such that their strain energy is zero. Accordingly, each reference compound describes one linear equation in the $\eta_k$'s, or in the current case, three unknowns. Three such equations define uniquely the numerical value of the three $\eta_k$'s. Equivalently, three reference compounds and their associated sets of $\nu_k$ (i.e., $\alpha$, $\beta$, and $\gamma$ values) are necessary and sufficient to define the needed heats of formation of the group increments $\{\eta_k\}$. The numbers $\eta_1$, $\eta_2$, and $\eta_3$ determined depend only on the reference compounds chosen, i.e., they are independent of one another.

[160] See, e.g., Section 5.B of this book.
[161] E. M. Engler, J. D. Andose, and P. von R. Schleyer, *J. Am. Chem. Soc.* **95**, 8005 (1973).
[162] Y. Kudo and S.-I. Sasaki, *J. Chem. Inf. Comput. Sci.* **16**, 43 (1973).
[163] J. F. Liebman and A. Greenberg, *Chem. Rev.* **76**, 311 (1976), Section XIV.
[164] Deborah Van Vechten and Joel F. Liebman, unpublished observations.

We may accordingly view them as coefficients of orthogonal unit vectors, i.e., describe $(\eta_1, \eta_2, \eta_3)$ as a three-component vector.

The method of group increments offers no guidelines as to how to choose the three reference compounds. Consider now what happens when the set of reference componds is chosen differently by "competitive" group-increment schemes. (We remind the reader that this is, in fact, the customary situation.) This can be expressed by saying that each set of reference compounds will correspond to a different set of $\eta$'s, a set we now label by the superscript $\eta^j$. For each $j$, the individual $\eta_k$'s, or more properly the individual $\eta_k^j$'s, are independent of the others. Are the various sets $\eta^j$ likewise independent of one another? The answer is a definitive no. From a knowledge of the choice of reference compounds of the competing schemes, one can arithmetically deduce the strain energy found for one scheme (1) from that of the other (2) by multiplying each $\nu_k \eta_k^{(1)}$ term by $\eta_k^{(2)}/\eta_k^{(1)}$. (Recall that the compound being considered determines the $\nu_k$ independent of the reference states.) However, even if one does not know the choice of reference compounds on which the schemes are based, it may be shown that there are three (no more and no less) linearly independent schemes, as that is the number of linearly independent 3-vectors. Fundamentally, there is no decision process based on the method of group increments to decide which schemes are the best. While it is perhaps obvious that cyclopropane is inappropriate as the generator of the —$CH_2$— group, decisions such as whether diamond or neopentane is more suitable for quaternary C cannot be made except by personal preference and prejudice.

There is a third benefit[163,164] of writing the compounds in the form of $(C)_\alpha (CH)_\beta (CH_2)_\gamma$. Recall the earlier discussions of strain energy per (carbon) atom, $SE_c$, and strain energy per C—C bond, $SE_b$. These two concepts appear useful, distinct, and complementary, but they sometimes lead to contradictory results. For the class of compounds being discussed in this section, $SE_c$ equals $SE/(\alpha + \beta + \gamma)$ while $SE_b = SE/[\frac{1}{2}(4\alpha + 3\beta + 2\gamma)] = SE/(2\alpha + \frac{3}{2}\beta +\gamma)$. Both quantities may be expressed as $SE/(a\alpha + b\beta + c\gamma)$. $\alpha$, $\beta$, and $\gamma$ depend only on the structure of the compound being considered; i.e., they are independent of each other. We may again accordingly view them as the coefficients of orthogonal unit vectors: that $\mathbf{C} \equiv (\alpha,\beta,\gamma)$ may be viewed as a three-component vector. Now if $\mathbf{S} \equiv (a,b,c)$ is known for the strain-energy scheme one wishes to consider—clearly $\mathbf{S} = (1,1,1)$ for the $SE_c$ scheme and $\mathbf{S} = (2, \frac{3}{2}, 1)$ for the $SE_b$ scheme—then the quantities to be computed for different compounds are $SE/(\mathbf{C} \cdot \mathbf{S})$. Moreover, one can consider $\mathbf{S}$ as a vector in a three-dimensional "scheme space." Since any vector in a three-dimensional space is expressible in terms of the three linearly independent basis vectors, we conclude that there are at most three independent schemes. There are two remaining questions. First, what is the third linearly independent scheme? Other than asking that $a, b$, and $c$ be all positive so as to preclude division by zero, there are no other guidelines. Second, which scheme should be chosen? The answer is "any," because they are fundamentally equally valid, even when they are seemingly contradictory.

The reader may be wondering whether the number and nature of the independent rings in the molecule can be used to lessen the ambiguities described above. We find from the simple formulas (13) and (13')[165–167] the number of independent rings; and for the compounds of interest in this

$$\text{faces} = \text{edges} - \text{vertices} + 1 \tag{13}$$
$$\text{rings} = \text{bonds} - \text{atoms} + 1 \tag{13'}$$

section use Eq. (13''). It is perhaps useful to consider the strain energy of

$$\text{rings} = (2\alpha + \tfrac{3}{2}\beta + \gamma) - (\alpha + \beta + \gamma) + 1 = \alpha + \tfrac{1}{2}\beta + 1 \tag{13''}$$

bicyclobutane vs. that arising from two cyclopropane rings [66.5 vs. 2(28.2) kcal/mole], or cubane vs. that arising from five (!) (independent) cyclobutane rings [157 vs. 5(27.4) kcal/mole]. The difference of these quantities may be labeled "superstrain." Indeed, the strain energy per ring appears to be a useful criterion. However, while the strain energy of bicyclo(4.1.0)heptane (30.3 kcal/mole) is comparable to that of the sum of its three- (28.2) and six- (1.4) membered rings, the strain energy per ring, $SE_r$, would give 15.1 kcal/ mole for the bicycle of interest. $SE_r$ thus appears to be conceptually useless.

In any discussion of rings in polycycles, it is important to know which rings are being considered. For example, in bicyclobutane there is a four-membered or cyclobutane ring as well, and cubane likewise has six nonindependent cyclobutane, 12 cyclohexane, and four cyclooctane rings. In general discussions of rings, it is necessary to be very careful to determine if one means dependent or independent. Which set of two and five rings, respectively, do we choose? One option, already implemented for nomenclature[167,168] and synthesis,[169 a,b] suggests two criteria for the set of rings: (a) it "must contain all the bonds which are members of any ring" and thus contain all of the atoms; and (b) "the sizes of the rings are as small as possible." We suspect that the sizes of the rings can be chosen by other criteria, such as "as large as possible" and "as close to six-membered, i.e., strain free, as possible."

[165] O. Ore, "Graphs and their Uses," Ch. 3 and 8. Random House, New York, 1963.

[166] D. H. Rouvray, *J. Chem. Educ.* **52**, 769 (1975).

[167] A. Zamora, *J. Chem. Inf. Comput. Sci.* **16**, 40 (1976).

[168] A. Zamora and T. Ebe, *J. Chem. Inf. Comput. Sci.* **16**, 36 (1976).

[169a] E. J. Corey and W. T. Wipke, *Science* **166**, 178 (1969).

[169b] E. J. Corey, W. T. Wipke, R. D. Cramer, III, and W. J. Howe, *J. Am. Chem. Soc.* **94**, 431 (1972).

We now briefly mention some other potential regularities in the structure and energetics of polycyclic hydrocarbons. (We intend to be neither exhaustive nor exhausting.) The first is "ring suturing,"[163] and we are reminded of the methods of "homodesmotic reactions," "group separation," and "minimal reference states," discussed in Section 1.A. The second is the "maxi-ring" hierarchy.[163] Intuitively, we expect some chemical properties such as ionization potential and ease of reaction with electrophiles to depend on the size of the largest ring. (Remember, an electron in a molecule does not know from which atom it came.) Before mentioning the remaining preliminary methods, we wish to admit that they were derived "merely" by looking at the *Journal of Chemical Information and Computer Science* and by adapting those authors' information-oriented approach to our thermochemical one. The first is the idea of "Exhaustive Enumeration of Unique Structures Consistent with Structural Information."[162,170a,b] These authors speak of a hierarchy of disjoint components $\{C_i\}$ such that

$$\bigcup_i C_i = \text{molecule} \tag{14}$$

$$C_i \cap C_j = \phi \text{ if } i \neq j \tag{14'}$$

From this approach and the "Code for Chemical Ring Compounds with Application of Fusion Lines, Suited for Calculation of their Physical Properties,"[171] it has become clear to the authors that, in general, organic chemists and thermochemists have been too limited in their choices of building blocks for molecules. The fifth and final approach entails the "Application of Artifical Intelligence for Chemical Inference" for the "Exhaustive Generation of Cyclic and Acyclic Isomers"[172] and "Computer Generation of Vertex Groups and Ring Systems."[173] "Vertex-graphs consist of the vertices ('nodes') which represent the points of ring fusion in pure cyclic molecules together with the paths ('edges') which interconnect these nodes." This approach legitimizes our interrelation of paddlanes, propellanes, bicycloalkenes, and cycloalkynes to be discussed in Section 6.B. We conclude this section by noting that we feel that "information science" has an honorable place in the "mental stockroom" of any organic chemist and/or thermochemist.

[170a] Y. Kudo and S.-I. Saskai, *J. Chem. Doc.* **14**, 200 (1974).

[170b] Y. Kudo, Y. Hirota, S. Aoki, Y. Takada, T. Taji, I. Fujioka, K. Higashino, H. Fujishima, and S.-I. Sasaki, *J. Chem. Inf. Comput. Sci.* **17**, 50 (1977).

[171] M. J. Romancec, *J. Chem. Doc.* **14**, 49 (1974).

[172] L. M. Masinter, N. S. Sridharan, J. Lederberg, and D. H. Smith, *J. Am. Chem. Soc.* **96**, 7702 (1974).

[173] R. E. Carhart, D. H. Smith, H. Brown, and N. S. Sridharan, *J. Chem. Inf. Comput. Sci.* **15**, 124 (1975).

CHAPTER 5

# Kinetic and Thermodynamic Stability

## 5.A Introduction

Throughout this book the term "stability" has been employed in a variety of ways. In this chapter we will attempt to clarify some of the ambiguities in our use of this word. We immediately recommend to the reader an excellent review article[1] which analyzes some of the problems raised in the use and abuse of this term. For example, the article notes that a species such as benzyl radical may be stabilized ($D[C_6H_5CH_2$—H] < $D[CH_3CH_2$—H], where $D$ is the dissociation enthalpy of the specified bond), yet transient, while 2,4,6-tri-*tert*-butylphenyl radical is destabilized ($D[2,4,6-(t\text{-Bu})_3C_6H_2$—H] > $D[CH_3$-CH$_2$—H]), but is relatively persistent (it has a relatively long half-life). In this instance, steric effects are the primary means for causing a destabilized radical to be persistent. In this book many cases have been cited in which inherently destabilized structures have been made amenable to observation and even isolation through protection of susceptible sites with bulky groups. Perhaps the most spectacular example to date is that of the cyclobutadiene system. Cyclobutadiene dimerizes at temperatures well below $-200°C$, while tri-*tert*-butylcyclobutadiene, which should be as antiaromatic as the parent molecule, can be observed via nmr in solution at $25°C$ (see Section 3.D.3). In the context of this book, we choose to define a destabilized molecule as one having (or anticipated to have) a higher enthalpy content than model calculations would indicate [e.g., comparison of $\Delta H_f^0$ (cyclobutadiene) with the sum of four standard $\Delta H_f^0[C_{olef.}(H)(C)]$ group increments.[2] Another example in

---

[1] D. Griller and K. U. Ingold, *Acc. Chem. Res.* **9**, 13 (1976).
[2] S. W. Benson, "Thermochemical Kinetics," 2nd Ed. Wiley, New York, 1976.

which steric constraints impart longevity to a destabilized system is the case of the bicyclic molecule **1**, whose encapsulated bridgehead proton resists abstraction, which would lead to an aromatic and less strained [8]paracyclophane derivative (**2**)[3] (see Section 3.H.3).

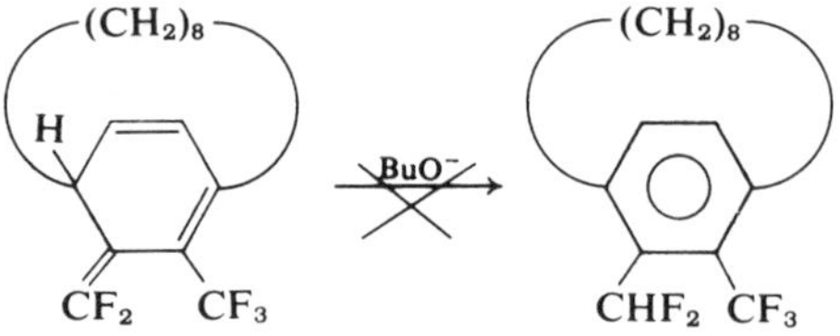

$$(1)$$

Another means of endowing a system with improved kinetic stability (persistence) is illustrated through comparison of the reactivities of 1-methylcyclopropene and 3,3-dimethylcyclopropene. Neat liquid 1-methylcyclopropene dimerizes within minutes at room temperature via an ene reaction [Eq. (2)],[4] but 3,3-dimethylcyclopropene, which is every bit as strained but lacks the requisite 3-hydrogen, can be heated for several days at 100°C in a sealed tube without significant decomposition.[5]

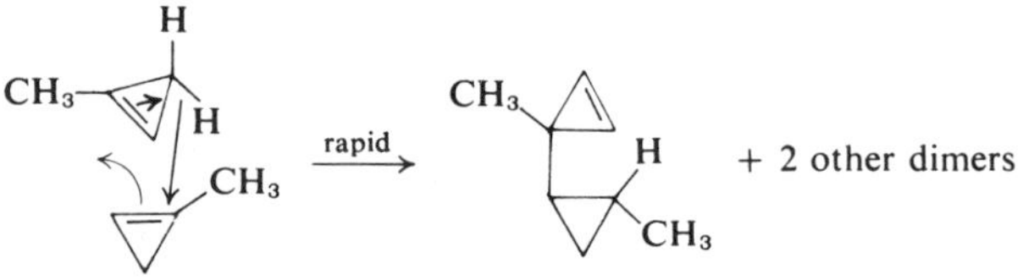

$$(2)$$

Rearrangements of strained systems in which anchimeric assistance plays a role are well documented.[6] The critical importance of the proper alignment of neighboring groups including heteroatoms, olefinic linkages, cyclopropane rings, aromatic rings, and carbon–carbon single bonds may be illustrated by noting the tendency of *exo*-twist-brendan-2-yl brosylate (**3**) to rearrange to the "brexyl" system (**4**) rather than the more stable "brendyl" cation (**5**) during acetolysis.[7] Here, departure of the leaving group is facilitated by concerted migration of the properly aligned *x* linkage in **3**.[7] Certain other molecules which can release a good deal of energy upon rearrangement are, nonetheless, long-lived because they lack facile configurations for anchimeric assistance.

[3] P. G. Gassman, S. R. Korn, and R. P. Thummel, *J. Am. Chem. Soc.* **96**, 6948 (1974).

[4] F. J. Weigert, R. L. Baird, and J. R. Shapley, *J. Am. Chem. Soc.* **92**, 6630 (1970).

[5] G. L. Closs, L. E. Closs, and W. A. Boll. *J. Am. Chem. Soc.* **85** 3796 (1963).

[6] See many review articles in G. A. Olah and P. von R. Schleyer, eds., "Carbonium Ions," Vols. 1–4. Wiley (Interscience), New York, 1968–1972.

[7] A. Nickon and R. C. Weglein, *J. Amer. Chem. Soc.*, **97**, 1271 (1975).

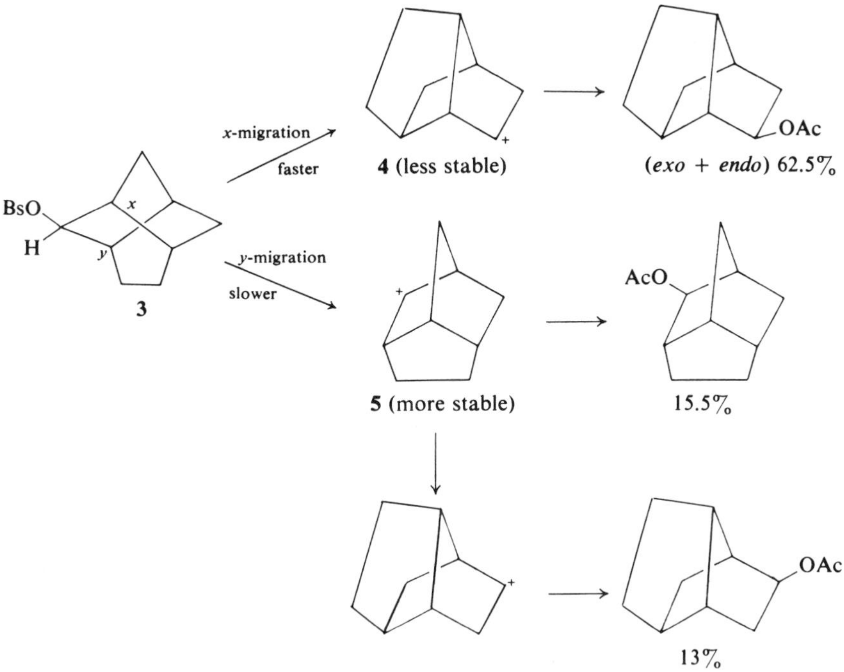

**Scheme 1**

In this chapter, some explicit consideration will be given to the electronic (electrostatic and resonance) effects of substituents upon strained rings. Substituents such as trifluoromethyl appear to promote changes in both the inherent strain in molecular systems and the reactivity of such molecules. We will also examine heteroanalogues of strained systems. While the slight changes in bond lengths and angles in a heterocycle such as ethylene oxide do not make it significantly more strained (destabilized) than cyclopropane, the heteroatom makes this ring highly reactive to electrophiles and nucleophiles. Three-membered rings containing heteroatoms may also open fairly readily to yield reactive 1, 3-dipolar species; this adds another dimension to their reactivity. Furthermore, constraint of a heteroatom such as tricoordinate nitrogen in an unnatural local geometry will exert some interesting effects upon its basicity and nucleophilicity, and such special effects are also of interest.

We will also spend some time discussing the role of the principle of the conservation of orbital symmetry and related theories in rationalizing and predicting the surprising persistence (kinetic stability) of some highly destabilized molecules. Although there are many such cases, we will emphasize the family of benzene valence isomers $(CH)_6$ which have provided some of the strongest support for this modern theory of organic chemistry. Shortly after the introduction of the conservation of orbital symmetry rules in 1965, it became apparent that transition metals could help molecules to seemingly violate these strictures. Starting about 1970, a secondary explosion of published work has occurred in this field that followed the initial interest in uncatalyzed reactions five years earlier. Aside from theoretical interest and synthetic utility, this combination of orbital symmetry-controlled and metal-catalyzed rearrangements may have practical application in various light-storage–latent-heat systems. For example, light will cause norbornadiene (6) to isomerize to the higher-energy quadricyclane (7), which is very stable to heat. Various transition metals, however, allow this molecule to rapidly revert to norbornadiene and release some of its stored energy. For example, the $Q$ value (a measure of energy storage capability) of 9 is about 8% when it is obtained photochemically from 8.[8] Molecule 9 is thermally stable to 295°C but can be converted (only slowly, unfortunately) into 8 with rhodium(I) catalysts.[8]

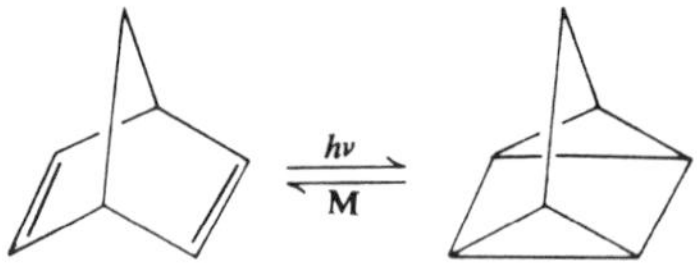

(3)

6    7

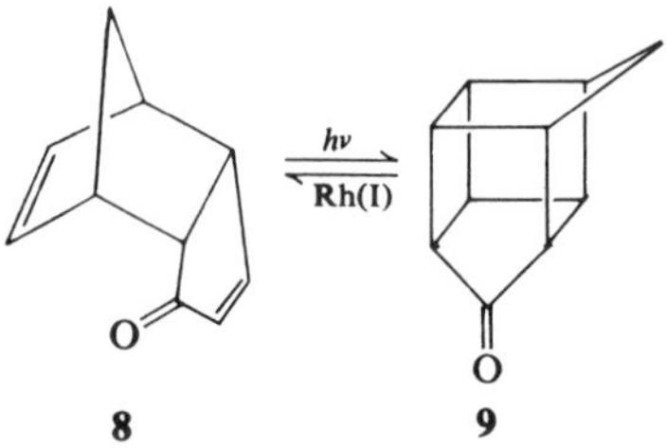

(4)

8    9

We will begin by discussing orbital symmetry and its role in the chemistry of benzene valence isomers, follow this by considering the effects of transition metals upon the structures and stabilities of strained molecules, and conclude the chapter by discussing heteroanalogues and substituent effects.

## 5.B    Orbital Symmetry and Stabilities of Benzene Valence Isomers

### 5.B.1    CONSERVATION OF ORBITAL SYMMETRY ILLUSTRATED

The reader has already been exposed to numerous highly strained molecules capable of releasing considerable energy in some seemingly facile ways.

[8] G. Jones, III, and B. R. Ramachandran, *J. Org. Chem.*, **41**, 798 (1976).

For example, while cyclobutane is somewhat more stable than two ethylenes [$\Delta G^0$(cyclo-$C_4H_8 \rightarrow 2C_2H_4$) $= +6.3$ kcal/mole; $\Delta H^0$(cyclo-$C_4H_8 \rightarrow 2C_2H_4$) $= +18.6$ kcal/mole], the analogous cycloreversion of quadricyclane (**7**) to norbornadiene (**6**) [Eq. (3)] is exothermic by $27.13 \pm 0.24$ kcal/mole at 25°C.[9] In spite of the considerable driving force for the latter reaction and a reaction pathway in which nuclear movement is minimized, temperatures around 200°C are required for this reaction. Similar considerations are evident in the analogous cycloreversion of cubane (**10**) to *syn*-tricyclo[4.2.0.0$^{2,5}$]octa-3,7-diene (**11**) (see discussion in Section 5.C.5.6). Explanations of these large activation barriers currently invoke the conservation of orbital symmetry and related approaches.[10–19]

$$\text{(5)}$$

**10**        **11**

The orbital symmetry approach may be employed to analyze the high barrier for cycloreversion of cyclobutane into two molecules of ethylene (or analogously **6** ⇌ **7** and **10** ⇌ **11**). Figure 5.1 is a correlation diagram for this ($_\sigma 2_s + {}_\sigma 2_s$) cycloreversion or ($_\pi 2_s + {}_\pi 2_s$) cycloaddition reaction. The notation $_\sigma 2_s$ refers to a σ bond between two atomic centers that interacts with another group (in this case one of the same type) in a suprafacial manner (that is, on the same side of the bonding surface). Similarly, $_\pi 2_s$ refers to a two-center π system interacting suprafacially, while substitution of "a" for "s" would indicate antarafacial interaction (opposite bonding surfaces). One assumes a

[9] K. B. Wiberg and H. A. Connon, *J. Am. Chem. Soc.* **98**, 5411 (1976).

[10a] R. B. Woodward and R. Hoffmann, *J. Am. Chem. Soc.* **87**, 395 (1965).

[10b] R. B. Woodward and R. Hoffmann, *Angew. Chem., Int. Ed. Engl.* **8**, 789 (1969).

[11] H. C. Longuet-Higgins and E. W. Abrahamson, *J. Am. Chem. Soc.* **87**, 2045 (1965).

[12] H. E. Zimmerman, *Acc. Chem. Res.* **5**, 393 (1972) and references cited therein.

[13] M. J. S. Dewar, *Angew. Chem., Int. Ed. Engl.* **10**, 761 (1971) and references cited therein.

[14] K. Fukui, *Acc. Chem. Res.* **4**, 57 (1971) and references cited therein.

[15] R. G. Pearson, *Acc. Chem. Res.* **4**, 152 (1971) and references cited therein.

[16] W. T. van der Hart, J. J. C. Mulder, and L. J. Oosterhoff, *J. Am. Chem. Soc.* **94**, 5724 (1972).

[17] N. D. Epiotis, *Angew. Chem., Int. Ed. Engl.* **13**, 751 (1974) and references cited therein

[18] E. A. Halevi, *Angew. Chem., Int. Ed. Engl.* **15**, 593 (1976) and references cited therein.

[19] R. E. Lehr and A. P. Marchand, "Orbital Symmetry—A Problem Solving Approach." Academic Press, New York, 1972, examines alternative views in an elementary, highly readable manner.

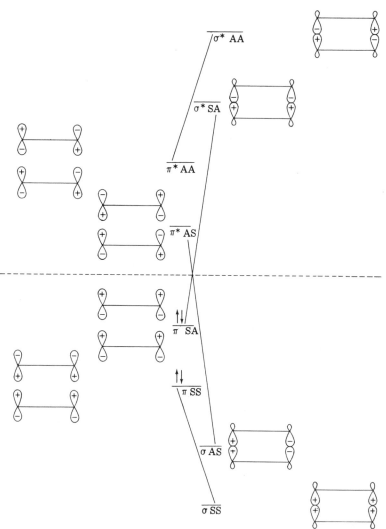

**Figure 5.1.** Correlation diagram for the orbitals comprising the bonds broken and formed in the thermally "forbidden" dimerization of ethylene to form cyclobutane. The orbitals are classified as symmetric (S) or antisymmetric (A) with respect to the planes which pass through each pair of C—C bonds in cyclobutane.

concerted reaction in which bond forming occurs simultaneously with, although not necessarily to the same extent as, bond breaking, and classifies the relevant orbitals in the reactants and products according to symmetry. If there is a smooth interconversion between bonding molecular orbitals in reactant and product, the reaction is termed "orbital symmetry allowed," bond forming and bond breaking occur concomitantly, and a lower energy barrier can usually be expected, in contrast to cases in which the initial step is simple bond breakage. Such concerted reactions occur with a high degree of stereospecificity. The Diels–Alder reaction is usually regarded as such a case. From Fig. 5.1 it is apparent that not all the ground-state molecular orbitals of cyclobutane correlate with ground-state molecular orbitals of ethylene molecules. This can be taken as an indication that this orbital-symmetry-"forbidden" thermal reaction should have a high barrier although it can and does occur. Photochemical interconversions between cyclobutane and two ethylenes in the suprafacial–suprafacial manner are orbital-symmetry-allowed.

Symmetry-forbidden concerted reactions do occur,[20] and symmetry-allowed concerted reactions sometimes have fairly high activation barriers if geometric and/or steric constraints are present. An example is the high-temperature pyrolysis of tricyclo[4.1.0.0$^{2,7}$]heptane (12), thought to initially yield the highly strained *trans*-cycloheptadiene 13, a species rapidly isomerizing to *cis*-bicyclo[3.2.0]hept-6-ene (14).[21] Woodward and Hoffmann[10b]

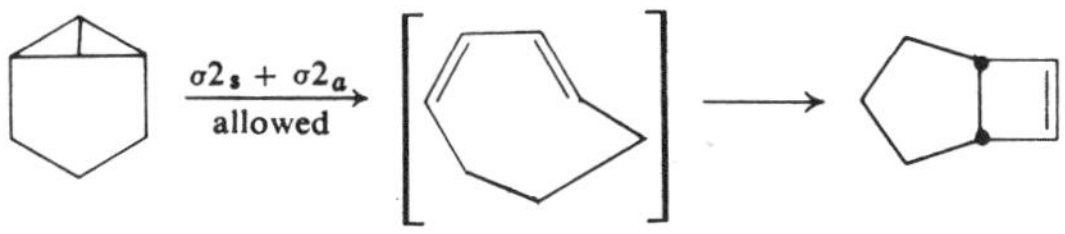

(6)

    12           13 (postulated)         14

note that while fairly smooth symmetry-allowed interconversions between 1,3,5,7-octatetraenes (15) and 1,3,5-cyclooctatrienes (16) occur, the analogous highly exothermic conversion of hydrocarbon 17[22] to 9,9'-bianthryl (18) is difficult because of geometric constraints which make 17 rather stable thermally.

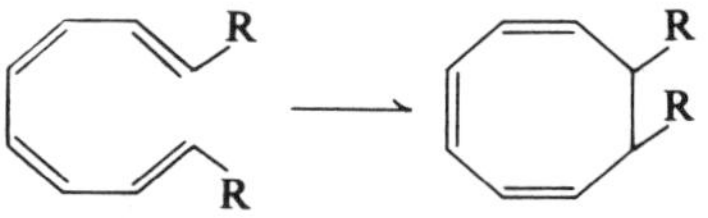

(7)

    15           16

[20] J. A. Berson, *Acc. Chem. Res.* **5**, 406 (1972).
[21] K. B. Wiberg and G. Szeimies, *Tetrahedron Lett.* p. 1235 (1968).
[22] N. M. Weinshenker and F. D. Greene, *J. Am. Chem. Soc.* **90**, 506 (1968).

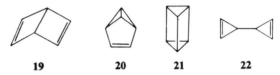

$$\textbf{17} \xrightarrow[t_{1/2}\ =\ 30\ \text{min}]{80°C,} \textbf{18} \qquad (8)$$

**17**                                          **18**

## 5.B.2 BENZENE VALENCE ISOMERS

Valence isomers[23] of benzenes[24–26] (**19–22**) are of historical[27] as well as modern theoretical, conceptual, and mechanistic interest. The recent syntheses

**19**          **20**     **21**     **22**

of **19–21** and derivatives of all four valence isomers are triumphs of modern synthetic artistry. Bicyclo[2.2.0]hexa-2,5-diene ("Dewar benzene," **19**) and tetracyclo[2.2.0.0.$^{2,6}$0$^{3,5}$]hexane ("Ladenberg benzene" or "prismane," **21**) are two of the most spectacular examples of molecules whose thermal stability is contrary to intuition but in accord with the theory of the conservation of orbital symmetry.[10a,b] As the result of their high strain energies, the valence isomers offer very feasible access (in terms of enthalpy) to some of the electronically excited states of the corresponding benzenoid compounds. Thus, "photochemistry without light" has been an active field of research for this class of compounds.[28]

Table 5.1 lists experimental and calculated relative enthalpies, respectively, of (CH)$_6$ derivatives and the parent compounds as well as observed thermal stabilities of **19–21**. The first substituted benzene valence isomer to be isolated was hexaphenyl-3,3′-bicyclopropenyl, obtained upon treatment of *syn*-triphenylcyclopropenyl bromide with zinc dust in benzene in what started

[23] L. T. Scott and M. Jones, Jr., *Chem. Rev.* **72**, 181 (1972).

[24a] E. E. van Tamelen, *Angew. Chem., Int. Ed. Engl.* **4**, 738 (1965).

[24b] E. E. van Tamelen, *Acc. Chem. Res.* **5**, 186 (1972).

[25] H. G. Viehe, *Angew. Chem., Int. Ed. Engl.* **4**, 746 (1965).

[26] D. Bryce-Smith and A. Gilbert, *Tetrahedron* **32**, 1309 (1976).

[27] W. Baker, *Chem. Brit.* **1**, 191 (1965).

[28] N. J. Turro, P. Lechtken, N. E. Schore, G. Schuster, H. C. Steinmetzer, and A. Yekta, *Acc. Chem. Res.* **7**, 97 (1974).

as an attempt to study a cyclopropenyl radical.[43] It is ironic that this molecule is a derivative of the $(CH)_6$ isomer of highest enthalpy (Table 5.1) and that 3,3'-bicyclopropenyl itself today remains the only $(CH)_6$ to elude synthesis. 1,2,5-Tri-*tert*-butyl-Dewar benzene (**24**) staggers its vicinal *tert*-butyl groups, and the realization that this would increase its stability relative to the crowded 1,2,4-tri-*tert*-butylbenzene (**23**) inspired its photochemical synthesis.[34] One year later, a "rational" synthesis of Dewar benzene (**19**) appeared.[35] This

(9)

23                    24

first example of a parent benzene valence isomer evidenced startling thermal stability. Conversion into benzene would at first glance appear to involve relatively small and smooth changes in the coordinates of the nuclei and disposition of electrons in this puckered molecule. Its reluctance to aromatize was a mystery at the time. However, orbital-symmetry arguments explain the high isomerization barrier by showing that concerted, thermally allowed, conrotatory ring closure of **19** would yield the unlikely *cis,cis,trans*-1,3,5-cyclohexatriene, while disrotatory ring closure of Dewar benzene to benzene does not correlate ground states and is therefore thermally forbidden (Fig. 5.2). A related explanation[13,44] is that the transition state for concerted conversion of Dewar benzene to benzene is antiaromatic (i.e., cyclobutadiene-like) and high in energy. Dewar benzene (along with benzvalene and fulvene) has also been obtained in very low quantum yield through vacuum uv (1650–2000 Å) irradiation of liquid benzene (benzene vapor yields only fulvene at these wavelengths).[45] It has been suggested[26] that the electronically excited state $(S_2)$ of benzene generated in the vacuum uv has considerable 1,4 bonding character which would facilitate the conversion to **19**. The earlier-cited approach of van Tamelen to obtain **24** from crowded benzenes has been employed to make other substituted Dewar benzenes[46,47] as well as the isomer **26** from 1,3,6,8-tetra-*tert*-butylnaphthalene (**25**), which has a severely repulsive *peri* interaction.[48] A different approach, involving reaction of 9,10-dichloro-9,9',10,10'-dianthracene with triphenylmethylsodium in benzene–

[43] R. Breslow and P. Gal, *J. Am. Chem. Soc.* 81 4747 (1959).

[44] R. Breslow, J. Napierski, and A. H. Schmidt, *J. Am. Chem. Soc.* **94**, 5906 (1972).

[45] H. R. Ward and J. S. Wishnok, *J. Am. Chem. Soc.* **90**, 1085, 5353 (1968).

[46] W. Schaefer, *Angew. Chem., Int. Ed. Engl.* **5**, 669 (1966).

[47] E. M. Arnett and J. M. Bollinger, *Tetrahedron Lett.* p. 3803 (1964).

[48] W. L. Mandella and R. W. Franck, *J. Am. Chem. Soc.* **95**, 971 (1973).

#### Table 5.1
#### Historical, Experimental, and Theoretical Data Pertaining to Benzene Valence Isomers $(CH)_6$

| Isomer | Initial publ. | | Relative enthalpies (kcal/mole) | | | | | Thermal stability of parent |
|---|---|---|---|---|---|---|---|---|
| | Substd. | Parent | Expt. | Expt.[30] | Theor.[31] (MINDO/2) | Theor.[32,33] (ab initio) | Bond-additiv. | |
| Benzene | | ca. 1825 | 0 | 0 | 0 | 0 | 0 | — |
| Dewar benzene (19) | 1962[34] | 1963[35] | 59.5[a] | 28.0[b] | 49.5 | 139.6 | 68[e] | $t_{1/2}^{RT}$ 2 days[35,36] |
| Benzvalene (20) | 1964[37] | 1967[38] | 67.5[d] | 34.4[b] | 58.9 | 100.1 | 62[f] | $t_{1/2}^{RT}$ 10 days[38] |
| Prismane (21) | 1965[c] 1966[40,41] | 1973[42] | 91.2[a] | 59.0[b] | 80.9 | 201.3 | 108[g] | stable$(RT)$, $t_{1/2}^{90°C}$ 11 hr[42] |
| 3,3'-Bicyclopropenyl (22) | 1959[43] | — | — | — | — | — | 119[h] | — |

[a] Hexamethyl series (Oth[29]).

[b] Hexakis(trifluoromethyl) series.

[c] A tri-*tert*-butyl prismane was obtained with 82% purity.[39]

[d] N. J. Turro, C. A. Renner, T. J. Katz, K. B. Wiberg, and H. A. Cannon, *Tetrahedron Lett.* p. 4133 (1976).

[e] $\Delta H_f$(19) = +88 kcal/mole: average of 1. and 2. below:
 1. 2 $\Delta H_f$(cyclobutene) + $\Delta H_f$(bicyclo[2.2.0]hexane) − 2 $\Delta H_f$(cyclobutane)
 2. $\Delta H_f$(1,4-cyclohexadiene) + $\Delta H_f$(bicyclo[2.2.0]hexane) − $\Delta H_f$(cyclohexane).

[f] $\Delta H_f$(20) = +82 kcal/mole: average of 3. and 4. below:
 3. $\Delta H_f$(2,4-dimethylbicyclobutane)(est) + $\Delta H_f$(cis-2-butene) − 2 $\Delta H_f$(ethane)
 4. $\Delta H_f$(cyclopentene) + $\Delta H_f$(bicyclobutane) − $\Delta H_f$(propane).

[g] $\Delta H_f$(21) = $\Delta H_f$(bicyclo[2.2.0]hexane) + 2 $\Delta H_f$(cyclopropane) + strain energy (cyclobutane) − 2 $\Delta H_f$(propane) = +128 kcal/mole.

[h] $\Delta H_f$(22) = +139 kcal/mole, average of 5. and 6. below:
 5. 2 $\Delta H_f$(3-methylcyclopropene)(est) − $\Delta H_f$(ethane)
 6. 2 $\Delta H_f$(cyclopropene) + $\Delta H_f$(bicyclopropyl) − 2 $\Delta H_f$(cyclopropane).

25                          26

(10)

27

ether, was used to obtain the "bridged Dewar anthracene" 27 ("9,10'-dehydrodianthracene").[49] This thermally stable molecule (m.p. 338°–342°C dec.) resists rearrangement to its impossibly strained [4]paracyclophane valence isomer. In contrast, tricyclo[5.2.2.0$^{1,7}$]undeca-8,10-diene (28) isomerizes rapidly, presumably through [5]paracyclophane (29),[50,51] unlike tricyclo[3.2.2.0$^{1,5}$]nona-6,8-diene (30) which, although more strained than 28,

---

[49] D. E. Applequist and R. Searle, *J. Am. Chem. Soc.* **86**, 1389 (1964).

[50] J. W. van Straten, I. J. Landheer, W. H. de Wolf, and F. Bickelhaupt, *Tetrahedron Lett.* p. 449 (1975).

[51] K. Weinges and K. Klessing, *Chem. Ber.* **109**, 793 (1976).

---

*Notes to Table 5.1*

[29] J. F. M. Oth, *Angew. Chem., Int. Ed. Engl.* **7**, 646 (1968); see also W. Adam and J. C. Chiang, *Int. J. Chem. Kinet.* **1**, 487 (1969).

[30] D. M. Lemal and L. H. Dunlap, Jr., *J. Am. Chem. Soc.* **94**, 6562 (1972).

[31] N. C. Baird and M. F. S. Dewar, *J. Am. Chem. Soc.* **91**, 352 (1969).

[32] G. Berthier, A. Y. Meyer, and L. Praud, *Jerusalem Symp. Quantum Chem. Biochem.*, **3**, 174 (1971); see also Ref. 33.

[33] M. D. Newton, J. M. Schulman, and M. M. Manus, *J. Am. Chem. Soc.* **96**, 17 (1974). In this article the geometries and wavefunctions of 19–21 are calculated but no values are given for the enthalpy differences between these isomers.

[34] E. E. van Tamelen and S. P. Pappas, *J. Am. Chem. Soc.* **84**, 3789 (1962).

[35] E. E. van Tamelen and S. P. Pappas, *J. Am. Chem. Soc.* **85**, 3297 (1963).

[36] E. E. van Tamelen, S. P. Pappas, and K. L. Kirk, *J. Am. Chem. Soc.* **93**, 6092 (1971).

[37] H. G. Viehe, R. Merenyi, J. F. M. Oth, J. R. Senders, and P. Valange, *Angew. Chem., Int. Ed. Engl.* **3**, 755 (1964).

[38] K. E. Wilzbach, J. S. Ritscher, and L. Kaplan, *J. Am. Chem. Soc.* **89**, 1031 (1967).

[39] K. E. Wilzbach and L. Kaplan, *J. Am. Chem. Soc.* **87**, 4004 (1965).

[40] D. M. Lemal and J. R. Lokensgard, *J. Am. Chem. Soc.* **88**, 5934 (1966).

[41] W. Schaefer, R. Criegee, R. Askani, and H. Gruener, *Angew. Chem., Int. Ed. Engl.* **6**, 78 (1967).

[42] T. J. Katz and N. Acton, *J. Am. Chem. Soc.* **95**, 2738 (1973).

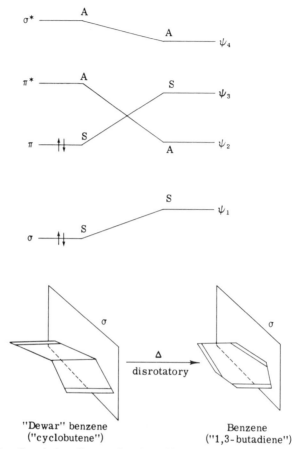

**Figure 5.2.** Correlation diagram for the orbitals comprising the bonds broken and formed in the thermally "forbidden" disrotatory ring opening of "Dewar" benzene to benzene analogous to the ring opening of cyclobutene to 1,3-butadiene.

lacks a viable aromatization pathway and is much more stable thermally (no reaction during flow pyrolysis at 300°C).[52] If we consider a "worst-case" calculation of the destabilization energy of [5]paracyclophane we might consider (a) a strain of 24 kcal/mole equal to twice the destabilization of the olefinic linkage in bicyclo[3.3.1]non-1-ene,[53] plus (b) loss of 36 kcal/mole of benzene resonance energy. The total destabilization of 60 kcal/mole (which has, however, neglected the pentamethylene chain) is still less than the

[52] I. J. Landheer, W. H. de Wolf, and F. Bickelhaupt, *Tetrahedron Lett.* p. 349 (1975).
[53] P. M. Lesko and R. B. Turner, *J. Am. Chem. Soc.* **90**, 6888 (1968).

enthalpy difference between Dewar benzene and benzene. Molecular-mechanics calculations predict that **29** is more stable than **28** by more than 25 kcal/mole.[54] Tricyclo[4.2.2.0$^{1,6}$]dec-7,9-diene (**31**)[55,56a] is also quite stable,

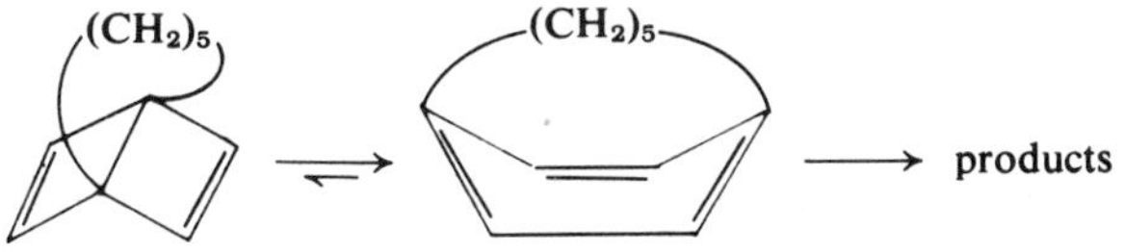

$$\text{(11)}$$

**28**            **29** (postulated)

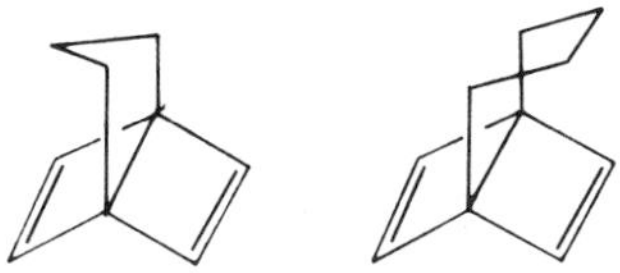

**30**            **31**

as is its analogue **27**, but can be induced in a flow system at 300°C to undergo a *retro*-Diels–Alder reaction to form ethylene and *p*-xylylene. The analogous reaction of **27** would generate benzyne and an extremely strained diolefin and does not occur. The Dewar isomer formed through photolysis of [6]-paracyclophane rearomatizes neatly ($E_{act} = 19.9$ kcal/mole).[56b] A twisted double bond in **33** speeds its aromatization to tetralin ($t_{1/2} = 58$ min),[56a] while **35** is so strained that it isomerizes too rapidly to be observed even at low temperatures.[52]

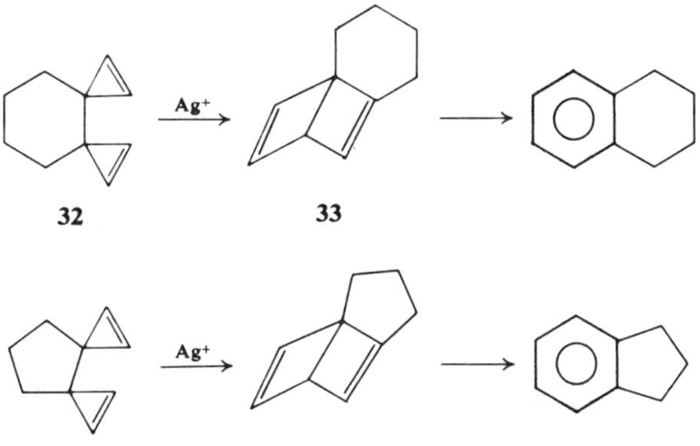

$$\text{(12)}$$

**32**            **33**

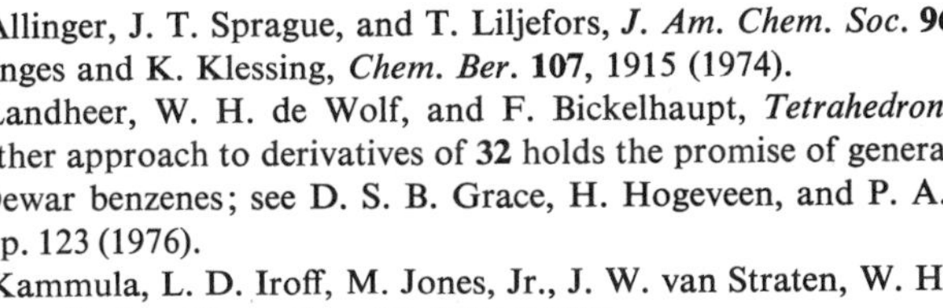

$$\text{(13)}$$

**34**            **35**

[54] N. L. Allinger, J. T. Sprague, and T. Liljefors, *J. Am. Chem. Soc.* **96**, 5100 (1974).

[55] K. Weinges and K. Klessing, *Chem. Ber.* **107**, 1915 (1974).

[56a] I. J. Landheer, W. H. de Wolf, and F. Bickelhaupt, *Tetrahedron Lett.* p. 2813 (1974). Another approach to derivatives of **32** holds the promise of generating (1,2)(3,4)-dibridged Dewar benzenes; see D. S. B. Grace, H. Hogeveen, and P. A. Wade, *Tetrahedron Lett.* p. 123 (1976).

[56b] S. L. Kammula, L. D. Iroff, M. Jones, Jr., J. W. van Straten, W. H. de Wolf, and F. Bickelhaupt, *J. Am. Chem. Soc.*, **99**, 5815 (1977).

Hexamethyl Dewar benzene, although not obtained photolytically from hexamethylbenzene,[40] is the most readily available hydrocarbon benzene valence isomer because it is obtained through alumina-catalyzed trimerization of 2-butyne.[46] Its central bond, at 1.63 Å, is one of the longest known carbon–carbon linkages.[57]

The first derivatives of benzvalene (tricyclo[3.1.0.0$^{2,6}$]hexene, 20) were obtained in photoisomerizations of substituted benzenes.[37,39] The parent hydrocarbon was produced in low yield (1% maximum) upon photolysis of liquid benzene at 2537 Å[39] and photolysis of the vapor in the presence of butene and neopentane at 2537 Å[58] (earlier reports of photolytic production of benzvalene in the vapor phase using vacuum uv apparatus are now known to be in error[59]). The production of benzvalene using the first absorption band of benzene has been rationalized in terms of a symmetry-allowed excited-state reaction[23] as well as through intermediacy of diradical 36 whose formation would be helped by the 1,3-bonding character of the $S_1$ state of benzene.[26] Benzvalene has also been produced in low yield during liquid-phase vacuum uv photolysis of benzene.[45] This compound can now be

$$\text{36} \qquad \qquad \qquad \text{Li}^+ + CH_2Cl_2 + CH_3Li \xrightarrow{CH_3OCH_3} \text{20} \qquad (14)$$

obtained in reasonable yield, but not in quantity (it is explosive in the pure form), through a "rational" synthetic pathway [Eq.(14)].[60] In isohexane solution benzvalene aromatizes very slowly ($t_{1/2}^{RT}$ 10 days[38]), but it decomposes explosively upon touch in even small quantities when pure.[60] Although the thermal aromatization of benzvalene has been classified as symmetry-forbidden and postulated to proceed via 36, ambiguities in the orbital-symmetry analysis have been discussed.[61a–d] The isomerization has been calculated to be thermally allowed using the criterion that bonding and anti-bonding orbitals do not cross during aromatization.[61a–d] The thermal stability of benzvalene in isohexane may be the result of a highly strained transition state operating even in the presence of such calculated noncrossing

[57] M. J. Cardillo and S. H. Bauer, J. Am. Chem. Soc. 92, 2399 (1970).

[58] L. Kaplan and K. E. Wilzbach, J. Am. Chem. Soc. 90, 3291 (1968).

[59] L. Kaplan and K. E. Wilzbach, J. Am. Chem. Soc. 89, 1030 (1967).

[60] T. J. Katz, E. J. Wang, and N. Acton, J. Am. Chem. Soc. 93, 3782 (1971).

[61a] M. J. S. Dewar and S. Kirschner, J. Am. Chem. Soc. 97, 2931 (1975).

[61b] N. J. Turro, C. A. Renner, T. J. Katz, K. B. Wiberg, and H. A. Cannon, Tetrahedron Lett. p. 4133 (1976).

[61c] U. Bruger and F. Mazenod, Tetrahedron Lett. p. 2885 (1976).

[61d] J. J. C. Mulder, J. Am. Chem. Soc., 99, 5177 (1977).

of excited- and ground-state molecular orbitals. Recent evidence (to be discussed later in this section) strongly supports this view, since thermally induced aromatization of benzvalene clearly does not produce excited-state molecules, even though the reaction is energy-sufficient.[61b]

Benzvalenes have been shown to be intermediates in a number of interesting photochemical reactions of benzene [Eqs. (15)[39], (16)[60,62,63], and (17)[64]]. However, a word of caution has appeared in the literature regarding the postulation of benzene valence isomer intermediates in cases where other mechanisms are possible.[65a,b] The degree of substitution required to absolutely

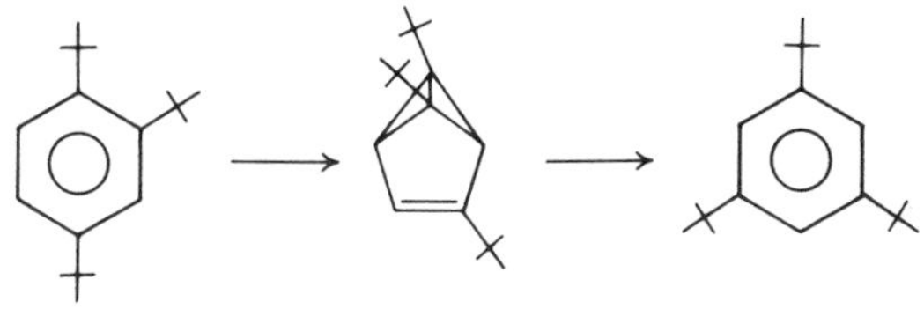

$$(15)$$

$$(16)$$

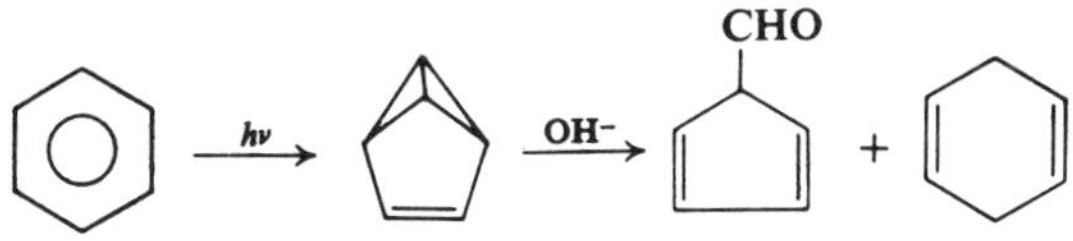

$$(17)$$

substantiate those mechanisms where benzvalene or other $(CH)_6$ are not isolable in the unsubstituted cases might invalidate extrapolations to the parent systems.

Although neither benzene nor hexamethylbenzene has been induced to isomerize photolytically to prismane and hexamethylprismane, other substituted benzenes are converted in this manner to the corresponding prismanes.[37,39] For example, photolysis at 2537 Å of both 1,2,4-tri-*tert*-butylbenzene and 1,3,5-tri-*tert*-butylbenzene produces a photoequilibrium having a tri-*tert*-butylprismane as the major isomer (ca. 65%).[39] This molecule also can be obtained through photolysis of **24**.[39] Hexamethylprismane (m.p. 89°–90.5°C) can be isolated following photolysis at 2537 Å of hexamethyl Dewar benzene.[40] The synthesis and behavior of prismane itself present numerous interesting

[62] L. Kaplan, J. S. Ritscher, and K. E. Wilzbach, *J. Am. Chem. Soc.* **88**, 2881 (1966).

[63] J. A. Berson and N. M. Hasty, Jr., *J. Am. Chem. Soc.* **93**, 1549 (1971).

[64] L. Kaplan, L. A. Wendling, and K. E. Wilzbach, *J. Am. Chem. Soc.* **93**, 3821 (1971).

[65a] J. A. Barltrop and A. C. Day, *J. Chem. Soc., Chem. Commun.* p. 177 (1975).

[65b] J. A. Barltrop, R. Carder, A. C. Day, J. R. Harding, and C. Samuel, *J. Chem. Soc., Chem. Commun.* p. 729 (1975).

aspects. Trost and Cory[66] were able to synthesize a logical precursor, diazo compound (37), but their attempts to photolyze this molecule and transform it to prismane at low temperatures were frustrated. Katz's group, following a suggestion by Turro, performed the successful photolysis of 37 at elevated temperature and was rewarded by the finding that prismane survived heat while in solution ($t^{90°}_{1/2}$ 11 hr).[67] It is interesting to note that Katz et al. used benzvalene as a starting material for prismane. The strain in prismane makes it explosive, but its slow aromatization in solution is again in line with

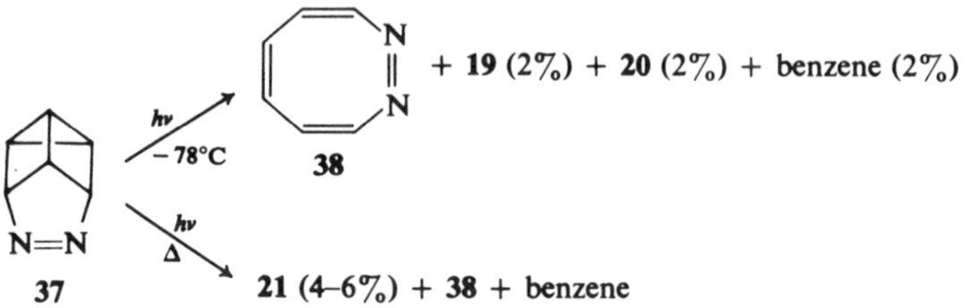

**Scheme 2**

orbital-symmetry predictions of a thermally forbidden transformation to benzene. For this reason, prismane has been likened to "an angry tiger unable to break out of a paper cage."[10b] (Energy-requiring conversion to 3,3'-bicyclopropenyl, and energy-releasing isomerization to Dewar benzene are also thermally forbidden, but a symmetry-allowed, although strained, exothermic pathway to benzvalene has been noted.[23]) The strain in prismane may be estimated at about 128 kcal/mole based upon the standard heat of formation of hexamethylprismane ( + 70.4 kcal/mole based upon the data in Table 5.1 and the standard heat of formation of hexamethylbenzene) and group incremental schemes. This is about equal to the sum of the strain energies of three cyclobutane and two cyclopropane faces.

3,3'-Bicyclopropenyl (22) is topologically unique among the benzene valence isomers in that it cannot be viewed as a multigraph resulting from pairwise connections of hexagon vertices[33] as the other four $(CH)_6$ can (benzene and 19–21 have "maxi-rings" of six; see Chapter 4). In Table 5.1, bond-additivity calculations are seen to suggest that 3,3'-bicyclopropenyl is less stable by 11 kcal/mole than prismane even though it is less strained than the latter (simply add the strain energies of two cyclopropene rings and compare that to the strain energies of three cyclobutane and two cyclopropane faces). Not surprisingly, derivatives of 22 aromatize, but they "take their time" because reasonable, energy-releasing entries onto the benzene manifold via concerted pathways are symmetry forbidden. For example, hexaphenyl-3,3'-bicyclopropenyl, the first member of this series, can be held for a few

[66] B. M. Trost and R. M. Cory, J. Am. Chem. Soc. 93, 5573 (1971).
[67] T. J. Katz and N. Acton, J. Am. Chem. Soc. 95, 2738 (1973).

**39**            **40** (*dl*)        **41** (*meso*)

seconds at its melting point (225°–226°C) before it resolidifies as hexaphenyl-benzene.[43] The simplest derivative presently characterized is 3,3'-dimethyl-3,3'-bicyclopropenyl (**39**),[68] which aromatizes to the three xylenes[69] but also suffers Cope rearrangement[69,70] to form the *dl* and *meso* isomers of 1,1'-dimethyl-3,3'-bicyclopropenyl (**40** and **41**). These last two compounds, as well as other 3,3'-bicyclopropenyls having 3-hydrogens, apparently aromatize more rapidly than they can dimerize by the ene reaction characteristic of simple cyclopropenes.

The observation that 2,2',3,3'-tetraphenyl-3,3'-bicyclopropenyl produced only 1,2,4,5-tetraphenylbenzene and 1,2,3,4-tetraphenylbenzene led Breslow *et al.* to postulate initial formation of an intermediate prismane and its subsequent isomerization to a Dewar benzene before aromatization occurs.[71] However, the distributions of benzene isomers from rearrangements of substituted 3,3'-bicyclopropenyls and the similarities between thermal- and silver-catalyzed reaction products led Weiss and Andrae to propose a vinyl-carbene mechanism[72] (see discussion of vinylcarbenes and their intermediacy in ring-opening of cyclopropenes in Section 3.D.2). Discovery of the Cope rearrangements of **39–41** appear to have caused these researchers to favor *anti*-tricyclohexyl diradicals as intermediates for these isomerizations as well as for aromatization.[70] Most recently, Bergman's group[73] has published persuasive evidence supporting the vinylcarbene hypothesis, which also has the virtue of relating thermal and silver-catalyzed (see Section 5.C.5c) re-arrangements of 3,3'-bicyclopropenyls. Their study involved the thermal aromatization of both **40** and **41**. According to both the prismane (Scheme 3) and the *anti*-tricyclohexyl diradical (Scheme 4) mechanisms, the *dl* (**40**) and *meso* (**41**) compounds should produce different extrapolated zero-time distributions of xylenes. The facts that **40** and **41** interconvert as well as provide

[68] W. H. de Wolf, W. Stol, I. J. Landheer, and F. Bickelhaupt, *Rec. Trav. Chim. Pays-Bas* **90**, 405 (1971).

[69] W. H. de Wolf, I. J. Landheer, and F. Bickelhaupt, *Tetrahedron Lett.* p. 179 (1975).

[70] R. Weiss and H. Kolbl, *J. Am. Chem. Soc.* **97**, 3224 (1975).

[71] R. Breslow, P. Gal, H. W. Chang, and L. J. Altman, *J. Am. Chem. Soc.* **87**, 5139 (1965).

[72] R. Weiss and S. Andrae, *Angew. Chem., Int. Ed. Engl.* **12**, 150, 152 (1973).

[73] J. H. Davis, K. J. Shea, and R. G. Bergman, *Angew. Chem., Int. Ed. Engl.* **15**, 232 (1976).

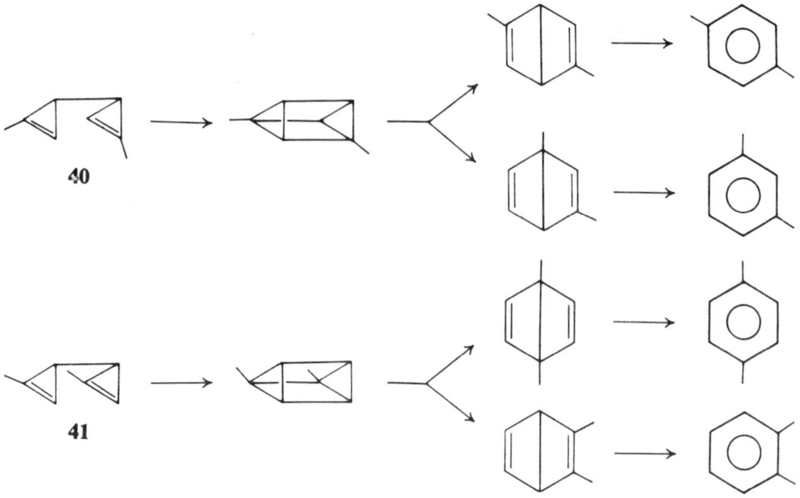

**40**

**41**

**Scheme 3**

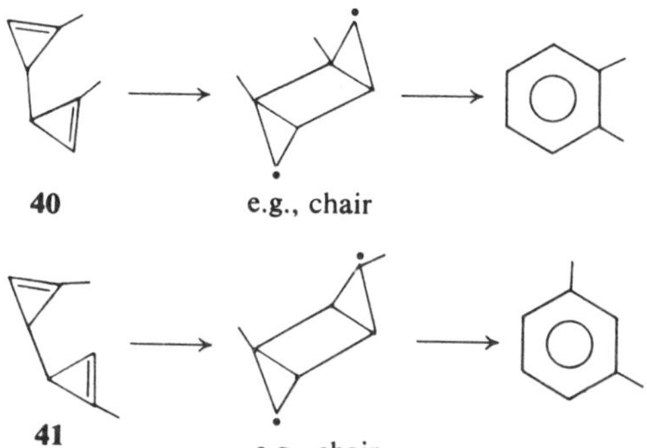

**40**      e.g., chair

**41**      e.g., chair

**Scheme 4**

within experimental error the same extrapolated zero-time isomer distribution are consistent with the formation of a vinyl carbene (or related diradical) (Scheme 5).[73] The observation that **39** produces only *o*- and *p*- but no *m*-xylene (within experimental error) at extrapolated zero-time also supports this mechanism.[73] The Dewar benzene intermediates proposed in Scheme 5 are well established in the aromatization mechanism of 3,3′-bicyclopropenyls[74] because their chemiluminescent "shadows" have been observed. Specifically, the very tiny steady-state concentration of dimethyl Dewar benzenes in the thermal aromatization of **39** is sensed by their conversion into a much

[74] N. J. Turro, G. B. Schuster, R. G. Bergman, K. J. Shea, and J. H. Davis, *J. Am. Chem. Soc.* **97**, 4758 (1975).

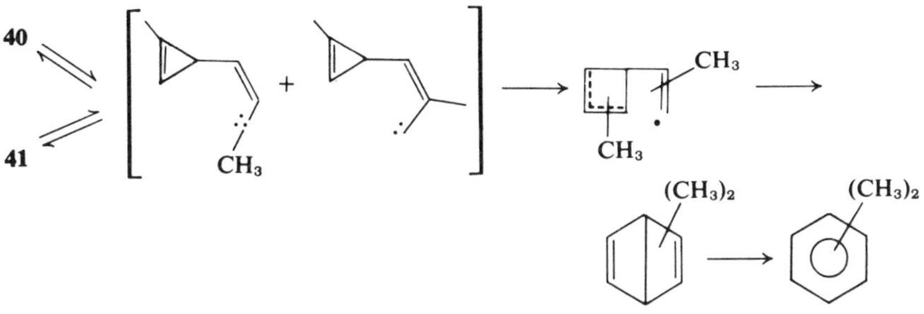

**Scheme 5**

lower concentration of short-lived (10 sec) triplet xylenes which cause fluorescence of 9,10-dibromoanthracene (DBA) placed in solution for this purpose.[75] The known thermal rearrangements of isolable Dewar benzenes to benzene triplets are reasonable because the difference in enthalpy between Dewar benzene and benzene (ca. 60 kcal/mole, see Table 5.1) and typical activation energies (ca. 25 kcal/mole) for aromatization provide access to the benzene triplet state (85 kcal/mole higher than the ground state).[75] Just as important is the orbital-symmetry-forbidden nature of the reaction, as previously noted, which precludes smooth interconversion between Dewar benzene and benzene electronic ground states.[61b] Figure 5.3 schematically depicts the energetics discussed here. The low yields of triplet benzenes produced from Dewar benzenes have been discussed in terms of an unfavorable $T \Delta S$ term,[76] but criticisms of these arguments have appeared.[77,78] It is interesting that the Dewar anthracene **42** does not generate energetically feasible electronically excited states of anthracene upon heating.[79] However, the excited singlet state of **42** exists long enough to rearrange to the excited singlet state of anthracene,[79] thereby furnishing one of the very rare cases in which an excited-state molecule, particularly a singlet, has rearranged. Related observations have been made in studies of the photochemistry of benzvalene,[80] **43**,[81] **44**,[82] **45**,[60] and **46**.[83a]

[75] P. Lechtken, R. Breslow, A. H. Schmidt, and N. J. Turro, *J. Am. Chem. Soc.* **95**, 3025 (1973).

[76] C. L. Perrin, *J. Am. Chem. Soc.* **97**, 4419 (1975).

[77] E. Lissi, *J. Am. Chem. Soc.* **98**, 3387 (1976).

[78] E. B. Wilson, *J. Am. Chem. Soc.* **98**, 3387 (1976).

[79] N. C. Yang, R. V. Carr, E. Li, J. K. McVey, and S. A. Rice, *J. Am. Chem. Soc.* **96**, 2297 (1974).

[80] C. A. Renner, T. J. Katz, J. Pouliquen, N. J. Turro, and W. H. Waddell, *J. Am. Chem. Soc.* **97**, 2568 (1975).

[81] G. D. Burt and R. Pettit, *Chem. Commun.* p. 517 (1965).

[82] R. N. McDonald, D. G. Frickley, and G. M. Moschi, *J. Org. Chem.* **37**, 1304 (1972).

[83a] T. J. Katz and K. C. Nicolaou, *J. Am. Chem. Soc.* **96**, 1948 (1974).

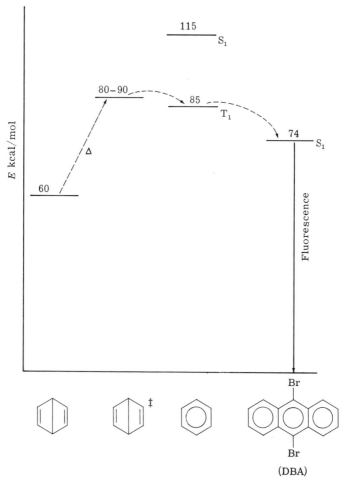

**Figure 5.3.** Energetics of the thermal production, from "Dewar" benzene, of electronically excited benzene, whose presence is sensed by the fluorescence of 9,10-dibromoanthracene (DBA). (See Lechtken *et al.*[75])

Because benzvalene is 67.5 kcal/mole less stable than benzene and has an activation barrier of 26.7 kcal/mole, the thermal aromatization satisfies the energy requirements for production of benzene triplet. However, no luminescence of DBA has been detected, which is consistent with smooth inter-conversion of ground states and the absence of triplet benzene.[61b] An optically active Dewar benzene has been recently reported which could in principle produce circularly polarized chemiluminescence.[83b] Since chemi-

[83b] J. H. Dopper, B. Greijdanus, and H. Wynberg, *J. Am. Chem. Soc.* **97**, 216 (1975).

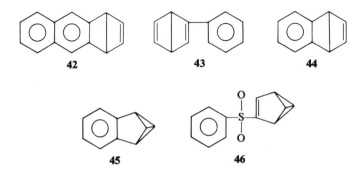

luminescence is only sensed indirectly through DBA, this effect may be lost unless a chiral benzene triplet trap is employed.

We conclude this section by noting that the most stable and available benzene valence isomers are those in the pertrifluoromethyl[84,85] and perpentafluoroethyl series.[85] Dewar $(CC_2F_5)_6$ is actually more stable than its distorted benzene isomer at temperatures above 280°C.[86a,b] These molecules will be discussed further in Section 5.E.3.

## 5.C Transition Metals and Strained Carbocyclic Molecules

Transition metals and their complexes catalyze the decomposition of many kinetically stable, high-energy molecules such as quadricyclane and Dewar benzene. On the other hand, they sometimes form stable complexes with inherently reactive chemical species such as carbenes (see below), cyclobutadiene,[87] benzyne (see below), cyclohexyne (see below), cyclopentadienone,[88] o-xylylene,[89] trimethylenemethane (see below), pentalene,[90] and $\Delta^{1,4}$-bicyclo[2.2.0]hexene (see below). The isomerization of cubane to cuneane as well as the synthetically useful olefin metathesis reaction are two examples of formal thermally forbidden conversions catalyzed by transition metals, and these will be considered in this section. Because transition metals allow

[84] D. M. Lemal, J. V. Staros, and V. Austel, *J. Am. Chem. Soc.* **91**, 3373 (1969).

[85] M. G. Varlowe, R. N. Haszeldine, and R. Hubbard, *J. Chem. Soc., Chem. Commun.* p. 202 (1969).

[86a] E. D. Clifton, W. T. Flowers, and R. N. Haszeldine, *J. Chem. Soc., Chem. Commun.* p. 202 (1969).

[86b] A.-M Debbagh, W. T. Flowers, R. N. Haszeldine, and P. J. Robinson, *J. Chem. Soc., Chem. Commun.* p. 323 (1975).

[87] J. D. Fitzpatrick, L. Watts, G. F. Emerson, and R. Pettit, *J. Am. Chem. Soc.* **87**, 3254 (1965).

[88] M. L. H. Green, L. Pratt, and G. Wilkinson, *J. Chem. Soc.* p. 989 (1960).

[89] W. R. Roth and J. D. Meier, *Tetrahedron Lett.* p. 2053 (1967).

[90] D. F. Hunt and J. W. Russell, *J. Am. Chem. Soc.* **94**, 7198 (1972).

thermally forbidden reactions to occur near ambient temperatures, direct measurements of the heats of these reactions can be obtained. Additionally, one can isolate or observe an intermediate which would rapidly isomerize under the normal uncatalyzed high-temperature conditions. Transition-metal studies also shed some light upon the validity of analogies between ethylene and cyclopropane or acetylene and cyclopropene.

A reader exploring the literature of this field is at first encouraged by the finding that the vast majority of relevant work, particularly in the realm of rearrangements, has been published since 1970. However, the rapid expansion of work since then and the multifarious reactions exhibited by even closely related compounds in the presence of different metals, upon change of ligands attached to a specific metal, and in different solvents make such an exploration challenging indeed. It is still an area in which much important work awaits completion. The task of preparing this section has been simplified by many excellent review articles, which will be cited when relevant. We especially recommend a recent article[91] which includes an introduction to transition-metal chemistry well suited to the organic chemist. This section will not consider transition-metal-catalyzed hydrogenations,[92] nor will it spend much time discussing certain other important reactions, such as carbonyl insertion. Its primary concern will be the special complexation properties of strained carbocyclic molecules, properties of transition metals, their role in stabilizing certain strained species, and their utility in inducing rearrangement, oligomerization, and other reactions.

Because we will be discussing transition-metal olefin complexes and because strained rings exhibit olefinic character, it is worthwhile to present a simplified picture of bonding in such complexes. Figure 5.4 furnishes a description of the Dewar–Chatt–Duncanson model[93a,b] of bonding in transition-metal olefin complexes. There are two components in this model: one in which the olefin acts as an electron donor through its occupied $\pi$ orbital and its interaction with empty metal valence-shell orbitals of proper symmetry, and one in which the $\pi^*$ orbital of the olefin accepts electrons from occupied metal orbitals of proper symmetry. Bonding involving the olefinic $\pi$ system is termed $\sigma$-type bonding because there is no node along the $z$ axis connecting metals and olefin, while the $\pi^*$ orbital is involved in $\pi$-type bonding because the $z$ axis is on a node. The relative importance of $\sigma$-type bonding (metal as acceptor) or $\pi$-type bonding (metal as donor) is, in part, indicated by the promotion energy (P.I.) and electron affinity (E.A.) of the metal.[94] It should

[91] K. C. Bishop, III, *Chem. Rev.* **76**, 461 (1976).

[92] R. L. Augustine, "Catalytic Hydrogenation." Dekker, New York, 1965.

[93a] M. J. S. Dewar, *Bull. Soc. Chim. Fr.* **18**, C71 (1951).

[93b] J. Chatt and L. A. Duncanson, *J. Chem. Soc.* p. 2939 (1953).

[94] The values for P.I. and E.A. are those found in Ref. 91.

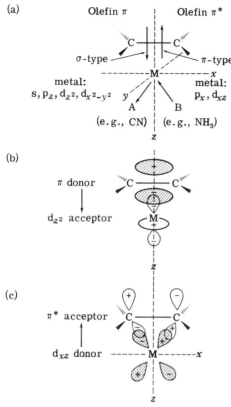

**Figure 5.4.** Dewar–Chatt–Duncanson picture of transition-metal–olefin bonding: (a) Schematic representation of donor–acceptor interactions including the effects of electron-donating ligands (A) and electron-accepting ligands (B). (b) Example of $\sigma$-type bonding (no node along the $z$ axis) involving olefin $\pi$ and metal $d_{z^2}$ orbitals. (c) Example of $\pi$-type bonding (node along $z$ axis) involving olefin $\pi^*$ and metal $d_{xz}$ orbitals.

be noted that these measured values refer to the uncomplexed metal or ion. For example, rhodium(I) should be a fairly strong donor (P.I. 1.6 eV) and a moderate acceptor (E.A. 7.31 eV); silver(I) should also be a moderate acceptor (E. A. 7.59 eV) and a weak donor (P.I. 9.9 eV). This latter property should tend to make silver(I) reluctant to participate in oxidative addition reactions[95] in which it would be oxidized to silver(III). Rhodium(I), however, should be a ready participant in such reactions. Platinum(II) (P.I. 3.39 eV; E.A. 19.42 eV and palladium(II) (P.I. 3.05 eV; E.A. 18.56 eV) obviously tend to participate both as donors and acceptors. Transition metals have been

[95] J. P. Collman, *Acc. Chem. Res.* **1**, 136 (1968).

employed to stabilize strained molecules, some of which, for example, a $Z,E,Z$-1,3,5-cycloheptatriene,[96a,b] have not been generated under other conditions.

### 5.C.1   ETHYLENE AND CYCLOPROPANE COMPLEXES

Reasoning that cyclopropane's ethylenic character (see Chapter 2) might allow it to form a $\pi$-complex such as **47**, Tipper[97] postulated that indeed such a complex having the three-membered ring intact was formed in the presence of $H_2PtCl_6$, because added cyanide regenerated free cyclopropane. A later study[98] indicated that the complex was oligomeric (initially postulated to be a polymer), and that addition of pyridine to Tipper's complex produced the ring-opened **48**, a product of oxidative addition. x-Ray studies confirmed

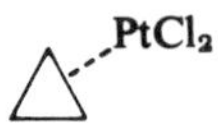

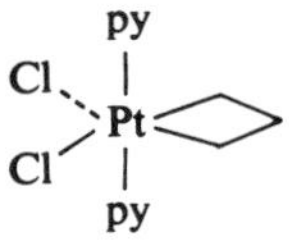

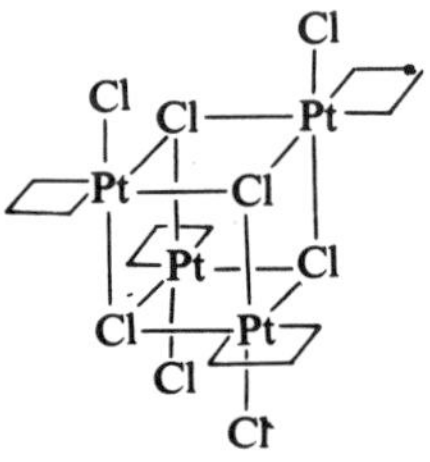

| **47** | **48** | **49** | **50** |

the structure of **48**,[99] and later also proved conclusively that Tipper's original complex was tetrameric and suggested opened cyclopropane rings (**49**).[100a] Such structures as **47** and **50** may be regarded as extreme canonical forms of the same complex, since there is no symmetry barrier between them.[100b] However, the long $C_1$—$C_3$ distance in **48** (2.54 Å) can be taken as evidence clearly supporting an insertion (oxidative addition) complex rather than a $\pi$-complex. The former is capable of easily regenerating the corresponding cyclopropanes upon addition of cyanide or other active ligands.[101a,b,102]

[96a] J. Browning, M. Green, J. L. Spencer, and J. G. A. Stone, *J. Chem. Soc., Dalton Trans.* p. 97 (1974).

[96b] J. Browning and B. R. Penfold, *J. Cryst. Mol. Struct.* **4**, 347 (1974).

[97] C. F. H. Tipper, *J. Chem. Soc.* p. 2045 (1955).

[98] D. M. Adams, J. Chatt, R. G. Guy, and N. Sheppard, *J. Chem. Soc.* p. 738 (1961).

[99] N. A. Bailey, R. D. Gillard, M. Keeton, R. Mason, and D. R. Russell, *Chem. Commun.* p. 396 (1966).

[100a] S. E. Binns, R. H. Cragg, R. D. Gillard, G. T. Heaton, and M. F. Pilbrow, *J. Chem. Soc. A* p. 1227 (1969).

[100b] For example, platinum(II) adds stereospecifically to cyclopropanes with retention at both carbons; see N. Dominelli and A. C. Oehlschlager, *Can. J. Chem.* **55**, 364 (1977).

[101a] W. J. Irwin and F. J. McQuillin, *Tetrahedron Lett.* p. 1937 (1968).

[101b] K. G. Powell and F. J. McQuillin, *Tetrahedron Lett.* p. 3313 (1971).

[102] R. Noyori and H. Takaya, *J. Chem. Soc. D* p. 525 (1969).

Evidence favoring the existence of the $\pi$-type platinum(II) complex **51** is based upon its elemental analysis and ir spectrum,[103] while the analogous silver(I) complex is highly unstable if it is indeed a discrete entity.[104] Although the isomerization of cyclopropane to propene requires a great deal of energy (see Section 2.B), Zeisse's dimer ($[C_2H_4PtCl_2]_2$) in chloroform or ethanol reacts with cyclopropane under gentle conditions to form propene $\pi$-complexes.[105] A suitably substituted cyclopropane will form a $\pi$-allyl complex of Pt(II) (e.g., **52**) which does not regenerate the three-membered ring upon addition of cyanide.[100a] Rhodium(I) also commonly inserts into strained bonds, and the reaction of dicarbonyl chlororhodium(I) dimer with liquid

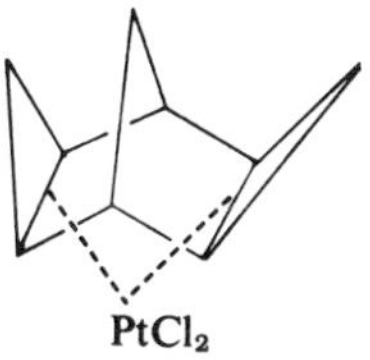

| 51 | 52 | 53 |

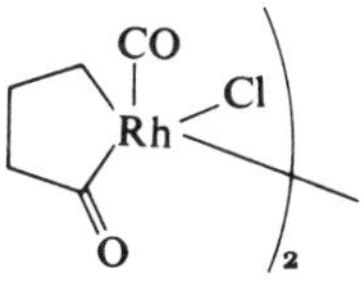

cyclopropane in a sealed tube at ca. 40°–50°C produced the carbonyl insertion product assigned structure **53**.[106]

## 5.C.2 CYCLOBUTANE COMPLEXES: THEIR RELEVANCE (IF ANY) TO OLEFIN METATHESIS

Cyclobutanes are less ethylenic than cyclopropanes and thus less reactive in the presence of transition metals. However, cyclobutane complexes (or "cyclobutanoids") have been postulated as intermediates in the olefin metathesis reaction (Eq. (19))].[107a,b108] A major problem with such a mechanism

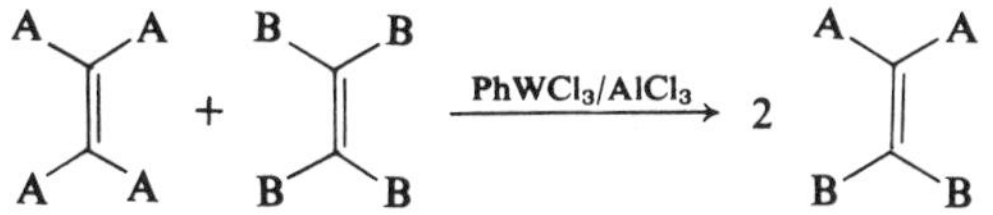

$$\tag{19}$$

is that only rarely have catalyzed dimerizations of ethylenes to cyclobutanes been observed, and until very recently only a 0.1% yield had been achieved

[103] H. C. Volger, H. Hogeveen, and M. M. P. Gassbeek, *J. Am. Chem. Soc.* **91**, 2137 (1969).

[104] B. C. Menon and R. E. Pinock, *Can. J. Chem.* **47**, 3327 (1969).

[105] D. B. Brown, *J. Organomet. Chem.* **24**, 787 (1970).

[106] D. M. Roundhill, D. N. Lawson, and G. Wilkinson, *J. Chem. Soc. A* p. 845 (1968).

[107a] N. Calderon, *Acc. Chem. Res.* **5**, 127 (1972).

[107b] N. Calderon, E. A. Ofstead, and W. A. Judy, *Angew. Chem., Int. Ed. Engl.* **15**, 401 (1976).

[108] R. J. Haines and G. J. Leigh, *Chem. Soc. Rev.* **4**, 155 (1975).

with a metathesis catalyst (a molybdenum compound).[109] However, the destabilized diene **54** has been induced to form cyclobutane **55**, and quadricyclane **56** converted to norbornadiene **57** under metathesis conditions; thus one may conclude that cyclobutanoid transition states for olefin metathesis might be viable under certain conditions.[110] Despite arguments* favoring

(20)

(21)

pairwise (combination of intact ethylenes) formation of a four-membered ring consisting of $sp^3$-hybridized carbon atoms stabilized by a transition metal,[111] pairwise formation of a cyclobutanoid transition state through relaxation of orbital symmetry constraints (see Section 5.B.1),[112a,b] or pairwise formation of a metallocyclopentane intermediate (i.e., cyclobutane metal-insertion product),[113a,b,114] most recent evidence supports a nonpairwise mechanism (Scheme 6) in a majority of cases.[114–117] Such evidence includes the isolation of catalytic metallocarbenes and metallocyclobutanes[114] as well as the *initial* formation of 2,10-tetradecadiene in a metathesis reaction employing a mixture of *cis*-cyclooctene, *cis*-2-butene, and *cis*-4-octene.[116] The

---

[109] G. S. Lewandos and R. Pettit, *Tetrahedron Lett.* p. 2281 (1971).

[110] P. G. Gassman and T. H. Johnson, *J. Am. Chem. Soc.* **98**, 861 (1976).

[111] G. S. Lewandos and R. Pettit, *J. Am. Chem. Soc.* **93**, 7087 (1971).

[112a] F. D. Mango and J. H. Schachtschneider, *J. Am. Chem. Soc.* **93**, 1123 (1971).

[112b] F. D. Mango, *Coord. Chem. Rev.* **15**, 109 (1975).

[113a] C. G. Biefield, H. A. Eick, and R. H. Grubbs, *Inorg. Chem.* **12**, 2166 (1973).

[113b] A. R. Graser, P. H. Bird, S. A. Bezman, J. R. Shapley, R. White, and J. A. Osborn, *J. Am. Chem. Soc.* **95**, 597 (1973).

[114] D. J. Cardin, M. J. Doyle, and M. F. Lappert, *J. Chem. Soc., Chem. Commun.* p. 927 (1972).

[115] T. J. Katz and J. McGinnis, *J. Am. Chem. Soc.* **97**, 1592 (1975).

[116] R. H. Grubbs, P. L. Burk, and D. D. Carr, *J. Am. Chem. Soc.* **97**, 3265 (1975).

[117] M. T. Mocella, M. A. Busch, and E. L. Muetterties, *J. Am. Chem. Soc.* **98**, 1284 (1976).

* See Addendum.

above $C_{14}$ hydrocarbon could only be formed as a secondary product if pairwise mechanisms were exclusively operative.[116] Similar conclusions were drawn in a study employing deuterated starting materials.[117]

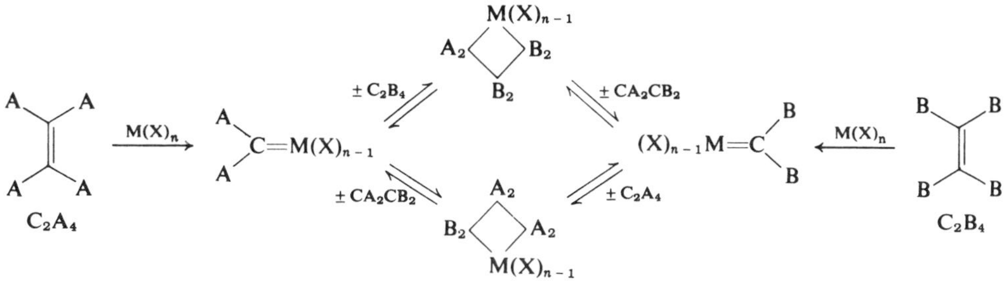

$$M(X)_{n-1}$$

$$A_2 \langle \quad \rangle B_2$$

$$B_2$$

$$\pm C_2B_4 \qquad \pm CA_2CB_2$$

$$A \quad \quad B$$
$$\underset{A}{\overset{A}{>}} C{=}M(X)_{n-1} \qquad (X)_{n-1}M{=}C \underset{B}{\overset{B}{<}}$$

$$\pm CA_2CB_2 \qquad A_2 \qquad \pm C_2A_4$$

$$B_2 \langle \quad \rangle A_2$$

$$M(X)_{n-1}$$

A₂C=CA₂  $C_2A_4$   →  M(X)ₙ

M(X)ₙ  ←  B₂C=CB₂  $C_2B_4$

**Scheme 6**

The metallocarbenes depicted in Scheme 6 are presumably generated in very low yield, but they are involved in chain reactions which provide for quantitative and rapid completion. Although nonpairwise mechanisms in olefin metathesis had been indicated and metallocarbenes shown to induce olefin metathesis, until recently there was no proof of the generation of a metallocarbene starting with olefins and a metathesis catalyst. Recognizing that alkylidene carbenes have considerable nucleophilic character, Gassman and co-workers completely halted olefin metathesis with various electrophilic olefins (e.g., Michael addition acceptors) and trapped metallocarbenes in the form of the very small amount of cyclopropane products they produced[118a-c] (Scheme 7). These authors also note that no similar trapping of a metallocar-

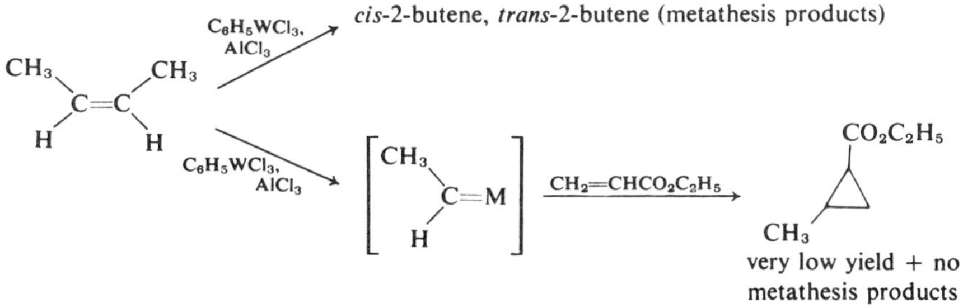

$$\underset{H}{\overset{CH_3}{>}}C{=}C\underset{H}{\overset{CH_3}{<}}$$

$\xrightarrow[AlCl_3]{C_6H_5WCl_3,}$ cis-2-butene, trans-2-butene (metathesis products)

$\xrightarrow[AlCl_3]{C_6H_5WCl_3,}$ $\left[ \underset{H}{\overset{CH_3}{>}}C{=}M \right]$ $\xrightarrow{CH_2{=}CHCO_2C_2H_5}$

$CO_2C_2H_5$ / $CH_3$ cyclopropane

very low yield + no metathesis products

**Scheme 7**

bene was achieved in the interconversion of **54** to **55**, thus lending further support to the view that at least this interconversion avoids the mechanism of

[118a] P. G. Gassman and T. H. Johnson, *J. Am. Chem. Soc.* **98**, 6055 (1976).
[118b] P. G. Gassman and T. H. Johnson, *J. Am. Chem. Soc.* **98**, 6057 (1976).
[118c] P. G. Gassman and T. H. Johnson, *J. Am. Chem. Soc.* **98**, 6058 (1976).

Scheme 6. Furthermore, retrocarbene addition in cyclopropanes is consistent with initial insertion of the metal (oxidative addition) to form a metallocyclobutanoid[116] (Scheme 8, see the earlier discussion of platinum insertion in cyclopropane).* The instability of cyclopropanes in the presence of this catalyst

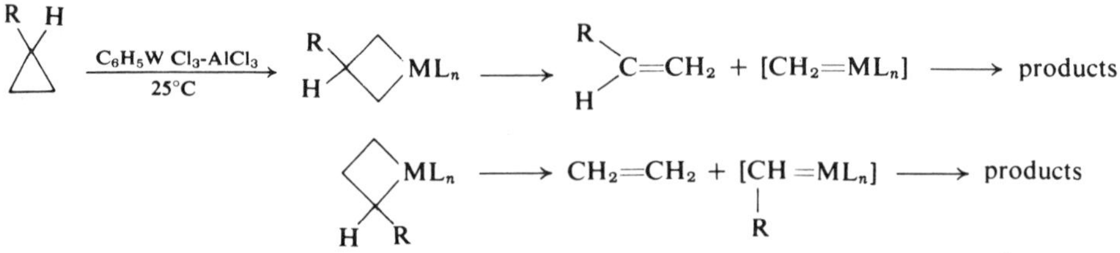

**Scheme 8**

explains why cyclopropane products normally are not observed during olefin metathesis.[118b] These results preluded successful cyclopropane–olefin cross metathesis.[118c] Other workers recently also have produced, isolated, and determined the structure of a hydrocarbon–carbene metal complex obtained from a cyclopropane (actually, bicyclopentane) system.[119]

### 5.C.3   TRANSITION-METAL-CATALYZED REARRANGEMENTS OF BICYCLIC AND SPIROCYCLIC HYDROCARBONS

#### 5.C.3.a   *Bicyclo[1.1.0]butanes: Simple and Bridged*

Although bicyclo[1.1.0]butane derivatives are highly strained, their thermal rearrangements (primarily to 1,3-butadienes) occur only at temperatures above 150° and require activation energies in excess of 40 kcal/mole,[120,121] despite the operation of a seemingly symmetry-allowed ($_\sigma 2_s + _\sigma 2_a$) and apparently concerted process[21,122] (see, however, arguments presented in refs. 61a–d). In striking contrast are the rapid rearrangements catalyzed by such transition metals as silver(I).[123a–c] For example, while tricyclo[4.1.0.0$^{2,7}$]-heptane (**58**, R = H) rearranges at high temperature to yield *cis*-bicyclo-[3.2.0]hept-6-ene, in the presence of a catalytic amount of AgBF$_4$ in chloroform, rearrangement to *cis,cis*-1,3-cycloheptadiene is complete in minutes at

[119] R. Aumann, H. Wormann, and C. Kruger, *Angew. Chem., Int. Ed. Engl.* **15**, 609 (1976).

[120] J. P. Chesick, *J. Phys. Chem.* **68**, 2033 (1964).

[121] H. M. Frey and I. D. R. Stevens, *Trans. Faraday Soc.* **61**, 90 (1965).

[122] G. L. Closs and P. E. Pfeffer, *J. Am. Chem. Soc.* **90**, 2452 (1968).

[123a] L. A. Paquette, *Acc. Chem. Res.* **4**, 280 (1971).

[123b] L. A. Paquette, *MTP Int. Rev. Sci., Org. Chem., Ser. 1* **5**, 127 (1973).

[123c] L. A. Paquette, *Synthesis* p. 347 (1975).

* See Addendum.

40°C via a *formal* thermally forbidden ($_\sigma 2_a + _\sigma 2_a$) pathway[124] (however, concerted metal-catalyzed decomposition is not indicated, as explained below). Silver(I) ion lowers the activation barrier for bicyclo[1.1.0]butane

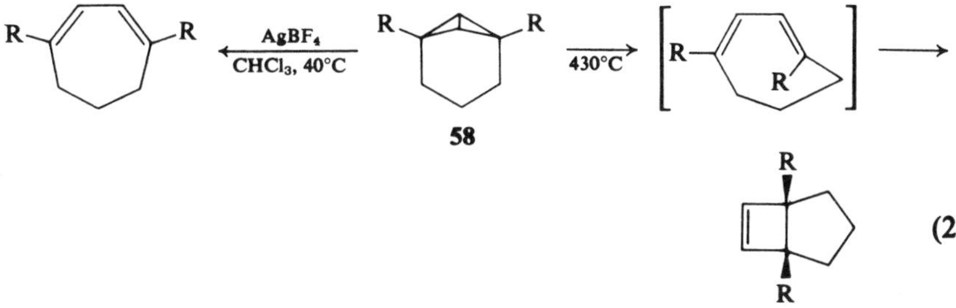

(22)

decomposition by more than 20 kcal/mole.[125] Silver(I) ion-catalyzed re-arrangements lead to four different skeletal bond reorganizations, depending upon the nature and placement of substituents in the starting material.[126] For example, while 2,6-dimethyltricyclo[4.1.0.0$^{2,7}$]heptane (**58**, R = $CH_3$) re-arranges to 1,4-dimethyl-*cis-cis*-1,3-cycloheptadiene, 1,7-dimethyltricyclo-[4.1.0.0$^{2,7}$]heptane (**59**) forms the methylenecyclohexane **60** and norcarene **61**,[127] and 1-methyltricyclo[4.1.0.0$^{2,7}$]heptane (**62**) rearranges to three products (**63–65**), the bicycloheptene being most abundant.[128a,b] The difference,

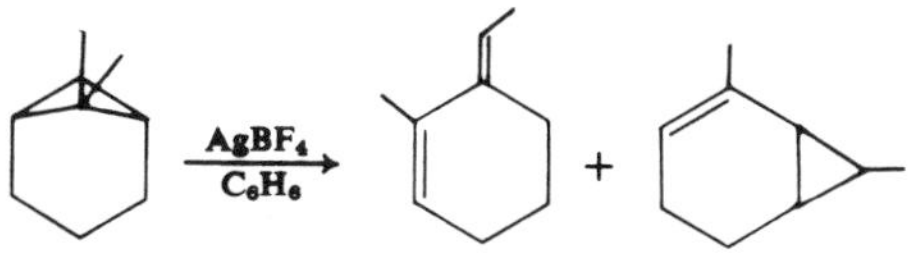

(23)

**59**        **60** (80%)        **61** (20%)

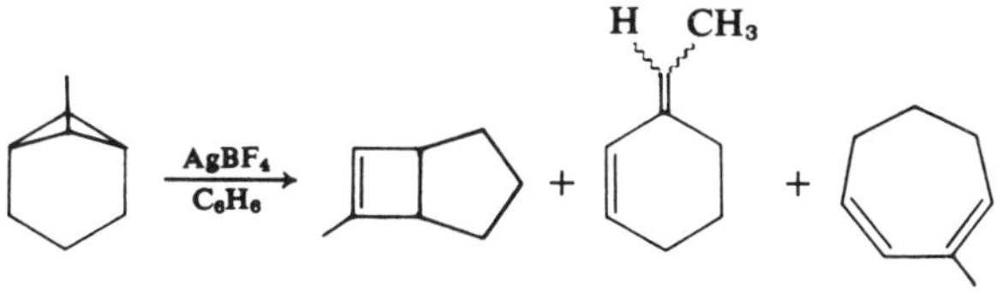

(24)

**62**        **63** (44%)        **64** (29%)        **65** (26%)

[124] L. A. Paquette, G. R. Allen, Jr., and R. P. Henzel, *J. Am. Chem. Soc.* **92**, 7002 (1970).

[125] L. A. Paquette, S. E. Wilson, and R. P. Henzel, *J. Am. Chem. Soc.* **93**, 1288 (1971).

[126] L. A. Paquette and G. Zon, *J. Am. Chem. Soc.* **96**, 203 (1974).

[127] L. A. Paquette, R. P. Henzel, and S. E. Wilson, *J. Amer. Chem. Soc.* **93**, 2335 (1971).

[128a] L. A. Paquette, S. E. Wilson, R. P. Henzel, and G. R. Allen, Jr., *J. Am. Chem. Soc.* **94**, 7761 (1972).

[128b] L. A. Paquette, R. P. Henzel, and S. E. Wilson, *J. Am. Chem. Soc.* **94**, 7780 (1972).

in product distribution in the Ag(I)-catalyzed rearrangements of **58** and **59**, for example, is ascribed to a changeover in mechanism. Compound **59** [and others having alkyl and certain other substituents at the bicyclobutane bridge-head(s)] is postulated to open to a 3° argentocarbonium ion (**66**)[128–132] which undergoes 1,2-hydride shift prior to forming products, while **58** and other derivatives lacking such bridgehead substituents avoid the less stable 2° argentocarbonium ion and proceed via a homoallylic (**67a**) or cyclopropyl-carbinyl (**67b**) carbonium ion.[123a–c,133] In neither of these mechanisms is the

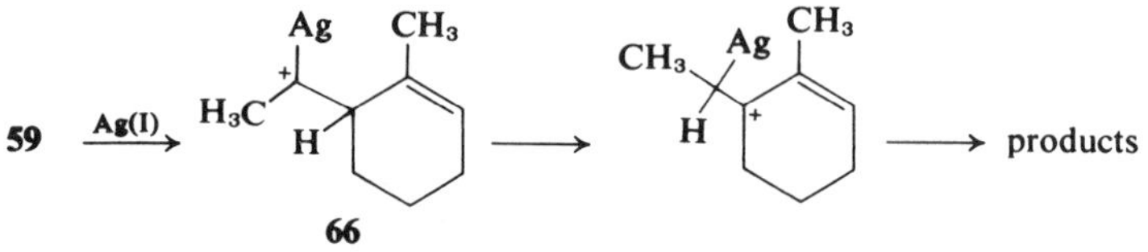

initial attack considered to occur at the highly strained bicyclobutyl central bond. A secondary "argentocarbonium ion" can be generated via Ag(I)-catalyzed decomposition of diazoalkene **68**, and the product expected from such an intermediate, 3-methylenecyclohexene, is found.[134] The steric effects produced by substituents in 1-substituted tricyclo[4.1.0.0$^{2,7}$]heptanes support the notion that 3-methylenecyclohexene derivatives are formed through

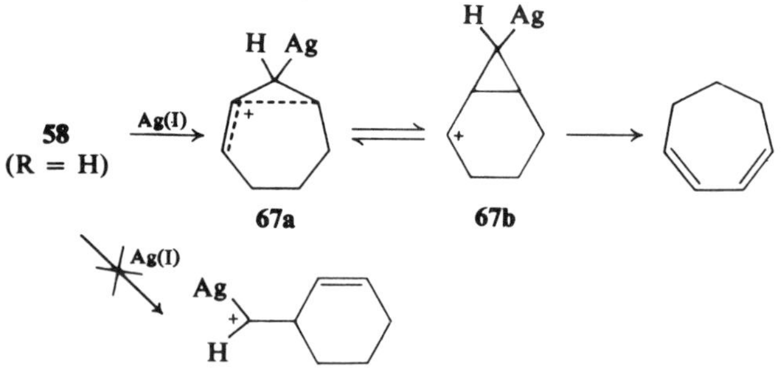

[129] The suggestion of metal-stabilized carbenes, which may be considered as one extreme canonical form of a species for which the metallocarbonium ion is another extreme, as intermediates in metal-catalyzed decompositions of bicyclobutanes may first be attributed to Gassman in connection with studies employing rhodium(I) (Ref. 130). Masamune first discussed these in connection with Pd(II) (Ref. 131), and Paquette in connection with Ag(I) (Refs. 128a,b). Noyori (Ref. 132) discusses the stabilities of such intermediates as a function of the transition metal.

[130] P. G. Gassman and T. J. Atkins, *J. Am. Chem. Soc.* **93**, 1042 (1971).

[131] M. Sakai, H. Yamaguchi, and S. Masamune, *Chem. Commun.* p. 486 (1971).

[132] R. Noyori, *Tetrahedron Lett.* p. 1691 (1973).

[133] M. Sakai, H. H. Westberg, H. Yamaguchi, and S. Masamune, *J. Am. Chem. Soc.* **93**, 4611 (1971).

[134] M. Sakai and S. Masamune, *J. Am. Chem. Soc.* **93**, 4610 (1971).

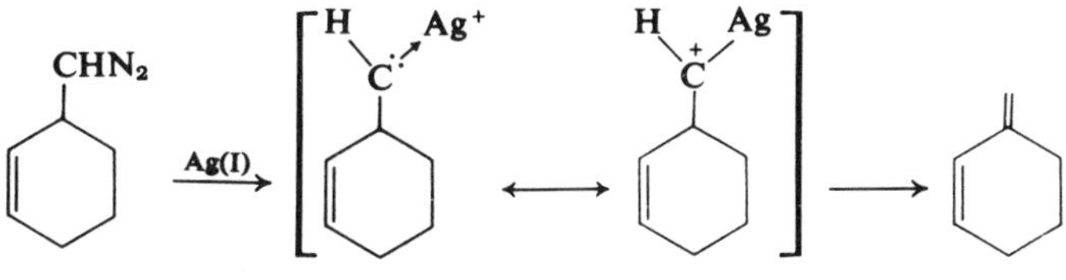

$$(27)$$

**68**

attack of Ag(I) at the more hindered edge bond with concomitant loss of the central bond (so-called $\beta$ pathway), while attack at the less hindered edge is implicated in the formation of bicyclo[3.2.0]hept-6-enes (so-called $\gamma$ pathway).[135a,b] (That is to say that bulky groups such as *tert*-butyl at the bicyclobutyl bridgehead increase the importance of the $\gamma$ pathway at the expense of the $\beta$ pathway.) The silver(I) chemistry of nonbridged substituted bicyclo[1.1.0] butanes is controlled by the same factors as those for the bridged system discussed above[123a–c,136] and can be calculated to release about $+25$ kcal/ mole.

Catalytic quantities of rhodium(I) dicarbonyl chloride dimer rapidly isomerize tricyclo[4.1.0.0$^{2,7}$]heptane to 3-methylenecyclohexene[130,137a,b] [recall that Ag(I) produces *cis,cis*-1,3-cycloheptadiene]. Apparently, the secondary rhodium–carbonium ion-carbenoid is more stable than its silver analogue. Similar behavior is exhibited by Pd(II),[137a,b,138] Ir(I),[137a,b] and Cu(I).[137a,b] The effects of these and other metals have been summarized elsewhere.[123b,137a,b]

Silver(I) ion promotes rearrangements in derivatives of strained systems having good leaving groups (e.g., **69**),[139] but this as well as related rearrangements[140,141] involve silver's role as complexing agent for the leaving group.

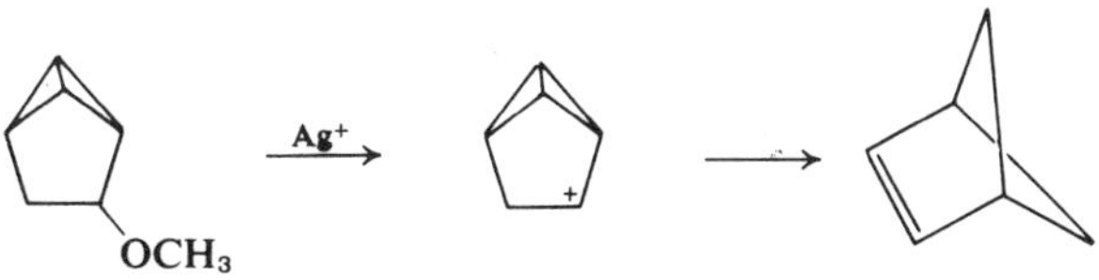

$$(28)$$

**69**    $Ag^+ \cdots OCH_3^-$

[135a] G. Zon and L. A. Paquette, *J. Am. Chem. Soc.* **96**, 215 (1974).

[135b] L. A. Paquette and G. Zon, *J. Am. Chem. Soc.* **96**, 224 (1974).

[136] M. Sakai, H. Yamaguchi, H. H. Westberg, and S. Masamune, *J. Am. Chem. Soc.* **93**, 1043 (1971).

[137a] P. G. Gassman, G. R. Meyer, and F. J. Williams, *J. Am. Chem. Soc.* **94**, 7741 (1972).

[137b] P. G. Gassman and T. J. Atkins, *J. Am. Chem. Soc.* **94**, 7748 (1972).

[138] M. Sakai, H. Yamaguchi, and S. Masamune, *Chem. Commun.* p. 486 (1971).

[139] F. Scheidt and W. Kirmse, *J. Chem. Soc., Chem. Commun.* p. 716 (1972).

[140] R. K. Murray, Jr. and K. A. Babiak, *Tetrahedron Lett.* p. 311 (1974).

[141] G. Zon and L. A. Paquette, *J. Am. Chem. Soc.* **96**, 5438 (1974).

**5.C.3.b**  *Bicyclo[2.1.0]pentanes (Simple and Bridged) and Higher Homologues*

Bicyclo[2.1.0]pentanes provide some interesting contrasts to the organometallic chemistry of the bicyclobutane system. Breakage of the central bond in bicyclo[2.1.0]pentane is accompanied by release of ca. 47 kcal/mole of strain energy, whereas loss of the bicyclobutane central bond provides about 35 kcal/mole. Thus, there is a greater thermodynamic driving force for this mode of reaction for the former compound. However, the central bond in bicyclo[1.1.0]butane is more strained, therefore more olefinic, and can be expected to react faster with electron-seeking metals. Thus, while bicyclobutanes are very reactive toward silver(I) ion, bicyclo[2.1.0]pentane and almost all of its derivatives are inert to this agent. However, rhodium(I) catalyzes rearrangement to cyclopentene [Eq. (29)].[142a,b,143] The observed

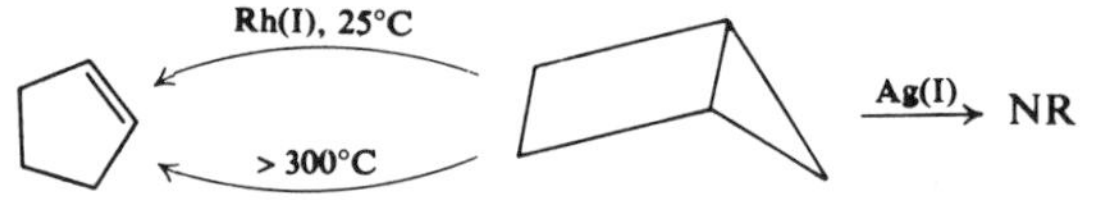

$$\text{(29)}$$

Rh(I)-catalyzed label-scrambling of deuterated bicyclo[2.1.0]pentanes[142a,b] and the interesting observation[143] that *endo*-5-methylbicyclo[2.1.0]pentane (**70**, R = CH$_3$) rearranges to 1-methylcyclopentene while the *exo* isomer does not rearrange are taken in support of a metal-hydride mechanism initiated by oxidative addition across the central bond [Eq.(30)].[142a,b,43] The isolation of iron complex **71** [Eq. (31)] supports the idea of initial attack by a transition metal upon the central bond of a bicyclopentane.[144] 1-Phenylbicyclo[2.1.0]-pentane does rearrange to isomeric cyclopentenes under Ag(I) catalysis,[145]

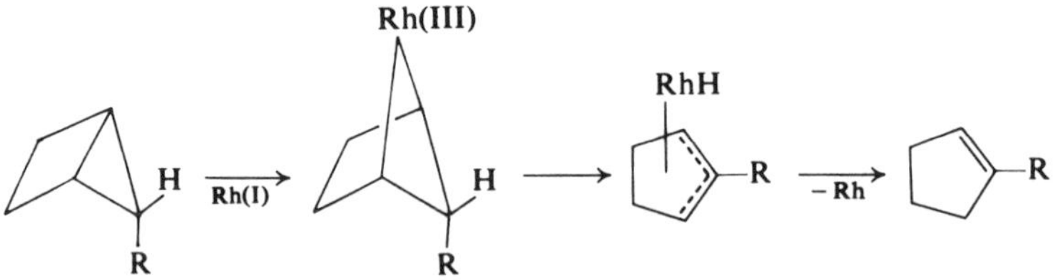

$$\text{(30)}$$

[142a] P. G. Gassman, T. J. Atkins, and J. T. Lumb, *Tetrahedron Lett.* p. 1643 (1971).

[142b] P. G. Gassman, T. J. Atkins, and J. T. Lumb, *J. Am. Chem. Soc.* **94**, 7757 (1972).

[143] K. B. Wiberg and K. C. Bishop, III, *Tetrahedron Lett.* p. 2727 (1973).

[144] R. Aumann and H. Averbeck, *Angew. Chem., Int. Ed. Engl.* **15**, 610 (1976).

[145] L. A. Paquette and L. M. Leichter, *J. Am. Chem. Soc.* **94**, 3653 (1972).

perhaps as the result of stabilization of positive charge by phenyl. [Zn(II) is also unreactive with the parent but catalyzes isomerization of the 1-phenyl derivative.[146] The phenyl group in **72**, however, appears to play a unique role

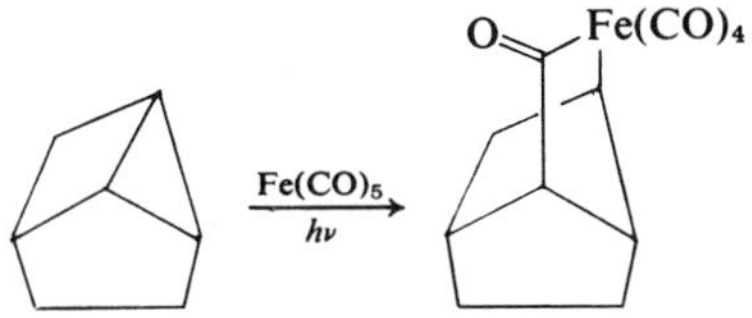

$$
\begin{array}{c}
\text{(structure)} \xrightarrow[h\nu]{\text{Fe(CO)}_5} \text{(structure with O=C–Fe(CO)}_4) \\
\mathbf{71}
\end{array}
\tag{31}
$$

in possibly complexing and guiding silver ion to form **73**, rather than a more stable benzylic carbonium ion, en route to the product (**74**). Somewhat

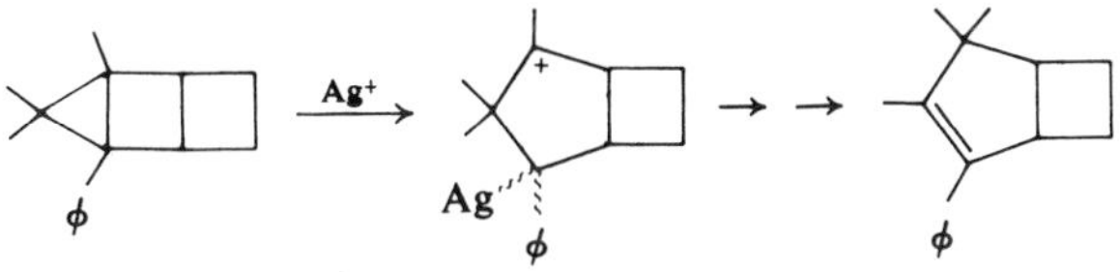

$$
\mathbf{72} \xrightarrow{\text{Ag}^+} \mathbf{73} \rightarrow \rightarrow \mathbf{74}
\tag{32}
$$

surprisingly, the spiro compound **75** is unreactive to silver(I) ion,[123a–c] in spite of its potential to create a cyclopropyl carbinyl carbonium ion (**76**) accompanied by release of about 55 kcal/mole of strain energy (see discussion of spiropentane below). It is interesting to remark that dehydronoriceane (**77**) rearranges to 2,4-ethenonoradamantane (**78**) with the use of Ag(I).[147] Although

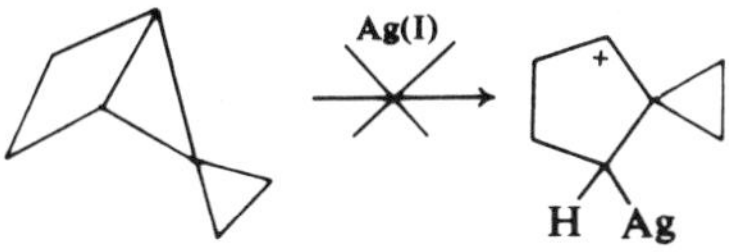

$$
\mathbf{75} \xrightarrow{\text{Ag(I)}} \mathbf{76}
$$

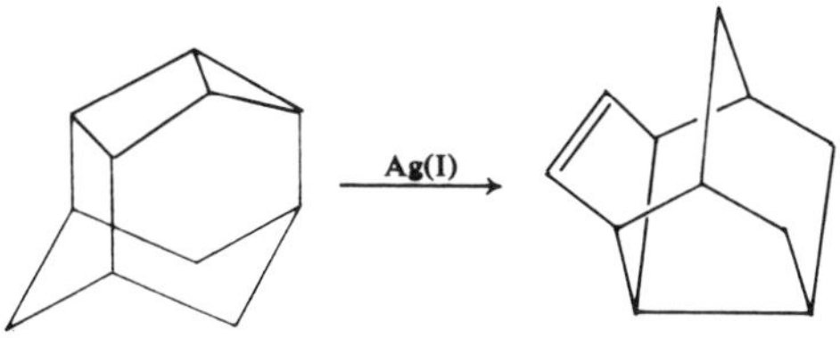

$$
\mathbf{77} \xrightarrow{\text{Ag(I)}} \mathbf{78}
\tag{33}
$$

[146] M. A. McKinney and S. K. Chou, *Tetrahedron Lett.* p. 1145 (1974).

[147] T. Katsushima, R. Yamaguchi, M. Kawanisi, and E. Osawa, *J. Chem. Soc., Chem. Commun.* p. 39 (1976).

there is a considerable thermodynamic driving force to rearrange to 2,4-ethenonoradamantane,[148] it appears that the rearrangement is facilitated by proximity and alignment of the migrating $C_4$—$C_9$ bond to the $C_3$—$C_5$ bond which is being lost (the bicyclopentane system in 77 is not significantly more strained than the parent molecule).[147] It might be interesting to see if a similarly placed C—C bond in endo-5-methylbicyclo[2.1.0]pentane accelerates rearrangement compared to the exo isomer. Bridging of the bicyclo[2.1.0]-pentane system by a single methylene group, as in tricyclo[2.2.0.0$^{2,6}$]hexane (79), either adds enough strain to facilitate the rearrangement shown[149] or again places the migrating bond in a favorable orientation (79a). Although the mechanism for rearrangement of 79 can be likened to that of 77, the bond alignment (see 79a) should not be as favorable for the former. Thermodynamic

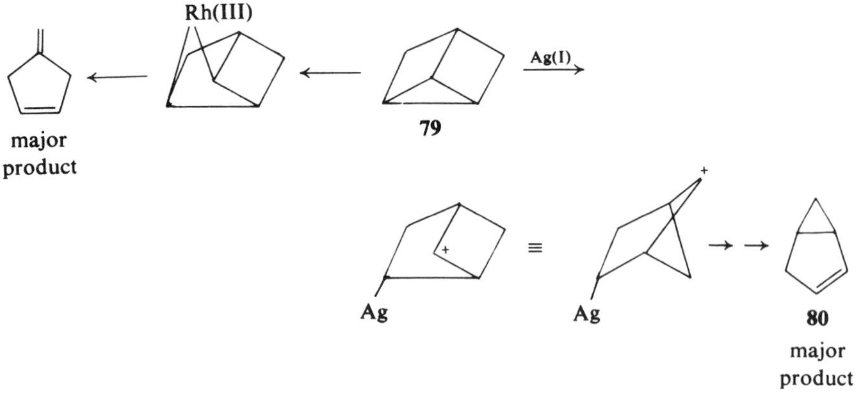

Scheme 9

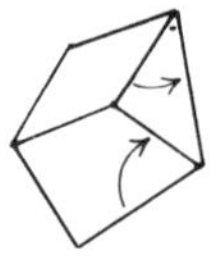

factors would not appear to play a major role because the release of strain in the transformation of 79 to bicyclo[3.1.0]hex-2-ene (80) (loss of two cyclobutane rings) should be about equal to that in the transformation of bicyclo[2.1.0]pentane to cyclopentane (loss of cyclopropane and cyclobutane rings). Again, rhodium(I) exhibits a tendency for initial oxidative addition to 71 and

---

[148] Based upon calculational results for the $C_{11}H_{16}$ tetracyclic manifold, see S. A. Godleski, P. von R. Schleyer, and E. Osawa, J. Chem. Soc., Chem. Commun. p. 38 (1976).
[149] R. J. Roth and T. J. Katz, J. Am. Chem. Soc. 94, 4770 (1972).

forms a different mixture of products.[149] [3.2.1]Propellane (**81**) (see Section 6B), avoids a bridgehead carbonium ion, and reacts with Ir(I) through postulated metallocarbonium ions to form two isomeric methylenecycloheptenes.[150] Bicyclo[3.1.0]hexane can also be induced to rearrange under

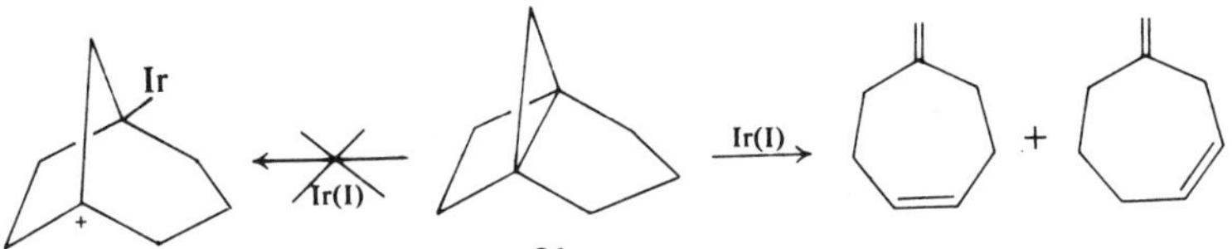

$$(34)$$

transition-metal catalysis, but about 100 times more slowly than bicyclo-[2.1.0]pentane.[151]

Transition metals catalyze cycloaddition of substituted ethylenes across the central bond of various bicyclo[2.1.0]pentanes,[152] while products arising from metal-catalyzed reaction between bicyclo[1.1.0]butanes and olefins are not the result of cycloadditions.[153] Although *syn*-tricyclooctane (**82**) reacts with

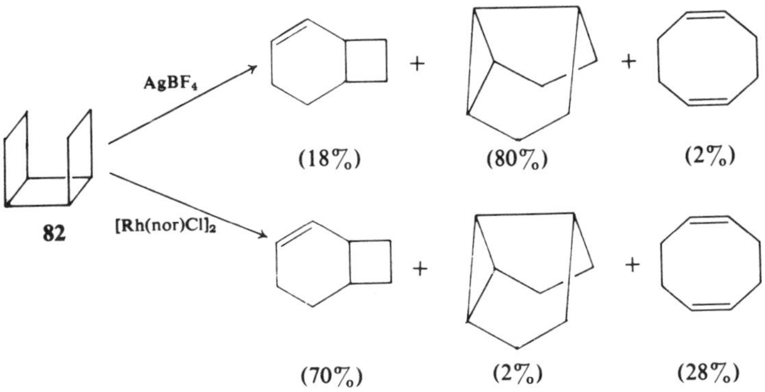

**Scheme 10**

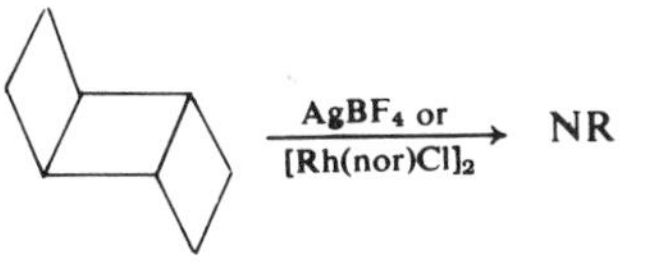

nor = norbornadiene

[150] P. G. Gassman and E. A. Armour, *Tetrahedron Lett.* p. 1431 (1971).
[151] See footnote 91 in Ref. 91.
[152] R. Noyori, T. Suzuki, and H. Takay, *J. Am. Chem. Soc.* **93**, 5896 (1971).
[153] R. Noyori, T. Suzuki, Y. Kumagia, and H. Takaya, *J. Am. Chem. Soc.* **93**, 5894 (1971).

both AgBF$_4$ and [Rh(nor)Cl]$_2$ to produce the products shown in Scheme 10, the *anti* isomer (**83**) is unreactive.[154] The fact that both **82** and **83** react with [Rh(CO)$_2$Cl]$_2$ to yield acylrhodium complexes suggests that steric effects do not explain this difference in reactivity.[154] (*Syn-* and *anti*-tricyclooctane will be discussed in more depth later, in the somewhat larger context of the rearrangements of cubanes and related systems.)

### 5.C.3.c  *Spiropentane*

Spiropentane (**84**) rearranges thermally to methylenecyclobutane with an activation energy of 55.5 kcal/mole[155] (a geometric isomerization which may be monitored using suitable substituted spiropentanes has a slightly lower activation barrier[156,157]). The inability of **84** to rearrange in the presence of

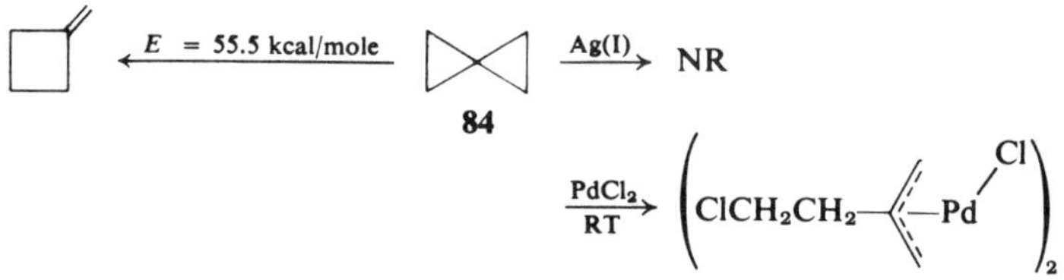

**Scheme 11**

silver(I) may be explained by reference to the results of *ab initio* calculations of the protonation pathways of spiropentane (Scheme 12).[158] Calculations based upon the ground-state structure of spiropentane suggest that edge attack (path A) is preferable to face attack (Path B) by a considerable energy margin (see Chapter 2 for analogous studies of protonated cyclopropane).

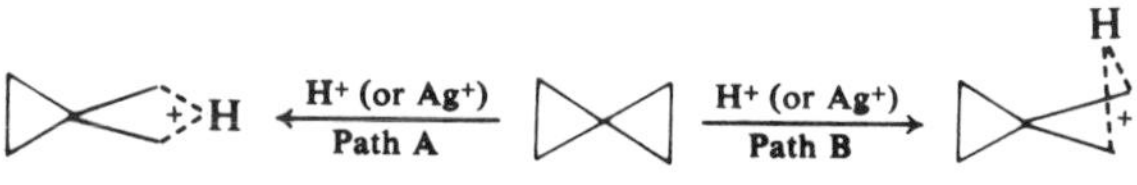

**Scheme 12**

This is, of course, a reflection of the electron densities in the hydrocarbon. Path A would lead to a structure in which the spiro cyclopropyl group cannot effectively stabilize developing positive charge through conjugation. Spiropentane is opened by Pd(II),[104,159a,b,160] and Rh(I)[160] to bisallyl complexes (spiro[2.4]heptane is inert under the published conditions[159a,b]), while it does

[154] J. Wristers, L. Brener, and R. Pettit, *J. Am. Chem. Soc.* **92**, 7499 (1970).
[155] M. C. Flowers and H. M. Frey, *J. Chem. Soc.* p. 5550 (1961).
[156] M. C. Flowers and A. R. Gibbons, *J. Chem. Soc. B* p. 612 (1971).
[157] J. J. Gajewski and L. T. Burka, *J. Am. Chem. Soc.* **94**, 8857 (1972).
[158] J.-M. Lehn and G. Wipff, *J. Chem. Soc., Chem. Commun.* p. 747 (1973).
[159a] A. D. Ketley and J. A. Braatz, *Chem. Commun.* p. 959 (1968).
[159b] A. D. Ketley, J. A. Braatz, and J. Craig, *Chem. Commun.* p. 1117 (1970).
[160] R. Rossi, P. Diversi, and L. Porri, *J. Organomet. Chem.* **47**, C21 (1973).

not react with Zeisse's acid $[HPtCl_3(C_2H_4)]$ in ethanol under conditions which cause cyclopropane to form a propene complex.[104] This may be a steric effect because 1,1-dimethylcyclopropane is also inert to Zeisse's acid.[104]

### 5.C.4 COMPLEXES OF STRAINED OLEFINIC AND AROMATIC MOLECULES

#### 5.C.4.a Cyclopropenes, Methylenecyclopropanes, Cyclobutenes, $\Delta^{1,4}$-Bicyclo[2.2.0]hexene, Norbornadiene

The enormous strain in cyclopropene and its simple derivatives causes these molecules to dimerize and polymerize spontaneously at temperatures below $0°C$.[161,162] Any reduction in the double bond character and concomitant $C_1$—$C_2$ stretch should relieve some strain in a cyclopropene. Transition-metal complexation induces this change, and one measure of it is the magnitude of cyclopropene's argentation constant[163] [silver ion is the metal, $K_{Ag} > 10^7$, see Eq. (35)] and its comparison with those of norbornene ($K_{Ag} = 0.268$) and the almost strainless cyclohexene ($K_{Ag} = 0.0184$). (No simple correlation between strain of cycloolefins and their complexing ability has yet appeared.[164])

$$K_{Ag} = \frac{[Ag\,(olefin)^+]_{aq}}{[Ag^+]_{aq}[olefin]_{CCl_4}} \tag{35}$$

Stable Pt(0) complexes have been obtained through displacement of ethylene from $Pt(C_2H_4)[P(C_6H_5)_3]_2$ by cyclopropene and its simple derivatives, and the intact three-membered rings in these complexes (e.g., **85**) maintain $C_1$—$C_2$ distances of about 1.50 Å, which is consistent with anticipated significant relief of strain.[165a,b] The low promotion energy (3.28 eV) and low electron

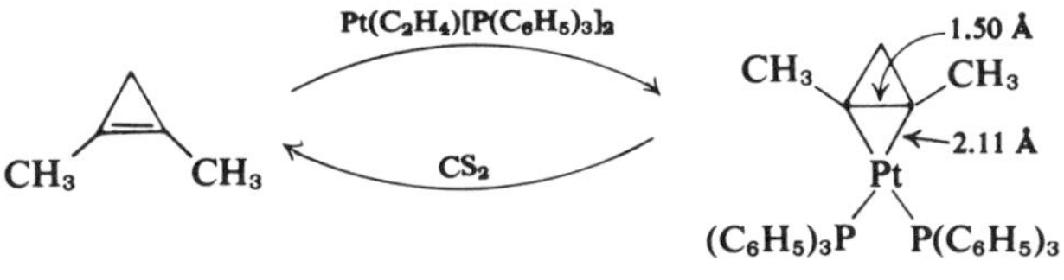

**85**

[161] K. B. Wiberg and W. J. Bartley, *J. Am. Chem. Soc.* **82**, 6375 (1960).

[162] J. G. Traynham and M. F. Sehnert, *J. Am. Chem. Soc.* **18**, 4024 (1956).

[163] F. R. Hartley, *Chem. Rev.* **73**, 163 (1973).

[164] M. Herberhold, "Metal $\pi$-Complexes," Vol. 2, Part 2, pp. 141–147. Elsevier, Amsterdam, 1974.

[165a] J. P. Visser, A. J. Shipperijn, J. Lukas, D. Bright, and J. J. de Boer, *Chem. Commun.* p. 1266 (1971).

[165b] J. J. de Boer and D. Bright, *J. Chem. Soc., Dalton Trans.* p. 662 (1975).

affinity (2.4 eV) of Pt(0) suggest that the dominant bonding here is of the $\pi$ type [Fig. 5.4(c)], and that $C_1$—$C_2$ bond lengthening is primarily due to population of the olefinic $\pi^*$ orbital. Happily, such stable complexes [m.p. (85) = 141°–144°C[165a,b]] release their cyclopropene ligands upon addition of carbon disulfide and thus function as cyclopropene carriers. In the presence of palladium(II) chloride, 1-methylcyclopropene and other simple derivatives are dimerized to *anti*-tricyclohexanes,[4] while this reagent causes certain tri- and tetra-substituted cyclopropenes to produce $\pi$-allyl complexes.[166a,b] In the presence of tetrakis(triphenylphosphine)palladium(0), a quantitative cyclotrimerization occurs which transforms 3,3-dimethylcyclopropene into hexamethyl-*trans*-$\gamma$-trishomobenzene (86).[167] Silver perchlorate catalyzes

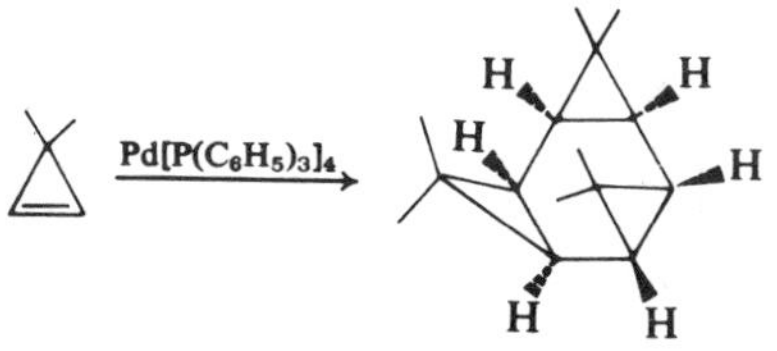

(37)

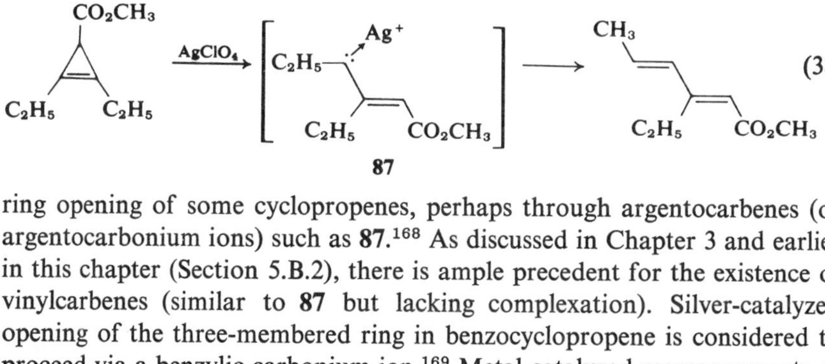

**86**

(38)

**87**

ring opening of some cyclopropenes, perhaps through argentocarbenes (or argentocarbonium ions) such as **87**.[168] As discussed in Chapter 3 and earlier in this chapter (Section 5.B.2), there is ample precedent for the existence of vinylcarbenes (similar to **87** but lacking complexation). Silver-catalyzed opening of the three-membered ring in benzocyclopropene is considered to proceed via a benzylic carbonium ion.[169] Metal-catalyzed rearrangements of 3,3'-bicyclopropenyl will be discussed later in this chapter.

Both methylenecyclopropane[170] and methylenecyclobutane[171] may be opened by Pd(II) to bisallyl complexes (**88** and **89**, respectively). Although

[166a] P. Mushak and M. A. Battiste, *J. Organomet. Chem.* **17**, 46 (1969).

[166b] M. A. Battiste, L. E. Friedrich, and R. E. Fiato, *Tetrahedron Lett.* p. 45 (1975).

[167] P. Binger, G. Schroth, and J. McMecking, *Angew. Chem., Int. Ed. Engl.* **13**, 465 (1974).

[168] J. H. Leftin and E. Gil-Av, *Tetrahedron Lett.* p. 3367 (1967).

[169] W. E. Billups, W. Y. Chow, and C. V. Smith, *J. Am. Chem. Soc.* **96**, 1979 (1974).

[170] R. Noyori and H. Takaya, *Chem. Commun.* p. 525 (1969).

[171] R. Rossi, P. Diversi, and L. Perri, *J. Organomet. Chem.* **31**, C40 (1971).

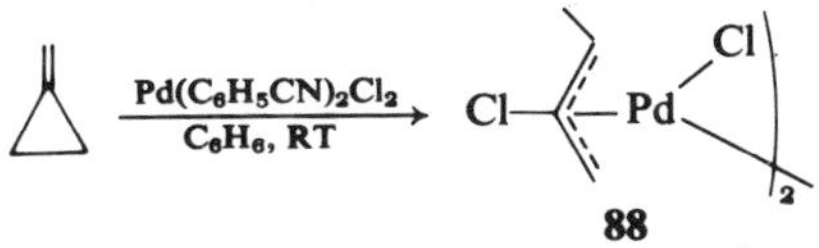

$$(39)$$

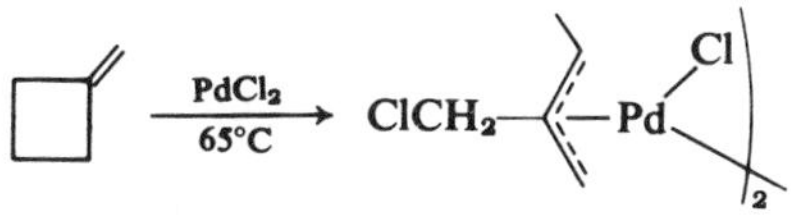

$$(40)$$

stable trimethylenemethane iron tricarbonyl (90) has been prepared via reaction of $Fe_2(CO)_9$ and 3-chloro-(2-chloromethyl)propene,[172] this complex is not obtainable through reaction of methylenecyclopropane and $Fe_2(CO)_9$, in which 1,3-butadiene iron tricarbonyl is produced instead.[173] However, 1-methylene-2-phenylcyclopropane and some other substituted methylenecyclo-

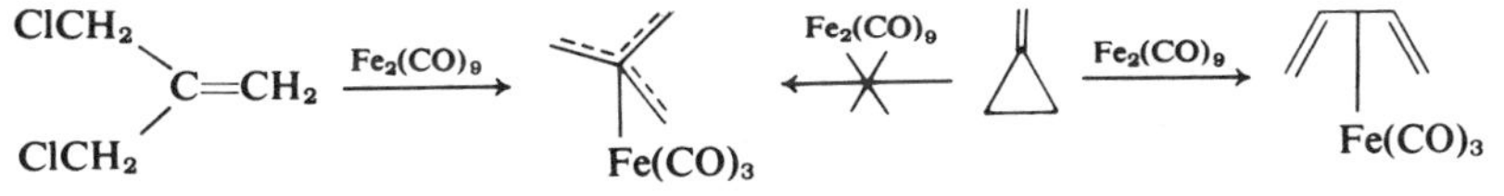

**Scheme 13**

propanes do open to the corresponding trimethylene methane complexes (e.g., 91) under comparable conditions. Resonance stabilization afforded by the phenyl group in 91 should exceed that in the corresponding 1,3-butadiene

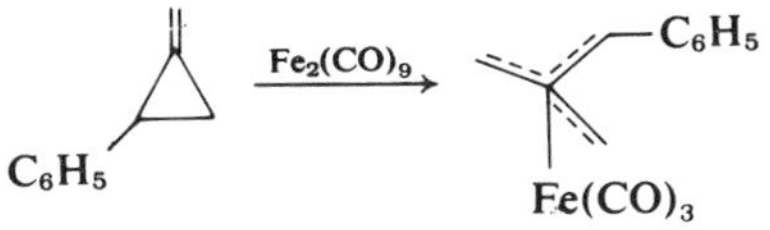

$$(41)$$

complex and will certainly be greater than in the starting material.[173] Zero-valent nickel catalyzes dimerization[174] and trimerization[175] of methylene-cyclopropane as well as cycloaddition of this molecule with substituted ethylenes[176] and norbornadiene.[177] Trimethylenemethane complexes of Ni(0) have been ruled out, and insertion of nickel into the ring occurs.[175–177]

[172] G. F. Emerson, K. Ehrlich, W. P. Giering, and P. C. Lauterber, *J. Am. Chem. Soc.* **88**, 3172 (1966).

[173] R. Noyori, T. Nishimura, and H. Takaya, *Chem. Commun.* p. 89 (1969).

[174] P. Binger, *Angew. Chem., Int. Ed. Engl.* **11**, 109 (1972).

[175] P. Binger and J. McMeeking, *Angew Chem., Int. Ed. Engl.* **12**, 995 (1974).

[176] R. Noyori, T. Odagi, and H. Takaya, *J. Am. Chem. Soc.* **92**, 5780 (1970).

[177] R. Noyori, T. Ishiganai, N. Hayashi, and H. Takaya, *J. Am. Chem. Soc.* **95**, 1674 (1973).

*Cis* and *trans* isomers of the dimethyl ester of Feist's acid form 1,3-butadiene iron tricarbonyl complexes in the presence of $Fe_2(CO)_9$,[178] stable complexes with Rh(I), Ir(I), Pt(0), and Pt(II), and an unstable Pd(II) complex.[179]

While cyclobutene should exhibit complexing ability intermediate between cyclopropene and cyclopentene, the argentation constants (in ethylene glycol) of 1-methylcyclobutene (0.54), 1-methylcyclopentene (1.9), and 1-methylcyclohexene (0.5)[163,180] do not satisfy this expectation. The most interesting aspect of cyclobutene chemistry is the ease with which metal species such as Ag(I), Cu(I), and Fe(0) promote orbital-symmetry-forbidden disrotatory ring opening[154,181a,b] in molecules such as **92** and Dewar benzene (to be discussed later).

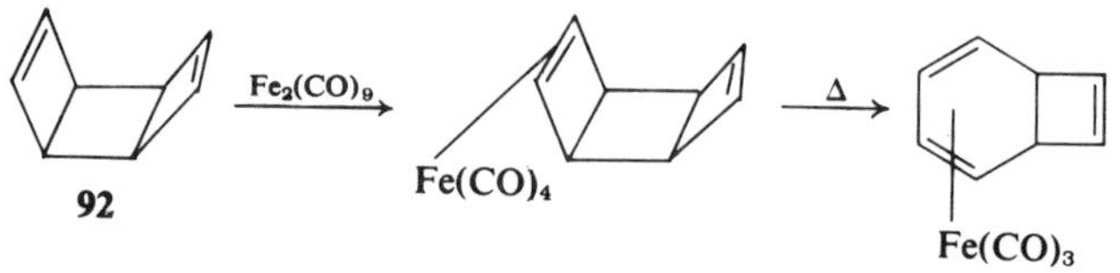

(42)

Just as cyclopropene can be stabilized in the form of its Pt(0) complex, $\Delta^{1,4}$-bicyclo[2.2.0]hexene (**93**) can also be "stored" in this manner and released upon addition of carbon disulfide.[182]

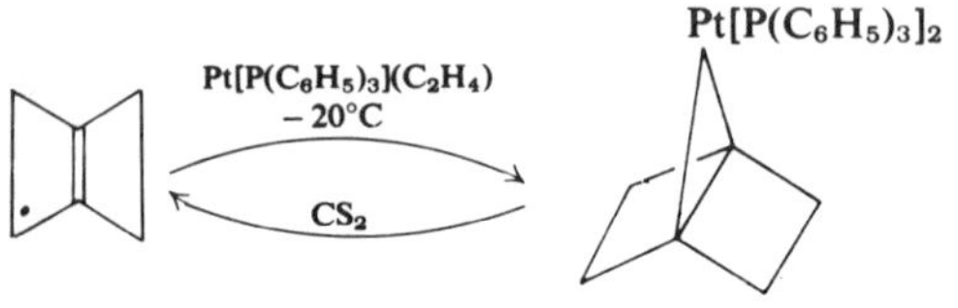

(43)

While unstrained olefins such as cyclohexene have a fairly weak tendency to react with iron carbonyls, the strain in norbornadiene is great enough to facilitate its dimerization as well as insertion of carbon monoxide [Eq. (44)].[183-187] Dimers and trimers may also be formed in the presence of metallic rhodium on charcoal[188a] or various compounds of Rh(I).[188b]

[178] T. H. Whitesides and R. W. Slaven, *J. Organomet. Chem.* **67**, 99 (1974).

[179] M. Green and R. P. Hughes, *J. Chem. Soc., Chem. Commun.* p. 686 (1974).

[180] M. A. Muhs and F. T. Weiss, *J. Am. Chem. Soc.* p. **84**, 4697 (1962).

[181a] W. Merk and R. Pettit, *J. Am. Chem. Soc.* **89**, 4788 (1967).

[181b] W. Slegeir, R. Case, J. S. McKennis, and R. Pettit, *J. Am. Chem. Soc.* **96**, 287 (1974).

[182] M. E. Jason, J. A. McGinnety, and K. B. Wiberg, *J. Am. Chem. Soc.* **96**, 6531 (1974).

[183] D. M. Lemal and K. S. Shim, *Tetrahedron Lett.* p. 368 (1961).

[184] C. W. Bird, D. L. Colinese, R. C. Cookson, J. Hudec, and R. O. Williams, *Tetrahedron Lett.* p. 373 (1961).

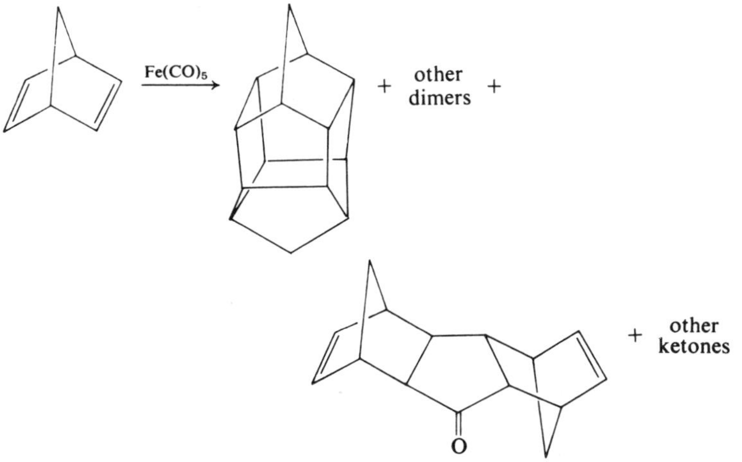

$$(44)$$

### 5.C.4.b  Twisted Olefinic Linkages

*Trans*-cyclooctene forms much more stable transition-metal complexes than its *cis* isomer (argentation constants in ethylene glycol: *trans*-cyclooctene, $K_{Ag} > 1000$; *cis*-cyclooctene, $K_{Ag} = 14.4$[180]) and, indeed, resolution of *trans*-cyclooctene was achieved through fractional crystallization of [*trans*-dichloro-(*trans*-cyclooctene)-(+ or −)α-methylbenzylamine]platinum(II).[189] Aqueous cyanide releases enantiomeric *trans*-cyclooctenes from their respective diastereomeric complexes. The structure of the Pt(II) complex **95**

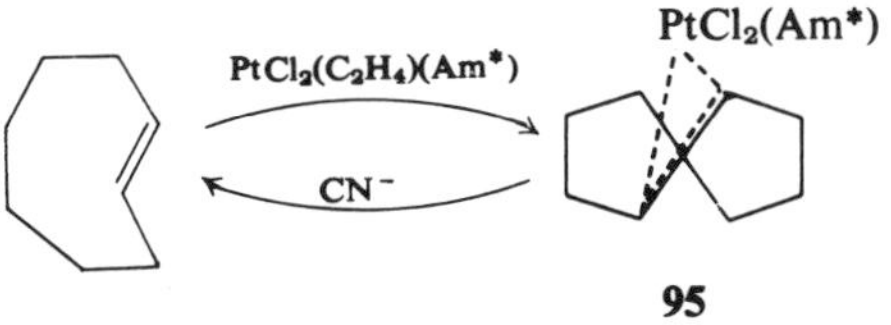

**95**

Am* = α-methylbenzylamine (optically active)     (45)

[185] S. C. Neely, D. van der Helm, A. P. Marchand, and B. R. Hayes, *Acta Crystallogr.*, Sect. B **32**, 335 (1976).

[186] E. Weissbergerf and P. Laszlo, *Acc. Chem. Res.* **9**, 209 (1976).

[187] H.-D. Scharf, G. Weisgerber, and H. Hüber, *Tetrahedron Lett.* p. 4227 (1967).

[188a] J. J. Mrowca and T. J. Katz, *J. Am. Chem. Soc.* **88**, 4012 (1966).

[188b] N. Acton, R. J. Roth, T. J. Katz, J. K. Frank, C. A. Maier, and I. C. Paul, *J. Am. Chem. Soc.* **94**, 5446 (1972).

[189] A. C. Cope, C. R. Ganellin, H. W. Johnson, Jr., T. U. Van Auken, and H. J. S. Winkler, *J. Am. Chem. Soc.* **85**, 3276 (1963).

indicates that the ring is in the crossed conformation, which, curiously, is also valid for the free ligand. The $C_1$—$C_2$ distance in **95** is 1.35 Å.[190] It is curious that platinum complexes do not appear to have been used to intercept *trans*-cycloheptene, $\Delta^{1,2}$-bicyclo[3.2.2]nonene, and similar bicyclic bridgehead olefins. However, transition-metal complexes of bridgehead olefins are being investigated,[191a] and indeed a Pt(0) complex of $\Delta^{1,8}$-bicyclo[4.2.1]nonene has been isolated.[191b] A nickel(0) complex of a $Z,E,Z$-1,3,5-cycloheptatriene has been isolated and its crystal structure analyzed.[96a,b]

### 5.C.4.c  *Complexes of Cycloalkynes, Benzyne, and Cycloallenes*

As noted in Chapter 3, cycloheptyne and cyclohexyne are transient intermediates. However, stable Pt(0) complexes have been prepared by generating these cycloalkynes from the corresponding 1,2-dibromocycloalkenes, using sodium amalgam in THF in the presence of tris(triphenylphosphine)-platinum(0).[192] Not surprisingly, acyclic alkynes cannot displace these ligands, but free reactive cycloalkynes can be regenerated upon addition of tetracyanoethylene (TCNE) [Eq. (46)].[192] As mentioned earlier in connection with its cyclopropene complex, Pt(0) should act primarily as an electron

donor to the $\pi^*$ orbital of an alkyne. Occupation of this orbital will weaken the acetylenic linkage as well as deform C—C—C angles. The $C_1$—$C_2$ distances in the cyclohexyne complex (**96**) [1.297(8) Å] and the cycloheptyne complex [1.283(5) Å], and the departures of the C—C—C angles from linear geometry (52.7° and 41.3°, respectively) are consistent with this view. The ligands thus seem to be more ethylenic than acetylenic.[193] Attempts to make stable silver and platinum complexes of benzyne have failed,[194–197] but a

[190] P. C. Manor, D. P. Shoemaker, and A. S. Parkes, *J. Am. Chem. Soc.* **92**, 5260 (1970).
[191a] Prof. Carol D. Meyer, personal communication to Arthur Greenberg.
[191b] J. R. Wiseman, *Int. Symp. Chem. Strained Rings*, Binghamton, New York, 1977.
[192] M. A. Bennett, G. B. Robertson, P. O. Whimp, and T. Yoshida, *J. Am. Chem. Soc.* **93**, 3797 (1971).
[193] G. B. Robertson and P. O. Whimp, *J. Am. Chem. Soc.* **97**, 1051 (1975).
[194] H. F. S. Winkler and G. Wittig, *J. Org. Chem.* **28**, 1733 (1963).
[195] S. L. Friedman, *J. Am. Chem. Soc.* **89**, 3071 (1967).
[196] C. D. Cook and G. S. Jauhal, *J. Am. Chem. Soc.* **90**, 1464 (1968).
[197] T. L. Gilchrist, F. J. Graveling, and C. W. Rees, *Chem. Commun.* p. 821 (1968).

nickel complex (97)[198,199] and osmium complexes[200a–d] have been characterized. Tetrafluorobenzyne, whose electronegative substituents should help $\pi$-type bonding (metal as electron donor), forms stable complexes in the presence of cobalt, iron, and nickel compounds.[201]

$$\text{(47)}$$

**97**

A variety of transition-metal allene complexes have been obtained in the pure state, and these are characterized by interaction of the metal with a single olefinic group accompanied by significant $C\!=\!C\!=\!C$ angle bending.[202,203] Thus, small cyclic molecules which already maintain a bent allenic linkage at some "cost" should be stabilized by interaction with transition metals. In fact, 1,2-cyclononadiene has been resolved in the form of its stable *trans*-dichloro(ethylene)($\alpha$-methylbenzylamine)platinum complex,[202] just as *trans*-cyclooctene was (see Section 5.C.4.b). (1,2-Cyclooctadiene)(bis-triphenylphosphine)platinum(0) (m.p. 151°–153°C) releases its strained allenic ligand upon addition of carbon disulfide. When this reaction is performed at $-60°C$, one can spectroscopically observe 1,2-cyclooctadiene.[203] However, 1,2-cycloheptadiene is understandably more "reluctant" to leave its stable complex (m.p. 156°–157°C), and once liberated at 35°C, it dimerizes more rapidly than it can form a platinum(0) complex.[203]

### 5.C.4.d  *Metal Complexes of Cyclophanes*

One might speculate briefly about what effect distortion of a benzene ring would have upon its ability to form metal complexes. Although one might

[198] E. W. Gowling, S. F. A. Kettle, and C. M. Sharples, *Chem. Commun.* p. 21 (1968).

[199] Another nickel-benzyne complex, see J. E. Dobson, R. G. Miller, and J. P. Wiggen, *J. Am. Chem. Soc.* **93**, 554 (1971).

[200a] C. W. Bradford, R. S. Nyholm, G. J. Gainsford, J. M. Guss, P. R. Ireland, and R. Mason, *J. Chem. Soc., Chem. Commun.* p. 87 (1972).

[200b] A. J. Deeming, R. S. Nyholm, and M. Underhill, *J. Chem. Soc., Chem. Commun.* p. 224 (1972).

[200c] G. J. Gainsford, J. M. Guss, P. R. Ireland, and R. Mason, *J. Organomet. Chem.* **40**, C70 (1972).

[200d] K. A. Azam and A. J. Deeming, *J. Chem. Soc., Chem. Commun.* p. 852 (1976).

[201] D. M. Roe and A. G. Massey, *J. Organomet. Chem.* **23**, 547 (1970).

[202] A. C. Cope, W. R. Moore, R. D. Bach, and H. J. S. Winkler, *J. Am. Chem. Soc.* **92**, 1243 (1970) and references 11–14 cited therein.

[203] J. P. Visser and J. E. Ramakers, *J. Chem. Soc., Chem. Commun.* p. 178 (1972).

naturally anticipate symmetric complexation of a metal ion such as Ag(I) with benzene (i.e., alignment with the sixfold symmetry axis), in fact this metal (and others) coordinates with a single pair of carbons on one benzene face.[204] x-Ray analysis indicates that the benzene ring of the silver perchlorate complex of benzene (98) is distorted so that the bonds coordinating to silver are 1.35 Å long, while uncoordinated bonds are 1.43 Å long.[204] Any tendency for [n]paracyclophanes to shorten and lengthen bonds in this manner (see Section 3.H.3) should increase the tendency to form complexes (perhaps in the manner of 99).

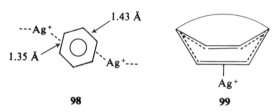

98                           99

### 5.C.5 TRANSITION-METAL-CATALYZED REARRANGEMENT OF POLYCYCLIC HYDROCARBONS

#### 5.C.5.a  Quadricyclanes and Related Molecules

The quantitative rearrangement of thermally stable quadricyclane (7) ($t_{1/2}^{140°C} > 14$ hr[205]) to norbornadiene is catalyzed by Rh(I), Pd(II), Pt(I), and Ag(I).[206,207] Initially, the rearrangement was thought to occur in a concerted manner,[206,207] but later evidence,[208] including the characterization of a rhodium acyl insertion product (100),[209] suggested that a nonconcerted

Rh(III)

7                           100

**Scheme 14**

[204] C. A. Fyfe, in "Molecular Complexes" (R. Foster, ed.), Part 1, pp. 270–272. Crane, Russak, New York, 1973 and references cited therein.
[205] G. S. Hammond, N. J. Turro, and A. Fischer, J. Am. Chem. Soc. 83, 4674 (1961).
[206] H. Hogeveen and H. C. Volger, J. Am. Chem. Soc. 89, 2486 (1967).
[207] H. Hogeveen and B. J. Nusse, Tetrahedron Lett. p. 159 (1974).
[208] T. J. Katz and S. Cerefice, J. Am. Chem. Soc. 91, 2405, 6519 (1969).
[209] L. Cassar and J. Halpern, Chem. Commun. p. 1082 (1970).

mechanism involving oxidative addition as the initial step occurred at least in some cases. When methanol is employed as solvent, Rh(I) still provides norbornadiene products quantitatively, but Ag(I) produces isomers (101) resulting from an apparent carbonium-ion mechanism [Eq. (48)].[207] A

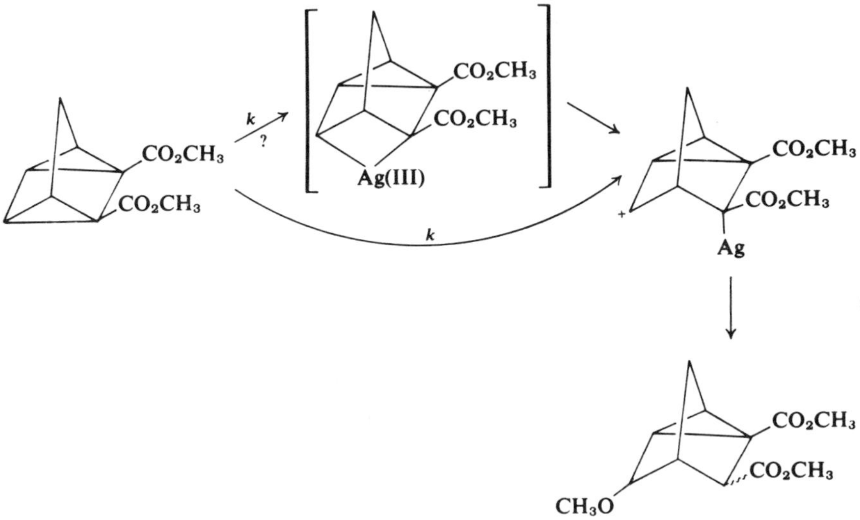

(48)

101

Michaelis–Menten treatment of kinetic data demonstrates that Rh(I) forms a preequilibrium complex before undergoing rate-limiting oxidative addition [Eq. (49)], while no such preequilibrium complex is found for Pd(II) prior to its rate-determining oxidative addition [Eq. (50)].[207] The oxidative addition of Ag(I) postulated at the time by these authors[207] and others is now widely

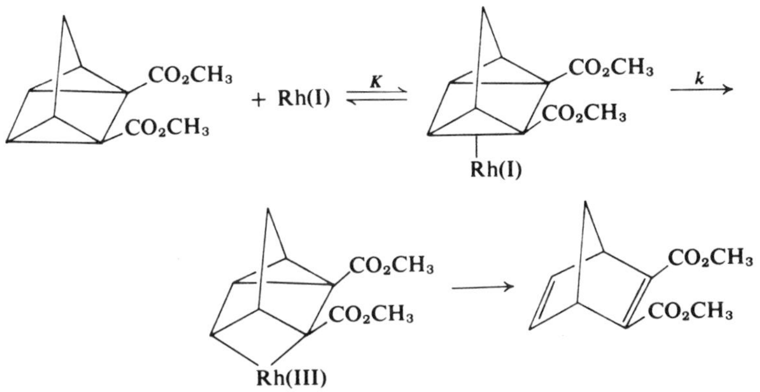

(49)

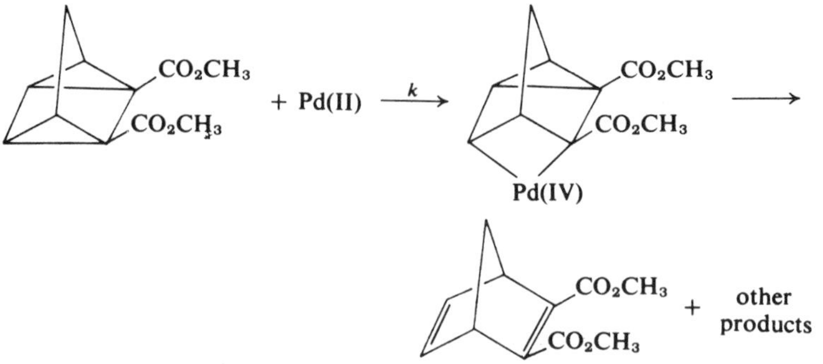

$$(50)$$

regarded as unlikely[123a–c,154,210] if one uses the promotion energies of silver(I) (9.94 eV) and rhodium(I) (1.6 eV) as indicators. It is also interesting that the normal products, norbornadiene and carbonium-ion-derived molecules, produced by silver trifluoroacetate in methanol are also accompanied by metallic silver and products of silver-promoted oxidation.[211] Oxidation of quadricyclane by silver(I) ion is rationalized through comparison of the ionization potential of the hydrocarbon (7.40 eV) with the electron affinity of silver(I) (7.57 eV).[211] Silver-promoted oxidation of cubane or tricyclo-[4.1.0.0$^{2,7}$]heptane does not occur because these molecules have ionization potentials of 8.74 and 8.15 eV, respectively.[211] Precipitation of metallic silver in a reaction involving silver nitrate (in methanol) and hexamethylprismane has also been noted.

The quadricyclane-to-norbornadiene isomerization, which is quite favorable thermodynamically ($\Delta H_r$(liq) $= -26.2$ kcal/mole[9]), occurs quite slowly as noted at the beginning of this section, and requires catalysis in order to run at a measurable rate at ambient temperatures. However, exo-tricyclo[3.2.1.0$^{2,4}$]-octane (102) is converted into bishomoprismane (103)[212] because the latter is more stable. The endo isomer 104 cannot be so transformed.[212] An

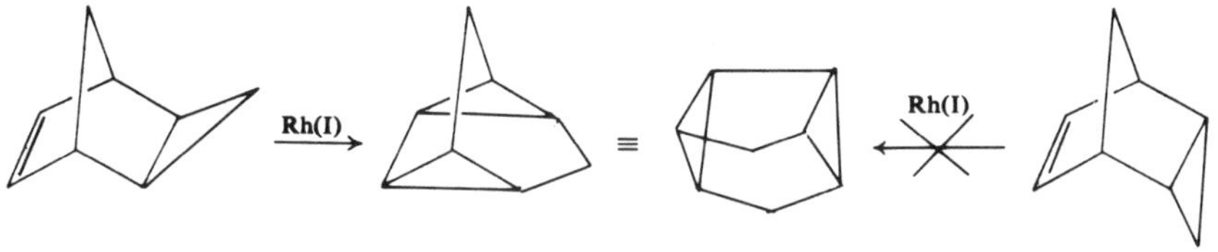

102                          103                                              104

**Scheme 15**

[210] P. E. Eaton, L. Cassar, R. A. Hudson, and D. R. Hwang, J. Org. Chem. 41, 1445 (1976).

[211] G. F. Koser and J. N. Faircloth, J. Org. Chem. 41, 583 (1976).

[212] H. C. Volger, H. Hogeveen, and M. M. P. Gassbeek, J. Am. Chem. Soc. 91, 218 (1969).

additional methylene bridge introduces more strain into **105** (homoanalogue of **103**) than in **106** (homoanalogue of **102**), and Rh(I) catalysis provides a means of interconverting **105** to **106**.[213a,b] However, **107** could not be isomerized to triasterane **108** in the presence of Rh(I), but Ir(I) converts it to **109**. The *exo,endo* isomer(**110**) which, unlike **107**, does not form a stable platinum

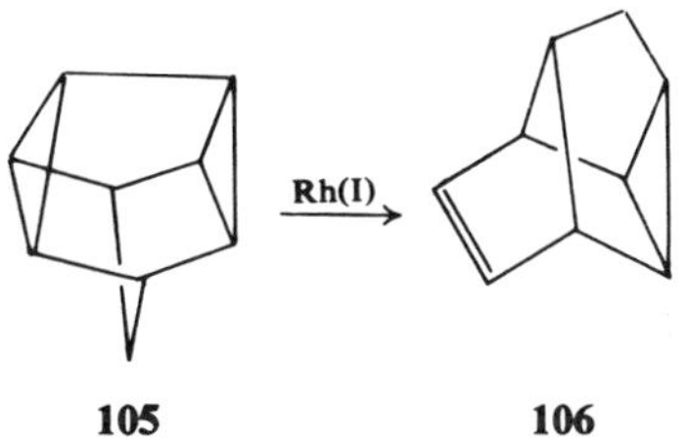

$$(51)$$

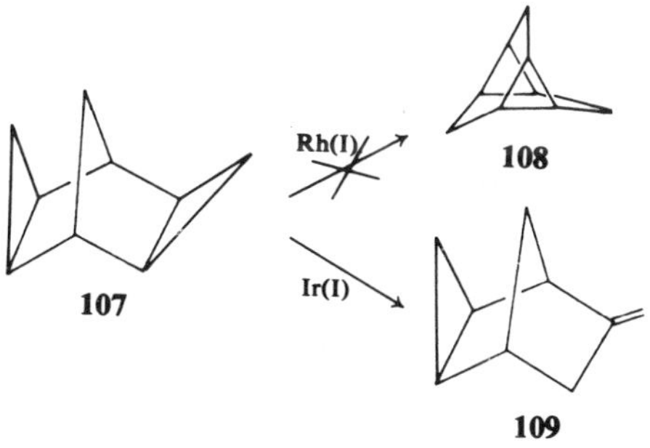

**105**          **106**

$$(52)$$

**107**          **108**

**109**

**110**

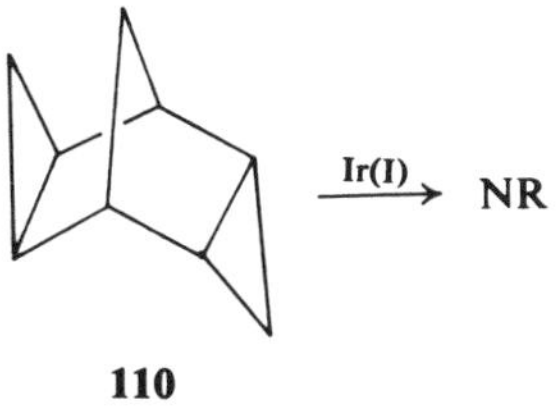

complex (see **51**), is inert to Ir(I).[102] The lack of reactivity of both **83** and **110** has been attributed by some to their inability to form bidentate complexes.[102,154]

## 5.C.5.b  *Cubanes and Related Systems*

Cubane (**10**) and related molecules have been perhaps the most explored of the polycyclic hydrocarbons in relation to their transition-metal chemistry.[123a–c,210] Silver-promoted rearrangements of bishomocubyl derivatives have been employed to provide access to the fascinating $(CH)_{10}$ molecules

[213a] T. J. Katz, J. C. Carnahan, Jr., and R. Boecke, *J. Org. Chem.* **32**, 1301 (1967).

[213b] T. J. Katz and S. Cerefice, *Tetrahedron Lett.* p. 2509 (1969).

snoutene, diademane, and triquinacene (see Chapter 4), while the diazabis-homocubyl system provided an entry to semibullvalene [$(CH)_8$] and some annelated semibullvalenes of theoretical importance.[123c] The first silver-promoted homocubyl rearrangement was unwittingly performed in the anticipated purification of **111** on a silica-gel column impregnated with silver nitrate.[214] Reports of silver(I), rhodium(I), palladium(II), and other transition-metal-catalyzed rearrangements of bishomocubanes,[215,216] homocubanes,[216]

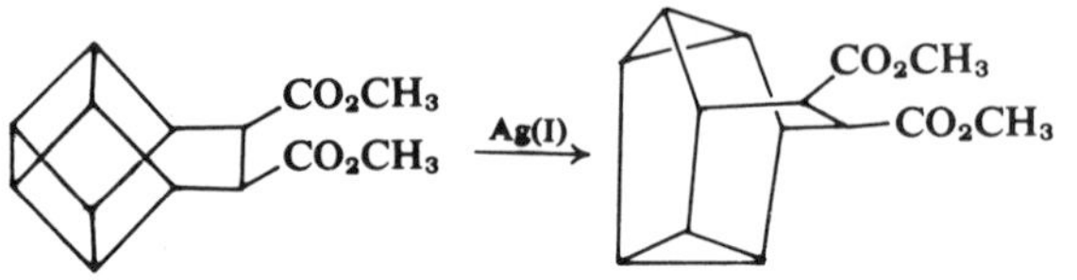

$$(53)$$

**111**                          **112**

cubanes,[217] secocubanes,[218] and *syn*-tricyclooctane (**82**)[154] appeared about this time. As with quadricyclane, rhodium(I)-catalyzed and silver(I)- [as well as Pd(II)-] catalyzed rearrangements of cubane follow markedly different

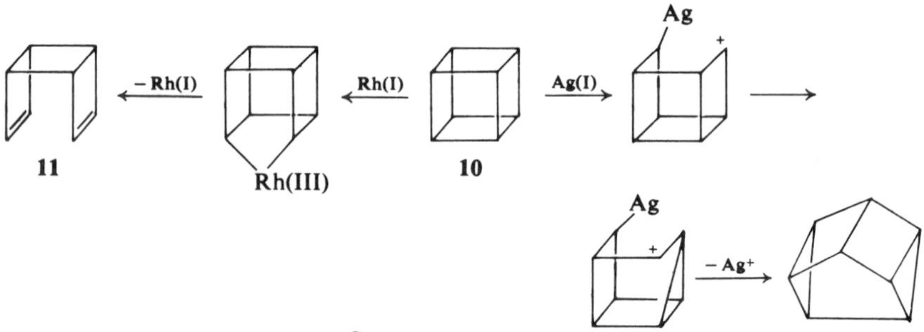

**Scheme 16**

pathways (Scheme 16).[91,123a–c] The rhodium-catalyzed rearrangement is a *formal* ($_\sigma 2_s + _\sigma 2_s$) thermally forbidden process, while the latter is a *formal* ($_\sigma 2_a + _\sigma 2_a$) process which is also thermally forbidden by the rules of orbital symmetry (the proved nonconcertedness of these reactions noted in Scheme 16 emphasizes the formalism noted above.) Isolation of an acyl rhodium complex, analogous to **100**, in the rhodium-induced rearrangement of cubane[217] is strong evidence favoring an oxidative-addition pathway for this

[214] R. Furstoss and J.-M. Lehn, *Bull. Soc. Chim. Fr.* p. 2497 (1966).

[215] W. G. Dauben, M. G. Buzzolini, C. H. Schallhorn, and D. L. Whalen, *Tetrahedron Lett.* p. 787 (1970).

[216] L. A. Paquette and J. C. Stowell, *J. Am. Chem. Soc.* **92**, 2584 (1970).

[217] L. Cassar, P. E. Eaton, and J. Halpern, *J. Am. Chem. Soc.* **92**, 3515, 6366 (1970).

[218] H. H. Westberg and H. Ona, *Chem. Commun.* p. 248 (1971).

metal[219] (but not necessarily others[210]). Although it was initially thought that silver played a similar role,[219] it is now known to behave as a Lewis acid.[123a-c,210] Parallel chemistries have been recorded for the rhodium- and silver-catalyzed rearrangements of homocubane (113, $X = CH_2$), bishomocubane (basketane 113, $X = CH_2CH_2$), secocubane (113, $X = $ no bridge or $H_2$), and syn-tricyclooctane (82)[91,123a-c,154,210,215-218] The pro-

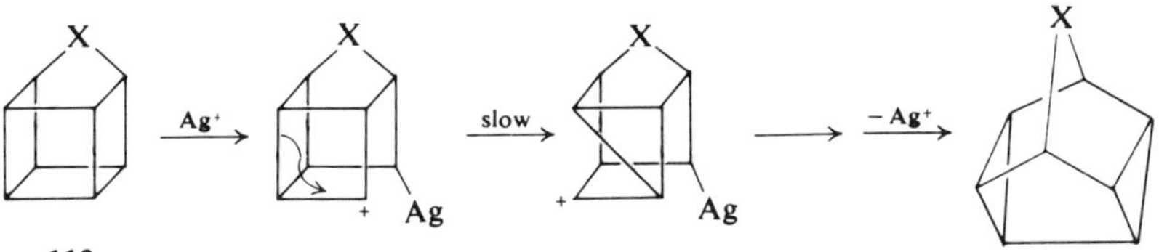

113                                                                    (54)

posed unifying mechanism[123a-c,210] involves initial argentation to form a cyclobutyl carbonium ion, rate-determining[210] rearrangement to a cyclo-propylcarbinyl carbonium ion via 1,2-migration of a bond *anti* to silver, and subsequent formation of the second three-membered ring with loss of silver ion [Eq. (54)]. Part of the driving force for this rearrangement is loss of the strain energy equivalent to a cyclobutane ring in the slow step.[210] The fact that *anti*-tricyclooctane (83) does not rearrange[154] can be attributed to a requirement of 1,2-migration of a bond *syn* to silver,[210] but might also have steric origins, or it may be electronic in origin.[220] Homopentaprismane (114) rearranges, as anticipated, to 115, but the expected silver-catalyzed rearrangement to 116 (obtained by an independent route[221]) does not occur.[210,221] Three explanations, all somewhat mutually dependent, have been offered.

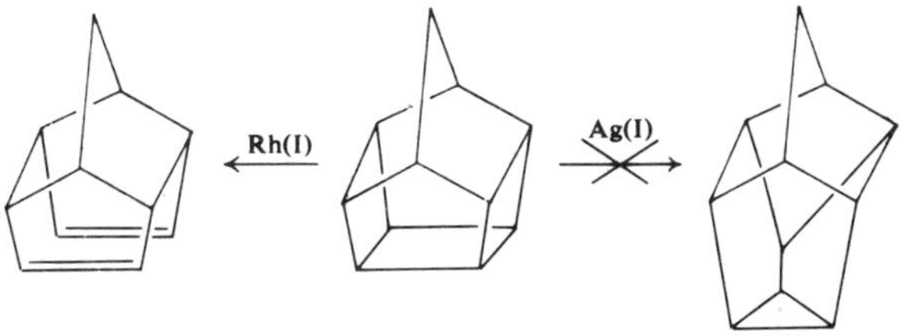

115                     114                     116                     (55)

First, it has been suggested that while changes in enthalpy of more than 30 kcal/mole have accompanied silver-catalyzed rearrangements in cubane and

[219] J. E. Byrd, L. Cassar, P. E. Eaton, and J. Halpern, *Chem. Commun.* p. 40 (1970).

[220] R. Gleiter, E. Heilbronner, M. Iteckman, and H.-D. Martin, *Chem. Ber.* **106**, 28 (1973).

[221] A. P. Marchand, T.-C. Chou, J. D. Ekstand, and D. van der Helm, *J. Org. Chem.* **41**, 1438 (1976).

related molecules, the estimated 24 kcal/mole strain relief upon conversion of **114** to **116** is an insufficient driving force.[221] It was also noted that the skeletal bonds in cubane are of greater p character and thus more susceptible to complexation than those in **114**.[209] Finally, the presumed rate-determining step in the silver-catalyzed rearrangement of **114** [assuming mechanism of Eq.(54)] would not be accompanied by net loss of the strain of a cyclobutane ring.[209]

### 5.C.5.c   *Benzene Valence Isomers*

The symmetry-forbidden ground-state aromatization of Dewar benzene has already been discussed (Fig. 5.2). Transition metals lower the activation barrier for this transformation {hexamethyl Dewar benzene → benzene: $E_{act}$(thermal) = 31.1 kcal/mole; $E_{act} \simeq 19.4$ kcal/mole [Rh(I) cat.][222]}. However, because Dewar benzenes are homoconjugated dienes, they sometimes form fairly stable transition-metal complexes. For example, a stable Pd(II) complex (**117**) has been characterized which releases Dewar benzene upon addition of pyridine.[223] The analogous Pd(II) complex of hexamethyl

$$\text{Pd}(C_6H_5CN)_2Cl_2/CH_2Cl_2 \atop \text{pyridine}$$

$$\text{PdCl}_2$$
**117**

(56)

Dewar benzene is stable in chloroform for about 20 minutes at 33°C, decomposes at 79°C, and will release the hydrocarbon upon addition of triphenylphosphine.[223] A chromium(0) complex rearranges at 115°C to produce the hexamethylbenzene chromium complex.[224] A Pt(II) complex of hexamethyl Dewar benzene decomposes at about 75°C.[225] Various effects of other metals on Dewar benzenes have been reported,[226] and some of these produce cyclopentadiene derivatives as well as aromatics.[227a,b]

In 1971, the rearrangement of 1,1′,2,2′-tetraphenyl-3,3′-bicyclopropenyl (**118**) to 1,2,4,5-tetraphenylbenzene was observed to be catalyzed by silver

[222] H. C. Volger and M. M. P. Gaasbeck, *Rec. Trav. Chim. Pays-Bas* **87**, 1290 (1968).

[223] H. Dietl and P. M. Maitlis, *Chem. Commun.* p. 759 (1967).

[224] E. O. Fischer, C. G. Kreiter and W. Berngruber, *Angew. Chem., Int. Ed. Engl.* **6**, 634 (1967).

[225] H. C. Volger and H. Hogeveen, *Rec. Trav. Chim. Pays-Bas* **86** 830 (1967).

[226] B. L. Booth, R. N. Haszeldine, and M. Hill, *Chem. Commun.* p. 1118 (1967).

[227a] P. V. Galakrishnan and P. M. Maitlis, *Chem. Commun.* p. 1303 (1968).

[227b] J. W. Kang, K. Mosley, and P. M. Maitlis, *Chem. Commun.* p. 1304 (1968).

perchlorate.[228] An interesting feature of this reaction is the intermediate formation of 2,3,5,6-tetraphenylbicyclo[2.2.0]hexa-2,4-diene (119), which has been isolated under these conditions. Large concentrations of this intermediate build up because its aromatization is much slower (see preceding discussion) than rearrangement of the corresponding bicyclopropenyl [Eq. (57)]. It should be noted that this rearrangement occurs more rapidly than the potential ene reaction. Evidence strongly supports a nonconcerted mechanism

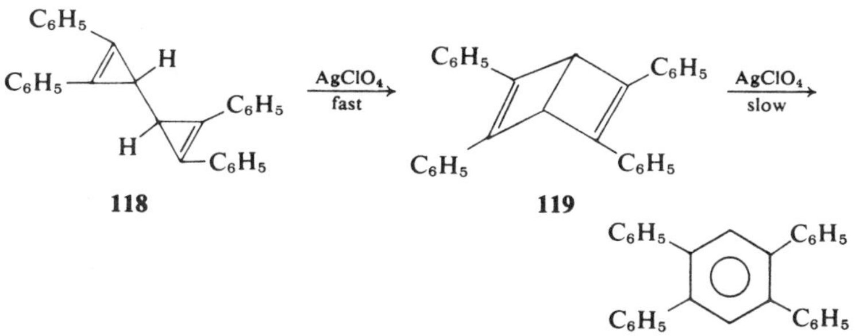

$$\tag{57}$$

involving the intermediacy of an argentocarbene-carbonium ion (120)[72] which is stabilized compared to the analogous intermediate in the uncatalyzed thermal reaction (see Scheme 5). Such a mechanism is consistent with

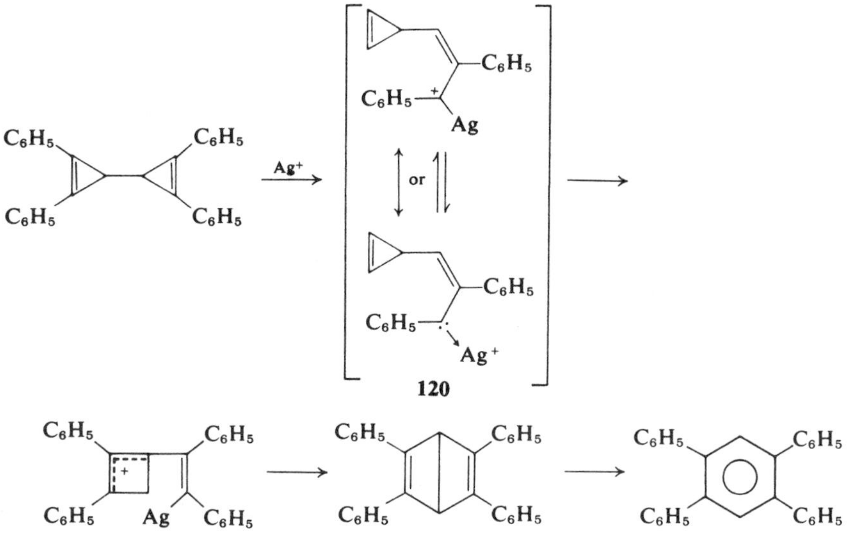

$$\tag{58}$$

[228] R. Weiss and C. Schlierf, *Angew. Chem., Int. Ed. Engl.* **10**, 811 (1971).

the substituent effects observed for the series: for example, the inertness of 3-cyano-2,2′,3,3′-tetraphenyl-3,3′-bicyclopropenyl to silver ion.[72] It is interesting to contrast the large concentrations of intermediate Dewar benzenes obtainable in this moderate-temperature silver-catalyzed rearrangement with the vanishingly small amounts indirectly detectable in the higher-temperature thermal pathway by which Dewar benzenes isomerize virtually as soon as they are formed.[74] This silver-catalyzed pathway to Dewar benzenes was employed to obtain **33** from the corresponding bicyclopropenyl.[56] Again, silver-catalyzed rearrangement of the Dewar benzene is much slower than rearrangement of the bicyclopropenyl isomer. Dewar benzene **35**, however, is only inferred as an intermediate en route to indane.[52]

Prismanes rearrange very rapidly and exothermically in the presence of transition-metal catalysts. For example, hexamethylprismane (**121**) in benzene rearranges at about 100°C ($\Delta H^{\ddagger} \simeq 34$ kcal/mole) to produce mostly hexamethylbenzene, while in the presence of Rh(I) in chloroform the half-life for rearrangement at $-30°$C is 40 minutes and the predominant product is hexamethyl Dewar benzene[229] (Scheme 17).

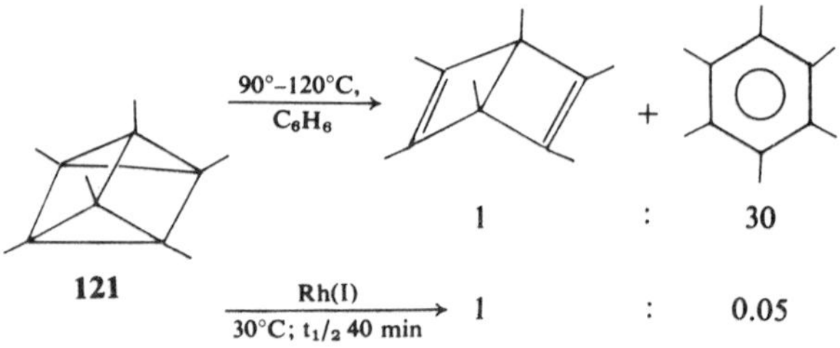

**Scheme 17**

Tri-*tert*-butylprismane (**122**) rearranges rapidly in the presence of a number of transition metals [e.g., $Ag^{+}$ in $CH_3OH$: $r_{1/2}^{30°} <$ min; $t_{1/2}^{25°-30°}$ Rh(I) in $CH_3OH \simeq 20$ min] to produce Dewar benzenes and benzvalenes which aromatize at a slower rate (Scheme 18).[230] Here, as in the bicyclopropenyl case, metals allow the isolation of an intermediate benzene valence isomer which could not normally be isolated in the thermal interconversion of a valence isomer to benzene. Silver-catalyzed rearrangement of benzvalene to benzene (see $Ag^{+}$-catalyzed rearrangements of bicyclobutanes, Section 5.C.3.a) has provided a direct value for $\Delta H_r$.[61b] There is a dichotomy in the rearrangements of benzvalene promoted by zero-valent metals. For example, Ag(0) readily provides benzene, while Cu(0) yields fulvene.[61c]

[229] H. Hogeveen and H. C. Volger, *Chem. Commun.* p. 1133 (1967).
[230] K. L. Kaiser, R. F. Childs, and P. M. Maitlis, *J. Am. Chem. Soc.* **93**, 1270 (1971).

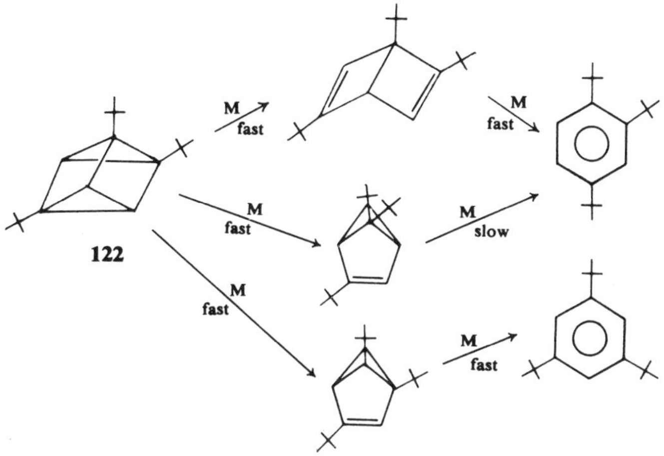

**Scheme 18**

## 5.D Strained Heterocyclic Molecules

### 5.D.1 FEATURES OF STRUCTURE AND STABILITY PECULIAR TO HETEROCYCLIC MOLECULES[231]

For the most part, the other sections of this book have dealt with strained hydrocarbons. In this section we examine some aspects of the effects of strain on the chemistry at a heteroatom as well as the effects of heteroatoms upon strained carbocyclic networks. The number of classes of heterocyclic molecules is so large and their chemistries so diverse that this section will merely be a survey concerned mostly with principles, rather than an exhaustive study of all types of strained systems. It is our intention to consider (a) the influence of distorted local geometries on the ionization potential, basicity, and complexing ability of constrained heteroatoms; (b) the strain energies of heterocyclic molecules in comparison to analogous hydrocarbons; (c) kinetic stability of heterocycles which are usually more reactive with electrophiles and nucleophiles than the analogous hydrocarbon; (d) special conjugative and inductive effects arising from nonbonded electrons as well as from the possible intervention of d orbitals of atoms such as Si, P, or S; (e) changes in the strain and reactivity of carbocyclic networks as the result of changes in bond lengths and angles introduced by heteroatoms; (f) special steric effects attributable to nonbonded electrons; (g) the potential role of intramolecular hydrogen bonding in stabilizing inherently strained molecules; and (h) the

---

[231] See also L. A. Paquette, "Heterocyclic Chemistry," Benjamin, New York, 1968.

relative ease of ring opening of certain three-membered heterocyclic rings due to the accessibility of stable ring-opened forms. A knowledge of strained heterocycles in many cases allows one to explain the chemistry of many important naturally occurring molecules as well as to examine the viability of numerous intermediates postulated for reaction mechanisms. This chapter will not include discussions of boranes, carboranes, and elemental sulfur rings.

At this time, molecular mechanics (see Chapter 1) has played a rather limited role in understanding heterocyclic chemistry. Calculations of certain sulfur-containing[232a-c] and oxygen-containing molecules[233] have appeared in the literature and these will be cited when appropriate. On the other hand, photoelectron spectroscopy has been particularly valuable in studies of heterocycles.[234a-b] The primary reason is that bands in the photoelectron spectrum primarily attributable to lone-pair electrons are often well separated from other bands and are sharp, rather than diffuse, as bands arising from carbocyclic systems tend to be. Sharp bands occur because the geometry of the radical cation produced by ionization of a lone-pair (nonbonding) electron is very similar to that of the neutral compound.[234a] In addition to the standard sources of thermochemical data cited in Chapter 1 and used implicitly throughout the book, we also employ here a compendium and discussion of thermochemical data for heterocyclic molecules.[235]

We might provide some examples of properties of heterocyclic molecules that diverge from those of hydrocarbons before beginning the survey of strained species. A useful system to examine for this purpose is the 1,3-dioxane ring. 1,3-Dioxane is some 8.3 kcal/mole more stable than 1,4-dioxane, an effect also apparent in acyclic 1,3-dioxa and 1,4-dioxa molecules, attributed to a stabilizing 1,3-resonance interaction rather than differences in ring strain.[235] Shortened C—O vs. C—C bond lengths cause certain 1,3-diaxial interactions to be particularly severe in the heterocycle. For example, the equatorial preference of a methyl group in cyclohexane ($\Delta G°$) is 1.7 kcal/mole, while that in 2-methyl-1,3-dioxane is greater than 3.6 kcal/mole[236]; corresponding equatorial preferences of the *tert*-butyl group on a cyclohexane ring and

[232a] N. Allinger and M. J. Hickey, *J. Am. Chem. Soc.* **97**, 5167 (1975).

[232b] N. L. Allinger, M. J. Hickey, and J. Kao, *J. Am. Chem. Soc.* **93**, 2741 (1976).

[232c] N. L. Allinger and J. Kao, *Tetrahedron* **32**, 529 (1976).

[233] N. L. Allinger and D. Y. Chung, *J. Am. Chem. Soc.* **98**, 6798 (1976).

[234a] E. Heilbronner, J. P. Maier, and E. Haselbach, *in* "Physical Methods in Heterocyclic Chemistry" (A. R. Katritzky, ed.), Vol. 6, p. 1. Academic Press, New York, 1974.

[234b] P. D. Mollere and K. N. Houk, *J. Am. Chem. Soc.* **99**, 3226 (1977).

[235] K. Pihlaja and E. Taskinen, *in* "Physical Methods in Heterocyclic Chemistry" (A. R. Katritzky, ed.), Vol. 6, p. 199. Academic Press, New York, 1974.

[236] E. L. Eliel and M. C. Knoeber, *J. Am. Chem. Soc.* **90**, 3444 (1968).

in the 2- position of a 1,3-dioxane ring are $>4.4$ kcal/mole and $>7.2$ kcal/mole, respectively.[236] On the other hand, the equatorial preference of the substituent in 5-methyl-1,3-dioxane is only 0.8–1.0 kcal/mole and in 5-*tert*-butyl-1,3-dioxane only 1.4–1.5 kcal/mole.[236] In these cases it appears that 1,3-diaxial alkyl–lone pair (oxygen) repulsion is less severe than alkyl–hydrogen repulsion. Assuming a model of a localized lone pair (i.e., "lobes"), one can argue that such lone pairs are "softer," "larger," "more diffuse," or "more easily polarized" than a bonded electron pair. However, the "size" varies with the environment, and a particularly large repulsion may appear when lone pairs are in close proximity. The relatively short methyl–methyl nonbonded distance in chair *trans*-2,2,4,6-tetramethyl-1,3-dioxane (**123a**) causes this molecule to predominantly adopt the twist-boat conformation (**123b**).[237a,b] Although the equatorial preference of hydroxyl bound to cyclohexane is 0.5–0.9 kcal/mole,[238] hydrogen bonding causes the group to be

**123a**                              **123b**

axial in 5-hydroxy-1,3-dioxane (**124**).[239,240] The methoxy group (equatorial preference 0.6 kcal/mole on cyclohexane[238]) is axial in 2-methoxy-1,3-dioxane[125] in order to minimize dipolar repulsion. This manifestation is usually referred to as the "anomeric effect."[241] For reasons which are still not clear, the substituent in 5-hydroxymethyl-1,3-dioxane (**126**) is axial but does not hydrogen bond to the ring's heteroatoms.[242,243]

As noted previously, molecular mechanics is presently playing a relatively minor role in the analysis of geometries and enthalpies of heterocyclic molecules. Calculations for many sulfur-containing compounds are in good agreement with known experimental structures and thermochemical data.[232a–c] However, the assumption of symmetrical electron density around sulfur is, as the authors admit,[232a] a very crude approximation, which will require

[237a] K. Pihlaja, *Acta Chem. Scand.* **22**, 716 (1968).

[237b] K. Pihlaja and S. Luoma, *Acta Chem. Scand.* **22**, 2401 (1968).

[238] J. A. Hirsch, *Top. Stereochem.* **1**, 210 (1967).

[239] N. Baggett, M. A. Bukhari, A. B. Foster, J. Lehman, and J. M. Weber, *J. Chem. Soc.* p. 4157 (1963).

[240] J. Gelas and R. Rambaud, *Bull. Soc. Chim. Fr.* p. 1300 (1969).

[241] E. L. Eliel and W. F. Bailey, *J. Amer. Chem. Soc.* **96**, 1798 (1974).

[242] E. L. Eliel and H. D. Banks, *J. Am. Chem. Soc.* **92**, 4730 (1970).

[243] R. Dratler and P. Laszlo, *Tetrahedron Lett.* p. 2607 (1970).

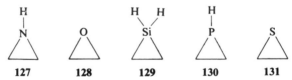

124                    125                    126

modification in molecules in which anisotropic electronic distribution is responsible for special behavior (e.g., intramolecular hydrogen bonding). While calculations upon oxygen-containing rings have also met with some success, there are marked discrepancies when two oxygen atoms are in the same molecule.[233]

### 5.D.2 SATURATED THREE-MEMBERED RINGS

#### 5.D.2.a   *One Heteroatom: Structure and Bonding*

The inherent strain in azirane (**127**) and oxirane (**128**) is about equal to that in cyclopropane (see Table 5.2). It is, therefore, not surprising that unimolecular thermal decomposition of oxirane to form acetaldehyde ($\Delta H_r^\circ = -27.3$ kcal/mole) has an activation energy ($E_{act} = 57.0$ kcal/mole[244]) comparable to that for the corresponding rearrangement of cyclopropane to propene ($\Delta H_r^\circ = -7.9$ kcal/mole; $E_{act} = 65.0$ kcal/mole[245a,b]). Obviously, the

heteroatom makes such heterocyclic rings highly reactive to both electrophiles and nucleophiles which initiate ring opening. Thus, nucleophiles to which tetrahydrofuran is essentially inert cleave carbon–oxygen bonds in oxiranes.

Heteroatoms in small heterocyclic rings affect bond angles, bond lengths, and bond strengths through a combination of factors, including their intrinsic hybridization, magnitude of covalent radii, angle-bending constants, nonbonded interactions, and long-range electronic effects. The geometries and electronic characteristics of three-membered heterocycles have been analyzed

[244] T. S. Chambers and G. S. Kistiakousky, *J. Am. Chem. Soc.* **56**, 399 (1934).
[245a] M. C. Neufeld and A. T. Blades, *Can. J. Chem.* **41**, 2956 (1963).
[245b] S. W. Benson, *J. Chem. Phys.* **40**, 105 (1964).

**Table 5.2**

**Selected Thermochemical and Structural Data for Three-Membered Heterocycles\***

| | Strain | $C_2$—$C_3$ | C—X | ∡ C—X—C | I.P. | I.P.($CH_3XCH_3$) |
|---|---|---|---|---|---|---|
| △ | 27.5[a] | 1.510 Å[b] | 1.510 Å | 60.0° | — | — |
| △ NH | 27.1[a] | 1.481[c] | 1.475 | 60.0° | 226[d,e] | 205.9 |
| △ O | 27.2[a] | 1.472[f] | 1.436 | 61.4° | 243.7[g,h] | 231.8 |
| △ $SiH_2$ | — | (1.520)[i] | (1.826) | (49.2°) | — | — |
| △ PH | — | 1.502[j] | 1.867 | 47.4° | — | — |
| △ S | 19.9[a] | 1.492[f] | 1.819 | 48.5° | 204.5[h,k] | 200.4 |

\* Strain energies and ionization potentials in kcal/mole.

[a] K. Pihlaja and E. Taskinen, in "Physical Methods in Heterocyclic Chemistry" (A. R. Katritzky, ed.), Vol. 6, p. 199. Academic Press, New York, 1974.

[b] O. Bastiansen, F. N. Fritsch, and K. Hedberg, *Acta Crystallogr.* **17**, 538 (1964).

[c] B. Bak and S. Skaarup, *J. Mol. Struct.* **10**, 385 (1971).

[d] H. Basch, M. B. Robin, N. A. Kuebler, C. Baker, and D. W. Turner, *J. Chem. Phys.* **51**, 52 (1969).

[e] N. Bodor, M. J. S. Dewar, W. B. Jennings, and S. D. Worley, *Tetrahedron* **26**, 4109 (1970).

[f] G. L. Cunningham, A. W. Boyd, R. J. Myers, W. D. Gwinn, and W. I. Le Van, *J. Chem. Phys.* **19**, 676 (1951).

[g] A. D. Baker, D. Betteridge, N. R. Kemp, and R. E. Kirby, *Anal. Chem.* **43**, 375 (1971).

[h] A. Schweig and W. Thiel, *Chem. Phys. Lett.* **21**, 541 (1973).

[i] Actually these are values for a dispiro derivative; G. L. Delker, Y. Wang, G. D. Stucky, R. L. Lambert, Jr., C. K. Haas, and D. Seyferth, *J. Am. Chem. Soc.* **98**, 1779 (1976).

[j] M. T. Bowers, R. A. Beaudet, H. Goldwhite, and R. Tang, *J. Amer. Chem. Soc.* **91**, 17 (1969).

[k] D. C. Frost, F. G. Herring, A. Katrib, and C. A. McDowell, *Chem. Phys. Lett.* **20**, 401 (1973).

in terms of an interaction between the heteroatom (or heterogroup) and ethylene (Fig. 5.5)[246–248] in a manner very reminiscent of the Dewar–Chatt–Duncanson model of bonding in transition-metal–ethylene complexes (Fig. 5.4). There are two components: (a) donation of electrons from an orbital of

[246] R. Hoffmann, H. Fujimoto, J. R. Swenson, and C.-C. Wan, *J. Am. Chem. Soc.* **97**, 7644 (1973).

[247] M.-M. Rohmer and B. Roos, *J. Am. Chem. Soc.* **97**, 202 (1975).

[248] G. L. Delker, Y. Wang, G. D. Stucky, R. L. Lambert, Jr., C. K. Haas, and D. Seyferth, *J. Am. Chem. Soc.* **98** 1779 (1976).

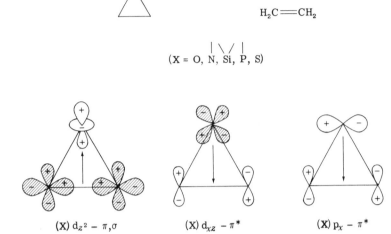

**Figure 5.5.** A picture of bonding in three-membered heterocycles invoking inter-action between ethylene and heteroatom (group) "fragments."

proper symmetry into "ethylene $\pi^*$," which would lengthen the carbon–carbon bond; and (b) donation of electrons from "ethylene $\pi$" into suitable heteroatom (heterogroup) orbitals, including d orbitals where accessible. This would also lengthen the bond distance (Table 5.2) in the series cyclo-propane, azirane, and oxirane and also in the series silirane (**129**), phosphirane (**130**), and thiirane (**131**), and can be explained in terms of heteroatom electro-negativity. Highly electronegative groups will manifest relatively little ten-dency to donate electrons to "ethylene $\pi^*$" whose occupation would lengthen $C_2$–$C_3$. Silirane is not yet known; the simplest known derivative is hexa-methylsilirane,[249] and the structure used in Table 5.2 is based upon the structure of the crystalline dispirocyclopropyl derivative **132**, which the authors realize is not the most suitable standard because the cyclopropyl groups interact electronically with the silirane ring.[248,250] The unstable mole-cule phosphirane (**130**) is a very recent addition to the three-membered heterocyclic ring series, having been obtained in lieu of the anticipated product of the reaction of phosphinide and 1,2-dichloroethane [Eq. (59)].[251a,b] While the tendency of nitrogen and oxygen to shorten the $C_2$–$C_3$

[249] D. Seyferth and D. C. Annarelli, *J. Am. Chem. Soc.* **97**, 2273 (1975).

[250] R. L. Lambert, Jr. and D. Seyferth, *J. Am. Chem. Soc.* **94**, 9246 (1972); see also P. D. Mollere and R. Hoffmann, *J. Am. Chem. Soc.* **97**, 3680 (1975).

[251a] R. I. Wagner, U.S. Patent, 3,086,056 (1963).

[251b] R. I. Wagner, L. D. Freeman, H. Goldwhite, and D. G. Rowsell, *J. Am. Chem. Soc.* **89**, 1102 (1967).

$$ClCH_2CH_2Cl \xrightarrow[NH_3]{NaPH_2} \xrightarrow[\quad]{H_2PCH_2CH_2PH_2} \mathbf{130}$$

 (59)

**132**

bond may be explained in the manner of Fig. 5.5, at least part of this feature is explicable in terms of decreasing covalent radii of the more electronegative atoms.

The decreased strain in thiirane (**131**) (Table 5.2) may be attributed to sulfur's large covalent radius, which lessens angle strain (C—C—C angle $\simeq 66°$); decreased nonbonded interactions ($C_2$—$C_3$ is 0.02 Å longer than in oxirane, $C_2$—X is 0.38 Å longer, and the sulfur lone pairs are more diffuse than those of oxygen); and the greater ease with which sulfur utilizes pure p orbitals to form ring bonds. Although thermochemistry is not yet available, similar aspects of molecular geometry are apparent in silirane and phosphirane (Table 5.2). The geometry of thiirane-1,1-dioxide (**133**) is especially interesting,

**133**

since its $C_2$—$C_3$ is much longer than one might predict after comparing thiirane and thiirane-1-oxide. Hoffmann et al.[246] have explained this via reference to calculations, excluding sulfur 3d orbitals, which show an occupied orbital on "$SO_2$" very close in energy to "ethylene $\pi^*$" and therefore capable of very effectively donating electrons into this orbital, thus increasing $C_2$—$C_3$. The sulfur lone-pair orbital in thiirane is calculated to be much lower in energy and therefore ineffective in donating electron density to $\pi^*$. Inclusion of sulfur 3d orbitals also explains the long $C_2$—$C_3$ bond in thiirane-1,1-dioxide, because the two electronegative oxygen atoms lower the 3d orbital energies and otherwise alter their properties to make them more effective electron acceptors.[246] (See a discussion of geometries of **131** and **133**.[252])

In the preceding paragraph we noted the effects of heteroatoms upon the molecular geometries and thermodynamics of rings. One may also consider how their presence in various three-membered rings affects the properties of the heteroatoms themselves: most notably basicity, nucleophilicity, and hydrogen-bonding properties. Because the ring orbitals are high in

[252] H. L. Ammon, D. Fallon, and L. A. Plastar, *Acta Crystallogr.*, *Sect. B* **32**, 2171 (1976).

p character, it is reasonable to assume that extraring orbitals include lone-pair orbitals have more s character than usual. The high ionization potentials of oxirane, azirane, and thiirane relative to those of dimethylether, dimethyl-amine, and dimethylsulfide are consistent with lone pairs having high s character in these strained molecules (Table 5.2).[234a] Additionally, the hetero-atoms in such rings should be relatively poor hydrogen-bond acceptors (but recall that trends in basicity and hydrogen-bond-acceptor tendencies need not correspond[253]).

Sulfur and phosphorus are capable of "expanding their octets" and forming "hypervalent"[254] compounds with higher coordination than in **127** and **128**. If one considers pentacoordinate phosphorus to be $sp^3d$-hybridized, then it becomes apparent that electronegative substituents should stabilize this hybridization state by reducing the energies of 3d orbitals. Thus, $PF_5$ is a well-known compound. However, attempts to make $P(CH_3)_5$ fail [Eq. (60)]

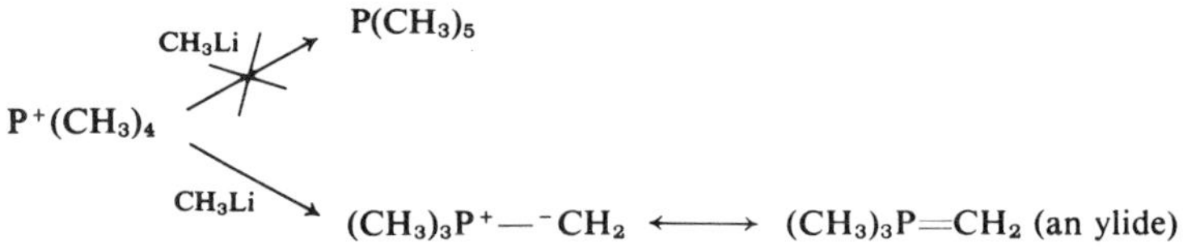

$$P^+(CH_3)_4 \quad \xrightarrow[\text{CH}_3\text{Li}]{\text{CH}_3\text{Li}} \quad \begin{array}{l} P(CH_3)_5 \\[2em] (CH_3)_3P^+\!\!-\!\!{}^-CH_2 \longleftrightarrow (CH_3)_3P{=}CH_2 \text{ (an ylide)} \end{array} \tag{60}$$

perhaps because of the reduced electronegativity of alkyl groups.[255] (Penta-phenyl phosphoranes are stable.[255]) Bis(trifluoromethyl)trimethylphosphorane is also known.[256] The only pentaalkyl (hydrocarbon) phosphorus derivative presently known is **135**.[255] Here angle strain has been utilized as the driving force to obtain **135**, since distortion of the would-be tetracoordinate phosphorus

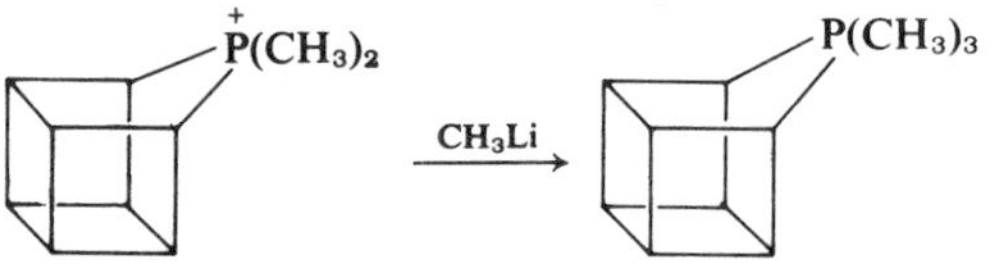

$$\text{134} \qquad \xrightarrow{\text{CH}_3\text{Li}} \qquad \text{135} \tag{61}$$

in **134** and in the potential ylide is greater than in **135**, where the ring can occupy apical and equatorial positions on phosphorus corresponding to an idealized C—P—C angle of 90°.[254] While no three-membered sulfuranes (tetracoordinate sulfur) or phosphoranes have been isolated, evidence favoring

[253] E. M. Arnett, E. J. Mitchell, and T. S. S. R. Murty, *J. Am. Chem. Soc.* **96**, 3875 (1974).

[254] J. Musher, *Angew. Chem., Int. Ed. Engl.* **8**, 54 (1969).

[255] E. W. Turnblum and T. J. Katz, *J. Am. Chem. Soc.* **93**, 4065 (1971) and references cited therein.

[256] K. I. The and R. G. Cavell, *J. Chem. Soc., Chem. Commun.* p. 716 (1975).

the intermediacy of **136** has appeared,[257] and the molecule **137** has been monitored by nmr, exhibiting some stability at $-5°C$.[258]

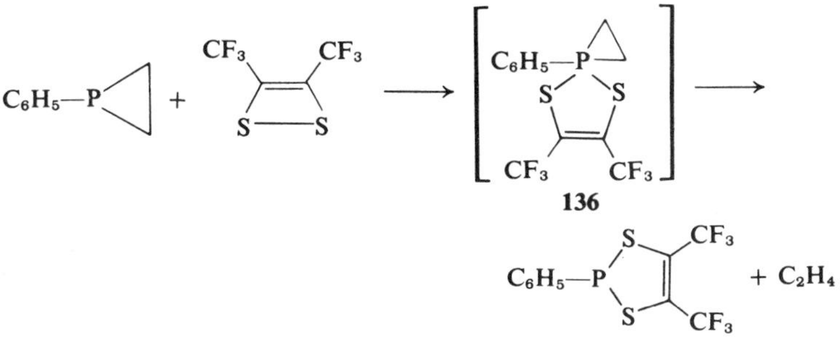

$$+ C_2H_4 \qquad (62)$$

Another approach to the analysis of the stability of three-membered rings might be afforded through comparison of equilibria corresponding to the tropylidene–norcaradiene equilibria isomerization [Eq. (63)]. Here it is

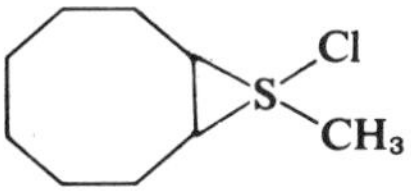

**137**

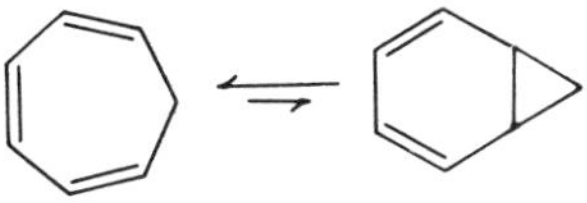

$$(63)$$

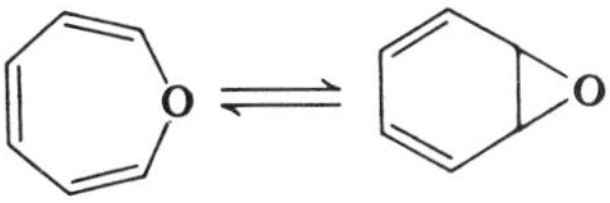

$$(64)$$

apparent that oxepin is actually almost as stable as its norcaradienelike isomer [Eq. (64)].[259] This might at first suggest some sort of stability for the oxepin ring relative to the cyclopropane ring, but it can also be argued that oxepin may have a small degree of antiaromatic destabilization.

### 5.D.2.b   *Three-Membered Rings Containing Two Heteroatoms and Three Heteroatoms*

When two heteroatoms are incorporated in a three-membered ring, a relatively weak heteroatom–heteroatom bond is present. By reference to Eqs.

[257] D. B. Denney and L. S. Shih, *J. Am. Chem. Soc.* **96**, 317 (1974).

[258] D. C. Owsley, G. K. Helmkamp, and M. F. Rettig, *J. Am. Chem. Soc.* **91**, 5239 (1969).

[259] E. Vogel and H. Gunther, *Angew. Chem., Int. Ed. Engl.* **6**, 385 (1967).

(65) and (66), it is a simple matter to demonstrate that such bonds, for example, N—N and O—O, are weak. The discrepancies calculated are

$$2 \, \Delta H_f(\text{CH}_3\text{NH}—\text{CH}_2\text{CH}_3) \longrightarrow \Delta H_f(\text{CH}_3\text{NH}—\text{NHCH}_3) +$$
$$\Delta H_f(\text{CH}_3\text{CH}_2—\text{CH}_2\text{CH}_3)$$

$$2(-10.5 \text{ kcal/mole, estd.}) \longrightarrow (+21.99 \text{ kcal/mole}) + (-30.36 \text{ kcal/mole}) \quad (65)$$

Discrepancy $= +12.6$ kcal/mole

$$2 \, \Delta H_f(\text{CH}_3\text{O}—\text{CH}_2\text{CH}_3) \longrightarrow \Delta H_f(\text{CH}_3\text{O}—\text{OCH}_3) + \Delta H_f(\text{CH}_3\text{CH}_2—\text{CH}_2\text{CH}_3)$$

$$2(-51.72 \text{ kcal/mole}) \longrightarrow (-30.1 \text{ kcal/mole}) + (-30.36 \text{ kcal/mole}) \quad (66)$$

Discrepancy $= +42.9$ kcal/mole

approximately attributable to bond weaknesses. In practice, this explains why 1,2-diaziridine (**138**) is unknown, although substituted derivatives have been characterized.[260] *trans*-Diaziridine has been calculated to be 7.8 kcal/mole more

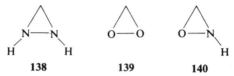

**138**                        **139**                        **140**

stable than the *cis* isomer,[261a] and this is about equal to the electronic destabilization induced by *cis* lone pairs on adjacent nitrogen atoms detectable by photoelectron spectroscopy.[234a] Parent dioxirane (**139**) has recently been detected spectroscopically in the vapor phase at $-150°\text{C}$.[261b] One may estimate a very crude heat of formation for **139** via Eq. (67), which assumes strain energy equal to that of cyclopropane, ignores the 1,3-dioxa resonance interaction mentioned in Section 5.D.1, and ignores the peroxide rotational barrier. (The last two features contribute energy terms of almost equal magnitude that are opposite in sign.) The calculation predicts that **139** is some 103 kcal/mole higher in energy than formic acid ($\Delta H_f = -90.6$ kcal/ mole); this is in excellent agreement with an *ab initio* calculation[261a] predicting a 98 kcal/mole energy difference. Valence-bond calculations suggest that **139** is almost 1 eV more stable than its open form, methylene peroxide ($\cdot\text{CH}_2—\text{OO}\cdot$), which has been postulated in the Criegee ozonolysis mechanism.[262] One might crudely estimate an N—O bond weakness (or discrepancy) by averaging the discrepancies of Eqs. (65) and (66). Applying the same method, a standard

[260] E. Schmitz, *Adv. Heterocyclic Chem.* **2**, 83 (1963).

[261a] W. A. Lathan, L. Radom, P. C. Hariharan, W. J. Hehre, and J. A. Pople, *Top. Curr. Chem.* **40**, 1 (1973).

[261b] Drs. Richard D. Suenram, Frank J. Lovas, John T. Herron, R. I. Martinez, and Robert E. Huie, unpublished results, personal communication to Joel F. Liebman. See J. T. Herron and R. E. Huie, *J. Am. Chem. Soc.*, **99**, 5430 (1977).

[262] W. R. Wadt and W. A. Goddard, III, *J. Am. Chem. Soc.* **97**, 3004 (1975).

enthalpy of formation of $+13$ kcal/mole may be estimated for oxaziridine (**140**). It is well known that oxaziridines are less stable than their ring-opened nitrone isomers[263] [**141** in Eq. (68)], and this is important in the rearrangements of nitrones.[264] Calculational evidence supporting the intermediacy of an oxathiirane has also appeared.[265]

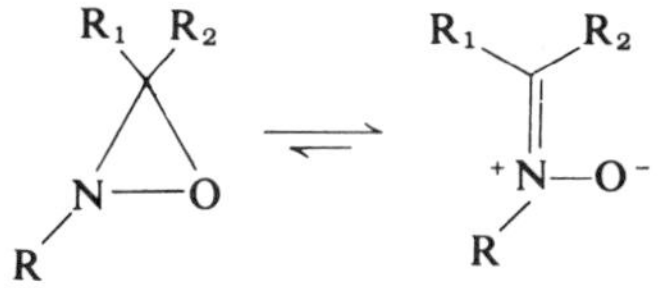

$$\text{(68)}$$

**141**

For the sake of completeness we note that sterically hindered oxadiaziridines (**142**, R = *tert*-butyl or *tert*-octyl)[266a,b] are thermally stable, as are certain thiadiaziridine-1,1-dioxides (**143**, R = *tert*-butyl or *tert*-octyl),[267,268] and that the former are viable intermediates in the isomerization of azoxyalkanes.[269]

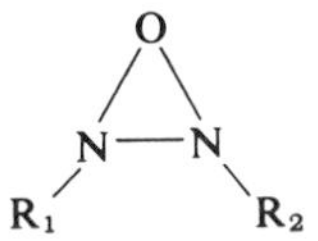

**142** 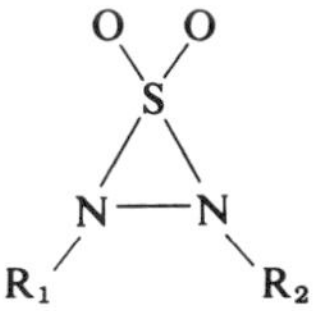          **143**

These molecules are "organic" by virtue of the ring substituents rather than the rings themselves, and in a sense they form a conceptual link between cyclopropane and such inorganic species as cyclic $O_3$.[261a] While $O_3$ is more stable in its ring-opened form, ozone,[261a] the very long N—N bond in **143** suggests an increasing ability to ring open.[267]

### 5.D.2.c  *Ring-Opened Forms of Three-Membered Rings*

The thermodynamically favorable ring opening of oxaziridine [Eq. (68)] is an extreme illustration of a general tendency of three-membered heterocycles to form open (1,3-dipolar) isomers. The nature and utility in organic synthesis

[263] M. F. Hawthorne and R. D. Strahm, *J. Org. Chem.* 22, 1263 (1957).

[264] M. Lamchen, *in* "Mechanisms of Molecular Rearrangements" (B. S. Thygarajan, ed.), Vol. 1, pp. 1–60. Wiley (Interscience), New York, 1968.

[265] J. P. Snyder, *J. Am. Chem. Soc.* 96, 5005 (1974).

[266a] S. S. Hecht and F. D. Greene, *J. Am. Chem. Soc.* 89, 6761 (1967).

[266b] F. D. Greene and S. S. Hecht, *J. Org. Chem.* 35, 2482 (1970).

[267] J. W. Timberlake and M. L. Hodges, *J. Am. Chem. Soc.* 95, 634 (1973); see also L. M. Trefonas and L. D. Cheung, *J. Am. Chem. Soc.* 95, 636 (1973).

[268] H. Quast and F. Kees, *Tetrahedron Lett.* p. 1655 (1973).

[269] J. Swigert and K. G. Taylor, *J. Am. Chem. Soc.* 93, 7337 (1971).

of such 1,3-dipoles have been reviewed elsewhere (Scheme 19).[270,271] Factors which facilitate the formation of ring-opened isomers include the ability of heteroatoms to inductively stabilize negative charge, their ability to stabilize adjacent positive charges via conjugation, and the prior-mentioned weakness of heteroatom–heteroatom bond in rings containing this structural feature.

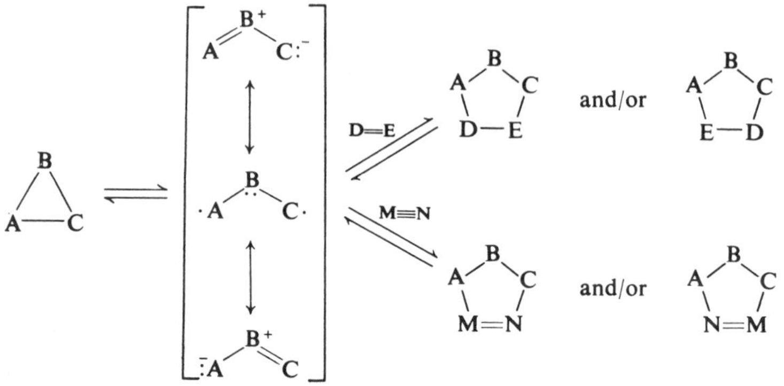

**Scheme 19**

Figure 5.6 depicts the relative energetics of isomeric aziridines and azomethine ylides (*cis* and *trans* isomers of both **144** and **145**, respectively), which are interrelated in Scheme 20.[272-274] The dipolar form is less than 8 kcal/mole higher in free energy, an amount great enough to preclude its direct observation in thermal equilibrium with its aziridine isomer. However, azomethine ylide (**145**) was generated in observable quantity through use of flash photolysis,[272] and its conversion to aziridine **144** was monitored to ingeniously produce the energy profile depicted in Fig. 5.6. Furthermore, in accord with the precepts of orbital symmetry (Section 5.B.1), aziridines open in a conrotatory manner thermally and in the disrotatory mode photochemically.[272-274] Thermal isomerization (see Scheme 20[272]) of *cis*-**144** to *trans*-**144** involves initial formation of *trans*-**145** and then slow bond rotation to form *cis*-**145** (this rotational barrier corresponds to the high point on the energy profile) and closure to *trans*-**144**. In the presence of dipolarophiles, such as dimethylacetylene dicarboxylate, rate-determining, stereospecific ring opening precedes 1,3- cycloaddition (Scheme 20).

Tetracyanoethylene oxide and other oxiranes having suitable substituents readily form their ring-opened isomers. The epimerization of *trans*-2,3-

---

[270] R. Huisgen, *Angew. Chem., Int. Ed. Engl.* **2**, 633 (1963).

[271] R. M. Kellogg, *Tetrahedron* **32**, 2165 (1976).

[272] H. Hermann, R. Huisgen, and H. Mader, *J. Am. Chem. Soc.* **93**, 1779 (1971).

[273] R. Huisgen, W. Scheer, and H. Huber, *J. Am. Chem. Soc.* **89**, 1753 (1967).

[274] R. Huisgen and H. Mader, *J. Am. Chem. Soc.* **93**, 1777 (1971).

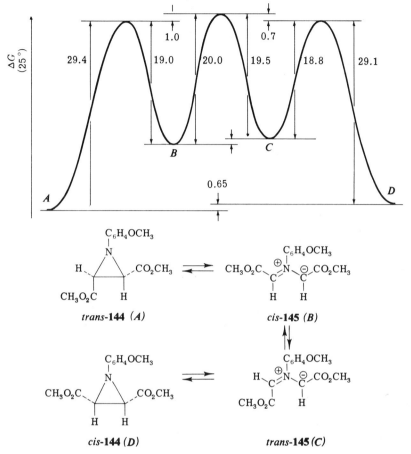

**Figure 5.6.** Reaction coordinate for the isomerization of aziridines through the corresponding azomethine ylides. (Courtesy of Professor R. Huisgen; see also Hermann et al.[272]).

dicyano-2,3-diphenyl oxirane (Scheme 21) requires an enthalpy of activation of about 27 kcal/mole.[275a] The rate-determining step in this reaction may well be interconversion between ring-opened isomers (carbonyl ylides) rather than ring opening.[275b] The thermal ring opening of oxaziridines [Eq. (68)] has been shown to be conrotatory, in accord with the predictions of orbital symmetry.[276]

[275a] H. Hamberger and R. Huisgen, *Chem. Commun.* p. 1190 (1971).

[275b] A. Dahmen, H. Hamberger, R. Huisgen, and V. Markowsky, *Chem. Commun.* p. 1192 (1971).

[276] J. S. Splitter, T.-M. Su, H. Ono, and M. Calvin, *J. Am. Chem. Soc.* **93**, 4075 (1971).

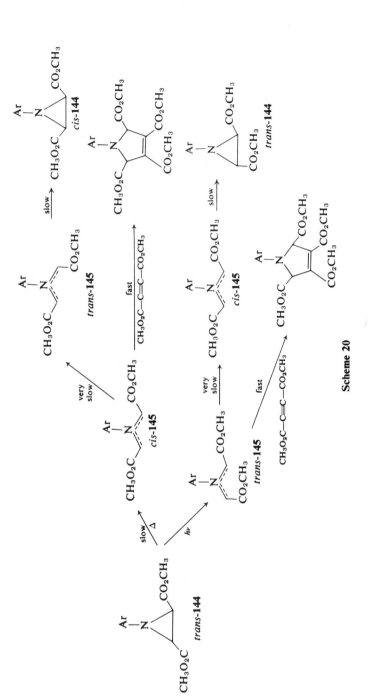

**Scheme 20**

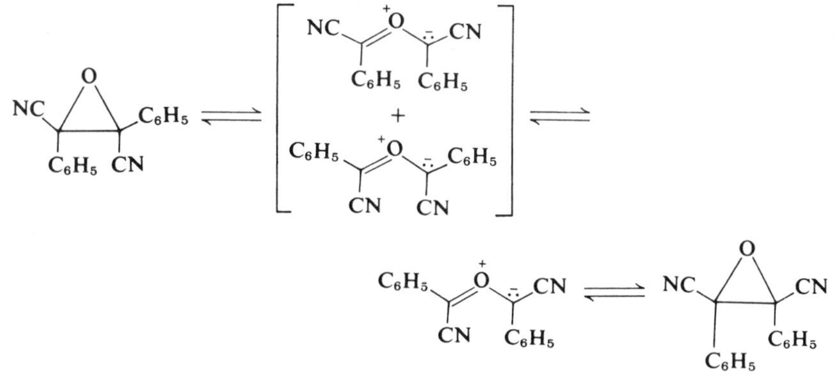

**Scheme 21**

Even appropriately substituted cyclopropanes can ring open to form 1,3-dipolar species. For example, racemization of **146** is thought to proceed via a zwitterionic intermediate **(147)**.[277]

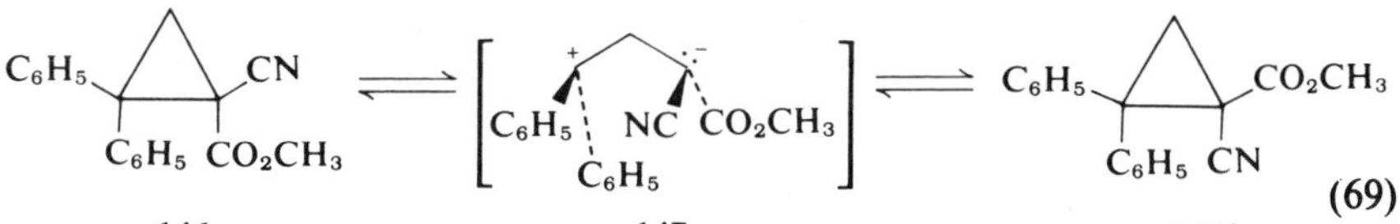

$$(69)$$

**146**            **147**           **146′**

### 5.D.2.d   *Positively Charged Three-Membered Rings*

As one would anticipate, protonation of a three-membered heterocycle will generate an ion that is extremely susceptible to nucleophilic attack. For example, amino acid **148** has not been isolated[278] because the zwitterion is extremely reactive, as are aziridinium salts in general (aziridinium trifluoroacetate is stable at 0°C, and the aziridinium ion can be monitored in $HSO_3F$-$SbF_5$-$SO_2$ solution at $-60°C$[279]). The measured gas-phase proton affinity of oxirane (183 kcal/mole) allows **149** to be assigned a standard enthalpy of

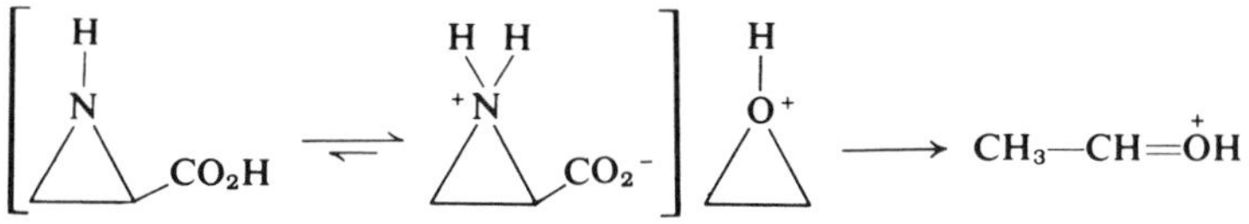

$$(70)$$

**148**           **149**           **150**

[277] E. W. Yankee, F. D. Badea, N. Howe, and D. J. Cram, *J. Am. Chem. Soc.* **95**, 4210 (1973) and references cited therein.

[278] Professor Kenneth B. Henery-Logan, personal communication to Arthur Greenberg.

[279] G. A. Olah and P. J. Szilagyi, *J. Am. Chem. Soc.* **91**, 2949 (1949).

formation some 27 kcal/mole higher than that of protonated acetaldehyde (150), to which 149 apparently rearranges with ease.[280] Thiirane-1-oxides which are thermally stable suffer two modes of unimolecular decomposition depending upon the presence and stereochemistry of methyl substituents.[281a] For example, the parent molecule (151) decomposes chelotropically [Eq. (70)],

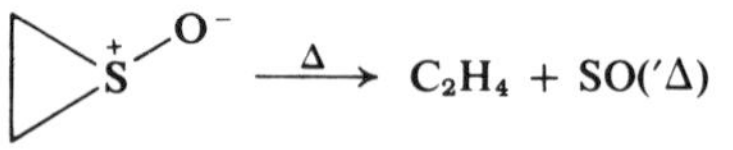

$$\triangleright\!\!\overset{+}{\underset{S}{}}\!\!\overset{O^-}{} \xrightarrow{\Delta} C_2H_4 + SO('\Delta) \tag{70}$$

151

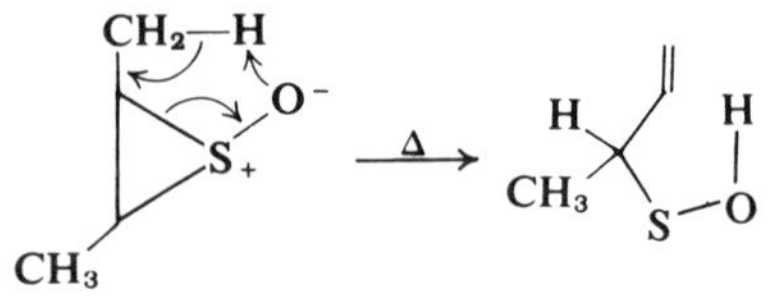

$$\tag{71}$$

152

but the presence of a suitably placed methyl substituent in 152 causes this molecule to tautomerize.[281a] Similarly, N-tert-butylaziridine-N-oxide (153)

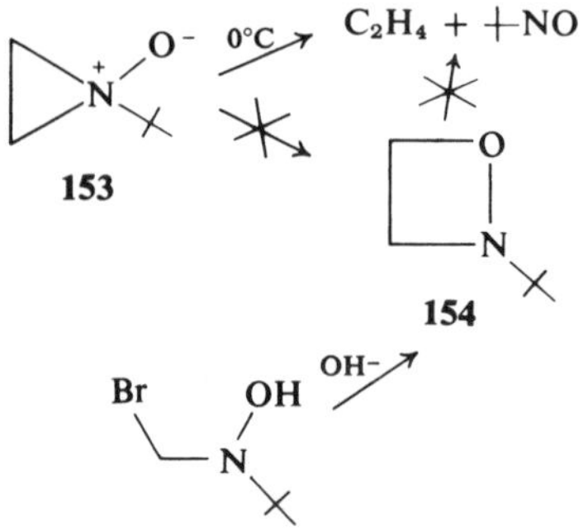

Scheme 22

154

can be obtained via ozonization of N-tert-butylaziridine and, while it is stable in solution to 0°C, will decompose above 0°C ($E_{act}$ = 22 kcal/mole) to produce ethylene and nitroso-tert-butane. (The intermediacy of the oxazetidine 154, synthesized independently, has been ruled out. See Scheme 22.)[281b] However, the methyl-substituted derivative 155 decomposes at $-30°C$ ($E_{act}$ = 15 kcal/mole) to form the rearranged N-hydroxy compound [Eq. (72)].[281b]

[280] J. L. Beauchamp and R. C. Dunbar, *J. Am. Chem. Soc.* **92**, 1477 (1970).

[281a] J. E. Baldwin, S. C. Choi, and G. Hofle, *J. Am. Chem. Soc.* **93**, 2810 (1971).

[281b] J. E. Baldwin, A. K. Bhatnager, S. C. Choi, and T. J. Shortridge, *J. Am. Chem. Soc.* **93**, 4082 (1971).

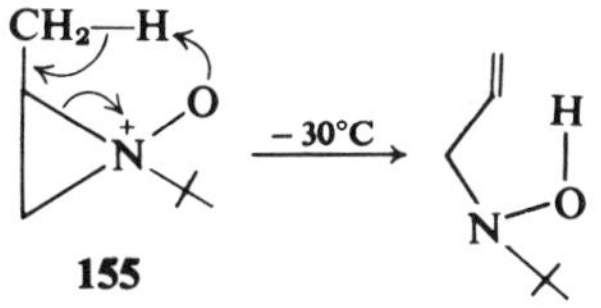

$$\text{155} \qquad \xrightarrow{-30°C} \qquad \tag{72}$$

Excellent evidence for the intermediacy of a perepoxide (155a) or its ring-opened form has been obtained from the reaction of camphenylidene–adamantane with singlet oxygen, in which a carbonium ion-like rearrangement occurs.[282a,b]

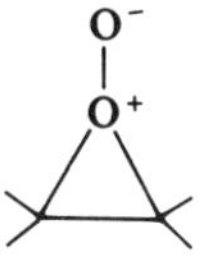

155a

In the realm of charged rings we mention the observation of bridged chloronium ions (e.g., 156), bromonium ions (e.g., 157), and iodinium ions via nmr in solution at low temperature.[283-285] 2-Fluoroethyl cations are found to prefer the acyclic form [283] in accord with calculations.[286] Thermochemical studies have established that three-membered ring bromonium ions are more stable than their chlorine analogues by about 7.5 kcal/mole,[287] a result explicable in terms of both the lower electronegativity of bromine and the relatively long C—Br bonds which reduce ring strain. Furthermore, the three-membered bromonium ion ring is less stable than its five-membered homologue (tetramethylenebromonium ion), but by only 10 kcal/mole in spite of an estimated 17 kcal/mole excess of strain relative to the larger ring (based upon sulfur analogues).[287] Gas-phase synthesis of tetramethylenechloronium ion has been reported.[288] In line with principles discussed in the Section 5.D.2.e, we note that two spiroadamantane rings provide sufficient steric hindrance to allow a bromonium ion to be isolated as a fairly stable salt (158).[289] The propadienylhalonium ions 159 have been generated in

[282a] F. McCapra and I. Beheshti, J. Chem. Soc., Chem. Commun. p. 517 (1977).

[282b] The perepoxide was calculated to be the initial adduct in an ethylene-singlet oxygen reaction; see, M. J. S. Dewar, A. C. Griffin, W. Thiel, and I. J. Turchi, J. Am. Chem. Soc. 97, 4439 (1975).

[283] G. A. Olah and J. M. Bollinger, J. Am. Chem. Soc. 89, 4744 (1967).

[284] G. A. Olah, "Halonium Ions," Part B. Wiley, New York, 1975.

[285] R. E. Peterson, Acc. Chem. Res. 4, 407 (1971).

[286] W. J. Hehre and P. C. Hiberty, J. Am. Chem. Soc. 96, 2665 (1974).

[287] J. W. Larsen and A. V. Metzner, J. Am. Chem. Soc. 94, 1614 (1972).

[288] C. C. Van de Sande and F. W. McLafferty, J. Am. Chem. Soc. 97, 2298 (1975).

[289] J. Strating, J. H. Wieringa, and H. Wynberg, Chem. Commun. p. 907 (1969).

acidic medium and observed via nmr at $-80°C$.[290] Having spoken about three-membered rings containing F, O, N, and C, we conclude Period II with references to three-membered rings containing B,[291a] Be$^+$,[291b] and Li$^+$.[291c]

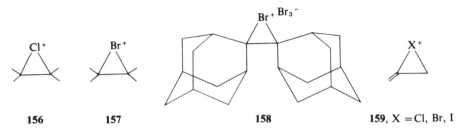

**156**            **157**                    **158**            **159,** X = Cl, Br, I

## 5.D.2.e  *Kinetic Stability Afforded by Bulky Substituents*

A major theme in this book has been the kinetic stability imparted to an inherently reactive linkage or functional group through suitable placement of bulky substituents. For example, oxiranes **160**[292a,b] and **161**,[293] aziridines

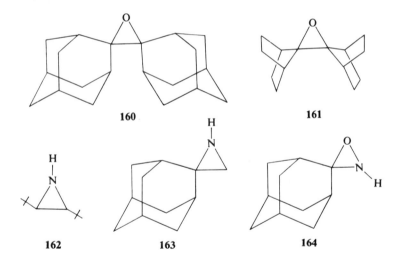

**160**                    **161**

**162**            **163**                    **164**

[290] J. M. Bollinger, J. M. Brinich, and G. A. Olah, *J. Am. Chem. Soc.* **92**, 4025 (1970).

[291a] S. M. van der Kerk, J. Boersma, and G. J. M. van der Kerk, *Tetrahedron Lett.* p. 4765 (1976).

[291b] W. C. Swope and H. F. Schaefer, III, *J. Am. Chem. Soc.* **98**, 7962 (1976).

[291c] R. H. Staley and J. L. Beauchamp, *J. Am. Chem. Soc.* **97**, 5920 (1975).

[292a] H. Wynberg, E. Boelema, J. H. Wieringa, and J. Strating, *Tetrahedron Lett.* p. 3613 (1970).

[292b] A. P. Schaap and G. R. Faler, *J. Am. Chem. Soc.* **95**, 3381 (1973).

[293] P. D. Bartlett and M. S. Ho, *J. Am. Chem. Soc.* **96**, 627 (1974).

$162^{294}$ and $163,^{295}$ and oxaziridine $164^{296}$ are all strikingly stable (recall also **158**). The first example of a three-membered heterocycle substituted with *cis-tert*-butyl groups is thiirane **165**. A combination of nonbonded repulsion and especially easy accessibility to one face of the ring makes this molecule much more reactive than its *trans* isomer, **166**.[297]

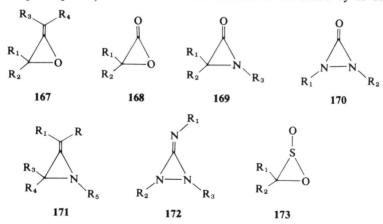

(73)

### 5.D.3 ALPHA-LACTONES, ALPHA-LACTAMS, RELATED HETEROCYCLES, AND OPEN ISOMERS

A family of three-membered rings bearing exocyclic double bonds can be mentioned at this point. Among these are allene oxides (**167**), alpha-lactones (**168**), alpha-lactams (**169**), diaziridinones (**170**), methylene aziridines (**171**), diaziridine imines (**172**), and alpha-sultines (**173**). Allene oxide (**167**, $R_1 = R_2 = R_3 = R_4 = H$) has not been observed and is calculated by *ab initio*

[294] J. C. Sheehan and J. H. Beeson, *J. Am. Chem. Soc.* **89**, 362 (1967).

[295] T. Saski, S. Eguchi, and Y. Hirako, *Tetrahedron* **32**, 437 (1976).

[296] E. Oliveros-Desherces, M. Riviere, J. Parello, and A. Lattes *Synthesis* p. 812 (1974).

[297] R. Raynolds, S. Zonnebelt, S. Bakker, and R. M. Kellogg, *J. Am. Chem. Soc.* **96**, 3146 (1974).

methods[298] as well as bond-additivity logic[299] to be about 20 kcal/mole less stable than its isomer, cyclopropanone. Substituted allene oxides protected by bulky alkyl groups are isolable,[300–302] the simplest to date being a *tert*-butyl derivative (**167**, $R_1$ = *t*-Bu, $R_2$ = $R_3$ = $R_4$ = H) stable in solution for 1.5 hours at 25°C.[302] These allene oxides are isomerized to the corresponding cyclopropanones. Only one alpha-lactone (**168**, $R_1$ = $R_2$ = $CF_3$) has been isolated up to this time.[303] Alkyl- and aryl-substituted alpha-lactones have been generated at very low temperatures and observed spectroscopically, but such species polymerize at temperatures well below $-30°C$[304–306] (Scheme 23). This may be rationalized by invoking large concentrations of various

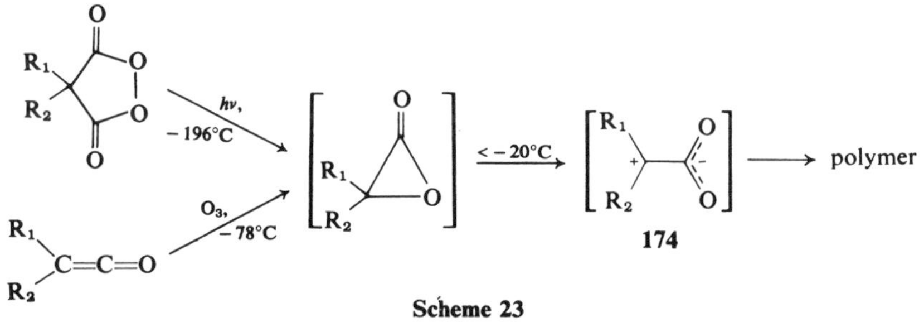

**Scheme 23**

ring-opened methylene carboxylates (**174**) (analogous to oxyallyls discussed in Chapter 3) or through operation of autocatalysis involving a tiny steady-state concentration of this species. Bond-additivity calculations[299] predict that alpha-lactone is only about 10 kcal/mole more stable than methylene carboxylate and that alkyl- and aryl-substituted molecules may be more stable in the open dipolar form.

The chemistry of alpha-lactams (**169**) has been reviewed elsewhere,[307] and discussions of methylene aziridines (**171**)[308] and diaziridine imines (**172**)[309] have also appeared. The thermal stabilities of certain sterically hindered

[298] A. Liberles, A. Greenberg, and A. Lesk, *J. Am. Chem. Soc.* **94**, 8685 (1972).

[299] J. F. Liebman and A. Greenberg, *J. Org. Chem.* **39**, 123 (1974).

[300] R. L. Camp and F. D. Greene, *J. Am. Chem. Soc.* **90**, 7349 (1968).

[301] J. K. Crandal and W. H. Machleder, *J. Heterocycl. Chem.* **6**, 777 (1969).

[302] T. H. Chan, B. S. Ong, and W. Mychajlowskij, *Tetrahedron Lett.* p. 3253 (1976).

[303] W. Adam, J.-C. Liu, and O. Rodriquez, *J. Org. Chem.* **38**, 2269 (1973).

[304] R. Wheland and P. D. Bartlett, *J. Am. Chem. Soc.* **92**, 6057 (1970).

[305] W. Adam and R. Rucktaschel. *J. Am. Chem. Soc.* **93**, 557 (1971).

[306] O. L. Chapman, P. W. Wojowski, W. Adam, O. Rodriquez, and R. Rucktaschel, *J. Am. Chem. Soc.* **94**, 8636 (1972).

[307] I. Lengyel and J. C. Sheehan, *Angew. Chem., Int. Ed. Engl.* **7**, 25 (1968).

[308] H. Quast, R. Frank, and E. Schmitt, *Angew. Chem., Int. Ed. Engl.* **11**, 329 (1972).

[309] H. Quast and E. Schmitt, *Angew. Chem., Int. Ed. Engl.* **8**, 448, 449 (1969).

diaziridinones (**170**) are particularly striking.[310a-c,311] The ir carbonyl absorption bands of alpha-lactams are about 20 cm$^{-1}$ higher than those of cyclopropanones, in contrast to the situation for unstrained lactams, whose carbonyl frequencies are 20–60 cm$^{-1}$ lower than those of the corresponding ketones.[310c] This is rationalized by invoking decreased contribution from resonance structure **175b**, which is more "strained" than **175a**. In contrast to common unstrained lactams, the local geometry at nitrogen in 1,3-diadamantyl-α-lactam (and probably most other derivatives) is markedly nonplanar.[312]

"relative strain" = 41 kcal/mole
**175a**

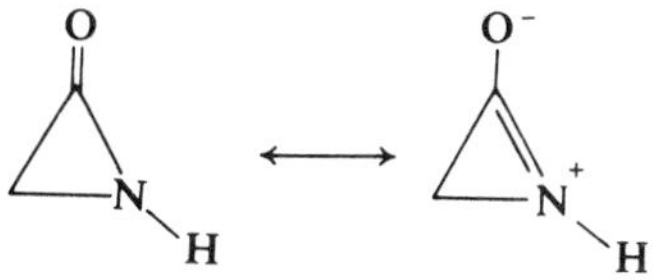

"relative strain" = 54 kcal/mole
**175b**

2,2,5,5-Tetramethyl-1,6-diazabicyclo[4.1.0]hepten-7-one (**176**)[311] contains the equivalent of *cis-tert*-butyl groups (note that *tert*-butyl and *tert*-octyl groups in other diaziridinones are *trans*;[310a-c] also see the previous discussion of **165**). Decarbonylation of **176** occurs spontaneously at room temperature ($t_{1/2}^{25°C} = 25\,hr$) in contrast to decarbonylation of *trans*-di-*tert*-butyldiaziridinone ($t_{1/2}^{180°C} \sim 2$ hr). This corresponds to a postulated destabilization energy of

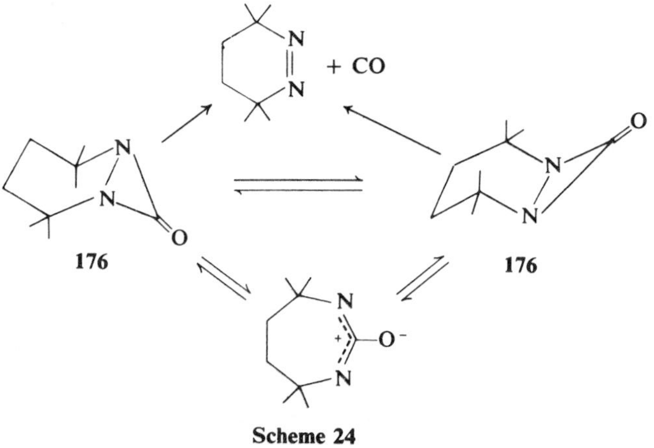

**Scheme 24**

[310a] F. D. Greene and J. C. Stowell, *J. Am. Chem. Soc.* **86**, 3569 (1964).

[310b] F. D. Greene, W. R. Bergmark, and J. G. Pacifici, *J. Org. Chem.* **34**, 2263 (1969).

[310c] F. D. Greene, J. C. Stowell, and W. R. Bergmark, *J. Org. Chem.* **34**, 2254 (1969).

[311] C. A. Renner and F. D. Greene, *J. Org. Chem.* **41**, 2813 (1976).

[312] A. H.-J. Wang, E. R. Talaty, and A. E. Depuy, Jr., *J. Chem. Soc., Chem. Commun.* p. 43 (1972).

10 kcal/mole in **176** attributed to nitrogen–nitrogen *cis* lone-pair repulsion.[311] Compound **176** is also much more susceptible to methanolysis than *trans*-di-*tert*-butyldiaziridinone and also exhibits a low activation energy ($\Delta G^{\ddagger} \leq$ 5–6 kcal/mole) for the conformational process depicted in Scheme 24.[311] The analogous conformational process, constituting racemization in *trans*-di-*tert*-octylaziridine, has an activation barrier of about 16 kcal/mole.[311]

### 5.D.4   SATURATED FOUR-MEMBERED RINGS

#### 5.D.4.a   *One Heteroatom: Structure and Stability*

Four-membered saturated ring systems, including oxetane (**177**), azetidine (**178**), and thietane (**179**), are less strained than their lower homologues and thus less reactive.[313] For example, oxetane is only $10^{-3}$ times as reactive to hydroxide as oxirane,[314] and unlike their lower homologues, azetidine and thietane are essentially inert to nucleophiles. The strain energies of **177–179** are roughly comparable to those of the three-membered heterocycles.[235] It is apparent that the hybridization of the heteroatoms is much closer to that in unstrained analogues than to three-membered ring systems. While the orbital-symmetry-forbidden ground-state conversion of cyclobutane to two ethylenes is appreciably endothermic ($\Delta H_r^0 = +18.1$ kcal/mole), the analogous thermally forbidden cycloreversion of oxetane is much less so ($\Delta H_r^0 = +5.75$ kcal/mole) mainly because of the stability of the carbonyl group in formaldehyde.

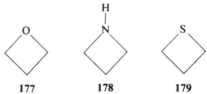

177            178            179

Of particular interest are the structures of these four-membered hetero-rings.[315] Cyclobutane is a puckered molecule [pucker angle $\phi$ (between planes including $C_1C_2C_3$ and $C_2C_3C_4$) = 35°] having an inversion barrier of about 1.5 kcal/mole [Fig. 5.7(a)]. The ground-state conformation can be considered to be a balance between angle distortion (minimum at the planar conformation, increased by puckering), nonbonded vicinal repulsions (decreased by puckering), and torsional energies.[315,316] Oxetane is a very flexible ring whose

---

[313] L. A. Paquette, "Modern Heterocyclic Chemistry," Ch. 3. Benjamin, New York, 1968.

[314] J. G. Pritchard and F. A. Long, *J. Am. Chem. Soc.* **80**, 4162 (1958).

[315] R. M. Moriarty, *Top. Sterochem.* **8**, 271 (1973).

[316] K. B. Wiberg and G. M. Lampman, *J. Am. Chem. Soc.* **88**, 4429 (1966).

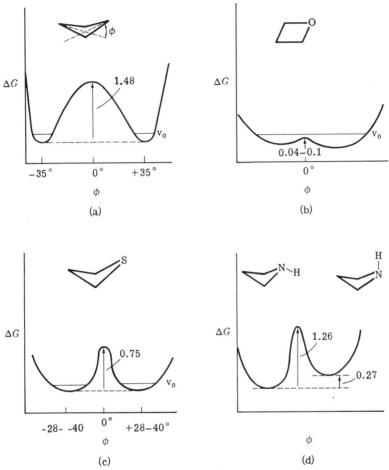

**Figure 5.7.** Inversion coordinate diagrams ($\phi$ is the pucker angle) for (a) cyclobutane, (b) oxetane (because the ground vibrational state is higher than the inversion barrier, the molecule is planar), (c) thietane, (d) azetidine (there are two conformers differing in stability). (All energy values in kcal/mole; see Moriarty.[315])

average geometry is planar because the molecules' ground vibrational state is higher than the 0.04–0.1 kcal/mole barrier separating puckered conformers [see Fig. 5.7(b)].[317,318] One can rationalize this by assuming decreased vicinal repulsions between hydrogens and oxygen lone pairs. However, thietane is appreciably puckered [Fig. 5.7(c)], and a nonbonded repulsion

[317] W. D. Gwinn, J. Zinn, and J. Fernandez, *Bull. Am. Phys. Soc.* **4**, 153 (1959).
[318] T. Veda and T. Shimanouchi, *J. Chem. Phys.* **47**, 4042 (1967).

rationale for this feature might be a bit difficult.[319] However, as noted previously, the natural acyclic C—S—C angle of 99°, as well as long C—S bonds, would lead one to expect that this molecule can tolerate puckering to a greater extent than oxetane can. Azetidine introduces an additional factor: two nonequivalent puckered conformers [Fig. 5.7(d)].[320] Silacyclobutane,[321] and derivatives of thietane-1-oxide,[322–324] thietane-1,1-dioxide,[322] and various phosphetanes[315] are puckered. One might also mention here the nmr observation at low temperature of a brometanium ion.[325] The lone-pair ionization potentials of **177–179** also reflect angular strain and the associated hybridization changes at the heteroatoms.[234b]

### 5.D.4.b    *Four-Membered Rings Including Two Heteroatoms*

While numerous possible systems in this classification might be explored, this section will concern itself only with 1,2-dioxetanes and related molecules, and oxaphosphetanes. Scheme 25 illustrates two methods employed to obtain 1,2-dioxetanes (**180**). The first clearly isolated representative of this system,

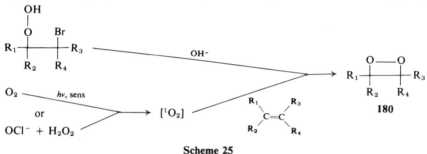

**Scheme 25**

3,3,4-trimethyl-1,2-dioxetane, was reported in 1968[326] and was obtained through decomposition of an α-halohydroperoxide.[326,327] Successful syntheses which employed singlet oxygen and an alkene[328,329] followed. Weak oxygen–oxygen bonds and strained rings make 1,2-dioxetanes high in energy. Their

[319] T. R. Borgers and H. L. Strauss, *J. Chem. Phys.* **45**, 947 (1966).

[320] L. A. Carreira and R. C. Lord, *J. Chem. Phys.* **51**, 2735 (1969).

[321] J. Laane and R. C. Lord, *J. Chem. Phys.* **48**, 1508 (1968).

[322] R. M. Dodson, E. H. Jancis, and G. Klose, *J. Org. Chem.* **35**, 2520 (1970).

[323] S. Allenmark, *Ark. Kemi* **26**, 73 (1960).

[324] S. Abrahamson and G. Rehnberg, *Acta Chem. Scand.* **26**, 494 (1972).

[325] J. H. Exner, L. D. Kershner, and T. E. Evans, *J. Chem. Soc., Chem. Commun.* p. 361 (1973).

[326] K. R. Kopecky and C. Mumford, *Can. J. Chem.* **47**, 709 (1969) and references cited therein.

[327] E. H. White, J. Wiecko, and D. F. Roswell, *J. Am. Chem. Soc.* **91**, 5194 (1969).

[328] P. D. Bartlett and A. P. Schaap, *J. Am. Chem. Soc.* **92**, 3223 (1970).

[329] S. Mazur and C. S. Foote, *J. Am. Chem. Soc.* **92**, 3225 (1970).

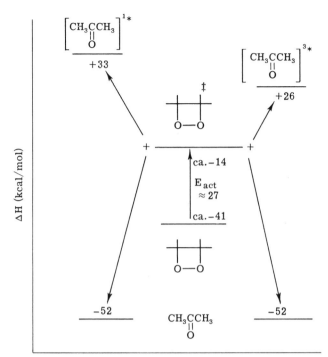

**Figure 5.8.** Enthalpy relations between tetramethyl-1,2-dioxetane and acetone monomers in their electronic ground and excited states. (Adapted from Turro *et al.* [330])

thermal stability in the presence of such formidable destabilization is noteworthy. Furthermore, many of the derivatives, for example 3,3,4,4-tetramethyl-1,2-dioxetane,[330] luminesce upon standing or heating. Luminescence, both fluorescence and phosphorescence, originates from electronically excited states of the carbonyl products produced by decomposition of 1,2-dioxetanes. The activation energy (ca. 27 kcal/mole) for unimolecular decomposition of 3,3,4,4-tetramethyl-1,2-dioxetane provides the activated complex with enough energy to produce a molecule of acetone in its ground state accompanied by a molecule of acetone in the triplet state ($S_0 \rightarrow T_1$, +78 kcal/mole) or excited singlet state ($S_0 \rightarrow S_1$, +85 kcal/mole)[330,331] (See Fig. 5.8). Accessibility of excited state molecules through thermal process allows the operation of "photochemistry in reverse"[330] or "photochemistry without light."[332] For example, 4,4-diphenylcyclohexadienone (**181**) can be

[330] N. J. Turro, P. Lechtken, N. E. Schore, G. Schuster, H.-C. Steinmetzer, and A. Yekta, *Acc. Chem. Res.* **7**, 97 (1974) and references cited therein.

[331] H. E. O'Neal and W. H. Richardson, *J. Am. Chem. Soc.* **92**, 6553 (1970).

[332] E. H. White, J. D. Miano, C. J. Watkins, and E. J. Breaux, *Angew. Chem., Int. Ed. Engl.* **13**, 229 (1974).

photoisomerized to **182** via its triplet state.[333] This isomerization can also be effected by brief heating of **181** in a solution containing trimethyl-1,2-dioxetane.[327] 3,3,4-Trimethyl-1,2-dioxetane will also initiate the isomerization of *trans*-stilbene to *cis*-stilbene—another normally photochemical reaction.[327]

The mechanism for thermal decomposition of 1,2-dioxetanes is one of keen interest for chemists because its intimate details might allow them to better understand the conversion of thermal energy to light. Initially, debate focused upon whether the decomposition was a concerted ($_\sigma 2_s + _\sigma 2_s$) cycloreversion, which is symmetry-allowed if an excited-state carbonyl product is formed, or if it involved initial cleavage of the oxygen–oxygen bond to form a diradical. Any fully satisfactory explanation must include reasoning consistent with preferred formation of triplet acetone in apparent contradiction to the rule of conservation of spin.[330] Bond-additivity calculations[331] are consistent with a nonconcerted mechanism generating a diradical expected to undergo intersystem crossing to produce a statistical 3:1 triplet:singlet mixture. Alternatively, the preferred formation of triplet acetone has been explained in terms of concerted bond cleavage accompanied by singlet-to-triplet conversion.[330] The recent finding that the barrier to thermal decomposition of **183** is 10 kcal/mole higher than that of 3,3,4,4-tetramethyl-1,2-dioxetane favors a nonconcerted mechanism, because one would anticipate that relief of nonbonded repulsion in the transition state for concerted decomposition would lower the activation barrier.[334,335] A mechanism in which formation of the diradical occurs in concert with spin inversion has been advanced to explain this result.[334] Compelling evidence for the intermediacy of 1,2-dioxetane itself comes from the observation of formaldehyde luminescence in a reaction between ethylene and singlet oxygen.[336a] An optically active dioxetane has been synthesized and produces circularly polarized light upon heating.[336b]

[333] H. E. Zimmerman and D. I. Schuster, *J. Am. Chem. Soc.* **84**, 4527 (1962).

[334] G. B. Schuster, N. J. Turro, H.-G. Steinmetzer, A. P. Schaap, G. Faler, W. Adam, and J. C. Liu, *J. Am. Chem. Soc.* **97**, 7110 (1975).

[335] G. Hohne, A. H. Schmidt, and P. Lechtken, *Tetrahedron Lett.* p. 3587 (1976).

[336a] D. J. Bogan, R. S. Sheinson, and F. W. Williams, *J. Am. Chem. Soc.* **98**, 1034 (1976).

[336b] H. Wynberg, H. Numan, and H. P. J. M. Dekkers, *J. Am. Chem. Soc.* **99**, 3870 (1977).

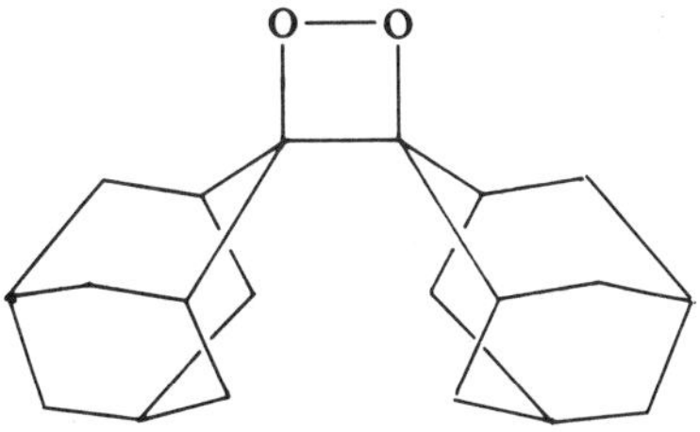

**183** 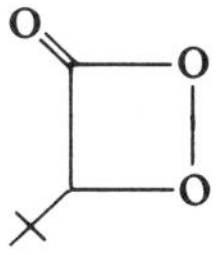　　　　　**184** 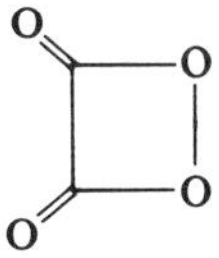　　　**185**

4-*tert*-Butyl-1,2-dioxetan-3-one (**184**) has been isolated and has a half-life of 58 minutes at room temperature,[337a,b] in good agreement with predictions based upon bond-additivity methods.[338] A number of other 1,2-dioxetan-3-ones have been observed spectroscopically; these thermally decompose to provide electronically-excited ketones.[337c] Moreover, they apparently arise from "perlactones" or ring-opened intermediates similar to **155a**.[337c] The luminescence of solutions of oxalates and hydrogen peroxide has been explained in light of the intermediacy of 1,2-dioxetan-3,4-dione (**185**), which has never been detected but is calculated to be capable of providing thermal access to triplet $CO_2$ but not the electronically excited state.[338] The 1,2-dioxetan-3-one system has been postulated in the luciferin derivative **186** thought to be the precursor of photo products producing firefly luminescence.[339]

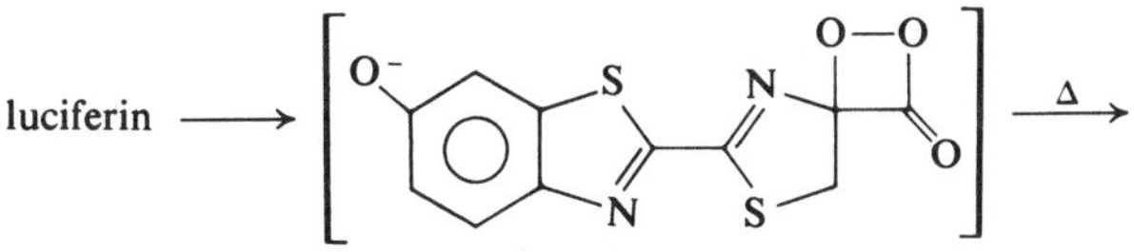

**186**

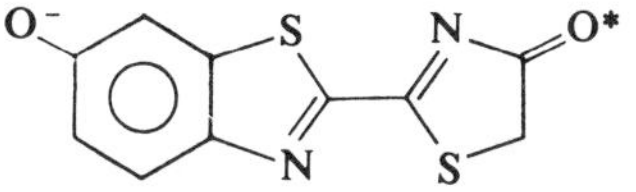

$+ CO_2$　　(75)

The 1,2-oxaphosphetane **187** can be generated under Wittig reaction conditions at $-70°C$ and observed up to $-15°C$, above which decomposition occurs.[340] These observations are used to argue that 1,2-oxaphosphetanes are

[337a] W. Adam and J. C. Liu, *J. Am. Chem. Soc.* **94**, 2894 (1972).

[337b] W. Adam and J. C. Liu, *Angew. Chem., Int. Ed. Engl.* **11**, 540 (1972).

[337c] N. J. Turro, Y. Ito, N.-F. Chow, W. Adam, O. Rodriguez, and F. Yany, *J. Am. Chem. Soc.* **99**, 5836 (1977).

[338] W. H. Richardson and H. E. O'Neall, *J. Am. Chem. Soc.* **94**, 8665 (1972).

[339] E. H. White, J. D. Miano, and M. Unbreit, *J. Am. Chem. Soc.* **97**, 198 (1975); see references cited therein summarizing arguments for and against this mechanism.

[340] E. Vedejs and K. A. J. Snoble, *J. Am. Chem. Soc.* **95**, 5778 (1973).

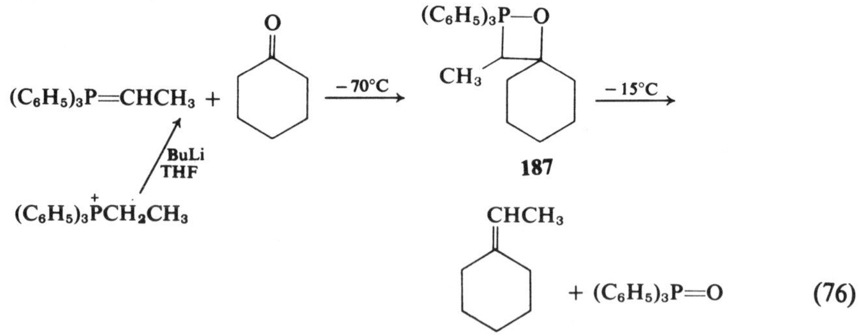

$$(C_6H_5)_3P=CHCH_3 +$$

CH_3

187

$$+ (C_6H_5)_3P=O \qquad (76)$$

$$(C_6H_5)_3\overset{+}{P}CH_2CH_3$$

more stable than betaines and thus more likely to be intermediates in Wittig reactions.[340] Another interesting observation is the thermal rearrangement of **188** to oxaphosphetane **189**.[341] This novel conversion of a five-membered

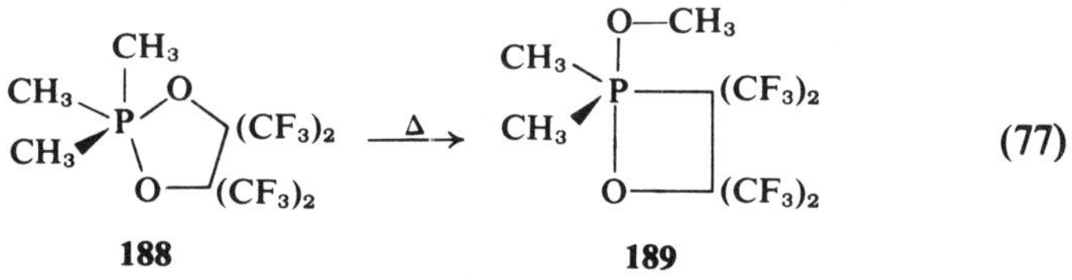

$$(77)$$

**188**                                    **189**

ring to a more stable four-membered ring is explained by noting (a) that **189** has both apical positions occupied by the more electronegative substituents (a most favorable arrangement), and (b) the ready accommodation of a four-membered ring, including a 90° apical–equatorial bond angle at phosphorus.[341]

**5.D.5** BICYCLIC AND SPIROCYCLIC HETEROCOMPOUNDS

**5.D.5.a** *Heteroatom Bridged Molecules*

We begin this topic by discussing the structure and properties of 7-oxabicyclo[2.2.1]heptane (**190**). The measured C—O—C angle in this molecule (ca. 95°)[342,343] is very similar to the corresponding bridge angle in norbornane (96°).[344] The strain in the molecule, curiously, is 3–6 kcal/mole less than that of norbornane.(In contrast, compare the strain energies of cyclopropane and oxirane, which are virtually equal.) Perhaps one means of rationalizing this difference is to note that the "flap" angle in **190** is about 113.5°,[342,343] while the corresponding angle in norbornane is about 108°.[344] Perhaps lessened

[341] F. Ramirez, C. P. Smith, and J. F. Pilot, *J. Am. Chem. Soc.* **90**, 6726 (1968).

[342] R. A. Creswell, *J. Mol. Spectrosc.* **56**, 133 (1975).

[343] K. Oyanagi, T. Fukuyama, K. Kuchitsu, and R. K. Bohn, *Bull. Chem. Soc. Jpn.* **48**, 751 (1975).

[344] J. F. Chiang, C. F. Wilcox, Jr., and S. H. Bauer, *J. Am. Chem. Soc.* **90**, 3149 (1968).

O
-95°

S

$\phi = 113.5°$ **190**

**191**

nonbonded $C_2$—$C_6$ (and/or the corresponding *endo* hydrogen) repulsions in **190** relative to norbornane are a factor. Another interesting observation is the anomalously high ionization potential of **190**.[345] The first ionization band (photoelectron spectroscopy) is unusually sharp and the first vertical ionization potential (9.57 eV) is equal to that in tetrahydrofuran, suggesting at first that 7-oxabicyclo[2.2.1]heptane is as "normal" as THF. However, **190** may be best regarded as a 2,5-dialkyltetrahydrofuran, which would lead one to expect it to have a lower ionization potential (i.e., the first vertical ionization potential of isopropyl ether is 9.20–9.25 eV[346]). These results have been explained in terms of a p-hybridized lone-pair orbital on oxygen which does not have the proper symmetry to mix with boat cyclohexane "ribbon orbitals."[347] Similar results are obtained for 7-thiabicyclo[2.2.1]heptane (**191**).[347] Incorporation of a suitably arranged acetal linkage into a small bicyclic framework may be accompanied by a large increase in reactivity. For example, 2,6-dioxabicyclo[2.2.1]heptane (**193**) is almost $10^6$ times more reactive than its acylic analogue.[348] On the other hand, **194** is hydrolyzed in acid about 15 times more slowly than triethoxymethane.[349]

**192**  **193**  **194**

The main consequence of cyclobutane fusion to the oxirane ring in **195** (m.p. 118°–120°C) is that heating of this molecule in solution at only 100°C produces the open singlet (presumably zwitterionic) ylide **196**.[350] Although ring fusion should thwart concerted thermally allowed conrotatory ring opening, relief of strain functions as a strong driving force. A purple solution

[345] A. D. Bain, J. C. Bunzli, D. C. Frost, and L. Weiler, *J. Am. Chem. Soc.* **95**, 291 (1973).

[346] J. C. Bunzli, D. C. Frost, and L. Weiler, *J. Am. Chem. Soc.* **95**, 7880 (1973).

[347] E. Heilbronner and A. Schmelzer, *Helv. Chim. Acta* **58**, 936 (1975).

[348] H. K. Hall, Jr. and Fr. DeBlauwe, *J. Am. Chem. Soc.* **97**, 655 (1975).

[349] H. K. Hall, Jr., Fr. De Blauwe, and T. Pyriad, *J. Am. Chem. Soc.* **97**, 3854 (1975).

[350] D. R. Arnold and L. A. Karnischky, *J. Am. Chem. Soc.* **92**, 1404 (1970).

of **196** may be maintained for 50 hours at 115°C and then cooled to produce **195** quantitatively. One can also react **196** with a dipolarophile and obtain cycloaddition products [Eq. (78)].[350] A 5-thiabicyclo[2.1.0]pentane has also

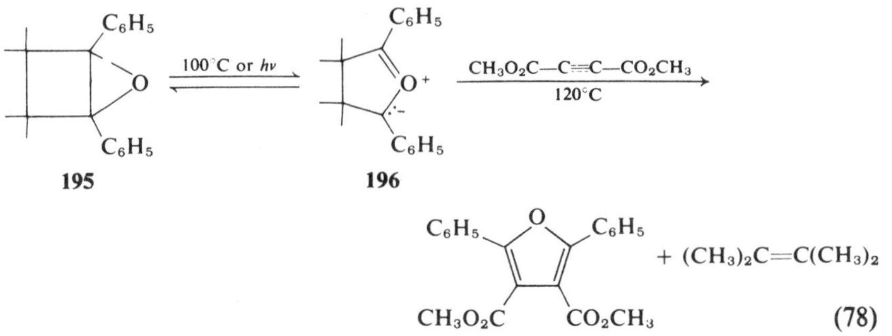

$$\text{(78)}$$

been characterized,[351] as has 6-thiabicyclo[2.1.1]hexane.[352] An attempt to prepare 5-thiabicyclo[1.1.1]pentane failed.[351] N-carboxyl esters of 5-aza-bicyclo[2.1.0]pentanes[353–355] are extremely unreactive with acetylene dicarb-oxylates, but this may be an electronic substituent effect rather than resistance to ring opening. Peroxidation of Dewar benzene produces the fused oxirane **197**, which rearranges thermally or photochemically to oxepineoxanorcara-diene.[356] 2-Azabicyclobutane (**198**)[357] and 2-oxabicyclobutane (**199**)[358–360]

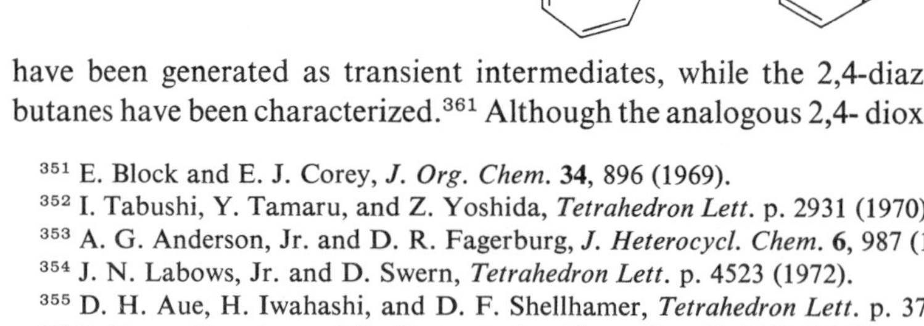

$$\text{(79)}$$

have been generated as transient intermediates, while the 2,4-diazabicyclo-butanes have been characterized.[361] Although the analogous 2,4- dioxabicyclo-

[351] E. Block and E. J. Corey, *J. Org. Chem.* **34**, 896 (1969).

[352] I. Tabushi, Y. Tamaru, and Z. Yoshida, *Tetrahedron Lett.* p. 2931 (1970).

[353] A. G. Anderson, Jr. and D. R. Fagerburg, *J. Heterocycl. Chem.* **6**, 987 (1969).

[354] J. N. Labows, Jr. and D. Swern, *Tetrahedron Lett.* p. 4523 (1972).

[355] D. H. Aue, H. Iwahashi, and D. F. Shellhamer, *Tetrahedron Lett.* p. 3719 (1973)

[356] E. E. van Tamelen and D. Carty, *J. Am. Chem. Soc.* **89**, 3922 (1967).

[357] D. H. Aue, R. B. Lorens, and G. S. Helwig, *Tetrahedron Lett.* p. 4795 (1973).

[358] J. Ciabattoni and J. P. Kocienski, *J. Am. Chem. Soc.* **91**, 6534 (1969).

[359] L. Friedrich and R. A. Cormier, *J. Org. Chem.* **35**, 450 (1970).

[360] L. E. Friedrich and R. A. Cormier, *Tetrahedron Lett.* p. 4761 (1971).

[361] A. A. Dudinskaya, L. I. Khmelnitski, I. D. Petrova, E. S. Baryshnikova, and S. S. Novikov, *Tetrahedron* **27**, 4053 (1971).

butane system has never been seen, its intermediacy has been postulated in the peroxy acid reaction of alkynes (see Scheme 26).

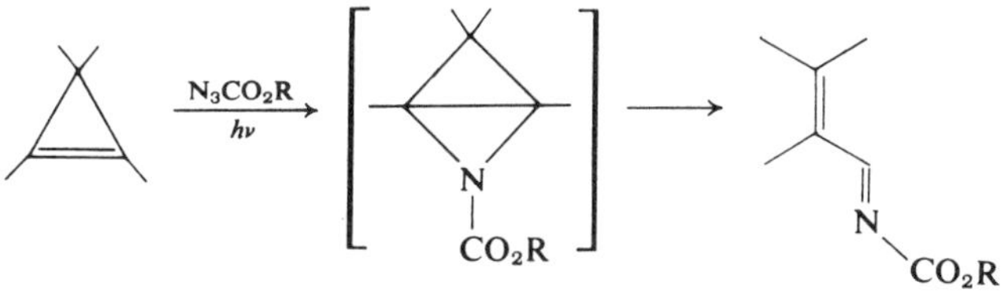

$$\text{(80)}$$

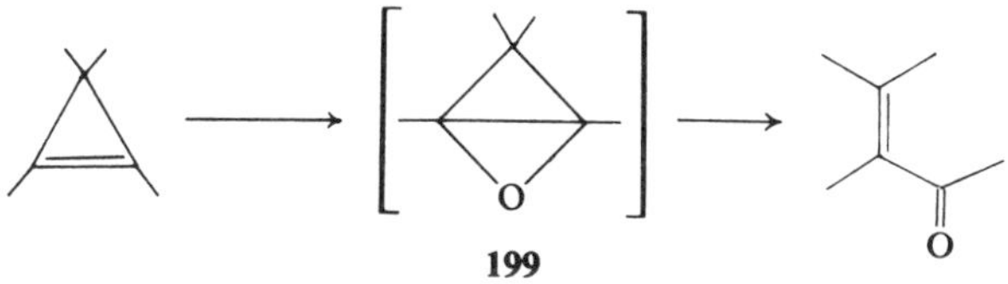

**198**

$$\text{(81)}$$

**199**

**5.D.5.b** *Nitrogen at the Bridgehead*

3-Phenyl-1-azabicyclobutane was actually the first reported heteroanalogue of bicyclobutane.[362,363] The vertical ionization potential of the parent molecule (**200**)[364] is relatively high (9.8 eV or 225 kcal/mole, compared to 197.4 kcal/mole for trimethylamine) and reflects the high s character (ca. 34%) assigned to the nitrogen lone-pair orbital.[365] 1-Azabicyclo[1.1.1]pentane has not yet been characterized, nor has its intermediacy been inferred. The postulated intermediacies of **202**,[366] oxygen analogue **203**,[367] and even **204**, the nmr of which is reported in $SO_2$ at $-20°C$,[368] are reported in the literature.

The vertical ionization potential [185.6 kcal/mole (8.05 eV)] of 1-azabicyclo-[2.2.2]octane (quinuclidine, **205**) is very similar to that of triethylamine, its acyclic model, although the lone-pair band width in the photoelectron spectrum of **205** is quite narrow, reflecting strain in its radical cation.[369] An interesting aspect of the photoelectron spectrum of 1,4-diazabicyclo[2.2.2]-octane (DABCO, **206**) is the splitting of the two bands attributed to nitrogen

[362] A. G. Hortman and D. A. Robertson, *J. Am. Chem. Soc.* **89**, 5974 (1967).

[363] A. G. Hortman and J. E. Martinelli, *Tetrahedron Lett.* p. 6205 (1968).

[364] W. Funke, *Angew. Chem., Int. Ed. Engl.* **8**, 70 (1969).

[365] D. H. Aue, H. M. Webb, and M. T. Bowers, *J. Am. Chem. Soc.* **97**, 4137 (1975).

[366] R. H. Higgins and N. H. Cromwell, *J. Am. Soc.* **95**, 120 (1973) and references cited therein.

[367] H. G. Richey, Jr., and D. V. Kinsman, *Tetrahedron Lett.* p. 2505 (1969).

[368] J. A. Deyrup and S. C. Clough, *J. Org. Chem.* **39**, 902 (1974).

[369] D. H. Aue, H. M. Webb, and M. T. Bowers, *J. Am. Chem. Soc.* **97**, 4136 (1975).

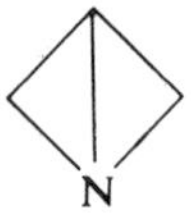

**200** 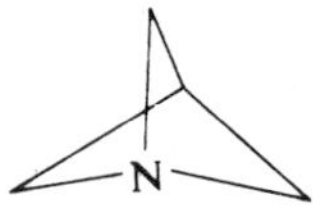          **201**

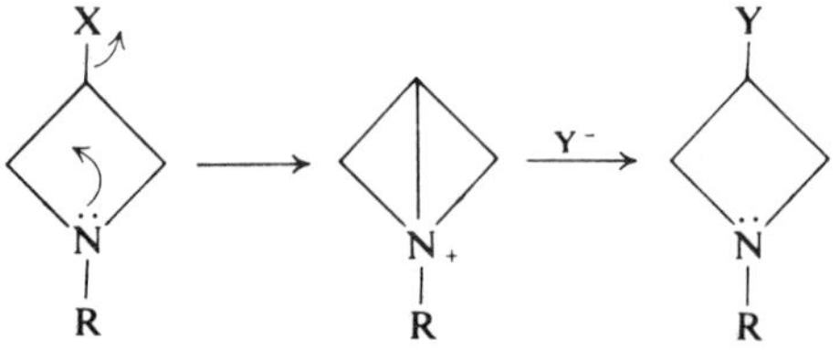

$$X \quad \longrightarrow \quad \overset{Y^-}{\longrightarrow} \quad Y \qquad (82)$$

**202**

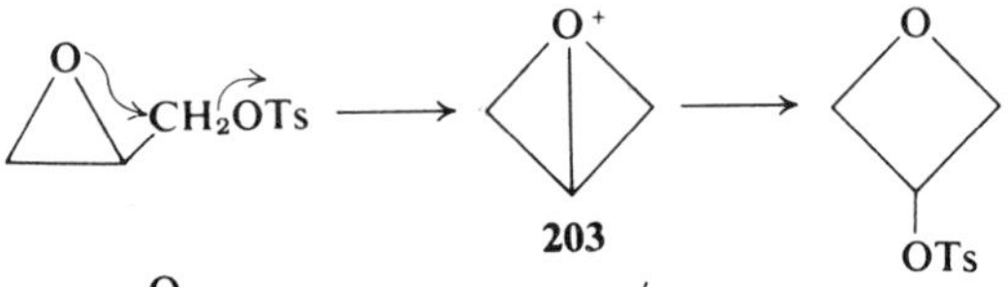

$$\longrightarrow \quad CH_2OTs \quad \longrightarrow \quad \longrightarrow \qquad (83)$$

**203**          OTs

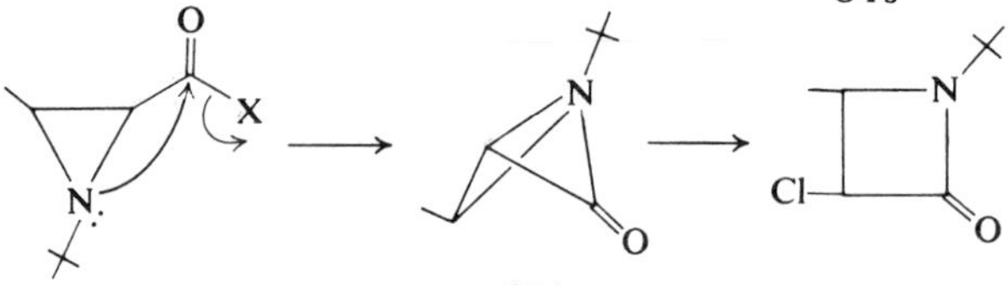

$$\longrightarrow \quad \longrightarrow \qquad (84)$$

**204**

lone pairs by some 51.5 kcal/mole (2.2 eV).[370] This is consistent with a splitting of the hypothetical noninteracting degenerate lone-pair orbitals as the result of "through-space" and "through-bond" interactions with the former dominating in this molecule.[370,371] Another novel situation arises in the analysis of the gas-phase ionization potential and basic properties of 1-azabicyclo[3.3.3]undecane (manxine, **207**).[372] This molecule maintains

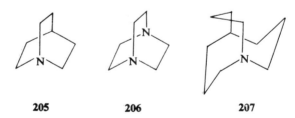

**205**              **206**              **207**

[370] E. Heilbronner and K. A. Muszkat, *J. Am. Chem. Soc.* **92**, 3818 (1970).

[371] R. Hoffmann, A. Imamura, and W. Hehre, *J. Am. Chem. Soc.* **90**, 1499 (1968).

[372] J. C. Coll, D. R. Crist, M. del C. G. Barrio, and N. J. Leonard, *J. Am. Chem. Soc.* **94**, 7092 (1972).

virtually planar geometry at the bridgehead nitrogen.[373] Its vertical ionization potential (161.7 kcal/mole) is more than 20 kcal/mole lower than that of its acyclic analogue, tri-*n*-propylamine, reflecting the almost pure p character of its lone pair.[369] This is in striking contrast to the high s character of the lone pair in **200**. However, the gas-phase proton affinity of manxine is actually slightly lower than that of tri-*n*-propylamine, reflecting the added strain in manxinium ion and showing that the relative weakness of manxine as a base in the gas phase is a manifestation of the dominance of strain over hybridization effects. Very similar considerations explain the important aspects of the first two ionization bands in **208** and **209**.[374] Compound **208**, which contains some structural features of both **206** and **207**, has its first two

ionization potentials, equal to 6.90 eV and 7.76 eV, yielding an average of 7.3 eV, which reflects the almost pure p character of the N lone pairs. An interesting feature is that the radical cation corresponding to the lowest band is considered to have inverted pyramidal geometry at the nitrogen bridgeheads.[374] The average ionization potential in **209** (8.2 eV) reflects essentially sp$^3$ hybridization of nitrogen.

### 5.D.5.c  *Silicon, Phosphorus, and Sulfur at the Bridgehead*

Nucleophilic substitution at the bridgehead of 1-chloro-1-silabicyclo-[2.2.1]heptane (**210**)[375] occurs much more rapidly than the corresponding reaction of the silaadamantyl molecule **211**.[376,377] This is the exact reverse of the reactivities of norbornane and adamantane systems (Section 3.B.2). The results suggest front-side displacement with the intermediacy of pentacoordinate silicon.[377] An electron-diffraction study of 1-methyl-1-silabicyclo-[2.2.1]heptane (**212**) supports this view because the geometry at the silicon bridgehead is very close to the trigonal pyramid expected for pentacoordinate

[373] A. H.-J. Wang, R. J. Missavage, S. R. Byrn, and J. C. Paul, *J. Am. Chem. Soc.* **94**, 7100 (1972).
[374] R. W. Alder, N. C. Goode, T. J. King, J. M. Mellor, and R. W. Miller, *J. Chem. Soc., Chem. Commun.* p. 173 (1976).
[375] L. H. Sommer and O. F. Bennet, *J. Am. Chem. Soc.* **79**, 1008 (1957).
[376] A. L. Smith and H. A. Clark, *J. Am. Chem. Soc.* **83**, 3345 (1961).
[377] G. D. Homer and L. H. Sommer, *J. Am. Chem. Soc.* **95**, 7700 (1973).

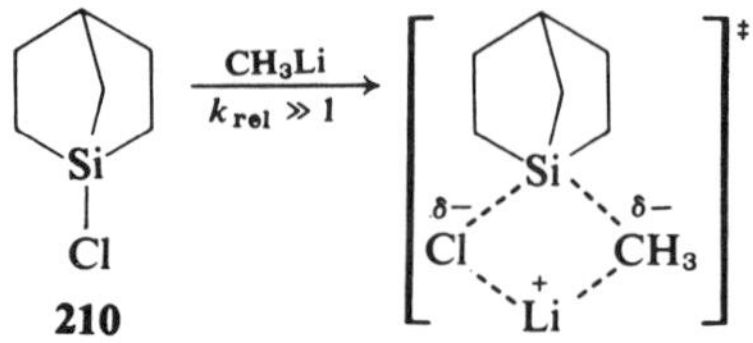

$$(85)$$

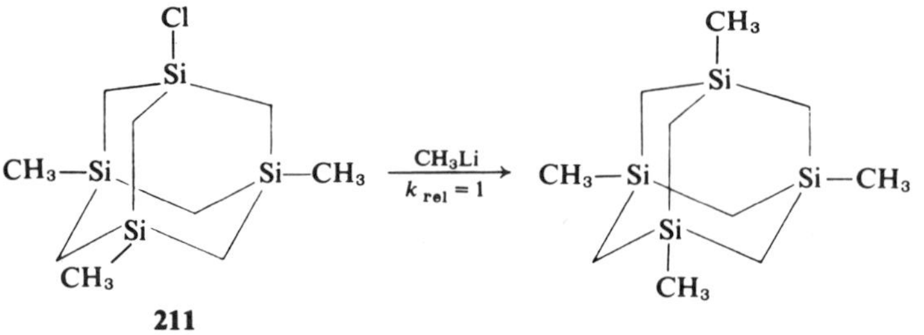

$$(86)$$

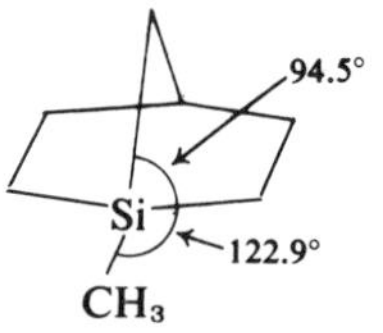

silicon.[378] Bridgehead phosphine oxides (e.g., 213) have been described.[379a,b] The nonbonded S—S distance in stable salts of dication 214 is ca. 0.5 Å shorter than the sum of van der Waals radii.[380] This is apparently related to the special stabilization noted elsewhere for bicyclo[2.2.2]octyl dication (see Section 3.B.2). Generation of the stable liquid tetramethylallene episulfide in

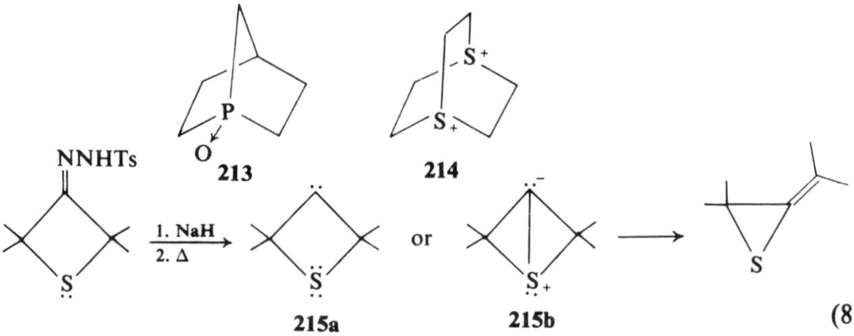

$$(87)$$

[378] R. L. Hilderbrandt, G. D. Homer, and P. Boudjouk, *J. Am. Chem. Soc.* **98**, 7476 (1976).

[379a] R. B. Wetzel and G. L. Kenyon, *J. Am. Chem. Soc.* **94**, 9230 (1972).

[379b] R. B. Wetzel and G. L. Kenyon, *J. Chem. Soc., Chem. Commun.* p. 287 (1973).

[380] E. Deutsch, *J. Org. Chem.* **37**, 3481 (1972).

the manner of reaction (87) supports the intermediacy of either thietanylidene **215a** or the bridged ylide **215b**.[381] The analogous oxygen-containing starting material follows a different course.[381]


### 5.D.5.d  *Spiro- and Trans-Fused Heterocycles*

Oxaspiropentane (**216**)[382] and some substituted derivatives[383a-c] are isolable, as is the azaspiro molecule **217**[384] and related species.[385a,b] A microwave determination of the structure of **216** implies that the oxygen substituent to the cyclopropyl ring appears to lengthen the $C_3$—$C_4$ bond through conjuga-

| **216** | **217** | **218** |

tive electron donation.[386] Spirodioxapentane **218** ($R_1 = R_2 = R_3 = R_4 = CH_3$) is a transient intermediate,[387a] but substitution by a bulky group allows another derivative of **218** ($R_1 = H$, $R_2 = t$-Bu, $R_3 = R_4 = CH_3$) to be isolated,[387b] As already mentioned in Section 5.D.2.a, spirocyclopropane rings stabilize silacyclopropanes (e.g., **132**).

No heterocyclic analogues of small *trans*-fused bicyclic hydrocarbons (Section 3.B.3) are known. The main reason, of course, is the nonavailability of *trans*-olefins smaller than *trans*-cyclooctane. Peracid oxidation of this olefin produces *trans*-9-oxabicyclo[6.1.0]nonane (**219**), which produces an unusual two-carbon ring-contraction product, 2-methylcyclohexane carboxaldehyde, along with other products in formic acid. The different products observed for the *cis* isomer are rationalized in terms of increased strain and

[381] A. G. Hortmann and A. Bhattacharjya, *J. Am. Chem. Soc.* **98**, 7081 (1976).

[382] J. R. Salaun and J. M. Conia, *Chem. Commun.* p. 1579 (1971).

[383a] J. K. Crandall and D. R. Paulson, *J. Org. Chem.* **33**, 991 (1968).

[383b] J. K. Crandall and D. R. Paulson, *Tetrahedron Lett.* p. 2751 (1969).

[383c] B. M. Trost and M. Bogdanowicz, *J. Am. Chem. Soc.* **95**, 5311 (1973).

[384] J. K. Crandall and W. W. Conover, *J. Chem. Soc., Chem. Commun.* p. 33 (1973).

[385a] D. H. Aue, R. B. Lorens, and G. S. Helwig, *Tetrahedron Lett.* p. 4795 (1973).

[385b] J. K. Crandall and W. W. Conover, *J. Org. Chem.* **39**, 63 (1974).

[386] W. D. Slafer, A. D. English, D. O. Harris, D. F. Shellhamer, M. J. Meshishnek, and D. H. Aue, *J. Am. Chem. Soc.*, **97**, 6638 (1975).

[387a] J. K. Crandall, W. H. Machleder, and S. A. Sojka, *J. Org. Chem.* **38**, 1149 (1973).

[387b] J. K. Crandall, W. H. Machleder, and M. J. Thomas, *J. Am. Chem. Soc.* **90**, 7346 (1968).

unusual spatial features in **219**.[388] The *cis* isomer is reduced by lithium aluminum hydride more rapidly than **219**, and this appears to be attributable to steric hindrance in the latter.[389]

$$\underset{\textbf{219}}{\text{[structure]}} \xrightarrow{\text{HCO}_2\text{H}} \text{[structure with CHO and CH}_3\text{]} + \text{other products} \qquad (88)$$

### 5.D.6  UNSATURATED HETEROCYCLES

#### 5.D.6.a  *Three-Membered Rings*

Numerous 3*H*-diazirine derivatives (**220**) are known,[390] and the parent (R = R′ = H) has been found experimentally to be 1–15 kcal/mole more stable than diazomethane[391] (cyclopropene is less stable than propyne or allene). The exothermic evolution of nitrogen from 3*H*-diazirines is "thermally

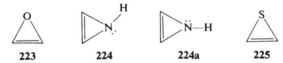

| **220** | **221** | **222** |

forbidden" and thus many of these compounds are thermally stable, although occasionally explosive upon impact. 1-Azirines (**221**) have been known since the early 1960's,[392] and even the spiro compound **222** has been characterized.[393]

| **223** | **224** | **224a** | **225** |

A great deal of interest has focused on the potentially antiaromatic heterocycles oxirene (**223**), 2-azirine (**224**), and thiirene (**225**), with the greatest

[388] A. C. Cope, A. Fournier, Jr., and H. E. Simmons, Jr., *J. Am. Chem. Soc.* **79**, 3905 (1957)

[389] M. Lj. Mihailovic, C. V. Andrejevic, J. Milanovic, and J. Jankovic, *Helv. Chim. Acta* **59**, 2305 (1976).

[390] E. Schmitz, *Proc. Int. Cong. Pure Appl. Chem. 23rd* **2**, 283 (1971).

[391] A. H. Laufer and H. Okabe, *J. Am. Chem. Soc.* **93**, 4137 (1971).

[392] For example, see G. Smolinsky, *J. Am. Chem. Soc.* **83**, 4483 (1961).

[393] H. J. Bestmann, T. Denzel, R. Kunstmann, and J. Lengyel, *Tetrahedron Lett.* p. 2895 (1968).

efforts being applied to studies of the first of these systems. Oxirenes have never been isolated or even observed spectroscopically. The initial evidence for formation of an oxirene (**226**) was the discovery of phenylacetic acid as a

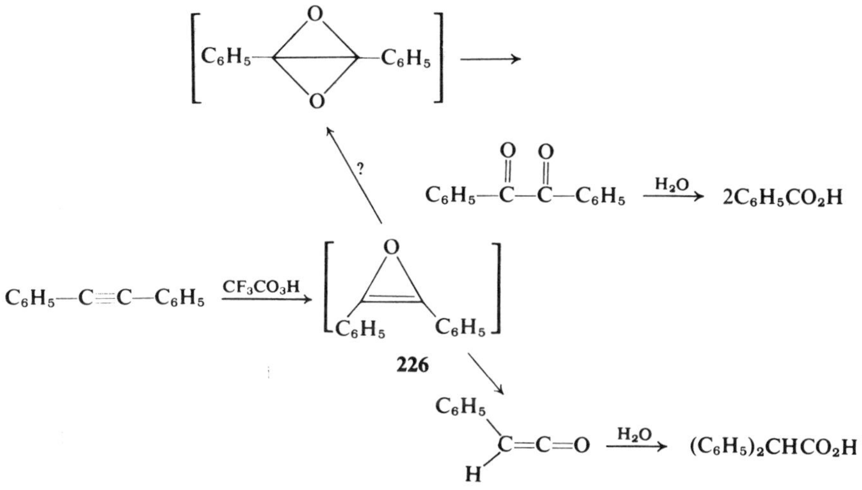

$$C_6H_5-C\equiv C-C_6H_5 \xrightarrow{CF_3CO_3H} \left[ \triangle_{C_6H_5 \quad C_6H_5}^{O} \right]$$

**226**

$$\left[ C_6H_5-\underset{O}{\overset{O}{\diamondsuit}}-C_6H_5 \right] \longrightarrow$$

$$C_6H_5-\overset{\overset{O}{\parallel}}{C}-\overset{\overset{O}{\parallel}}{C}-C_6H_5 \xrightarrow{H_2O} 2C_6H_5CO_2H$$

$$\underset{H}{\overset{C_6H_5}{\diagdown}}C=C=O \xrightarrow{H_2O} (C_6H_5)_2CHCO_2H$$

**Scheme 26**

reaction product in the oxidation of diphenylacetylene by peroxytrifluoro-acetic acid (Scheme 26).[394,395] Labeling studies suggest the formation of oxirenes in photochemical and thermal Wolff rearrangements (Scheme 27) in the vapor phase and to a somewhat lesser extent in solution.[396–398b]

One might argue that such labeling experiments merely provide evidence for two rapidly interconverting α-ketocarbenes without the intermediacy of an oxirene or with the oxirene as a transition state lacking "chemical significance." Here, in principle, is a fruitful field for predictions based upon theoretical chemistry. Unfortunately, there is substantial disagreement among theoretical calculations when applied to this problem. For example, MINDO/3

[394] R. N. McDonald and P. A. Schwab, *J. Am. Chem. Soc.* **86**, 4866 (1964).

[395] J. K. Stilk and D. D. Whitehurst, *J. Am. Chem. Soc.* **86**, 4871 (1964).

[396] For general review, see H. Meier and K.-P. Zeller, *Angew. Chem., Int. Ed. Engl.* **14**, 32 (1975).

[397a] I. Csizmadia, J. Font, and O. P. Strausz, *J. Am. Chem. Soc.* **90**, 7360 (1968).

[397b] D. E. Thornton, R. K. Gosavi, and O. P. Strausz, *J. Am. Chem. Soc.* **92**, 1768 (1970).

[397c] G. Frater and O. P. Strausz, *J. Am. Chem. Soc.* **92**, 6654 (1970).

[397d] J. Fenwick, G. Frater, K. Ogi, and O. P. Strausz, *J. Am. Chem. Soc.* **95**, 124 (1973).

[398a] S. A. Matlin and P. G. Sammes, *J. Chem. Soc., Chem. Commun.* p. 11 (1972).

[398b] S. A. Matlin and P. G. Sammes, *J. Chem. Soc., Perkin Trans. 1* p. 2623 (1972).

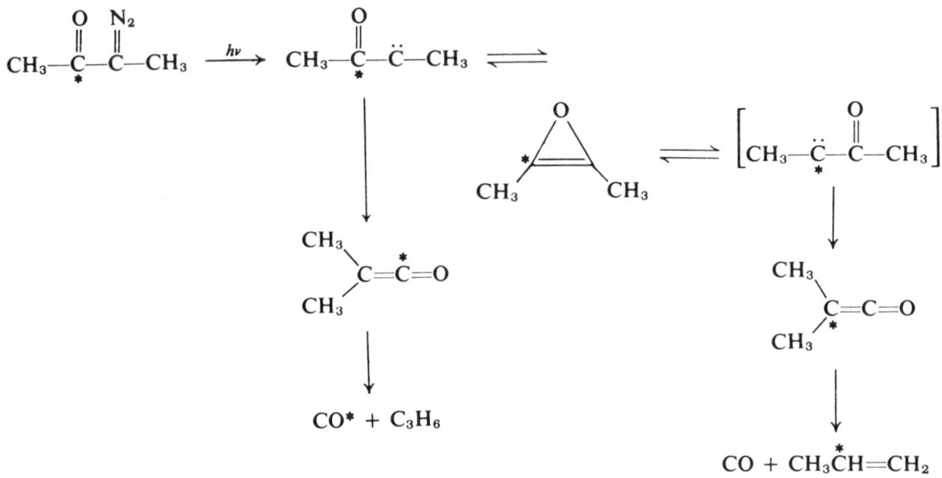

**Scheme 27**

calculations[399] and an *ab initio* study[400] both predict that oxirene is more stable than α-ketocarbene. The MINDO/3 study also suggests the presence of an appreciable activation energy (ca. 18 kcal/mole) for isomerization to ketene, thus implying the potential isolability of oxirene. However, a larger basis-set *ab initio* computation[401] supports earlier results of extended Hückel theory that found the open α-ketocarbene to be more stable than oxirene by more than 11 kcal/mole and further calculated a barrier for isomerization to ketene of only about 7 kcal/mole. This would suggest that oxirene is most likely to be observed at very low temperatures. Calculations also find little antiaromatic character in this molecule. Finally, we should mention here the apparent trapping of an oxirene [Eq. (89)], suggesting to us that these species are not merely transition states and are chemically significant.[402]

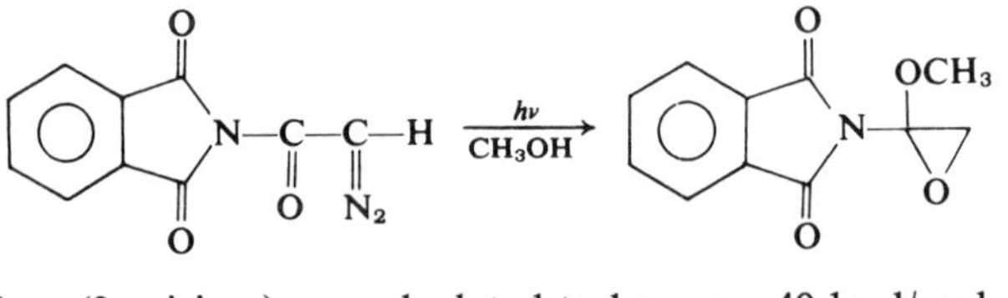

(89)

1*H*-Azirines (2-azirines) are calculated to be some 40 kcal/mole less stable than 2*H*-azirines (1-azirines), and the tautomerization interconverting them

[399] M. J. S. Dewar and C. A. Ramsden, *J. Chem. Soc., Chem. Commun.* p. 688 (1973).

[400] A. C. Hopkinson, *J. Chem. Soc., Perkin Trans. 2*, p. 794 (1973).

[401] O. P. Strausz, R. K. Gosavi, A. S. Denes, and I. G. Csizmadia, *J. Am. Chem. Soc.* **98**, 4784 (1976).

[402] H. Meier and K.-P. Zeller, *Angew. Chem., Int. Ed. Engl.* **14**, 32 (1975), see p. 40 and footnote 232 therein.

should occur readily. Not surprisingly, none of the 2-azirines has been observed even spectroscopically. Again, indirect chemical evidence (Scheme 28) supports the intermediacy of these species (which should be negligibly antiaromatic simply as the result of nonplanarity at nitrogen).[403a-d] It should

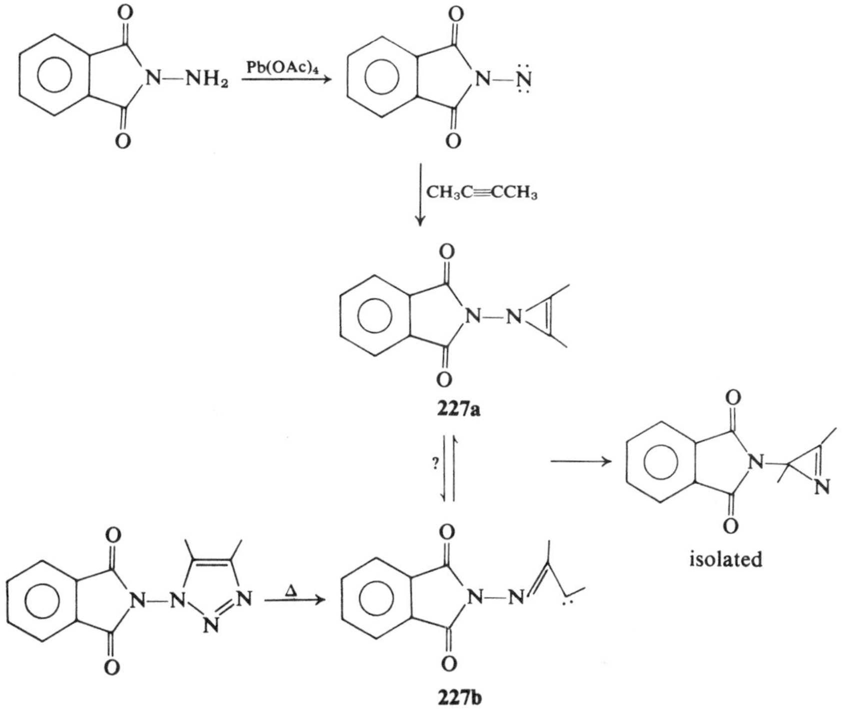

**227a**

**227b**

isolated

**Scheme 28**

be noted that there is apparent precedent for an iminocarbene (analogous to **227b**) in the pyrolytic formation of styrene and hydrogen cyanide from 2-phenyl-3-methyl-2H-azirine[404] [Eq. (90)] (see earlier sections in this chapter and in Chapter 3 for proposed formation of vinylcarbenes from cyclopropenes and 3,3'-bicyclopropenyls with or without the aid of transition

[403a] D. J. Anderson, T. L. Gilchrist, and C. W. Rees, *J. Chem. Soc., Chem. Commun.* p. 147 (1969).

[403b] D. J. Anderson, T. L. Gilchrist, G. E. Gymer, and C. W. Rees, *J. Chem. Soc., Chem. Commun.* p. 1518 (1971).

[403c] T. L. Gilchrist, G. E. Gymer, and C. W. Rees, *J. Chem. Soc., Chem. Commun.* p. 1519 (1971).

[403d] T. L. Gilchrist, G. E. Gymer, and C. W. Rees, *J. Chem. Soc., Perkin Trans. 1* p. 1 (1975).

[404] L. A. Wendling and R. G. Bergman, *J. Am. Chem. Soc.* **96**, 308 (1974).

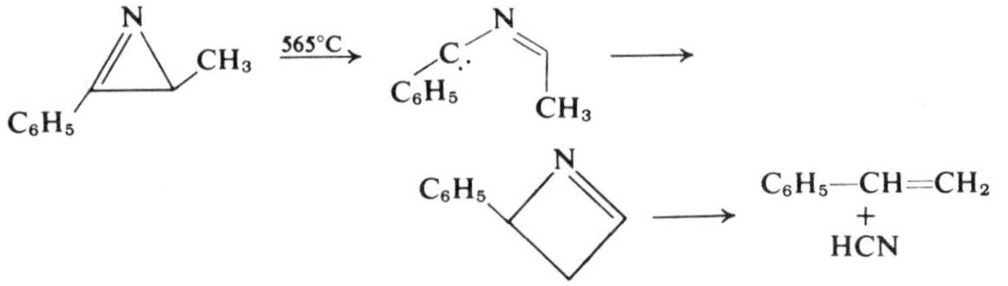

$$C_6H_5 \overset{N}{\triangle} CH_3 \xrightarrow{565°C} \overset{N}{\underset{C_6H_5 \quad CH_3}{C}} \longrightarrow$$

$$C_6H_5 \overset{N}{\underset{}{\square}} \longrightarrow \begin{array}{c} C_6H_5-CH=CH_2 \\ + \\ HCN \end{array} \qquad (90)$$

metals). One might imagine a possible means of isolating a $1H$-azirine derivative through suitable steric constraints, namely, as **228** (a hydrocarbon analogue is isolable, see Chapter 3). Carbon-13 labeling studies strongly suggest the intermediacy of $1H$-benzazirine (**229**).[405]

Unlike its oxygen and nitrogen analogues, thiirene has actually been monitored (by kinetic mass spectroscopy), exhibiting a half-life of about 2 seconds under the low-pressure conditions of the mass spectrometer (dimethylthiirene has a half-life of about 7 seconds).[406] The increased stability of thiirene relative to oxirene is attributed to a decreased tendency to form thioketene compared to ketene, because the carbon–sulfur double bond is not as stable as a carbon–oxygen double bond.[406] Generation of thiirenes from 1,2,3-thiadiazoles in the presence of $Fe_2(CO)_9$ results in the trapping of the open forms of these species.[407,408a] The parent molecule thiirene appears to

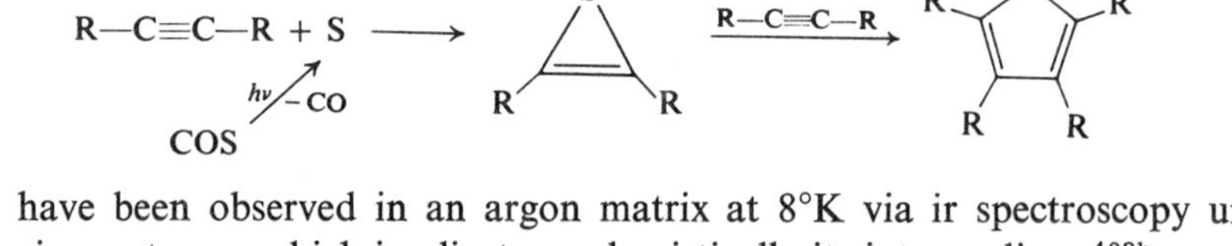

$$\overset{N}{\underset{R}{\square}} \overset{?}{\underset{\Delta}{\longrightarrow}} \overset{}{\square}N-R \overset{}{\xrightarrow{\times}} \overset{R}{\square}N \qquad \overset{}{\bigcirc}NH \qquad (91)$$

$$\underset{\textbf{228}}{} \qquad \qquad \underset{\textbf{229}}{}$$

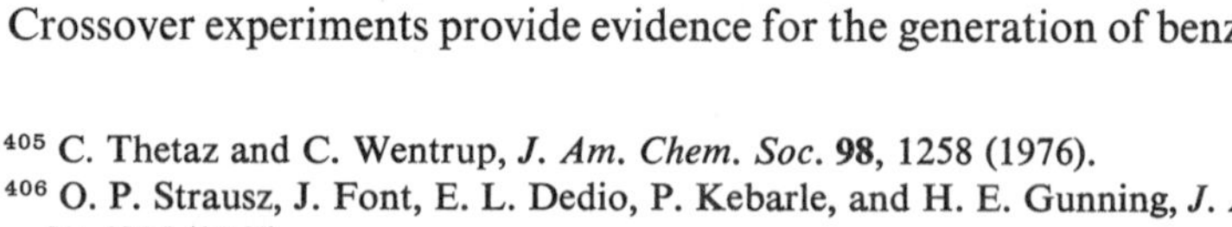

$$R-C\equiv C-R + S \longrightarrow \overset{S}{\underset{R}{\triangle}}\overset{}{R} \xrightarrow{R-C\equiv C-R} \overset{R}{\underset{R}{\square}}\overset{S}{\underset{R}{\square}}R \qquad (92)$$

$$\overset{hv}{\underset{COS}{\nearrow}}{}_{-CO}$$

have been observed in an argon matrix at 8°K via ir spectroscopy under circumstances which implicate mechanistically its intermediacy.[408b]

Crossover experiments provide evidence for the generation of benzothiirene

[405] C. Thetaz and C. Wentrup, *J. Am. Chem. Soc.* **98**, 1258 (1976).

[406] O. P. Strausz, J. Font, E. L. Dedio, P. Kebarle, and H. E. Gunning, *J. Am. Chem. Soc.* **89**, 4805 (1967).

[407] P. G. Mente and C. W. Rees, *J. Chem. Soc., Chem. Commun.* p. 418 (1972).

[408a] G. N. Schrauzer and H. Kisch, *J. Am. Chem. Soc.* **95**, 2501 (1973).

[408b] J. Laureni, A. Krantz, and R. A. Hajdu, *J. Am. Chem. Soc.* **98**, 7892 (1976).

(231) via the α-thiaketocarbene 230, but α-ketocarbene 232 does not isomerize to a benzooxirene.[409]

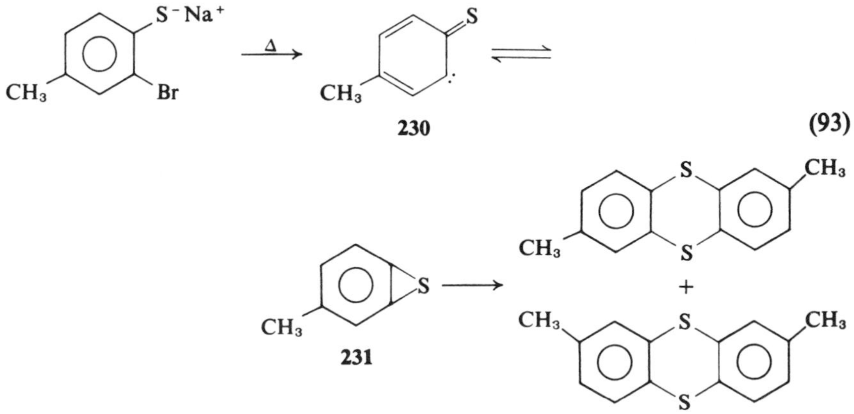

(93)

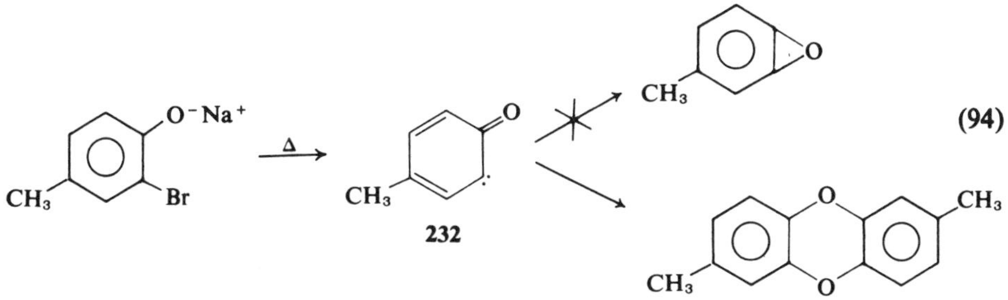

(94)

Tetramethylsilacyclopropene (233), formed from attack of dimethylsilylene on 2-butyne, is stable for weeks at room temperature in the absence of air, which will in turn destroy the molecule in less than one minute.[410] Thiirene oxide 234 is quite stable perhaps because of d-orbital participation, thought to lend some aromatic character.[411]

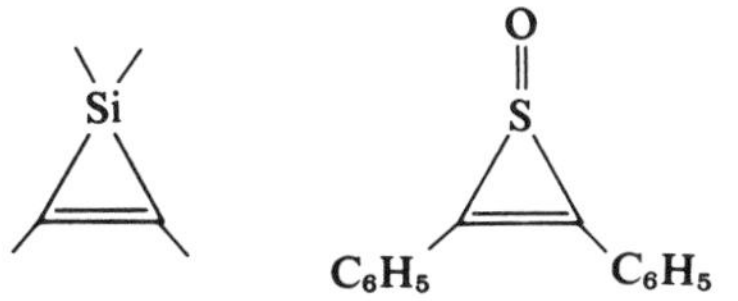

[409] J. I. G. Cardogan, J. T. Sharp, and M. J. Trattles, *J. Chem. Soc., Chem. Commun.* p. 900 (1974).

[410] R. T. Conlin and P. P. Gaspar, *J. Am. Chem. Soc.* **98**, 3715 (1976). For another stable silacyclopropene, see D. Seyferth, D. C. Annarelli, and S. C. Vick, *J. Am. Chem. Soc.* **98**, 6382 (1976).

[411] L. A. Carpino and H.-W. Chen. *J. Am. Chem. Soc.* **93**, 785 (1971).

### 5.D.6.b    Unsaturated Four-Membered Heterocycles

The first oxetene to be observed in solution was reported in 1968;[412] and tetramethyloxetene (235) was the first derivative reported to be stable at room temperature.[413a,b] This molecule is also formed from an oxetanylidene

(95)

235

236                 237                 238                 239

analogous to 215a.[381] One might argue that the inherent strain of the molecule is not the major factor in its instability; rather, the stability of the carbonyl group of the ring-opened isomer [Eq. (95)] is reflected in a low barrier for this transformation.[413b] A somewhat less stable oxetene derivative was produced via photochemical cycloaddition of 2-butyne and benzaldehyde.[413c] Benzo-thietedioxide 236 can be heated to 200°C before it ring opens,[414] while 237 is less stable but isolable[415] [in both cases, ring opening analogous to that in Eq. (95) involves loss of aromaticity.] Benzopropiolactone 238 has been observed in an argon matrix at 8°K.[416]

The first representative (239) of the dithiete family was initially thought to be aromatic, consistent with its modest thermal stability.[417] However, gas-phase electron-diffraction studies establish that there is little sulfur–sulfur double-bond character in this molecule.[418] Dioxetene 241 has been implicated as an intermediate in the chemiluminescent reaction of the strained cyclo-alkyne 240 with both singlet and triplet oxygen.[419] Here the higher-energy $\pi$

[412] J. M. Holovka, P. D. Gardner, C. B. Strow, M. L. Hall, and T. V. Van Auken, J. Am. Chem. Soc. 90, 5041 (1968).

[413a] L. E. Friedrich and G. B. Schuster, J. Am. Chem. Soc. 91, 7204 (1969).

[413b] L. E. Friedrich and G. B. Schuster, J. Am. Chem. Soc. 93, 4602 (1971).

[413c] L. E. Friedrich and J. D. Bower, J. Am. Chem. Soc. 95, 6869 (1973).

[414] D. C. Dittmer and T. R. Nelson, J. Org. Chem. 41, 3044 (1976).

[415] E. Voigt and H. Meier, Angew. Chem., Int. Ed. Engl. 15, 117 (1976).

[416] O. L. Chapman, C. L. McIntosh, J. Pacansky, G. V. Calder, and G. Orr, J. Am. Chem. Soc. 95, 4061 (1973).

[417] C. G. Krespan, B. C. McKusick, and T. L. Cairns, J. Am. Chem. Soc. 82, 1515 (1960).

[418] K. L. Hencher, Q. Shen, and D. G. Tuck, J. Am. Chem. Soc. 98, 899 (1976).

[419] N. J. Turro, V. Ramamurthy, K.-C. Liu, A. Krebs, and R. Kemper, J. Am. Chem. Soc. 98, 6758 (1976).

orbital of the strained cycloalkyne (Section 3.G) induces the formation of an oxygen complex in which triplet–singlet intersystem crossing is greatly facilitated. Spin inversion, not ring opening of **241**, is the rate-determining step of reaction (**96**). Reactions related to this one appear to hold the potential for generating singlet oxygen from triplet oxygen by purely thermal means.[419]

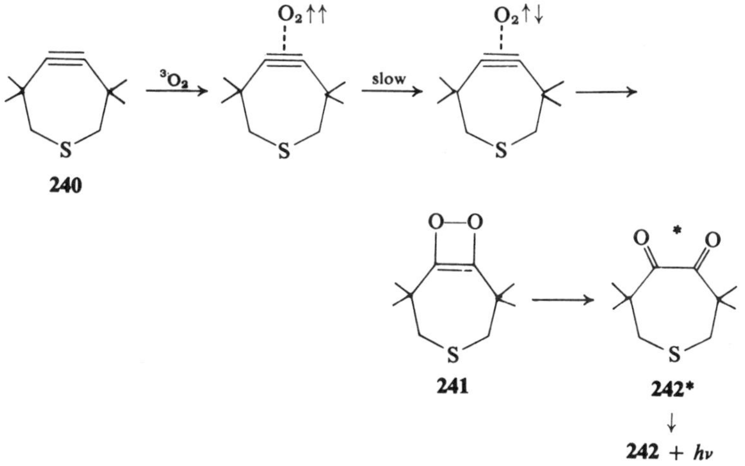

$$240 \xrightarrow{^3O_2} \cdots \xrightarrow{slow} \cdots \longrightarrow$$

$$(96)$$

**241** $\longrightarrow$ **242\*** $\downarrow$ **242** + $h\nu$

### 5.D.7  TWISTED $\pi$ SYSTEMS

The molecules **243**[420,421] and **244**[422] are isolable analogues of *trans*-cyclo-octene and bicyclo[3.3.1)non-1-ene, respectively. Molecules **245**[423] and **246**[424]

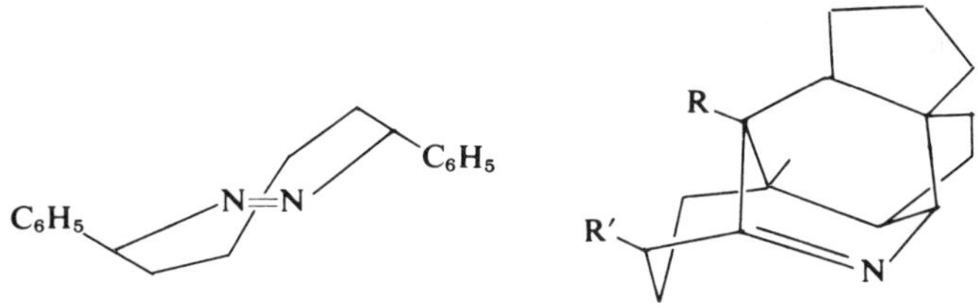

**243**                                        **244**

are transient species formed via the bridgehead nitrenes of adamantane and various tryptycenes, respectively. When tricoordinate nitrogen is in the

[420] Compound **243** was originally thought to be a *cis*-diazo isomer. The *syn*-diphenyl and unsubstituted compounds are also known. See Ref. 421.

[421] G. Vitt, E. Haedicke, and G. Quinkert, *Chem. Ber.* **109**, 518 (1976).

[422] M. Toda, Y. Hirata, and S. Yamamura, *Chem. Commun.* p. 1597 (1970).

[423] H. Quast and P. Eckert, *Justus Liebigs Ann. Chem.* 1727 (1974).

[424] H. Quast and P. Eckert, *Angew. Chem., Int. Ed. Engl.* **15**, 168 (1976).

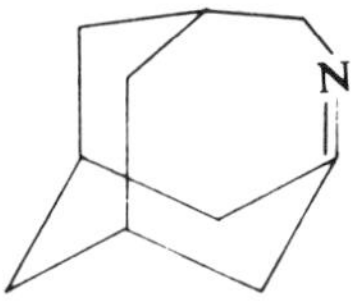

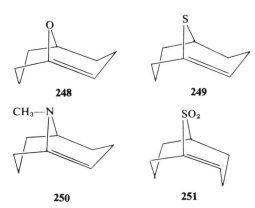

**245** 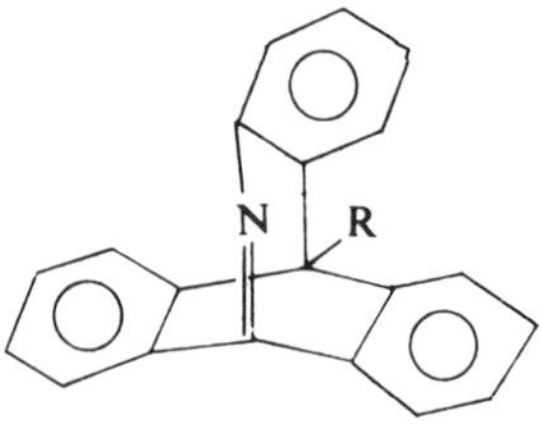          **246** 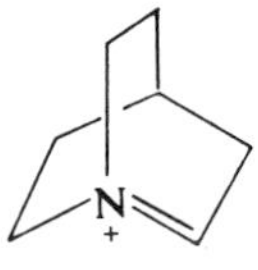          **247**

2-position of a 1-halobicyclo[2.2.2]octane system, resonance stabilization of the corresponding bridgehead carbonium ion produces a net $10^5$-fold increase in solvolysis rate (a $10^5$-fold decrease normally would be attributed to nitrogen's inductive effect).[425] This agrees with the results obtained in generating 1-azabicyclo[2.2.2]oct-2-yl cation (**247**),[426] which is more stable than the 1-azabicyclo[2.2.1]hept-2-yl cation.[427] 9-Oxabicyclo[3.3.1]non-1-ene **248** exhibits spectral characteristics atypical of normal enol ethers, and is about 4000 times less reactive to electrophilic addition than bicyclo[3.3.1]-non-1-ene, in marked contrast to the usual increased reactivity of enol

**248**          **249**

CH₃—N          SO₂

**250**          **251**

ethers.[428] 9-Thiabicyclo[3.3.1]non-1-ene (**249**)[429] also exhibits atypical spectral properties, as does nitrogen analogue **250**.[430] This latter compound suffers alkylation at nitrogen with methyl iodide in contrast with typical enamines which are carbon alkylated.[430] All of these special properties arise because the bicyclic skeleton strongly decreases the resonance interactions

[425] T. A. Wnuk and P. Kovacic *J. Am. Chem. Soc.* **97**, 5807 (1975).

[426] C. A. Grob and A. Sieber, *Helv. Chim. Acta* **50**, 2531 (1967).

[427] P. G. Gassman, R. L. Cryberg, and K. Shudo, *J. Am. Chem. Soc.* **94**, 7600 (1972).

[428] C. B. Quinn and J. R. Wiseman, *J. Am. Chem. Soc.* **95**, 1342 (1973).

[429] C. B. Quinn and J. R. Wiseman, *J. Am. Chem. Soc.* **95**, 6120 (1973).

[430] H. O. Krabbenhoft, J. R. Wiseman, and C. B. Quinn, *J. Am. Chem. Soc.* **96**, 258 (1974).

present in monocyclic and acylic analogues. Studies employing photoelectron spectroscopy also point up the relatively small $n-\pi$ interactions in **248** and **249**.[431] The very highly strained molecule $E$-9-thiabicyclo[3.3.1.]non-1-ene-9, 9-dioxide (**251**) has been generated under kinetic control conditions (conditions of thermodynamic control yield the $Z$ isomer) and trapped.[432]

Bridgehead amides such as 2-quinuclidone (1-azabicyclo[2.2.2]octan-2-one) (**252**) are of interest as potential indicators of the stability of the analogous olefins. 2-Quinuclidones are isolable and display ir spectra atypical of

252a          252b          253          254

amides and explicable in terms of reduced resonance contribution of **252b**.[433] The bridgehead amides **253** and **254**,[434] neither of which has been reported, should have virtually zero dipolar resonance stabilization of the type shown in **252b**. These compounds, once obtained, may help to shed some light on the question of the role of resonance in the thermodynamic and chemical stabilities and physical properties of amides. A thermochemical study on the as yet unknown 1-azabicyclo[3.3.1]nonan-2-one would be interesting because its strain could be compared to that of bicyclo[3.3.1]non-1-ene, whose enthalpy of formation is known (Section 3.E.3). 1-Azabicyclo[3.3.3]-undecan-2-one should also be interesting because of the likely presence of a planar nitrogen.

$\beta$-Lactones and $\beta$-lactams are usually isolable molecules which are, as expected, much more reactive than their larger homologues. The strain energy of $\beta$-lactone (propiolactone) has been evaluated at about 22.8 kcal/mole (after taking the ester resonance energy into account).[235] An interesting aspect of $\beta$-lactam chemistry is a published correlation between nonplanarity of the bridgehead nitrogens in various penicillins (**255**) and $\Delta^3$-cephalosporins (**256**), their ease of hydrolysis, and biological activity.[435] Nonplanarity of the

[431] C. Batich, E. Heilbronner, C. B. Quinn, and J. R. Wiseman, *Helv. Chim. Acta* **59**, 512 (1976).

[432] C. B. Quinn, J. R. Wiseman, and J. C. Calabrese, *J. Am. Chem. Soc.* **95**, 6121 (1973).

[433] H. Pracejus, M. Kehlen, H. Kehlen, and M. Matschiner, *Tetrahedron* **21**, 2257 (1965).

[434] H. K. Hall, Jr., *J. Am. Chem. Soc.* **80**, 6412 (1958).

[435] R. M. Sweet and L. F. Dahl, *J. Am. Chem. Soc.* **92**, 5489 (1970).

bridgehead nitrogen would decrease amide resonance stabilization and render this linkage more reactive.

Heteroatoms and groups have successfully spanned the 1 and 8 (*peri*) positions of naphthalenes. 1,8-Naphthalene thiete (**257**),[436a,b] the related sulfoxide (**258**)[436a,b] and sulfone,[437] as well as the analogous hydrocarbon (Section 3.H.1) have all been obtained through recourse to extrusion reactions. Evidence has been published which supports the transient existence of **260**.[438] More recently, silete **261** has been synthesized via a reaction between dichlorosilane and 1,8-dilithionaphthalene.[439] Apparently, the long Si—C bonds allow formation of **261** by nonextrusion means.

**257**, X = S
**258**, X = SO
**259**, X = SO$_2$
**260**, X = O
**261**, X = SiH$_2$

**5.D.8**   DISTORTED AROMATIC RINGS, HETARYNES, CYCLOPHANES,
              AND POLYCYCLES

A variety of hetarynes have been implicated as intermediates.[440,441] We will very briefly discuss 2-pyridyne and 3-pyridyne, the most important six-membered hetarynes, and 2-thiophyne, the only known five-membered hetaryne.

Of the two possible pyridynes, **263** and **265**, simple resonance arguments predict that the 2,3 isomer (**263**) is the most stable.[440,441] However, under the usual conditions of nucleophilic aromatic substitution, a 3-halopyridine will

---

[436a] J. Meinwald and S. Knapp, *J. Am. Chem. Soc.* **96**, 6533 (1974).

[436b] J. Meinwald, S. Knapp, S. K. Obendorf, and R. E. Hughes, *J. Am. Chem. Soc.* **98**, 6643 (1976).

[437] R. W. Hoffmann and W. Sieber, *Angew. Chem., Int. Ed. Engl.* **4**, 786 (1965).

[438] D. C. DeJongh and G. N. Evenson, *Tetrahedron Lett.* p. 4093 (1971).

[439] L. S. Yang and H. Schechter, *J. Chem. Soc., Chem. Commun* p. 775 (1976).

[440] R. W. Hoffmann, "Dehydrobenzenes and Cycloalkynes," Ch. 6. Academic Press, New York, 1967.

[441] H. J. den Hertog and H. C. van der Plas, *in* "Chemistry of Acetylenes" (H. G. Viehe, ed.), p. 1149. Dekker, New York, 1969.

produce 3,4-dehydropyridine (3-pyridyne, **265**) (Scheme 29). This is rationalized by assuming that conditions of kinetic control, favoring formation of **264**, are operative.[440,441] 3-Pyridyne has also been generated in the vapor state following pyrolysis of pyridine-3-diazonium-4-carboxylate, and its lifetime is comparable to that of benzyne.[442]

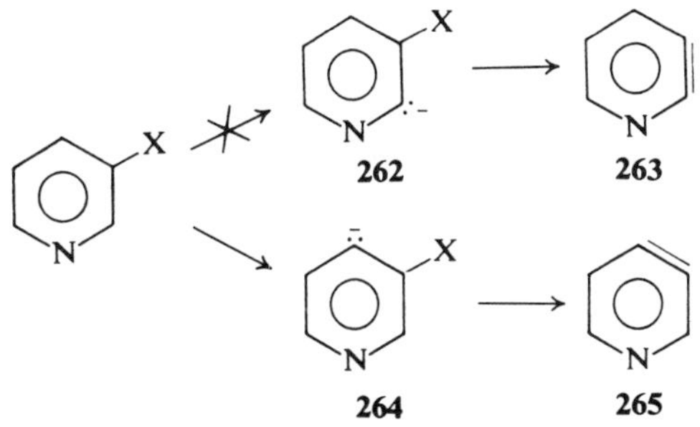

**Scheme 29**

The first five-membered hetaryne to be unambiguously identified as an intermediate was 2,3-thiophyne (**266**) [Eq. (97)].[443] An aromatic substitution

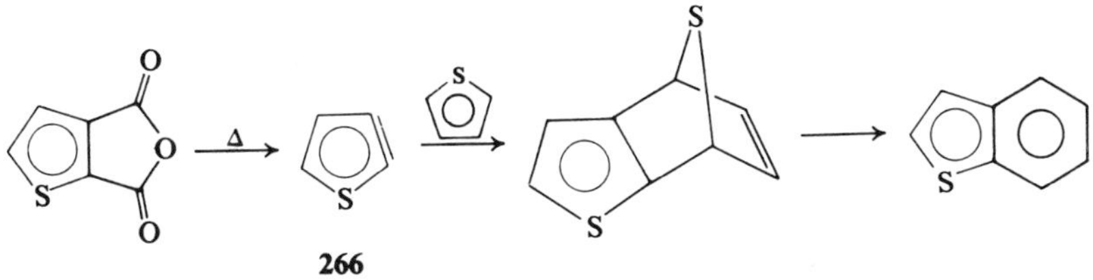

(97)

reaction involving 2-bromothiophene produced the *cine* substitution compound, 3-aminothiophene, as the dominant product.[444] Usually, *cine* substitution is acceptable as evidence for the intermediacy of an aryne. However, careful analysis of the reaction precluded the intermediacy of **266**. Efforts to generate 3,4-thiophyne in a manner analogous to reaction (97) have been unsuccessful.[445]

Systematic investigations of highly distorted heterophanes have not appeared to date. Some interesting aspects of heterocyclophane chemistry include (a) a tendency to Diels–Alder reactions [Eq. (98)],[446] and (b) some

[442] J. Kramer and R. S. Berry, *J. Am. Chem. Soc.* **94**, 8336 (1972).
[443] M. G. Reinecke and J. G. Newson, *J. Am. Chem. Soc.* **98**, 3021 (1976) and references cited therein.
[444] M. G. Reinecke and H. W. Adickes, *J. Am. Chem. Soc.* **90**, 511 (1968).
[445] B. E. Ayres, S. W. Longworth, and J. E. W. McOmie, *Tetrahedron* **31**, 1755 (1975).
[446] D. J. Cram and G. R. Knox, *J. Am. Chem. Soc.* **83**, 2204 (1961).

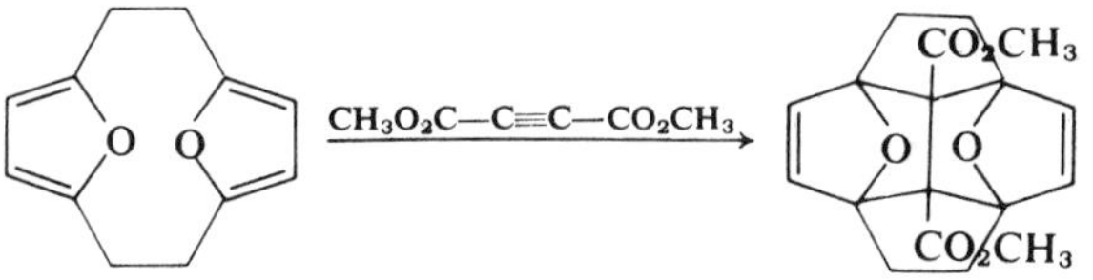

$$CH_3O_2C—C\equiv C—CO_2CH_3 \tag{98}$$

unusual conformational properties, such as the finding of perpendicular rings in [2.2](2,6)pyridinoparacyclophane-1,9-diene (267).[447a,b] Anomalous reactivity (e.g., 268 does not form an N-oxide under conditions in which the 2,6-decamethylene compound does; isolation of the stable salt 269)[448] is observed in certain instances.

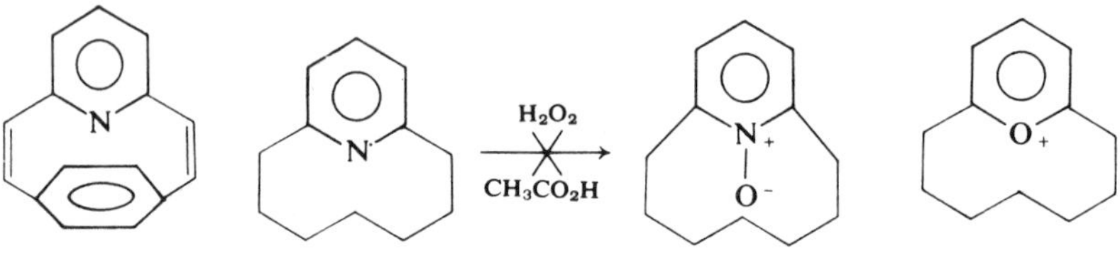

A number of valence isomers (analogues of Dewar benzene, prismane, etc.) of nitrogen-[449a–451] and oxygen-containing[452,453] heterocycles have been reported in the literature. Heterocyclic adamantanes, twistanes, and related species have been amply reviewed.[454] We note briefly published details of interconversions of oxaadamantanes, oxatwistanes, and oxaisotwistanes via carbonium ions and oxiranium ions.[455]

[447a] V. Boekelheide, K. Galuszko, and K. S. Szeto, J. Am. Chem. Soc. 96, 1578 (1974).
[447b] L. H. Weaver and B. W. Mathews, J. Am. Chem. Soc. 96, 1581 (1974).
[448] H. Nozaki, S. Fujita, and T. Mori, Bull. Chem. Soc. Jpn. 42, 1163 (1969).
[449a] C. G. Allison, R. D. Chambers, Y. A. Cheburkov, J. A. H. McBride, and W. K. R. Musgrave, J. Chem. Soc., Chem. Commun. p. 1200 (1969).
[449b] R. D. Chambers, W. K. R. Musgrave, and K. C. Srivastana, J. Chem. Soc., Chem. Commun. p. 264 (1970).
[450] R. N. Haszeldine, M. G. Barlow, and J. G. Dingall, J. Chem. Soc., Chem. Commun. p. 1580 (1970).
[451] K. E. Wilzbach and D. J. Rauch, J. Am. Chem. Soc. 92, 2178 (1970).
[452] J. A. Barltrop, K. Dawes, A. C. Day, and A.-J. H. Summers, J. Chem. Soc., Chem. Commun. p. 1240 (1972).
[453] J. W. Pavlik and E. L. Clennan, J. Am. Chem. Soc. 95, 1697 (1973).
[454] R. C. Fort, Jr. "Adamantane: The Chemistry of Diamond Molecules," Ch. 6. Dekker, New York, 1976.
[455] P. Ackermann, R. E. Portmann, and C. Ganter, Helv. Chim. Acta 59, 2515 (1976).

## 5.E Substituted Derivatives of Strained Molecules

### 5.E.1 INTRODUCTION

In earlier sections of this book, the hydrogen atom and various alkyl groups were implicitly considered as the only substituents on the various strained molecular skeletons discussed. However, it is obvious that an enormous number of substituted derivatives exist, at least in principle, and we might ask ourselves whether the structures, enthalpies, and reactivities follow the same order as do the parent molecules. (The term "substituted derivatives" is taken to mean replacement of H by a univalent or polyvalent atom as well as replacement of carbon by a polyvalent atom. The latter cases are termed heterocyclic compounds, and have been discussed in the sections immediately preceding this one.) For example, while the $\Delta H_f^0$ values are known for the three isomeric bicyclooctanes, what can be said about the 1-halo derivatives? Can one assume that replacement of H by F will not measurably alter strain? In the analysis of the relative stabilities of isomeric 1-azabicycloundecanes one might expect a relatively greater stability for 1-azabicyclo[3.3.3]undecane than would be found for the corresponding hydrocarbon, because tricoordinate nitrogen can stand planar geometry better than tetracoordinate carbon (see Section 5.D.5.b). Usually it is assumed that the $\Delta \Delta H_f^0$ of a pair of isomers is essentially unchanged if a carbon group is replaced by another group of equal covalence.[456] For example, $CH_3$ may be replaced by H or Br, $-CH_2-$ by $-O-$,[457] $>CH-$ by $>N-$, and $>C<$ by $>Si<$. Equivalently, isomerization of monosubstituted hydrocarbons[458] is considered equivalent to the less selective and often more difficult Lewis-acid-catalyzed reaction with unsubstituted hydrocarbons.[459]

Substituent effects should be most significant when the substituent (a) strongly interacts conjugatively with the rest of the molecule, (b) is extremely electron-withdrawing or -donating electrostatically, or (c) is large and so introduces considerable added strain. For example, we may contrast the $\Delta \Delta H_f^0$ of the isomeric 2-substituted allyl and 1-substituted cyclopropyl

[456] C. R. Somayajulu and B. J. Zwolinsky, *J. Chem. Soc., Faraday Trans. 2* **70**, 973 (1974).

[457] For extensive discussion of the stereochemical and thermochemical consequences of replacing $CH_2$ by O, see J. Dole, *Tetrahedron* **36**, 1603 (1973).

[458] Equilibration of bicyclooctanes; see P. von R. Schleyer, K. R. Blanchard, and C. D. Woody, *J. Am. Chem. Soc.* **85**, 1358 (1963).

[459] Recall the ubiquity of adamantane syntheses from other hydrocarbons; see Section 4.A.

cations: Taking a composite of theoretical results, we find the stability of the cyclopropyl species increases in the order H, $CH_3$, F, OH, $NH_2$, and $O^-$.[460a,b]

## 5.E.2 SUBSTITUTED ETHYLENES AND CYCLOPROPANES

In a simplistic manner, the reader will recall that cyclopropanes enjoy some ethylenic character (Chapter 2), and therefore olefinic substituent effects might be expected to be reflected, if dampened, in the cyclopropane series. Reasoning that an $sp^2$-hybridized carbon is more electronegative than an $sp^3$-hybridized carbon, one would predict that electron-withdrawing substituents would destabilize olefins. Alternatively, using Bent's logic,[461] one would say that the carbon orbital directed to an electronegative substituent should be of high p character, in opposition to the natural requirement of relatively high s character in the olefinic carbon orbital. Unfortunately, such uncomplicated substituent effects are not observed. On the basis of studies of the equilibrium represented by Eq. (99), a list of double-bond stabilization parameters (DBSP) has been tabulated including —$OCH_3$ (+5.2 kcal/mole), —F (3.3 kcal/mole), —$CH_3$ (3.2 kcal/mole), —CN (2.3 kcal/mole), —Cl (1.8 kcal/mole), and —H (0.0 kcal/mole).[462] When two methyl groups are attached to ethylene, the total DBSP is about equal to the sum of the individual contributions. However, 1,2-dimethoxyethylene is actually destabilized relative to ethylene even though methoxyethylene is considerably stabilized.[463]

$$
\begin{array}{c}
H \qquad\qquad Y \\
\diagdown \qquad\qquad \diagup \\
C{=}C \qquad \rightleftarrows \qquad C{=}C \\
\diagup \qquad\qquad \diagdown \\
XCH_2 \qquad\qquad H \qquad\qquad H \qquad\qquad CH_2Y
\end{array}
\qquad (99)
$$

There are very few structural and even fewer thermochemical data on substituted cyclopropanes; Hoffmann, through reference to the Walsh description of cyclopropane (Fig. 2.3), predicted that $\pi$-acceptor substituents at the 1- position of a cyclopropane would strengthen and shorten the $C_2$—$C_3$ bond while weakening and lengthening adjacent bonds.[464a,b] (Note that cyclopropanone has been treated as a cyclopropane disubstituted with

[460a] L. Radom, J. A. Pople, and P. von R. Schleyer, *J. Am. Chem. Soc.* **95**, 8193 (1973). (X = H, $CH_3$, F, OH, $NH_2$).

[460b] J. F. Liebman and A. Greenberg, *J. Org. Chem.* **39**, 123 (1974). (X = $O^-$)

[461] H. A. Bent, *J. Chem. Phys.* **33**, 1258, 1259, 1260 (1960).

[462] J. Hine, "Structural Effects on Equilibria in Organic Chemistry," pp. 270–276. Wiley, New York, 1975.

[463] E. Taskinen, *Tetrahedron* **32**, 2327 (1976).

[464a] R. Hoffmann, *Tetrahedron Lett.* p. 2907 (1970).

[464b] R. Hoffmann and W.-D. Stohrer, *J. Am. Chem. Soc.* **93**, 6941 (1971).

"O".[464c]) This prediction has been supported by a number of experimental findings (see compounds **270–276** in Table 5.3), including an x-ray crystallographic structure of 7,7-dicyano-2,5-dimethylnorcaradiene.[465] Note that no prediction is made about the overall relative stability of the ring. Hoffmann's theoretical description predicts that $\pi$-donating substituents should lengthen all three bonds, and here the experimental findings have not been supportive. 1,1-Dichlorocyclopropane (**279**, Table 5.3) does, however, manifest this effect, and isodesmic-type calculations show that geminal dichloro groups do appear to destabilize the ring (by $+5.2$ kcal/mole). However, 1-chlorocyclopropane's ring dimensions (see **278**, Table 5.3) are about the same as those in cyclopropane. Lengthening of the $C_1$—$C_2$ bond with no change of $C_2$—$C_3$ in cyclopropylamine (**280**, Table 5.3) is rationalized by the investigators through a picture of $\pi$ donation to an antibonding orbital of cyclopropane other than the $1A_2'$, to which the lone pair is orthogonal (see Fig. 2.3). 1,1-Difluorocyclopropane (**277**, Table 5.3) departs markedly from geometry predicted for a ring substituted by $\pi$ donors or $\pi$ acceptors. This same feature is also evident in 3,3-difluorocyclopropene, in which $C_1$—$C_2$ is longer by 0.032 Å and $C_2$—$C_3$ is shorter by 0.071 Å than the corresponding bonds in cyclopropene.[466] Aside from the 1,1-dichloro compound, most substituents other than F and Li seem to have minimal effects upon the thermodynamic stability of the cyclopropane ring. The strain in 1,1-dichlorocyclopropane may be equated to $32$–$33 \pm 3$ kcal/mole. The thermodynamic stability of hexafluorocyclopropane appears to deviate markedly from the hydrocarbon, and this will be covered in Section 5.E.3.

Interestingly enough, bridgehead substitution of strong $\pi$ acceptors on bicyclobutane appears to measurably stabilize this ring system relative to acyclic analogues (Table 5.4). The acrylonitrile comparison in Table 5.5 would appear to suggest the presence of a ring-strain-reducing effect, rather than an exocyclic hybridization effect. A rationalization of this stabilizing effect might be that cyano withdraws electrons from the HOMO of bicyclobutane, a bonding $\pi$ orbital across the central bond, and lengthens this bond. It is well known that this bond is short (1.497 Å, see Section 3.B.1) and that bicyclobutane has an increment of more than 10 kcal/mole extra strain compared to two cyclopropanes. If the extra increment is the result of this short fusion bond, then lengthening of the bond, perhaps to normal cyclopropane length, might remove the increment.

There are interesting chemical consequences arising from substituent effects as noted in Eqs. (100)–(105), which support the general views stated above

[464c] C. A. Deakyne, L. C. Allen, and V. W. Laurie, *J. Am. Chem. Soc.* **99**, 1343 (1977).
[465] C. J. Fritchie, Jr., *Acta Crystallogr.* **20**, 27 (1966).
[466] K. R. Ramaprasad, V. W. Laurie, and N. C. Craig, *J. Chem. Phys.* **64**, 4832 (1976).

**Table 5.3**

**Geometries and Thermochemical Data for Some Substituted Cyclopropanes**

| X, Y | $r_{12}$ (Å) | $r_{23}$ (Å) | $\Delta\Delta H_1$(kcal/mole)(A) ($c$-$C_3H_4XY$ − $c$-$C_3H_6$) | $\Delta\Delta H_2$(kcal/mole)(B) ($CH_3CXYCH_3$ − $C_3H_8$) | Diff. = (A − B) |
|---|---|---|---|---|---|
| X = Y = H | 1.510[a] | 1.510[a] | | | |
| | 1.514[b] | 1.514[b] | | | |
| | 1.528[c] | 1.500[c] | | | |
| | 1.535 (av.)[f] | 1.485[e] | +30.4[d] | +31.0 | −0.6 |
| 270, X = H; Y = CN | | 1.462[f] | | | |
| 271, X = Y = CN | | 1.500[g] | | | |
| 272, X = Y = $CO_2H$ | | 1.492[g] | | | |
| 273, X = H; Y = $NO_2$ | | 1.488[g] | | | |
| 274, X = H; Y = $CO_2H$ | | 1.490[g] | | | |
| 275, X = H; Y = $COCH_3$ | | | | | |
| 276, X = H; Y = CHO | | | | | |
| 277, X = Y = F | 1.464[h] | 1.553[h] | | | |
| 278, X = H; Y = Cl | 1.513[i] | 1.515[i] | | | |
| 279, X = Y = Cl | 1.532[i] | 1.534[i] | −11.4 ± 0.5[k] | −16.6 ± 2.5 | +5.2 |
| 280, X = H; Y = $NH_2$ | 1.535[i] | 1.513[i] | +5.4[m] | +4.8 | +0.6 |
| 281, X = H; Y = $CH_3$ | | 1.510[g] | −6.5 | −7.3 | +0.8 |

[a] O. Bastiansen, F. N. Fritsch, and K. Hedberg, *Acta Crystallogr.* **17**, 538 (1964).

[b] W. J. Jones and B. P. Stoicheff, *Can. J. Phys.* **42**, 2259 (1964).

[c] R. Pearson, Jr., A. Choplin, and V. W. Laurie, *J. Chem. Phys.* **62**, 4859 (1975).

[d] H. K. Hall, Jr. and J. H. Baldt, *J. Am. Chem. Soc.* **93**, 140 (1971).

[e] R. Pearson, Jr., A. Choplin, V. W. Laurie, and J. Schwartz, *J. Chem. Phys.* **62**, 2949 (1975).

[f] M. A. M. Meester, H. Schenk, and C. H. MacGillary, *Acta Crystallogr.*, *Sect. B* **27**, 630 (1971).

[g] R. E. Penn and J. E. Boggs, *J. Chem. Soc., Chem. Commun.* p. 666 (1972).

[h] A. T. Perretta and V. W. Laurie, *J. Chem. Phys.* **62**, 2469 (1975).

[i] R. H. Schwendeman, G. D. Jacobs, and T. M. Krigas, *J. Chem. Phys.* **40**, 1022 (1964).

[j] W. Flygare, A. Narath, and W. Gwinn, *J. Chem. Phys.* **36**, 200 (1962).

[k] $\Delta H_f°$(1,1-dichlorocyclopropane, liq.) = −6.5 kcal/mole [V. P. Kolesov, E. M. Tomareva, V. M. Shostakovskii, O. M. Nefedov, and S. M. Skuratov, *Zh. Fiz. Khim.* **44**, 1548 (1970); *Chem. Abstr.* **73**, 55506g (1970)]. Employed Trouton's rule (b.p. = 75°C) to obtain $\Delta H_f°$(g) = +1.3 ± 0.5 kcal/mole.

[l] M. D. Harmony, R. E. Bostrom, and D. K. Hendrickson, *J. Chem. Phys.* **62**, 1599 (1975).

[m] $\Delta H_f°$(aminocyclopropane, liq.) = +10.95 kcal/mole [W. D. Good and R. T. Moore, *J. Chem. Thermodyn.* **3**, 701 (1971)]. Employed Trouton's rule (b.p. = 50°C) to obtain $\Delta H_f°$(g) = +18.1 kcal/mole. Standard enthalpy of formation of methylcyclopropane [$\Delta H_f°$(g) = +6.25 kcal/mole obtained from published standard enthalpy of combustion; W. D. Good, *J. Chem. Thermodyn.* **3**, 539 (1971) and application of Trouton's rule (b.p. 5°C)].

## Table 5.4

**Thermochemical Effects of Substituents at the Bridgehead Positions of Small Bicyclo[n.1.0]alkanes on Ethylene and Cyclopropane***

| | $\Delta H_f$(kcal/mole)[a] | | | | $\Delta H_f$(kcal/mole) | | |
| Group | Group X | Group H | Diff. A | Model | Model X | Model H | Diff. B |
|---|---|---|---|---|---|---|---|
| H₂C=CH— | | | | | | | |
| △ | (X = CN)42.95[b] | 12.50 | (30.5) | Et- | (X = CN) 12.4[c] | -20.24 | (32.6) |
| 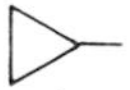 ◁▷ | (X = CN)43.17[b] | 12.74 | (30.4) | i-Pr- | (X = CN) 5.62 | -24.83 | (30.5) |
| 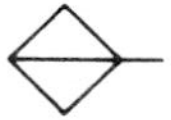 ◁▷ | (X = CN)72.78[b] | 51.90 | (20.9) | t-Bu | (X = CN) -0.79[b] | -32.41 | (31.6) |
| ◁▷ | (X = CO₂CH₃)-39.34[b] | 51.90 | (-91.2) | t-Bu | (X = CO₂CH₃)-118.16[b] | -32.41 | (-85.8) |
| 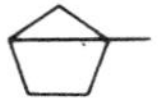 ⬠ | (X = CN)65.02[b] | 37.6 | (27.4) | t-Bu | (X = CN)-0.79[b] | -32.41 | (31.6) |
| 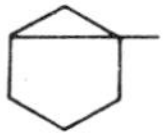 ⬡ | (X = CN)33.95[b] | 9.07 | (24.9) | t-Bu | (X = CN)-0.79 | -32.41 | (31.6) |

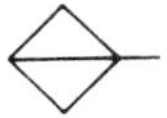

* If Diff. A is less (or more negative) than Diff. B, the substituent is said to stabilize the olefinic, cyclic, or bicyclic group to which it is attached.

[a] Unless noted otherwise, $\Delta H_f$ values taken from S. W. Benson, F. R. Cruickshank, D. M. Golden, G. R. Haugen, H. E. O'Neal, A. S. Rodgers, R. Shaw, and R. Walsh, *Chem. Rev.* **69**, 279 (1969).

[b] H. K. Hall, Jr. and J. H. Baldt, *J. Am. Chem. Soc.* **93**, 140 (1971).

[c] Estimated value, see ref. *a* above.

329

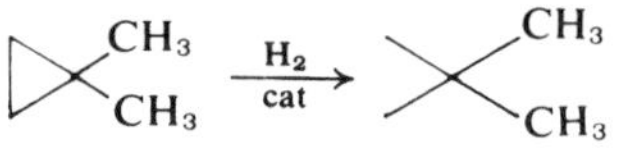

$$\text{(100)}$$

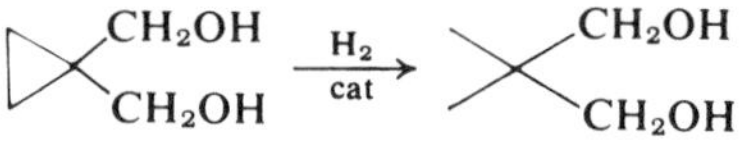

$$\text{(101)}$$

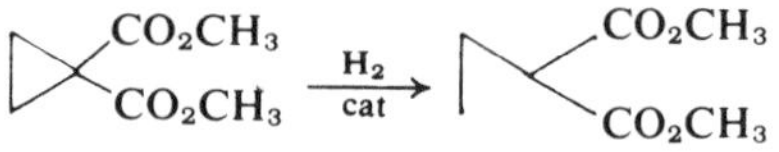

$$\text{(102)}$$

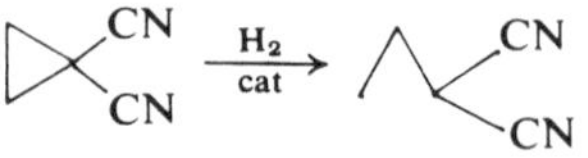

$$\text{(103)}$$

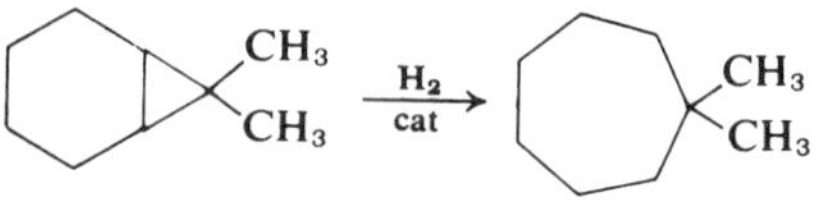

$$\text{(104)}$$

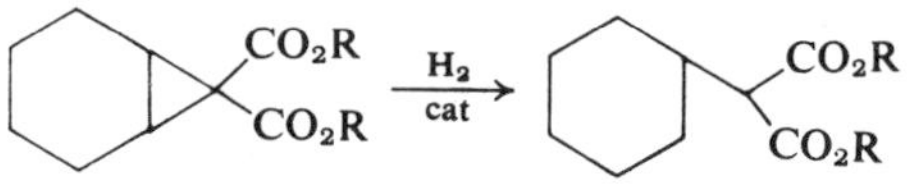

$$\text{(105)}$$

(all these reactions were performed at relatively low temperatures to mitigate loss of selectivity).[467-469] According to Hoffmann's predictions, $C_1$—$C_2$ in 1,2-dicyanocyclopropane should be particularly weak while the remaining two bonds should be normal. While this compound is inert to Pt(0), 1,1,2,2,-tetracyanocyclopropane readily forms an insertion product [Eq. (106)].

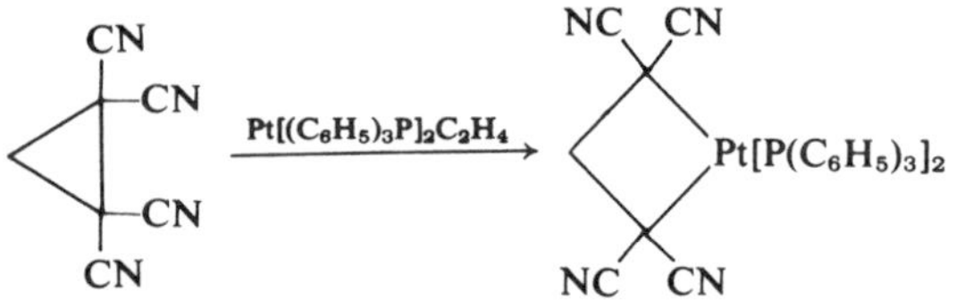

$$\text{(106)}$$

Because the dicyano compound does not react, and, further, a 1,1,2,2-tetra-cyanocyclobutane derivative also inserts, release of strain alone would not appear to be the major driving force. It might also be argued that the weakened substituted strained bond is the cause. However, here it would seem that the four electron-withdrawing cyano groups function to make the moiety to which they are attached more susceptible to relatively nucleophilic metal species, such as the one above.[470a] Curiously, the relatively low barrier for

[467] C. Groger and H. Musso, *Angew. Chem., Int. Ed. Engl.* **15**, 373 (1976).

[468] W. Cocker, P. V. R. Shannon, and P. A. Staniland, *Chem. Commun.* p. 254 (1965).

[469] H. Musso, *Chem. Ber.* **101**, 3710 (1968).

[470a] R. Ros, M. Lenarda, N. B. Pahor, M. Calligaris, P. Delise, L. Randaccio, and M. Graziani, *J. Chem. Soc., Dalton Trans.* p. 1937 (1976).

isomerization of *cis*- to *trans*-1,2-dicyanocyclopropane has been explained only in terms of stabilization of the transition state, not as the result of a weakened $C_1$—$C_2$ bond.[470b]

It has been argued that electronegative substituents might simply inductively withdraw electron density from bonding cyclopropane orbitals and thus increase C—C bond lengths, and experimental data supporting this view have been published.[471] Some of the data in Table 5.2 are consistent with this idea. However, the C—C bond lengths in hexafluorocyclopropane (1.505 Å) are actually shorter than those in cyclopropane.[472] The strain in hexafluorocyclopropane appears to be more than twice as high as in cyclopropane.[473] There is, however, a reasonable degree of uncertainty in the standard value for a —$CF_2$— increment. While MINDO/3 calculations[474] suggest that hexafluorocyclopropane is half as strained as cyclopropane, problems here include calculational difficulties with fluorines and the implicit assumption that tetrafluoroethylene is as "normal" as ethylene, which it is not.

Finally, we note that, in contrast to various acrylic esters, dimethyl-1,1-cyclopropane dicarboxylate requires vigorous, prolonged conditions in order to undergo Michael addition. However, forced conjugation in **282** augmented perhaps by some spiro-induced lengthening of the $C_2$—$C_3$ bond, make this compound a reactive synthon.[475]

$$(107)$$

**282**

Further support for Hoffmann's predictions about the effects of $\pi$-acceptor substituents in the stabilities of three-membered rings can be obtained through examination of substituted cycloheptatriene (tropylidene)-norcaradiene systems, semibullvalenes, and bullvalenes.[476] Cycloheptatriene exhibits no detectable norcaradiene isomer, but the 7,7-dicyano derivative exists entirely in the latter form (**283**).[477] Here, shortening of the 1,6 bond (bond opposite the carbon bearing the substituents) stabilizes the norcaradiene structure. It is interesting to note briefly here that the electronegativity of the

[470b] W. von E. Doering, G. Horowitz, and K. Sachdev, *Tetrahedron* **33**, 279 (1977).

[471] J. W. Lauher and J. A. Ibers, *J. Am. Chem. Soc.* **97**, 561 (1975).

[472] J. F. Chiang and W. A. Bernett, *Tetrahedron* **27**, 975 (1971).

[473] W. A. Bernett, *J. Org. Chem.* **34**, 1772 (1969).

[474] R. C. Bingham, N. J. S. Dewar, and D. H. Lo, *J. Am. Chem. Soc.* **97**, 1307 (1975).

[475] S. Danishefsky and R. K. Singh, *J. Am. Chem. Soc.* **97**, 3239 (1975).

[476] L. A. Paquette and W. E. Volz, *J. Am. Chem. Soc.* **98**, 2910 (1976) and references cited therein.

[477] E. Ciganek, *J. Am. Chem. Soc.* **87**, 652 (1965).

$\rangle$C(CN)$_2$ group is about equal to that of —O— and that, as mentioned in Section 5.D.2.a, the norcaradienelike structure is significantly stabilized in this system. (Cyano groups cannot, of course, act as lone-pair electron donors, as divalent oxygen can[478].) The monocyano derivative exists entirely in the cycloheptatriene form 284,[479] while 285b, detected by nmr, is about 0.9 kcal ($\Delta G^0_{-150°C}$) less stable than 285a.[480] While the 7,7-bistrifluoromethyl compound is seen only in the cycloheptatriene structure 286,[481] both structures are seen for the 7-cyano-7-trifluoromethyl derivative 287.[482] The strongly

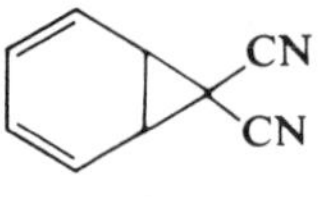

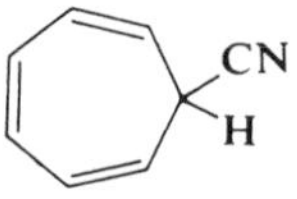

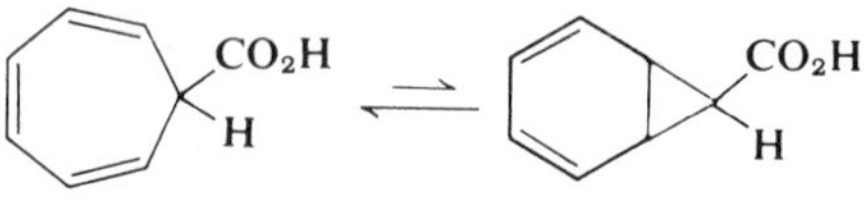

π-withdrawing cyano group causes semibullvalene tautomer 288a to be favored over 288b in solution ($K_{eq} = 2.05 \times 10^{-2}$).[476] In the crystalline form, only 288a is observed. The cyclopropane ring bond opposite the carbon bearing the cyano group is somewhat shortened relative to the corresponding bond in semibullvalene, as anticipated, but the adjacent bonds are also shortened.[276] Nmr observations on the piperidino compound 289 provide the conclusion that tropylidene, and norcaradiene isomers are of comparable stability, in contrast to Hoffmann's predictions[464a,b] for an electron-releasing substituent.[483a] However, 5-methoxysemibullvalene's structure (analogous to 288b) supports these predictions.[483b]

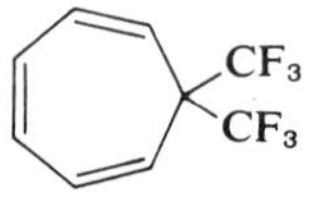

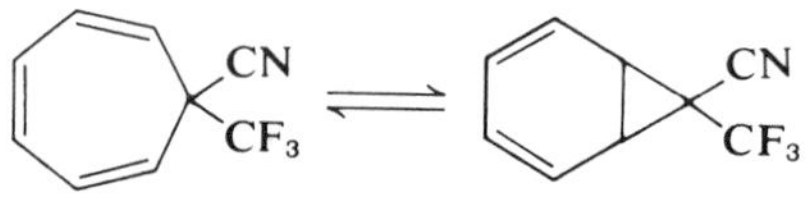

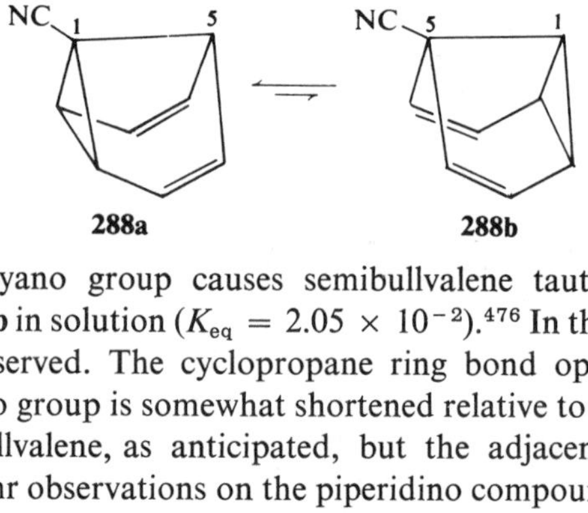

[478] K. Wallenfels, K. Friedrich, J. Rieser, W. Ertel, and H. K. Thieme, *Angew. Chem., Int. Ed. Engl.* **15**, 260 (1976).

[479] C. H. Bushweller, M. Sharpe, and S. J. Weininger, *Tetrahedron Lett.* p. 453 (1970).

[480] R. Wehner and H. Guenther, *J. Am. Chem. Soc.* **97**, 923 (1975).

[481] D. M. Gale, W. J. Middleton, and C. G. Krespan, *J. Am. Chem. Soc.* **88**, 3617 (1966).

[482] E. Ciganek, *J. Am. Chem. Soc.* **87**, 1149 (1965).

[483a] S. W. Staley, M. A. Fox, and A. Cairncross, *J. Am. Chem. Soc.*, **99**, 4524 (1977).

[483b] R. W. Hoffmann, N. Havel, and F. Frickel, *Angew. Chem., Int. Ed. Engl.*, **16**, 475 (1977).

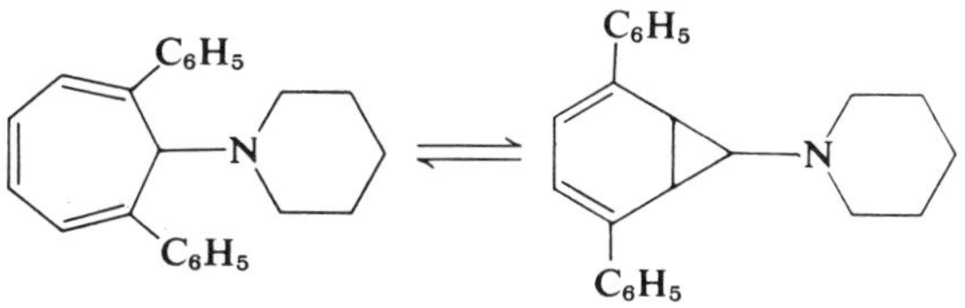

289a          289b

### 5.E.3 FLUORO- AND PERFLUOROALKYL SUBSTITUTION

As a substituent, fluorine is a superficially simple case because its steric requirements are about the same as those of hydrogen[484] and it should be involved in relatively little conjugative interactions with $\pi$ systems.[485] However, as the most electronegative element, fluorine often exerts a striking influence on the properties of systems to which it is attached. For example, 1-norbornene (290), a highly strained bridgehead olefin (Section 3.E.3), is seemingly stabilized by fluorine substitution at the other bridgehead.[486] Resonance contributors may be drawn in which fluorine stabilizes the bridgehead electrostatically (290b), conjugatively (290c), or hyperconjugatively (290d). Studies of the acidities of various fluorinated bicyclic compounds have

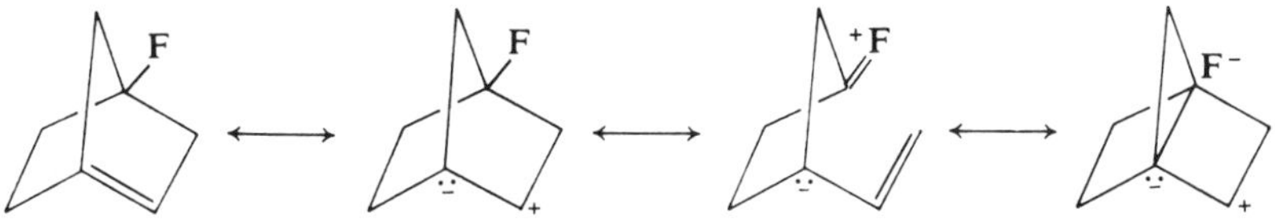

290a          290b          290c          290d

led Streitwieser *et al.* to discount the importance of such structures as 290d and to argue that electrostatic stabilization is the major interaction.[487] Varying the bridgehead substituent from F to $CH_3O$, $(CH_3)_2N$, and aryl might prove useful in the determination of the relative importance of the various resonance structures.

Thermochemical data on perfluorinated molecules generally are lacking. We recall that a reference $-CF_2-$ has been constructed [$\Delta H_f^0$ ($-CF_2-$) = $-98.1$ kcal/mole] which has allowed estimates of strain in tetrafluoroethylene (41.2 kcal/mole), hexafluorocyclopropane (68.7 kcal/mole), and octafluorocyclobutane (32.0 kcal/mole).[473] We again remind the reader that the C—C

[484] W. A. Sheppard and C. M. Sharts, "Organic Fluorine Chemistry," pp. 17–40, Benjamin, New York, 1969, present a thorough treatment of the C—F bond including a subsection on the CF₃ group and C—CF₃ bonding.

[485] P. Politzer and J. W. Timberlake, *J. Org. Chem.* **37**, 3447 (1972).

[486] S. F. Campbell, J. M. Leach, J. C. Tatlow, and K. N. Wood, *J. Fluor. Chem.* **1**, 103 (1971).

[487] A. Streitwieser, Jr., D. Holtz, G. R. Ziegler, J. O. Stoffer, M. L. Brokaw, and F. Guibé, *J. Am. Chem. Soc.* **98**, 5229 (1976).

bonds in hexafluorocyclopropane (1.505 Å) are shorter than those in the hydrocarbon.[472] It has been suggested that the $\pi$ bond in $C_2F_4$ is relatively weak in comparison to that in $C_2H_4$, and that geminal fluorines do not "like" being on a formally sp²-hybridized carbon, as they appear to force F—C—F angles close to 109°.[473] However, removal of a $\pi$ electron from $C_2F_4$ requires approximately the same energy as removal from $C_2H_4$.[488] This would naïvely suggest that $C_2F_4$ should be less strained than $C_2H_4$ because its $\pi$ bond is comparable and its $\sigma$ bond stronger than those in $C_2H_4$. The reader may recall similar arguments comparing $C_2H_2$, $C_2H_4$, and $C_2H_6$ in Chapter 2. Common to both cases is a sense of frustration with difficulties in employing ionization-potential logic in this manner. We note briefly additional seeming anomalies involving fluorine, including the dominance of the bridgehead isomer in fluorobullvalene (291) as opposed to the chloro- and bromobullvalenes 292 and 293 in which the halogens are at vinylic positions,[489] and the

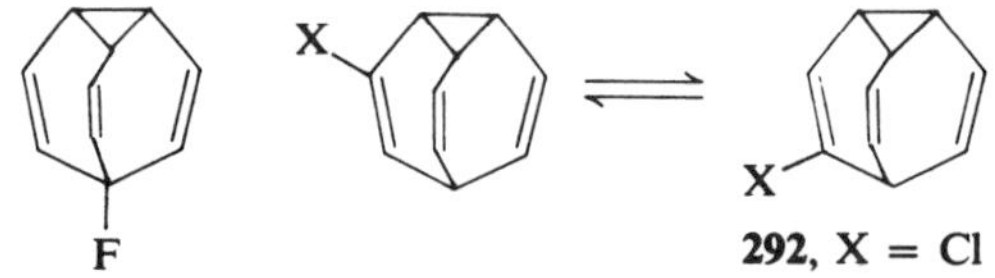

F
**291**

**292**, X = Cl
**293**, X = Br

F   F
**294**

X   X
**295**, X = Cl
**296**, X = Br

contrast in the structure of 294 vs. 295 and 296.[490] Table 5.5 presents some differences between hydrocarbon and fluorocarbon chemistry, and we

[488] This phenomenon occurs for numerous unsaturated species that are mostly or exhaustively fluorinated, perfluorinated and thus goes under the name "The Perfluoro Effect:" C. R. Brundle, M. B. Robin, N. A. Kuebler, and H. Basch, *J. Am. Chem. Soc.* **94**, 1451 (1972); C. R. Brundle, M. B. Robin, and N. A. Kuebler, *J. Am. Chem. Soc.* **94**, 1466 (1972). However, we believe it is a characteristic of linear or planar species with occupied $\pi$ orbitals and general fluorine substitution (J. F. Liebman and P. Politzer, unpublished results). For example, the $\pi$ ionization potentials of HF and $F_2$ are 16.01 and 15.69 eV, respectively; see J. Berkowitz, W. A. Chupka, P. Guyon, J. Holloway, and R. Spohr, *J. Chem. Phys.* **54**, 5165 (1971).

[489] J. F. M. Oth, R. Merenyi, H. Rottle, and G. Schroder, *Tetrahedron Lett.* p. 3941 (1968).

[490] V. Rautenstrauch, H.-J. Scholl, and E. Vogel, *Angew. Chem., Int. Ed. Engl.* **7**, 288 (1968).

## Table 5.5

### Some Contrasts between Hydrocarbon and Fluorocarbon Chemistries

| Carbon skeletons | | Relative stability | | |
| --- | --- | --- | --- | --- |
| Compd. A | Compd. B | H case | F case | Ref. |
| C≡C—C | C—C≡C | A < B | A > B | a |
| 2C≡C | □ | A < B | A < B | b |
| ▷ (methylenecyclopropane) | △ | A < B | A > B | c |
| C≡C—C≡C | □ | A > B | A < B | d |
| C=C—C=C | C—C≡C | A < B | A > B | e |
| C—C=C—C | C=C=C | A > B | A < B | e |
| C—C=C—C | C=C—C | A > B | A < B | e |
| C—C=C—C | C—C≡C | A > B | A > B | f |
| (methylenecyclobutane) | □ | A < B | A < B | g |
| C—C—C—C—C | □□ | A > B | A < B | h |

[a] R. E. Banks, M. G. Barlow, W. D. Davies, and R. N. Haszeldine, J. Chem. Soc. C p. 1104 (1969).

[b] The reader should recall the earlier strain energy per $CH_2$ vs. per $CF_2$ discussion of ethylene, cyclopropane, and cyclobutane and their perfluorinated analogues.

[c] B. E. Smart, J. Am. Chem. Soc. 96, 927 (1974).

[d] J. L. Anderson, R. E. Putnam, and V. H. Sharkey, J. Am. Chem. Soc. 83, 382 (1961). These authors additionally showed that $F_2C$=CHCH=$CF_2$ is less stable than the cyclobutene. In addition, $F_2C$=C($CF_3$)C($CF_3$)=$CF_2$ is almost thermoneutral with respect to isomerization to the cyclobutene: J. P. Chesick, J. Am. Chem. Soc. 88, 4800 (1966).

[e] W. T. Miller, W. Frass, and P. R. Resnick, J. Am. Chem. Soc. 83, 1767 (1961).

[f] R. E. Banks, A. Braitwaite, R. N. Haszeldine, and D. R. Taylor, J. Chem. Soc. C p. 454 (1969).

[g] Dr. Bruce Smart, personal communication. The same trend is also found for H and F cases of methylenecyclopentane and 1-methylcyclopentene (footnote f).

[h] A. H. Fainberg and W. T. Miller, J. Am. Chem. Soc. 79, 4170 (1957).

recommend Sheppard and Sharts,[484] Banks,[491] and Chambers[492] as thorough reviews of organofluorine chemistry.

We noted in Section 5.B.2 that $CF_3$ groups stabilize the strained hexaperfluoromethylbenzene valence isomers relative to the "Kekulé" or normal $(CCF_3)_6$.[493] In fact, at temperatures over 280°C, "Dewar $(CC_2F_5)_6$" is more stable than hexaperfluoroethylbenzene.[494a,b] The "perfluoroalkyl effect,"[493] in which groups such as $CF_3$ appear to provide both thermodynamic and kinetic stability to strained rings, has provided some striking departures from hydrocarbon chemistry. For example, hexakis(perfluoromethyl)-3,3'-bicyclopropenyl, the thermodynamically least stable $(CCF_3)_6$, is the most stable kinetically, aromatizing with a half-life of more than two hours at 360°C![495] Octaperfluoromethylcubane is greatly stabilized and octaperfluoromethylcyclooctatetraene amazingly stabilized compared to their hydrocarbon counterparts.[496] Perfluorotetramethyl (Dewar thiophene) is also anomalously stable, having a half-life for aromatization of 5.1 hours at 160°C, in contrast to bicyclo[2.1.0]pent-2-ene, which rearranges rapidly ($t_{1/2} = 4$ h at 34°C) to the nonaromatic compound cyclopentadiene.[497a,b]

We end this discussion by noting a remarkable reaction reported to generate a 1,4-trimethylene-bridged Dewar benzene [**297**, see Eq. (108)] via trimerization of a highly fluorinated cyclopentyne.[498] The $^{13}C$ nmr data cited, however, are somewhat ambiguous in their support of this structure (actually one of three possible isomers) for the $C_{27}F_{42}$ fluorocarbon found.[498]

$$(108)$$

**297**

[491] R. E. Banks, "Fluorocarbons and Their Derivatives," 2d Ed. McDonald Tech. Sci., London, 1970.

[492] R. D. Chambers, "Fluorine in Organic Chemistry." Wiley, New York, 1973.

[493] D. M. Lemal and L. H. Dunlap, Jr., *J. Am. Chem. Soc.* **94**, 6562 (1972).

[494a] E. D. Clifton, W. T. Flowers, and R. N. Haszeldine, *J. Chem. Soc., Chem. Commun.* p. 1216 (1969).

[494b] A.-M. Debbagh, W. T. Flowers, R. N. Haszeldine, and P. J. Robinson, *J. Chem. Soc., Chem. Commun.* p. 323 (1975).

[495] M. W. Grayston and D. M. Lemal, *J. Am. Chem. Soc.* **98**, 1278 (1976).

[496] L. L. Pelosi and W. T. Miller, *J. Am. Chem. Soc.* **98**, 4311 (1976).

[497a] C. Y. Kobayashi, I. Kumadaki, A. Ohsawa, and Y. Sekine, *Tetrahedron Lett.* p. 1639 (1975).

[497b] C. H. Bushweller, J. A. Ross, and D. M. Lemal, *J. Am. Chem. Soc.* **99**, 629 (1977).

[498] B. L. Dyatkin, N. I. Delyagin, E. I. Mysov, and J. L. Kunyants, *Tetrahedron* **30**, 4031 (1974).

### 5.E.4 LITHIUM AS A SUBSTITUENT

With the possible exception of alkyl groups, all the substituents considered up to this point clearly act as $\sigma$ acceptors. Withdrawal of $\sigma$ electrons from Walsh orbital $\psi_1$ (Fig. 2.3) decreases bonding character in a cyclopropane ring and thus increases its destabilization energy or strain.[499] Employing this point of view, the high strain energy in hexafluorocyclopropane is explicable in terms of $\sigma$ withdrawal as the dominant substituent effect, but the basis of the perfluoroalkyl effect remains ambiguous. However, what would be the effect of a strong $\sigma$ donor on the ring strain of cyclopropane as well as on the thermochemical properties of other strained molecules? Lithium is a $\sigma$ donor and $\pi$ acceptor, and the discussion below will provide evidence supporting substantially decreased strain in lithiated derivatives.[499]

Lithium's unusual properties as a substituent are reflected in the structures of a number of lithiated compounds. For example, experimental[500] and theoretical[501] studies of $Li_2O$ find this molecule to be linear instead of exhibiting the bent geometry of $H_2O$. While HF dimer is known to be bridged by a single hydrogen, LiF dimer is bridged by both lithiums.[502,503] Dilithio-acetylene[504] is also calculated to be doubly bridged (298).[505a] The known compound $C_3Li_4$ is calculated to possess an extraordinary triply bridged structure.[505b] Lithium substituents are predicted to stabilize planar tetracoordinate carbon, and $CLi_2F_2$, 1,1-dilithiocyclopropane, as well as 3,3-dilithiocyclopropene, are all calculated to favor a planar tetracoordinate carbon over the tetrahedral-type structure.[506] (See Chapter 6 for a much more detailed discussion of bonding in hypothetical compounds which contain planar tetracoordinate carbon.) We note that some experimental studies of polylithiated acyclic hydrocarbon species have been published recently[507,508a,b], but these have not included structural data.

[499] Dr. J. D. Dill, personal communication to Arthur Greenberg and Joel F. Liebman.

[500] A. Buchler, J. L. Stauffer, W. Klemperer, and L. Wharton, *J. Chem. Phys.* 39, 2299 (1963).

[501] R. J. Buenker and S. D. Peyerimhoff, *J. Chem. Phys.* 45, 3682 (1966).

[502] M. J. Linevsky, *J. Chem. Phys.* 34, 587 (1961).

[503] P. A. Kollman, J. F. Liebman, and L. C. Allen, *J. Am. Chem. Soc.* 92, 1142 (1970).

[504] H. H. Inhoffen, H. Pommer, and E. G. Meth, *Justus Liebigs Ann. Chem.* 565, 45 (1949).

[505a] Y. Apeloig, P. von R. Schleyer, J. S. Binkley, J. A. Pople, and W. L. Jorgensen, *Tetrahedron Lett.* p. 3293 (1976).

[505b] E. D. Jemmis, D. Poppinger, P. von R. Schleyer, and J. A. Pople, *J. Am. Chem. Soc.* 99, 5696 (1977).

[506] J. B. Collins, J. D. Dill, E. D. Jemmis, Y. Apeloig, P. von R. Schleyer, R. Seeger, and J. A. Pople, *J. Am. Chem. Soc.* 98, 5419 (1976).

[507] E. K. S. Liu and R. J. Lagow, *J. Am. Chem. Soc.* 98, 8270 (1976).

[508a] W. Priester, R. West, and T. L. Chwang, *J. Am. Chem. Soc.* 98, 8413 (1976).

[508b] W. Priester and R. West, *J. Am. Chem. Soc.* 98, 8421, 8426 (1976).

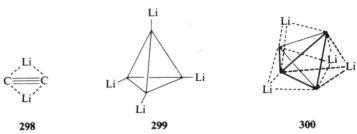

**298**                    **299**                    **300**

*Ab initio* calculations (via the Gaussian 70 package)[499,500] support the qualitative description of lithium as a strain-reducing substituent. Thus, the calculated (corrected) strain energy of lithiocyclopropane is only 18.5 kcal/mole, almost 10 kcal/mole less than that of the parent. While the strain energy in bicyclobutane is about 66.5 kcal/mole, the values in the 2-lithio and 1-lithio (bridgehead) derivatives are 11.5 kcal/mole and 25.0 kcal/mole lower, respectively.[509] Bicyclobutanes are known to be lithiated by butyllithium at the bridgehead in preference to the bridge,[510,511] reflecting the high s character of the exocyclic bridgehead carbon orbital, and the above-cited results are consistent with this property. 1,3-Dilithiobicyclobutane is calculated to have only about 24 kcal/mole of strain. Similarly, the strain in lithiospiropentane is predicted to be more than 10 kcal/mole lower than that in spiropentane, and the 1,3-dilithio derivative some 18 kcal/mole lower. Lithiotetrahedrane is calculated to be about 37 kcal/mole less strained than the parent hydrocarbon, and successive substitution of hydrogen by lithium reduces strain, finally reaching a value of only about 26–27 kcal/mole in tetralithiotetrahedrane. It has thus been proposed that tetralithiotetrahedrane may be a reasonable synthetic target, perhaps via photochemical dimerization of $C_2Li_2$ as well as a variety of carbene routes.[499] The geometry of tetralithiotetrahedrane is calculated as "face-lithiated" (**300**) in preference to "corner-lithiated" **299**.[509] In order to further investigate the isolability of **300** a detailed study of the $C_4Li_4$ energy surface must be undertaken.[499]

The *ab initio* calculations cited above also indicate that 1-fluorocyclopropane and 1,2,3-trifluorocyclopropane are slightly less strained and slightly more strained, respectively, than cyclopropane itself.[509]

**5.E.5**   CARBONIUM IONS AND CARBANIONS: ARE VACANT ORBITALS
AND LONE PAIRS IDEALIZED SUBSTITUENTS?

A novel way of describing a carbonium ion would be to consider the vacant orbital as an idealized electron-withdrawing ($\pi$ and $\sigma$) substituent.

[509] P. von R. Schleyer and J. D. Dill, unpublished calculations.
[510] G. L. Closs and R. B. Larabee, *Tetrahedron Lett.* p. 287 (1965).
[511] H. M. Cohen. *J. Organomet. Chem.* **9**. 375 (1967).

Similarly, the lone pair in a carbanion would be treated as an idealized electron-releasing ($\pi$ and $\sigma$) substituent.[512] Carbanions in which a metal ion such as $Li^+$ would play merely the role of solvated counterion are different from the lithiated compounds discussed in the previous section, where lithium plays an active role as $\sigma$ donor and $\pi$ acceptor. However, extreme caution should be exercised with this approach, since carbonium ions and carbanions are only tricoordinate. Furthermore, while bond-angle constraints keep the geometries of norbornane, 1-norbornyl cation, and 1-norbornyl anion fairly similar, one cannot have the same expectation for cyclopropane and its ions.

We remind the reader of the seemingly straightforward relationship between $J(^{13}C—H)$ and acidity: "highly strained rings have a proclivity commensurate with the degree of internal stresses present for acidity . . . and large $J(^{13}C—H)$ coupling constants."[513] Although there are conceptual problems with and empirical departures from this statement, as will be discussed below, we believe it is a good general regularity to be used with appropriate caution. Its basic utility arises because an exocyclic carbon orbital of abnormally high s character will hold lone pairs of electrons especially well. However, despite the same formal C—H hybridization, cubane is $10^3$ times as acidic as cyclopropane.[514] What tacitly is assumed in the above relationship is that (a) hybridization of the carbanion is the dominant effect, and/or (b) differences in strain between RH and $R:^{(-)}$ are negligible, or that there is a negligible difference in strain between (RH and $R:^{(-)}$) vs. (R'H and $R':^{(-)}$). If $\sigma$-releasing substituents reduce strain, as postulated in the previous section, then perhaps this effect is greater in the more strained cubyl anion than in cyclopropyl anion. Also assumed in such intermolecular comparisons is the unimportance of differential solvent effects. While we would have expected differences in hydrocarbon solvation energies to be small, it is not obvious that the differential solvation energies of carbanions should also be small. In general, however, carbanions appear to be weakly solvated.[515a,b] However, gas-phase basicity differences are most desirable but remain unmeasured.

Another interesting comparison is the isoelectronic one between saturated hydrocarbons and carbanions vs. ammonium ions and the corresponding

---

[512] This logic was earlier introduced by H. A. Bent, *Chem. Rev.* **61**, 275 (1961).

[513] L. N. Ferguson, "Highlights of Alicyclic Chemistry," Part 1, p. 94. Franklin Publ., Palisades Park, New Jersey, 1973.

[514] T.-Y. Luh and L. M. Stock, *J. Am. Chem. Soc.* **96**, 3712 (1974).

[515a] A. Streitwieser, Jr., R. A. Caldwell, and W. R. Young, *J. Am. Chem. Soc.* **91**, 529 (1969).

[515b] A. Streitwieser, Jr. and W. R. Young, *J. Am. Chem. Soc.* **91**, 529 (1969). The authors state: "The transition state (of deprotonation) has substantially the structural character of the carbanion intermediate."

amines. At first, there are obvious similarities. Thus, aziridinium ion is much more acidic than cyclohexane. However, the gas-phase basicity of aniline is surprisingly higher than that of ammonia,[516] but toluene is more acidic than water.[517a,b] Thus, it is clear that simple comparisons between the acidities of ammonium ions and alkanes, even in the gas phase, must be made with caution. Furthermore, recalling the considerable discrepancies which arise in the general comparison of gas-phase and aqueous basicities of amines,[518a–c] it is apparent that studies in pure or mixed solvents will be accompanied by ambiguities in interpretation. As previously noted in Section 5.D.5.b, manxinium ion is a stronger acid (gas-phase) than tri-*n*-propylamine because of a crucial balance between two very sizeable (ca. 20 kcal/mole) and opposing influences: hybridization and strain. To what extent can we extrapolate to predict the (gas-phase) acidity of the bridgehead hydrogen in manxane? Slight differences in the strain and hybridization effects in the 1-manxyl anion could completely reverse the situation and make manxane a weaker acid than the model compound.

Another problem in considering the stabilities of various carbanions as well as their comparison with amines is the ambiguous role of steric effects. That such influences are important may be illustrated by citing the classical study which found that while the Brønsted basicities of 1-azabicyclo[2.2.2]-octane (quinuclidine) and triethylamine are equal, the latter is a weaker Lewis base.[519] This is understood by recognizing that quinuclidine's lone pair is less crowded because the attached alkyl groups are "tied back," and while this makes it more susceptible to Lewis acids than triethylamine, the proton affinities of the two bases are the same. Nyholm–Gillespie or valence shell electron pair repulsion[520] logic considers lone pairs to be larger than bond pairs, and lone pairs of anions to be larger still. Therefore, loss of a proton should be hindered in a sterically encumbered environment on these grounds alone. Unfortunately, in condensed media, such crowding will hinder the approach of bases as well as molecules of solvation, and no simple separation of these three complementary influences is evident. Nevertheless, the above-cited greater acidity of cubane in comparison with cyclopropane can be

[516] E. M. Arnett, *Acc. Chem. Res.* **6**, 404 (1973) and references cited therein.

[517a] J. I. Brauman and L. K. Clair, *J. Am. Chem. Soc.* **93**, 4315 (1971).

[517b] D. K. Bohme, E. Lee-Ruff, and L. B. Young, *J. Am. Chem. Soc.* **94**, 5153 (1972).

[518a] M. T. Bowers, D. H. Aue, H. M. Webb, and R. T. McIver, Jr., *J. Am. Chem. Soc.* **93**, 4314 (1971).

[518b] E. M. Arnett, R. M. Jones, III, M. Taagepera, W. G. Henderson, J. L. Beauchamp, D. Holtz, and R. W. Taft, *J. Am. Chem. Soc.* **94**, 4724 (1972).

[518c] F. M. Jones, III and E. M. Arnett, *Prog. Phys. Org. Chem.* **11**, 263 (1974).

[519] H. C. Brown, *J. Chem. Soc.* p. 1248 (1956).

[520] R. J. Gillespie, *J. Chem. Educ.* **40**, 295 (1963).

explained on purely steric grounds if one assumes that tying back of all three groups affords a "more comfortable" environment, sterically, for the lone pair in cubyl anion. While we remain too ignorant as to how to quantify lone-pair room in order to compare the acidities of cubane and cyclopropane, it is intuitively obvious that either species (as well as ethylene) should be more acidic than propane or cyclobutane. Recent quantum-chemical calculations have been performed to clarify the effect of structural distortion on the acidity of methane, ethylene, and ethane.[521a,b] Greater geometry variation and interrelation with the results of molecular-mechanics studies are desirable.

All the above considerations also arise when considering carbonium ions. These positive ions are produced in considerably more polar media than corresponding carbanions with the same carbocyclic skeleton. As such, solvation energies are expected to be much higher for the carbonium ions of interest. Furthermore, when comparing bridgehead carbonium ions (see Section 3.B.2) with acyclic species, two opposing trends in solvation arise. While bridgehead carbonium ions are only solvated on one face, their pyramidal character should localize charge and cause tighter solvation.[522] (1-Manxyl cation's solvation energy would be an interesting subject for investigation.) Quite surprisingly, good correlations are found with the calculated differences in strain energy of the hydrocarbon and cation (see Section 3.B.2). Analogous correlations exist between the gas-phase appearance potential of the cation (derived from the bromide) and the above-calculated difference.[523]

[521a] A. Streitwieser, Jr. and H. P. Owens, *Tetrahedron Lett.* p. 5221 (1973).

[521b] A. Streitwieser, Jr., H. P. Owens, R. A. Wolf, and J. E. Williams, *J. Am. Chem. Soc.* **96**, 5448 (1974).

[522] Experimentally, *tert*-butyl cation is solvated more strongly than 1-adamantyl cation by at least 4 kcal/mole; see R. D. Wieting, R. H. Staley, and J. L. Beauchamp, *J. Am. Chem. Soc.* **96**, 7552 (1974); **97**, 924 (1975).

[523] For a more complete discussion of this topic, see Section XVI.C in J. F. Liebman and A. Greenberg, *Chem. Rev.* **76**, 311 (1976).

## CHAPTER 6

# A Potpourri of Pathologies

### 6.A    When is Tetracoordinate Carbon Tetrahedral?

We have discussed very few strained species that have the possibility of containing any strictly tetrahedral carbon atoms. As noted in the discussion of adamantanes (Section 4.A), a necessary but not sufficient condition is to have four identical groups on the carbon atom of interest. Of course, there are species such as spiropentane, **1**, that seemingly fit our criterion yet

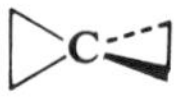

**1**

patently do not have strict tetrahedral geometry. There are yet other species with four identical substituents on a given carbon that are chiral.[1] Such a compound is 6,6-vespirene (**2**).[2a,b] However, there are more subtle examples of distortion from the idealized shape. For example, a highly mathematical stereochemical analysis of tetraaryl methanes[3] showed by dynamic stereochemistry and permutation analysis that even tetraphenylmethane will not have precisely the archetypal tetrahedral angle. (The symmetry of the phenyl groups is incompatible with threefold symmetry of the regular tetrahedron.) Subsequent experiments by three research groups[4] confirmed this theoretical

---

[1] M. Farina and C. Morandi, *Tetrahedron* **30**, 1819 (1974).

[2a] G. Haas and V. Prelog, *Helv. Chim. Acta* **52**, 1202 (1960).

[2b] G. Haas, P. B. Hulbert, W. Klyne, V. Prelog, and G. Snatzke, *Helv. Chim. Acta* **54**, 491 (1971).

[3] M. G. Hutchings, J. G. Nourse, and K. Mislow, *Tetrahedron* **30**, 1525 (1974).

[4] A. Robbins, G. A. Jeffrey, J. P. Chesick, J. Donohue, F. A. Cotton, B. A. Frenz, and C. A. Murillo, *Acta Crystallogr. Sect. B* **31**, 2395 (1975).

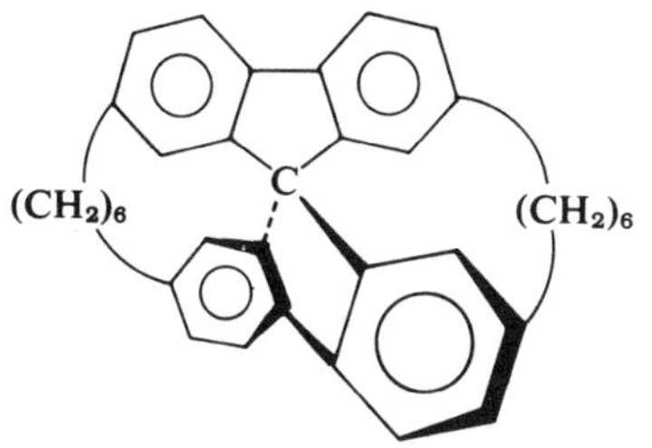

**2**

prediction. These experimentalists also discussed the isoelectronic $(C_6H_5)_4E$, where $E = B^-$, Si, Ge, or Sn. We note that additional complications arise[5] when dealing with the ionic $E = B^-$ case. If we wish to consider extremely accurate experiments and fundamental theory, even methane itself is not strictly tetrahedral, as it has a nonzero dipole moment![6a,b] However, the distortion and the resultant dipole moment are negligible in magnitude compared to other compounds of interest. In particular, the dipole moment of methane[6a] was given as $5.38 \pm 0.10 \times 10^{-6}$D, in contrast to $0.084 \pm 0.01$D for the nonpolar propane.[7] We do not know the dipole moment of ethane: it is too complicated a species for the type of study reported in Ozier;[6a] yet, indubitably, the numerical value is too small for most conventional measurement techniques reported in Nelson *et al.*[7] We will not discuss the origin of the deformation that gives rise to the dipole moment, as it not only transcends the models of molecular structure discussed earlier in this text but also, verily, any other. Suffice it to say, it arises from a coupling of rotational, vibrational, and electronic modes. As such, we will blithely consider methane to be tetrahedral, although we may still ask why. We refer the reader to Section 2.A on cyclopropane for discussions of the simpler 1-carbon compounds of interest. We conclude this section by noting that $CH_4^+$ is definitely and markedly nontetrahedral and indeed, there are two "isomers."[8]

## 6.B Inverted Tetrahedra, Propellanes, Buttaflanes, and Paddlanes

Let us return to polycyclic species, as there is little interest among organic chemists in methane. In particular, we now consider a selected group of

[5] M. Di Vaira, *Inorg. Chem.* **14**, 1442 (1975).

[6a] I. Ozier, *Phys. Rev. Lett.* **27**, 1327 (1971).

[6b] W. Holt, M. C. L. Gerry, and I. Ozier, *Phys. Rev. Lett.* **31**, 1033 (1973).

[7] R. D. Nelson, Jr., D. R. Lide, Jr. and A. A. Maryott, *Natl. Stand. Ref. Data Ser. Natl. Bur. Stand.* **NSRDS–NBS 10**, (1967).

[8] J. P. Arents and L. C. Allen, *J. Chem. Phys.* **53**, 73 (1970).

hydrocarbons, the [$m.n.p$]propellanes[9a–13d] (3). These compounds are "defined as tricyclic systems conjoined 'in' or 'by' a carbon–carbon single bond."[13a–d] For sufficiently large values of $m$, $n$, and $p$ (in practice $m, n \geq 4, p \geq 2$), these species behave essentially normally, as described in the initial articles[9a–12] and by Ginsburg, a founder of and the major chronicler of these species.[12–13d] We do not mean to say that these species are without idiosyncrasies or without interest. For example, one should not ignore the use of "propellanes as substrates for stereochemical studies"[13c] or as probes of inter-ring inter-actions.[13a–15c] However, as these features usually are hard to interrelate with strain-energy considerations, we shall omit discussion of the larger propellanes in our text.

We commence our discussion with the smallest propellane: tricyclo-[1.1.1.0$^{1,3}$]pentane (4) or [1.1.1]propellane (i.e., 3, $m = n = p = 1$). This species is experimentally unknown, although formally related metal analogues

3                                    4

[9a] G. Snatzke and G. Zanati, *Justus Liebigs Ann. Chem.* **684**, 62 (1965).

[9b] F. Nernd, K. Janowsky, and D. Frank, *Tetrahedron Lett.* p. 2979 (1965).

[10a] R. L. Cargill, M. E. Beckham, A. E. Siebert, and J. Dorn, *J. Org. Chem.* **30**, 3647 (1965).

[10b] R. L. Cargill, J. R. Damewood, and M. M. Cooper, *J. Am. Chem. Soc.* **88**, 1330 (1966).

[11a] J. J. Bloomfield and A. Mitra, *Chem. Ind.* (*London*), p. 2012 (1966).

[11b] J. J. Bloomfield and J. R. S. Ireland, *Tetrahedron Lett.* p. 2971 (1966).

[11c] J. J. Bloomfield and J. R. S. Ireland, *J. Org. Chem.* **31**, 2017 (1966).

[12] J. Altman, E. Badad, J. Itzchaki, and D. Ginsburg, *Tetrahedron, Suppl.* **8**(1), 279 (1966).

[13a] D. Ginsburg, *Acc. Chem. Res.* **2**, 121 (1969).

[13b] D. Ginsburg, *Acc. Chem. Res.* **5**, 249 (1972).

[13c] D. Ginsburg, *Tetrahedron* **30**, 1487 (1974).

[13d] D. Ginsburg, "Propellanes: Structure and Reactions." Verlag Chemie, Weinheim, 1975.

[14a] J. J. Bloomfield and R. E. Moser, *J. Am. Chem. Soc.* **90**, 5625 (1968).

[14b] S. C. Neely, C. Fink, C. Van Der Helm, and J. J. Bloomfield, *J. Am. Chem. Soc.* **93**, 4903 (1971).

[14c] D. Dougherty, J. J. Bloomfield, G. R. Newkome, J. F. Arnett, and S. P. McGlynn, *J. Phys. Chem.* **80**, 2212 (1976).

[15a] L. A. Paquette and G. L. Thompson, *J. Am. Chem. Soc.* **94**, 7118 (1972).

[15b] L. A. Paquette and G. L. Thompson, *J. Am. Chem. Soc.* **94**, 7127 (1972).

[15c] R. Gleiter, E. Heilbronner, L. A. Paquette, G. L. Thompson, and R. E. Wingard, Jr., *Helv. Chim. Acta* **29**, 565 (1973).

exist, $5^{16a}$ and $6.^{16b,c}$ In both cases, the geometry has been experimentally determined, and the bond lengths and angles are compatible with the propellane description. While [1.1.1]propellane, 4, remains only a theoretician's reality, there are nevertheless several features of general interest that transcend this

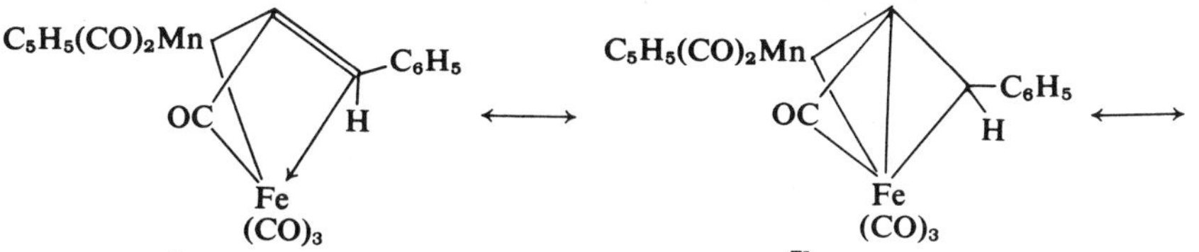

**5a**      **5b**

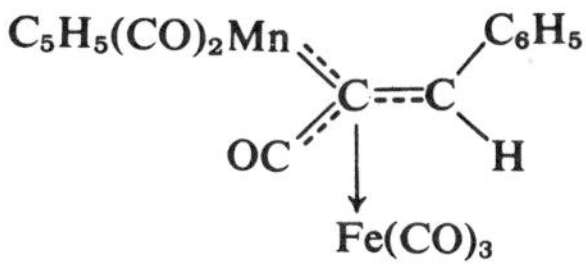

**5c**

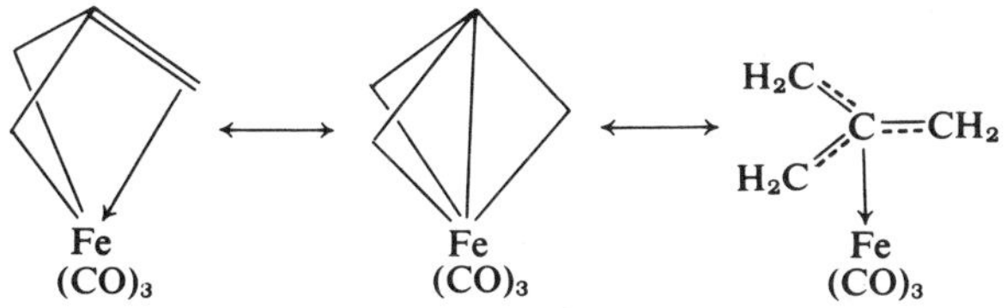

**6a**      **6b**      **6c**

species. First, "with the structural formula taken literally, ... (it would) contain four carbon–carbon bonds to the molecule side of a plane containing the bridgehead carbon."[17] This arrangement of having all four bonds lying in the same hemisphere has also been referred to as "'inverted' tetrahedral geometry at a bridgehead carbon,"[18] a feature observed experimentally using low temperature x-ray crystallography on a derivative of the considerably more "normal" [3.2.1]propellane.[18] Second, there is the possibility of "bond-stretch isomerism" for this and other small propellanes.[17,19–21] ["Bond-

[16a] V. G. Adrianov, Y. T. Struchkov, N. E. Kolobova, A. B. Antonaova, and N. S. Obezyuk, *J. Organomet. Chem.* **122**, C33 (1976).

[16b] G. F. Emerson, K. Ehrlich, W. P. Giering, and P. C. Lauterbur, *J. Am. Chem. Soc.* **88**, 3172 (1966).

[16c] A. Almenningen, A. Haaland, and K. Wahl, *Acta Chem. Scand.* **23**, 1145 (1969).

[17] M. D. Newton and J. M. Schulman, *J. Am. Chem. Soc.* **94**, 773 (1972).

[18] K. B. Wiberg, G. J. Burgmaier, K. W. Shen, S. J. LaPlaca, W. C. Hamilton, and M. D. Newton, *J. Am. Chem. Soc.* **94**, 7402 (1972).

[19] W. D. Stohrer and R. Hoffmann, *J. Am. Chem. Soc.* **94**, 4391 (1972).

[20a] M. D. Newton and J. M. Schulman, *J. Am. Chem. Soc.* **94**, 4391 (1972).

stretch isomerism" is the somewhat fancy name for the case of two minima in the potential-energy curve associated with a simple bond-length variation (i.e., $V(R)$ and $R$). It generally arises from a disallowed or forbidden curve crossing of two states with the same total symmetry. In the case of interest, the two states arise from $\sigma^2$ and $\sigma^{*2}$ configuration.] In fact, bond-stretch isomerism is a very rare phenomenon and seemingly occurs only for the [2.2.2]propellane and not the [1.1.1] species of interest.[19,21] There is, of course, a bond-stretched excited state that is more properly referred to as the diradical of bicyclo[1.1.1]pentane. The geometry of this species mimics the parent hydrocarbon[20a,b] and so conforms to the general rule that the geometry of a mono (or di) radical is usually approximately the same as the species with one (or two) more electrons or hydrogen atoms appended to it.[22] Third, on the basis of the above-cited quantum-chemical calculations and associated bond- and strain-energy analyses,[20a,b] the strain energy of [1.1.1]propellane interpolates that of bicyclobutane (see Section 3.B.1) and tetrahedrane (see Section 3.C). That is, the compound of interest has a strain energy between an experimentally well-documented species and one whose synthesis represents an unresolved challenge to the experimentalist. Fourth, these theoretical studies give a C—C bond of 1.6 Å, somewhat longer than normal but seemingly not too stretched (see Section 1.B). However, it is apparently a "nonbonding, or possibly antibonding interaction."[17,20a,b] We may understand this finding in terms of charge densities, orbital occupancies, and nonbonded repulsions. However, the inherent complexity bodes poorly for understanding this molecule in terms of "molecular mechanics" (Section 1.B) or any of the qualitative and semiquantitative schemes discussed for the simpler cyclopropane in Sections 2.A.1 through 2.A.6. Finally, we recognize [1.1.1]propellane as a bridged bicyclobutane and so admit the possibility of an "allowed"[23] rearrangement to the corresponding butadiene. In particular, we may inquire as to the exothermicity of reaction (1). There are no experimental data on species 7. However, it may be crudely estimated by any of the

$$\mathbf{4} \longrightarrow \mathbf{7} \qquad \Delta H_r = ? \qquad (1)$$

[20b] M. D. Newton, *in* "Modern Theoretical Chemistry" (H. F. Schaefer, ed.), Vol 4, pp. 223–268. Plenum, New York, 1977.

[21] J. J. Dannenberg and T. M. Procliv, *J. Chem. Soc., Chem. Commun.* p. 291 (1973).

[22] J. F. Liebman and J. S. Vincent, *J. Am. Chem. Soc.* **97**, 1373 (1975).

[23] R. B. Woodward and R. Hoffmann, "The Conservation of Orbital Symmetry," pp. 76–78. Academic Press, New York, 1971.

following bond-additivity schemes [(2a)–(2c)]. (The numbers underneath are the heats of formation in kcal/mole.) All are experimental, except the

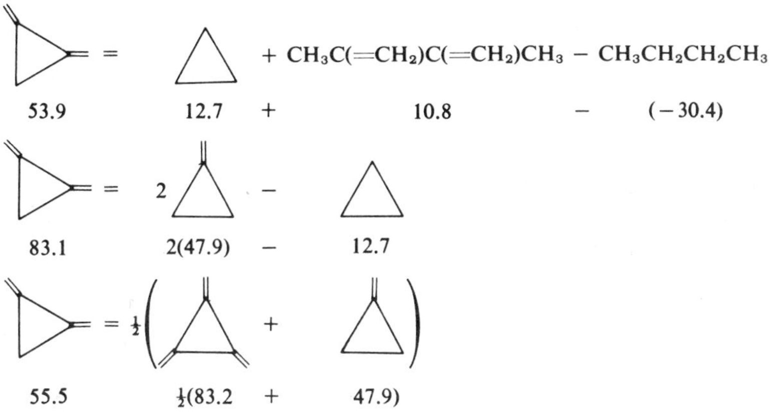

$$\text{(propellane)} = \triangle + CH_3C(=CH_2)C(=CH_2)CH_3 - CH_3CH_2CH_2CH_3 \qquad (2a)$$

| 53.9 | 12.7 | + | 10.8 | − | (−30.4) |

$$\text{(propellane)} = 2\,\triangle - \triangle \qquad (2b)$$

| 83.1 | 2(47.9) | − | 12.7 |

$$\text{(propellane)} = \tfrac{1}{2}\left(\triangle + \triangle\right) \qquad (2c)$$

| 55.5 | ½(83.2 | + | 47.9) |

trismethylene–cyclopropane or [3]radialene, which is taken from semi-empirical quantum-chemical calculations.[24] Even assuming that the highest value given above is correct, the propellane of interest is unstable by some 30 kcal/mole. This bodes very poorly for the synthesis of the compound of interest as well as other [n.1.1]propellanes except for those species in which n is so large that the compound is without interest.

There is, however, one way by which a derivative of small [n.1.1]propellanes may be synthesizable. We recall that 2,4-disubstituted bicyclobutanes with both substituents *exo* or *endo* thermally rearrange to form *cis,trans*-1,4-disubstituted butadienes.[23] (This assertion is sufficient to explain the high thermal stability of a bicyclobutane so joined by a —CH=CH— group, i.e., benzvalene—see Section 5.B.2.) Accordingly, connecting the "1" bridges in these propellanes (see 8) may stabilize it in that a small *cis,trans*-cyclodiene (9) must be formed. The strain in the propellane is played against the strain in the resultant diene and so it is possible that the former will be isolable [Eq. (3)]. Or course, the best X for producing the destabilized product is "nothing," but it is doubtful that the parent polymethylene–tetrahedrane (8, X = nil) will be synthesizable.

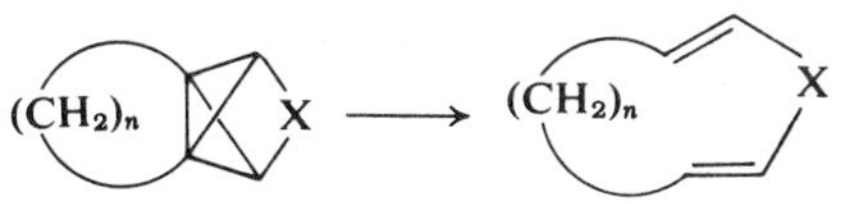

$$(CH_2)_n \quad \text{8} \quad X \longrightarrow (CH_2)_n \quad \text{9} \quad X \qquad (3)$$

[24] R. C. Bingham, M. J. S. Dewar, and D. H. Lo, *J. Am. Chem. Soc.* 97, 1294 (1975).

In principle, [2.1.1]propellane could arise from photochemical rearrangement of 1,2-dimethylenecyclobutane. Unfortunately, none of the desired product was observed in photolyzing the latter.[25] It is interesting to note that $\Delta^{1,3}$bicyclo[1.1.0]butene, **10**, has been calculated[26a] to have a puckered geometry similar to that of cyclobutane and the saturated bicyclo[1.1.0]-butane. This would suggest the possibility of relatively facile addition across the double bond to form [2.1.1] and other small propellanes. However, the strain energy in **10** is sufficiently formidable as to preclude optimism on our part, although a short-lived derivative of **10** has been implicated as an intermediate.[26b] Larger analogues of this olefin are known, in particular, **11**[27]–**15** (see Section 3.D.2). There is a remarkable spectrum of relative stabilities associated with these species. Compound **12** (R = CH$_3$) is stable for five

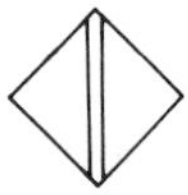

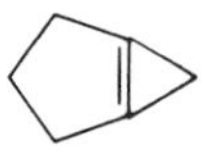

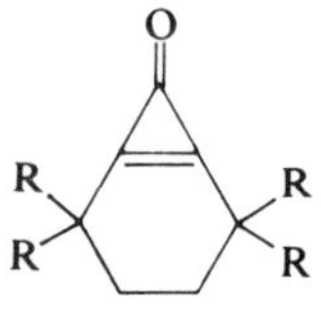

hours at 100°C[28] but is apparently unisolable without the four methyl groups (i.e., R = H).[29a,b] In contrast, **13**, with the stabilizing cyclopropenone ring replaced by a dimethylcyclopropene, can be isolated.[29a,b] Analogous stability patterns are found for **14** and **15**,[30a,b] with the former only having been observed in solution at −60°C. It would appear that compounds **12** (R = CH$_3$) and **13**–**15** are all potential precursors of [n.1.1]propellanes, but somehow n = 4 and 5 seem rather large here.

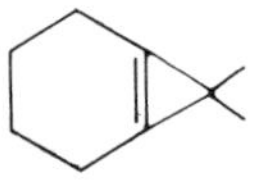

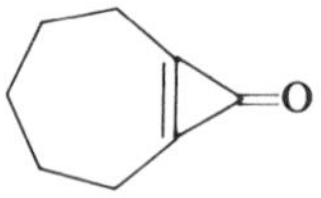

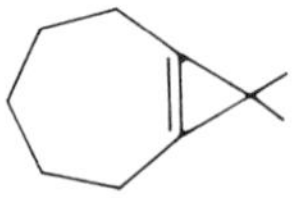

[25] P. A. Kelso, A. Yeshurun, C. N. Shih, and J. J. Gajewski, *J. Am. Chem. Soc.* **97**, 1513 (1975).

[26a] W. J. Hehre and J. A. Pople, *J. Am. Chem. Soc.* **97**, 6941 (1975).

[26b] G. Szeimies, J. Harnisch, and O. Baumgärtel, *J. Am. Chem. Soc.* **99**, 5183 (1977).

[27] Prof. Philip Warner, personal communication to Joel F. Liebman.

[28] M. Suda and S. Masamune, *J. Chem. Soc., Chem. Commun.* p. 504 (1974).

[29a] R. Breslow, J. Posner, and A. Krebs, *J. Am. Chem. Soc.* **85**, 234 (1963).

[29b] R. Breslow, L. J. Altman, A. Krebs, E. Mohacsi, I. Murata, R. A. Peterson, and J. Posner, *J. Am. Chem. Soc.* **87**, 1326 (1965).

[30a] G. L. Closs and W. A. Boll, *J. Am. Chem. Soc.* **85**, 3904 (1963).

[30b] G. L. Closs, W. A. Boll, H. Heyn, *J. Am. Chem. Soc.* **90**, 173 (1968).

We now consider [n.2.1]propellanes commencing with $n = 2$, e.g., **16**. (The value $n = 1$ has been neglected, as [1.2.1] and [2.1.1]propellane are synonymous.) Species **16** remains spectroscopically uncharacterized, although synthetic studies are suggestive of its intermediacy.[31–34c] Indeed, a recent study[34a–c] reports a finite-lived species that reacts chemically in a manner consistent with that expected for [2.2.1]propellane. Encouragingly, these last authors' approach gave a good yield of the larger, and earlier isolated[18,31–36b] [3.2.1]propellane (**17**).

Species **16** is an eminently reasonable species. The estimated strain energy[17,21] is 85 kcal/mole or, equivalently, 12 kcal/mole per carbon atom and ca. 9.4 kcal/mole per C—C bond. It is important to emphasize that this does not mean either synthesis and/or isolation will be facile. The organometallic analogue with the "1" bridge replaced by platinum, i.e., an olefin–Pt $\pi$ complex, **18**, corresponding to olefin **19**, has been isolated.[37a] (The crystal structure of **18** has also been determined.[37b]) We recall that the original Walsh model for cyclopropane[38] (see Section 2.A.4) described this species as an analogous complex of ethylene and $CH_2$. (Also see Sections 2.B.1 and 2.B.2 for discussions of related complexes of ethylene with $CH_3^+$ and with $CH_2^+$ and $CH_2^-$, respectively, and Chapter 5 for metal–olefin complexes in general as well as for three-membered heterocycles.) This suggests a possible correlation between such cyclopropanes and other [m.n.1]propellanes and olefin–platinum complexes. Can one derive a correlation between the stabilities of

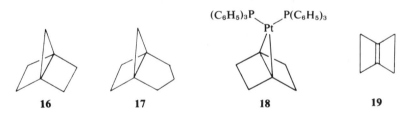

$(C_6H_5)_3P$   $P(C_6H_5)_3$

**16**　　　　**17**　　　　**18**　　　　**19**

[31] C. F. Wilcox, Jr. and C. Leung, *J. Org. Chem.* **33**, 577 (1968).
[32] M. R. Rifi, *Collect. Czech. Chem. Commun.* **36**, 932 (1971).
[33] K. B. Wiberg and G. J. Burgmaier, *J. Am. Chem. Soc.* **94**, 7396 (1972).
[34a] K. B. Wiberg, W. E. Pratt, and W. F. Bailey, *J. Am. Chem. Soc.* **99**, 2297 (1977).
[34b] K. B. Wiberg and G. J. Burgmaier, *Tetrahedron Lett.* p. 317 (1969).
[34c] K. B. Wiberg, B. C. Lupton, and G. J. Burgmaier, *J. Am. Chem. Soc.* **91**, 3372 (1969).
[35] D. H. Ave and R. N. Reynolds, *J. Am. Chem. Soc.*, **95**, 2027 (1973).
[36a] P. G. Gassman, A. Topp, and J. W. Teller, *Tetrahedron Lett.* p. 1093 (1969).
[36b] P. G. Gassman and E. A. Armour, *Tetrahedron Lett.* p. 1431 (1971).
[37a] M. E. Jason, J. A. McGinnerty, and K. B. Wiberg, *J. Am. Chem. Soc.* **96**, 6531 (1974).
[37b] M. E. Jason and J. A. McGinnerty, *Inorg. Chem.*, **14**, 3024 (1975).
[38] A. D. Walsh, *Nature (London)* **159**, 165 (1947).

these two classes of compounds? The transfer of $CH_2$ from ethylene to cyclopropene is exothermic (i.e., the conceptual reaction cyclopropane + cyclopropene → ethylene + bicyclobutane is energetically favorable.) However, in the absence of thermochemical data on both **18** and **19**, we cannot appraise the validity of such a supposed correlation. It would appear that **19** would be the perfect precursor for [2.2.1]propellanes by carbene addition to the double bond. However, in the three years since the initial synthesis of **19**,[37,39,40] no success has been achieved in this seemingly simple reaction.[41] However, we are optimistic as to the eventual isolation of **19** and indeed are tempted to say it has been observed.[34a-c] We additionally recall a resonance structure containing [2.2.1]propellane has been suggested to explain hyperfine interactions in 1-halonorbornyl radicals,[42] (**20**). We recall our earlier and related argument for the stabilization of 1-norbornene (**21**) by fluorination at the other bridgehead[43] (see Section 3.E.3 and 5.E.3). Finally, we note

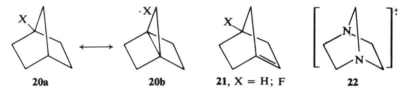

**20a**              **20b**         **21, X = H; F**      **22**

theoretical[44] studies on the radical cation of 1,4-diazabicyclo[2.2.1]heptane, **22**. This species is isoelectronic to the radical anion of the propellane of interest. Noting the stability of the bicyclo[2.2.2]octane homologue of **22**,[44-45b] species **22a** makes us wonder about the stability of the radical anions of small propellanes in general. If they are sufficiently stable, synthesis of

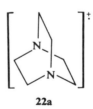

**22a**

[39] J. Casanova and H. R. Rogers, *J. Org. Chem.* **39**, 3803 (1974).

[40] K. B. Wiberg, W. F. Bailey, and M. E. Jason, *J. Org. Chem.* **39**, 3804 (1974).

[41] Prof. Kenneth B. Wiberg, personal communication to Joel F. Liebman.

[42] D. E. Wood, W. F. Bailey, K. B. Wiberg, and W. E. Pratt, *J. Am. Chem. Soc.* **99**, 268 (1977).

[43] S. F. Campbell, J. M. Leach, J. C. Tatlow, and K. N. Wood, *J. Fluorine Chem.* **1**, 103 (1971).

[44] R. Hoffmann, A. Imamuea, and W. J. Hehre, *J. Am. Chem. Soc.* **90**, 1499 (1968).

[45a] E. Heilbronner and K. A. Muszkrat, *J. Am. Chem. Soc.* **92**, 3818 (1970).

[45b] G. W. Eastland and M. C. R. Symons, *Chem. Phys. Lett.* **45**, 422 (1977).

[2.2.1]propellane via electrochemical or alkali-metal reduction may be characterized by a lessened likelihood of success than that hypothesized "merely" by strain-energy considerations.

We now turn to [3.2.1]propellane (17) and its derivatives.[18,31–36b] We earlier cited the crystal structure of the 8,8-dichloro derivative,[18] and noted that the expected structural features were observed. The strain energy of 17 has been estimated as at least 60 kcal/mole[35–36b] by comparison with the 8-oxa analogue (60 kcal/mole) and with bicyclo[2.1.0]pentane (57 kcal/mole). Despite its inherent instability, the half-life of 17 in diphenyl ether at 192°C is 20 hours.[33] These authors offer multiple reasons for this surprisingly high thermal, kinetic stability:

> Fragmentation of the cyclobutane ring to two double bonds in a concerted process is electronically forbidden while because of the rigid structure, involving first the central bond would probably have the electronic characteristics of a concerted process. Similarly, cleavage of the central bond to a diradical cannot be followed by a hydrogen migration found with most cyclopropanes because this will lead to a double bond at the bridgehead. Cleavage of one of the other cyclopropane bonds would have to be followed by an alkyl shift and such shifts are not common in free radical reactions.

In support of these explanations, [3.2.1]propellane polymerizes on standing in the liquid state,[33] while in the vapor phase, where decomposition may be expected to be intramolecular, the "forbidden" product 1,3-dimethylene-cyclohexane is formed.[46] As further support of the inherent thermodynamic instability of the propellane, we recall the spontaneous reaction with oxygen to form a copolymer and with bromine even at $-60°C$.[33,35–36b] This bromination reaction was conducted in methylene chloride and yielded not only 1,5-dibromobicyclo[3.2.1]octane but the 1-bromo-5-chloro species by chlorine abstraction from the solvent. Analogous iodination and iodochlorination were observed using iodine in methylene chloride. This is indeed a rather remarkable reaction in that there are very few carbon–carbon bonds, no less single bonds, to which iodine will add. Perhaps the most remarkable reaction is with bromotrichloromethane. In the absence of both light and oxygen, i.e., free-radical reaction initiators, 1-bromo-5-trichloromethylbicyclo[3.2.1]-octane was still formed. This parallels the numerous cases where $CBrCl_3$ adds to an olefin,[47] but even here it is striking that some catalysis is not required. We close our discussion of these halogenation reactions with a suggestion on how to determine the heat of formation of 17 and other reactive propellanes: React the compound with $Br_2$ or $O_2$ to form an intermediate product, and measure the heat of reaction in a calorimeter. Then, using appropriate

[46] D. H. Aue and R. N. Reynolds, *J. Org. Chem.* **39**, 2315 (1974).
[47] C. Walling and E. S. Huyser, *Org. React.* **13**, 91 (1963).

calorimetric techniques, burn either the dibromide or peroxide to determine its heat of formation. In principle, this can be a "one-pot" thermochemical study for which a simple thermochemical cycle then gives the desired heat of formation of the original compound of interest. Indeed, why should the formation of the peroxide prevent one from determining the heat of combustion, and thereby heat of formation of 17, directly? Reaction with acetic acid to form 1-bicyclo[3.2.1]octyl acetate[33] is also rapid. (This provides yet another approach for the two-step calorimetric determination of the strain energy of 17.) There are no conclusive mechanistic data for this reaction. Recent suggestive evidence exists[48] to disqualify a destabilized (bridgehead) 1-bicyclo[3.2.1]octyl cation (see Section 3.B.2) and a simple six-membered transition state, 23. {The latter mimics the highly strained tricyclo[4.3.2.1^{1,6}] dodecane (23a) called [4.3.2.1]paddlane (see Section 6.B.2).} It appears[49] that four molecules of acetic acid react with one of the propellane in the probably cyclic transition state.

23                                23a

We now turn our attention to the hetero analogue of 17, 8-oxa[3.2.1]-propellane.[35] Like the parent carbocycle it is highly thermally stable, having a half-life of 4 hours at 192°C in diphenyl ether. It is not obvious whether the fivefold difference in stability of 17 and the 8-oxa derivative (24) is significant. (By contrast, we recall the marked difference in the stability of cyclopropene and its antiaromatic oxa derivative, oxirene—see Section 5.D.6.) Few reactions of 24 have been reported. Perhaps the most noteworthy is the reaction with LiAlH₄. Twelve days of reaction in ether result in but 41% conversion to the alcohol, bicyclo[3.2.0]heptan-1-ol. While this is understandable in terms of the traditional inertness of epoxides of tetrasubstituted olefins, it is still surprising that such a strained molecule as this propellane can be so inert. In addition, a 1,3-ethano bridged[3.2.1]propellane,[46] 25a and its 8-oxa derivative 25b have recently been reported. While they may also be recognized as methano[4.2.1]propellanes, the methano and ethano bridges conspire to generally increase the reactivity to mimic a "[2½.2.1]propellane."[48] An interesting reaction of species 17 and 25a is with dimethylacetylene-dicarboxylate at room temperature.[46] Rather than forming some [3.2.2.1]-paddlane (cf. 26), the cyclopropane ring opens to form products reminiscent

[48] Prof. Donald H. Aue, personal communication to Joel F. Liebman.
[49] P. G. Gassman, *Acc. Chem. Res.* **4**, 126 (1971).

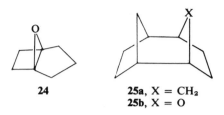

**24**

**25a, X = CH₂**
**25b, X = O**

of a carbene + olefin, and a trimethylene diradical (see Section 2.B.2). In particular, **25** forms **27** and **28**, while **25a** forms the ethano analogues **29** and **30**. (More recent experiments[48] suggest considerable diradical character in

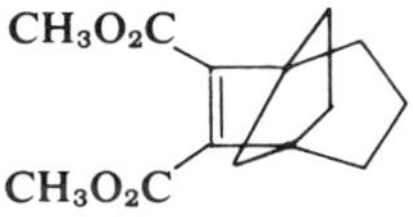

**26**

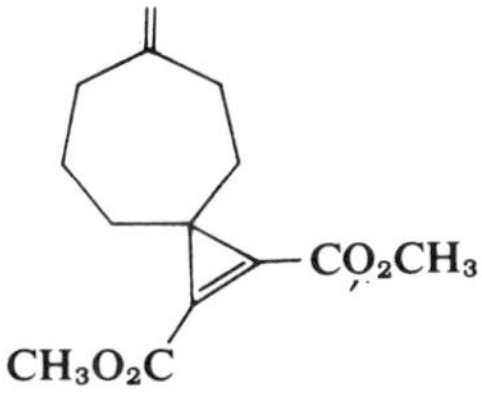

**27**

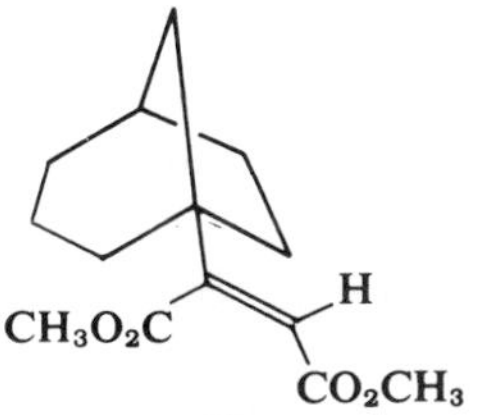

**28**

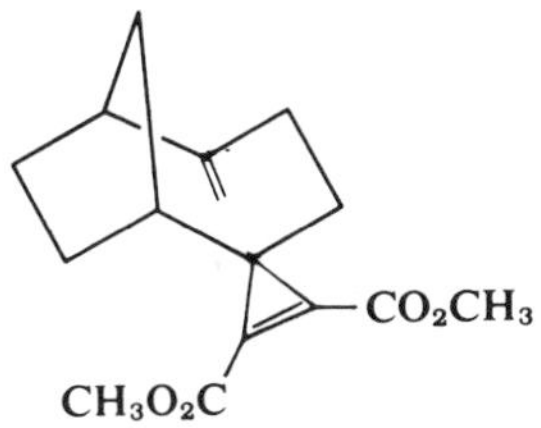

**29**

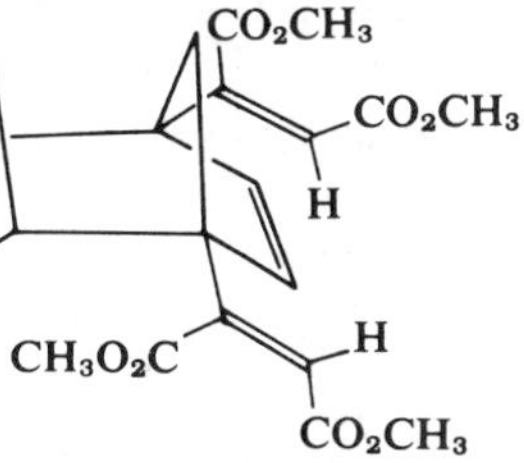

**30**

the reactions lending to **27–30**.) It is interesting that while the reactions of **17** and **25a** are nearly quantitative, the ratios of **27:28** and **29:30** are markedly different: In the former case, it is 1:7, while for the latter it is 3:2. This may be understood in terms of the instability, or more properly the high strain energy, of syntricyclo[4.2.1.1$^{2,5}$]decane, **31**, and its derivatives. Further documentation of this comes from the relative endothermicity of the addition of acetic acid to **25a** relative to the less strained and usually less reactive **17**.

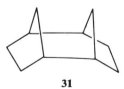

**31**

We now consider [*n*.2.1]propellanes for still larger *n*. These compounds may be expected to behave increasingly normally with increasing *n*. Nonetheless, as they remain bicyclo[2.1.0]pentanes regardless of the value of *n*, we expect these species still to be reactive if not also unusual. Most of our discussion will center around the chemistry of the olefinic derivative of [4.2.1]propellane, **32**. It has been found that addition of $Br_2$ across the "conjoining" single bond to form 1,6-dibromobicyclo[5.2.1]nonane **33** is

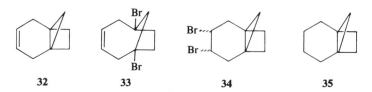

three times faster than the addition to the double bond to form the dibromopropellane, **34**.[50] Corresponding competition reactions with $Cl_2$ have seemingly not been reported. However, the reaction with iodobenzene dichloride, $C_6H_5ICl_2$, seemingly yields only products from the addition to the double bond.[51]

It is not surprising that acetic acid prefers to add across the strained cyclopropane ring rather than to the double bond.[51] (Few olefins directly add carboxylic acids even though the resultant ester is energetically favored.) However, even here there are seeming anomalies. For example, we understand why the reactions of [4.2.1]propellanes **32** and **35** are slower than those of [3.2.1]propellane **17** in terms of the relative strain energies. This would suggest that **32** should react faster than **35**, while in fact it is some 12 times slower. Literature explanations suggest "a combination of inductive and conformational factors."[50] We now present an alternative possibility: We hypothesize that there is an "aromatic" transition state[52,53] for the addition of acetic acid to small propellanes. The double bond, say in **32**, increases the number of "active" electrons by two. This converts the "aromatic" transition state to one that is "antiaromatic" and, accordingly, the reaction rate is slowed. A most interesting study would thus be with the corresponding propelladiene, **36**, where rate acceleration would be expected. It is also of interest that olefin **32** reacts readily with *m*-chloroperbenzoic acid to form the *syn*- and *anti*-epoxides with seemingly no competition or subsequent addition

[50] P. Warner and R. LaRose, *Tetrahedron Lett.* p. 2141 (1972).

[51] Unpublished research of K. Redecker and W. Grimme, cited in Ginsburg (Ref. 13d), p. 37.

[52] H. E. Zimmerman, *Acc. Chem. Res.* **4**, 272 (1971).

[53] M. J. S. Dewar, *Angew. Chem., Int. Ed. Engl.* **10**, 761 (1971).

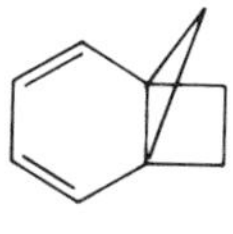

**36**

to the "conjoining" carbon–carbon bond by either the peracid or the by-product *m*-chlorobenzoic acid.[51] Still more amazing is the rapid rearrangement of **37**, the 7,7-dichloro analogue of **32**, to the bridgehead olefin **38** via the doubly Bredt-violating allyl cation **39**[52] (see Section 3.E.3). Direct comparison with the analogous 7,7-dichloropropellane (**40**) and propelladiene (**41**) would be instructive because we recall that the analogous and earlier discussed dichloro[3.2.1]propellane **42**[18] is stable to rearrangement. We believe the rearrangement of **37** to **38** is so rapid, even in nonpolar media, because the

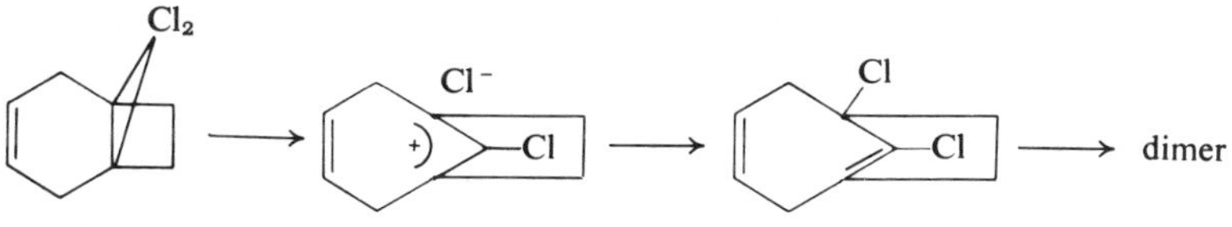

**37**   **39**   **38**   dimer

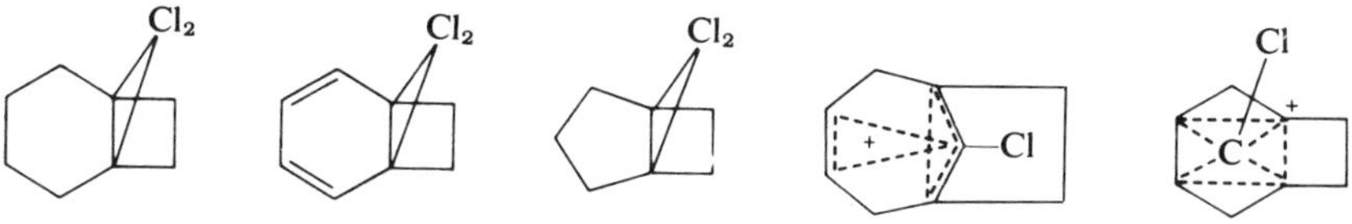

**40**   **41**   **42**   **39a**   **39b**

intermediate cyclopropyl cation can be stabilized by interaction with the olefinic cloud. This cation may be reformulated as a bishomocyclopropenyl cation **39a**, or even as a square-pyramidal $C_5H_5^+$ derivative **39b** (see Section 6.D). Useful information, but as yet unreported, is the energy of activation of this rearrangement. (The original authors[54] estimated the rearrangement of **37** to **38** to be ca. 10 kcal/mole exothermic.) It would also be of interest to consider other mono- and dihalo analogues of **32**. The difluoro analogue has been synthesized via a thermolysis reaction[51] and so is presumably thermally stable. Comparative ease of ionization of C—Cl groups over corresponding C—F groups is well documented. We need cite only the relative hydrolysis rates of benzyl chloride and fluoride.[55a–c] Perhaps more related, but regret-

---

[54] P. Warner, R. LaRose, C.-M. Lee, and J. C. Clardy, *J. Am. Chem. Soc.* **94**, 7607 (1972).

[55a] C. G. Swain and R. E. T. Spalding, *J. Am. Chem. Soc.* **82**, 6104 (1960).

[55b] R. E. Parker, *Adv. Fluorine Chem.* **3**, pp. 63–91 (1963).

[55c] J. F. Liebman and B. B. Jarvis, *J. Fluorine Chem.* **5**, 41 (1975).

tably not quite comparable, is the fact that while 7,7-difluorobenzocyclo-propene (**43**) is a covalent species (with a high dipole moment),[56] 1,4-diphenyl-7,7-dichlorobenzocyclopropene, **44**, exists as an ionic salt, **44a**[57] (see Section 3.H.A). Relative rate studies for the ionization of other halo derivatives of

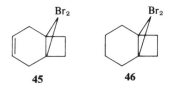

**43**                **44**                **44a**

**32**, **35**, and **36** would be of interest. While we hesitate to extrapolate from the [4.3.1] to [4.2.1]propellanes, we nonetheless note that the dibromopropellene **45** solvolyzes considerably more slowly than the corresponding propellane **46**.[58a,b] (Comparable data are found for monobromo[4.3.1]propellene and propellane.) We now leave [*n*.2.1]propellanes with the awareness that, regardless of the value of *n*, these species will always be derivatives of bicyclo-[2.1.0]pentane and so be highly strained.

**45**                **46**

We now consider [*n*.2.2]propellanes and commence with $n = 2$ (**47**). The possibility of bond-stretch isomerism[19-21] remains experimentally unproved but provides a good explanation for both the numerous failures[59-65] and occasional successes[66] in the synthesis of this compound. It is extremely interesting, at least to the authors, that Ginsburg[59] and Dannenberg[60]

[56] R. Pozzi, K. R. Ramaprasad, and E. A. C. Lucken, *J. Mol. Struct.* **28**, 111 (1975).

[57] B. Holton, A. D. Woodhouse, H. M. Hugel, and D. P. Kelly, *J. Chem. Soc., Chem. Commun.* p. 247 (1974).

[58a] P. Warner, J. Fayos, and J. Clardy, *Tetrahedron Lett.* p. 4473 (1973).

[58b] Prof. Phillip Warner, personal communication to Joel F. Liebman.

[59] See Ginsburg (Ref. 13d), pp. 43-51.

[60] J. J. Dannenberg, *Angew. Chem. Int. Ed. Engl.* **15**, 519 (1976).

[61] See Ginsburg (Ref. 13d), p. 50.

[62] K. B. Wiberg, G. A. Epling, and M. Jason, *J. Am. Chem. Soc.* **96**, 912 (1974).

[63] J. J. Dannenberg, T. M. Procliv, and C. Hutt, *J. Am. Chem. Soc.* **96**, 913 (1974).

[64] J. J. Gajewski, L. K. Hoffman, and C. N. Shih, *J. Am. Chem. Soc.* **96**, 3705 (1974).

[65] See Dannenberg (Ref. 60), pp. 520-521.

[66] P. E. Eaton and G. H. Temme, *J. Am. Chem. Soc.* **95**, 7508 (1973).

strongly disagree on what constitutes a successful synthesis of **47**. In particular, Ginsburg[61] describes the results of Wiberg et al.[62] and Dannenberg et al.[63] as "disappointing" and hedges about those of Gajewski et al.[64] while Dannenberg[65] is, in strong counterpoint, seemingly exuberant about the successes of these same references. We take something of a compromise view. Species **47** remains unknown in that it has not been characterized by either physical methods or even spectroscopy. However, to the extent that chemical reactions may be used as diagnostics to suggest the intermediacy of a species (as in the case of the smaller [2.2.1]propellane, **16**), then indeed [2.2.2]propellane has been synthesized. (See Wiberg et al.[34a–c] for a so-defined synthesis of both propellanes.)

Fortunately for the experimental and theoretical chemist alike, a substituted derivative of **47**, N,N-dimethyl[2.2.2]propellane-2-carboxamide, **48**, has been

**47**  **48**  **48a**  **48b**

isolated[66] in good yield (40%), and its nmr and ir spectra taken. No thermochemical measurements of its stability have been reported, and we can only chronicle an energy of activation of ca. 22 kcal/mole for its facile rearrangement to the isomeric **48a** and **48b**. Theoretical studies[19–21] seem to have emphasized the "bond-stretch isomerism" and rearrangement to 1,4-dimethylene cyclohexanes, but deemphasized estimates of the strain energy. References 20 and 20c are somewhat unique in that these authors presented a value for this quantity, 90 kcal/mole. These authors computed the strain energy from the energy difference for the reaction of the propellane with two ethane molecules to form three molecules of cyclobutane. (An alternative estimate by the same authors involved fusing three planar cyclobutane rings and yielded a value of 87 kcal/mole) "Molecular mechanics" (see Chapter 1) gave ca. 74 kcal/mole strain,[67] while the use of "distorted methane" logic[68] yielded the comparable value of 80 kcal/mole. Semiempirical calculations (MINDO/3)[24] have been reported on the ionization potential and heat of formation of **47**. Using the first approach of Newton and Schulman,[20a,b] the above heat of formation, and the correspondingly calculated quantities for

[67] E. M. Engler, J. D. Andose, and P. von R. Schleyer, *J. Am. Chem. Soc.* **95**, 8005 (1973).

[68] K. B. Wiberg, G. B. Ellison, and J. J. Wendolowski, *J. Am. Chem. Soc.* **98**, 1212 (1976).

cyclobutane and ethane[24] and the experimental strain energy of cyclobutane, we derive a value for the strain energy in excess of 110 kcal/mole. Only experiment will resolve the apparent discrepancy in the calculated heat of formation of [2.2.2]propellane.

The high strain energy in [2.2.2]propellanes is accompanied by a low energy of activation for rearrangement.[66] (The calculated value for the parent **47**[20a,b] is 29 kcal/mole, in close agreement to the earlier-cited[66] value of 22 kcal/mole for the carboxamido derivative **48**.) It is therefore not surprising that most syntheses of [2.2.2]propellanes failed. Without any intent to embarrass the authors, we cite additional unsuccessful attempts to make (a) the parent, by $SO_2$ extrusion[69] from 3-thia[3.2.2]propellane dioxide; (b) the 2-methylthio derivative, by a Stevens-like rearrangement[70a,b] of 3-methyl-3-thionia[3.2.2]propellanium tetrafluoroborate[69] (**49**); (c) a dimethylthio derivative, by a double-Stevens-like rearrangement of a dithionia[3.2.2]-propellanium species[69]; (d) the 2-keto derivative, by addition of ketene to the strained olefin, **19**[71]; (e) the dimer of the same olefin,[71] **19**, pentacyclo-[4.2.2.2.$^{2,5}$0.$^{1,6}$0$^{2,5}$]dodecane (**50**) by simple thermal dimerization; and (f) the tetrabenzo analogue of **50**, compound **51**, by photolysis of the strained olefin, **52**.[72] The product from reactions (a)–(e) was the related 1,4-bismethylene cyclohexane. It may be argued that these five reactions were all sufficiently exothermic that the energy generated to produce the propellane was also

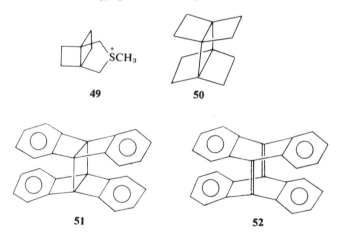

**49**          **50**

**51**                    **52**

[69] I. Lantos and D. Ginsburg, *Tetrahedron* **28**, 2507 (1972).

[70a] S. H. Pine, *Org. React.* **18**, 403 (1974).

[70b] B. M. Trost and L. S. Melvin, Jr., "Sulfur Ylides: Emerging Synthetic Intermediates," Ch. 7. Academic Press, New York, 1975.

[71] K. B. Wiberg and M. E. Jason, *J. Am. Chem. Soc.* **98**, 3393 (1976).

[72] R. L. Viavettene, F. D. Greene, L. D. Cheung, R. Majeste, and L. M. Trefonas, *J. Am. Chem. Soc.* **96**, 4342 (1974).

sufficient to cyclorevert the newly formed cyclobutane to the diolefin; While this explanation may seem to be faulted by recalling the synthesis of **48** via the highly exothermic Wolff rearrangement, we remember that this reaction was done at $-70°C$.[66] Perhaps these other studies should be redone at lower temperatures.

The synthesis of [2.2.2]propellanes will still not be easy. We recall that alkali-metal and electrochemical reduction[73,74] of 9,10-dihalotryptycenes **52a** did not yield tribenzo[2.2.2]propellene (**53**) but rather the mono- and didehalogenated tryptycenes (**54**). The formation of **53** may be understood in

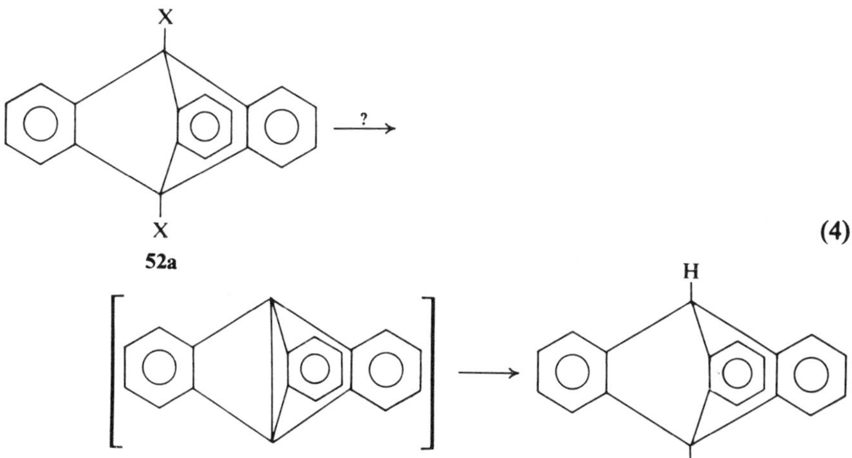

(4)

**52a**

**53**

**54**

terms of the high strain in the hypothesized **52**. Indeed, it looks considerably more strained than [4.2.2]propella-3,7-diene (**55**), erstwhile described as "the most highly strained propellane known to date" (1971).[75] To the extent that this species is considerably more strained than the saturated propellane **55a**, intuitively the unannelated [2.2.2]propellatriene (**56**) should be considerably more strained than its saturated analogue, **47**. Corresponding to the ring

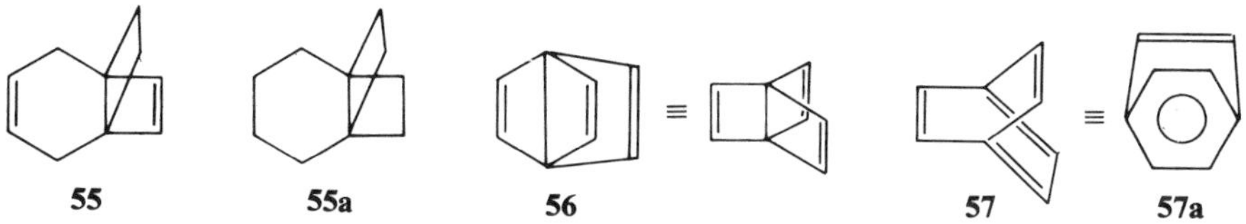

**55**      **55a**      **56**      **57**      **57a**

[73] G. Markl and A. Mayl, *Tetrahedron Lett.* p. 1817 (1974).
[74] H. Bohm, J. Kalo, C. Varnitzky, and D. Ginsburg, *Tetrahedron* **30**, 217 (1974).
[75] L. A. Paquette and R. W. Hauser, *J. Am. Chem. Soc.* **93**, 4522 (1971).

opening of **47** to the bis-1,4-dimethylene cyclohexane, the ring opening for **55** would yield the extremely strained **56**, and so this mode of decomposition is not to be expected. {We recognize **57** as an olefinic derivative of [2]paracyclophane (**57a**), and recall the earlier discussion of the relative stabilities of Dewar and benzenoid paracyclophanes (see Section 3.H.3 and 5.B.2). In addition, returning to **52**, the corresponding structure related to **56** would have had one benzene ring converted into an *o*-quinoid form, a process expected to be highly energetically disfavored.[76]} However, there is a simple process that allows the conversion of **52** into **53**: the addition of one or two electrons to form tryptycenyl- anion, **58a**, or dianion **58b**. Protonation, gain of another electron, and reprotonation, or diprotonation then yields tryptycene. We

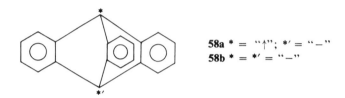

$$\textbf{58a} \; * = \text{``}\uparrow\text{''}; \; *' = \text{``}-\text{''}$$
$$\textbf{58b} \; * = *' = \text{``}-\text{''}$$

remind the reader that while bridgehead cations are destabilized (see Section 3.B.2), there is no such obstacle to bridgehead radicals and anions.[77] As such, tryptycyl radical[78a] and anion[78b] are stable but probably highly reactive species, as they lack the special stability of triphenylmethyl radical and anion where the benzene rings can be conjugated with the "odd carbon." Even in the absence of any resonance or inductive effects of the benzene rings, [2.2.2]propellane still forms radical anions and dianions,[31–34c] as evidenced by the products formed on reduction of 1,4-dihalobicyclo[2.2.2]octane. We had earlier cited discussions of 1,4-diazabicyclo[2.2.2]octane and its radical cation[44,45b] (species **22a**), and now remind the reader that these are isoelectronic to the [2.2.2]propellane dianion and radical anion. We recognize the existence of [2.2.2]propellane dication, earlier named bicyclo[2.2.2]octane dication (**59**)[79] (see Section 2.C.2). The interesting suggestion has been made[60] that this stable cation should be reducible to the neutral propellane. Semiempirical quantum-chemical calculations have been performed on [2.2.2]-propellane, its radical anion and radical cation, and its dication and dianion,

---

[76] E. Clar, "The Aromatic Sextet." Wiley, New York, 1972.

[77] E. D. Jemmis, V. Buss, P. von R. Schleyer, and L. C. Allen, *J. Am. Chem. Soc.* **98**, 6483 (1976).

[78a] P. D. Bartlett and F. D. Greene, *J. Am. Chem. Soc.* **76**, 1088 (1954).

[78b] A. Streitwieser, Jr., R. A. Caldwell, and M. R. Granger, *J. Am. Chem. Soc.* **86**, 3578 (1964).

[79] G. A. Olah, G. Liang, P. von R. Schleyer, E. M. Engler, M. J. S. Dewar, and R. C. Bingham, *J. Am. Chem. Soc.* **95**, 6829 (1973).

at a uniform calculational level.[80] The same authors also performed studies on isoelectronic heteroanalogues, such as **60** and **61** and the above diaza-compound.[80] *Ab initio* calculations have also been reported on species **61**.[81]

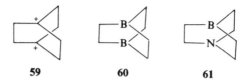

 **59**                    **60**                    **61**

We now turn briefly to the reaction chemistry of [2.2.2]propellanes, in particular that of the *N,N*-dimethyl carboxamide derivative, **48**. This species readily reacts with $Br_2$ at $-70°C$ but is seemingly inert to atmospheric oxygen even at room temperature.[66] This is in contrast to the isomeric [3.2.1]propellane **17**, which readily reacts with both substances at $-60°C$.[33,36a,b] This result cannot be understood either in terms of the relative strain energies of the propellanes ([2.2.2] > [3.2.1]), or the relative strain energies of the bicyclooctane intermediates or products ([2.2.2] $\approx$ [3.2.1]). One may invoke, *a posteriori*, the relative reactivities of compounds containing three- and four-membered rings in general (See Section 2.B.1). However, this argument is weakened by the near equality of the strain energies of the parent cyclopropane and cyclobutane (see Table 3.1.). Neither can we extricate ourselves by the invoking of strain energy per carbon atom or strain energy per C—C bond in the current case of interest, as the parent [2.2.2] and [3.2.1]propellanes have the same number of relevant atoms and bonds. We conclude by noting that while it is not initially surprising that the [3.2.1]propellane is more reactive, it is surprising that a simple explanation remains evasive.

Turning now to the higher [*n*.2.2]propellanes, we find that most of them are characterizable as one of three major "varieties." The first are the saturated compounds, and in particular, we chronicle $n = "2\frac{1}{2}"$ (species **62**),[38,39] and $n = 3$ and 4 (species **63**). Thermally stable in excess of 160°C, they nonetheless still react (slowly) with $Br_2$ at room temperature to yield the bridgehead dibromides of the corresponding bicyclo[*n*.2.2]alkanes.[82] For us, at least, the major interest in this class of compounds is that they provide a systematic synthetic entry into the more strained [2.2.2]propellanes. The second class of compounds are hetero derivatives, such as species **49**, already cited in unsuccessful attempts to make the [2.2.2]propellane. The third class

[80] J. J. Dannenberg and T. M. Procliv, unpublished results, personal communication from Prof. J. J. Dannenberg to Joel F. Liebman.

[81] M. N. Paddon-Row, L. Radom, and A. R. Gregory, *J. Chem. Soc., Chem. Commun.* p. 427 (1976).

[82] P. E. Eaton and N. Kyi, *J. Am. Chem. Soc.* **93**, 2786 (1971).

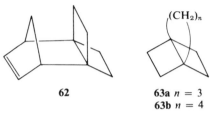

**62**

**63a** $n = 3$
**63b** $n = 4$

of compounds is the dienes, or the Dewar-benzene analogues of [$n$]para-cyclophanes. Here, $n = 3$[83a] and $4$[83b,84] have both been isolated and characterized (species **64**).

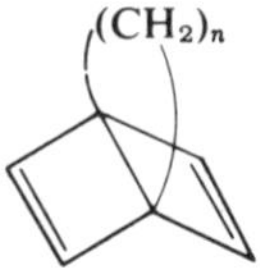

$(CH_2)_n$

**64a** $n = 3$
**64b** $n = 4$

As may be expected, [$n$.3.1]propellanes are more "normal," as the minimum value of $n$ is 3. However, as these species must by definition remain cyclo-propanes and bicyclo[3.1.0]hexanes, we are not surprised by their reactivity. In particular, [3.3.1]propellane reacts with $Br_2$ at $-78°C$ and at $100°C$ has a half-life of ca. eight hours with acetic acid.[85] To us, the most interesting [3.3.1]propellane is 1,3-dehydroadamantane (**65**) and its 5-substituted (**66**) and 5,7-disubstituted derivatives (**67**).[86a–c] The parent hydrocarbon, **65**,

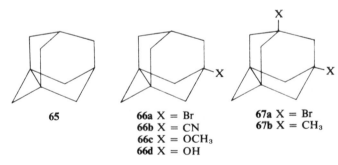

**65**

**66a** X = Br
**66b** X = CN
**66c** X = OCH$_3$
**66d** X = OH

**67a** X = Br
**67b** X = CH$_3$

[83a] I. J. Landheer, W. H. de Wolf, and F. Bickelhaupt, *Tetrahedron Lett.* p. 349 (1975).
[83b] I. J. Landheer, W. H. de Wolf, and F. Bickelhaupt, *Tetrahedron Lett.* p. 2813 (1974).
[84] K. Weinges and K. Klessing, *Chem. Ber.* **107**, 1915 (1974).
[85] P. Warner, R. LaRose, and T. Schleis, *Tetrahedron Lett.* p. 1409 (1974).
[86a] R. E. Pincock and E. J. Torupka, *J. Am. Chem. Soc.* **91**, 4593 (1969).
[86b] R. E. Pincock, J. Schmidt, M. B. Scott, and E. J. Torupta, *Can. J. Chem.* **50**, 3958 (1972).
[86c] W. B. Scott and R. E. Pincock, *J. Am. Chem. Soc.* **95**, 2040 (1973).

readily adds acetic acid, water (in an acid-catalyzed reaction), oxygen (to form the peroxide), iodine, and bromine. The last reaction (in ether) gives both 1,3-dibromoadamantane and 1-bromo-3-ethoxyadamantane, the latter product arising from an unstable but isolable oxonium salt, **68**.[86b] The most convenient derivative of **65** to study is the cyano species, **66b**. This compound may easily be made by reaction of **66a** with NaCN at −35°C, a surprisingly low temperature, as most bridgehead halides are very sluggish in their reactions with nucleophiles. (Analogous displacement reactions have been described with methoxide and hydroxide ions, but the resultant **66c** and **66d** rearrange prior to isolation.) Compound **66b** is very convenient to work with and has, via x-ray crystallography[87] been shown to have a "conjoining" C—C bond length of 1.64 Å. Species **66b** readily polymerizes to form a poly-1,3-adamantylene, **69**, with liberation of ca. 29 kcal/mole. This value is sufficiently

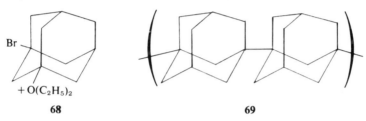

Br

$+ O(C_2H_5)_2$

**68**                                **69**

close to that of the hypothetical polymerization of cyclopropane that the strain energy has been equated for 1,3-dehydroadamantanes and cyclopropanes.[85] However, this equality is invalidated by the knowledge[87] that adamantane itself is strained and that, *a fortiori*, so are the 1,1′-biadamantyl units found in **69**. This leads to a value of ca. 42 kcal/mole for the strain energy in **65**. We additionally note that species **65–67** have two "chair" [3.1.0]bicyclohexane rings. We recall that this bicycloalkane prefers a boat conformation[88] and has a strain energy of 33 kcal/mole.[89]

Ideally, one can imagine the reductive ring closure of species **67a** to the tetradehydroadamantane, **70**, by analogy to the synthesis of **65** and other propellanes cited herein. When this reaction was performed by reducing 1,3,5,7-tetrabromoadamantane **71**, only **65** was produced. This mimics the cleavage of some other small propellanes already cited. It remains, however, unestablished whether this is "merely" a consequence of a high strain in **70** or as we conjecture, a low-lying empty orbital which facilitates the reduction. Other [*n*.3.1]propellanes are known (e.g., species **45** and **46**), but considerations other than strain seem to dominate their chemistry. Numerous other

[87] C. S. Gibbons and J. Trotter, *Can. J. Chem.* **51**, 87 (1973).

[88] R. L. Cook and T. B. Malloy, Jr., *J. Am. Chem. Soc.* **96**, 1703 (1974).

[89] P. von R. Schleyer, J. E. Williams, Jr., and K. R. Blanchard, *J. Am. Chem. Soc.* **92**, 2377 (1974).

and larger propellanes have been described (see Ginsburg[13d]), but most of these species provide little additional insight into the strain of saturated alicyclic species. This does not belittle their intriguing structures; for example, we may cite species 72[90] (cf. 63 and 65), 73[91] (cf. 50 and 51), and 74[92] (cf. 65, 72, and both [3.3.3] and [4.4.3]propellanes, which we draw as 75a and 75b). Nor do we mean to belittle the interrelationships with other strained species.

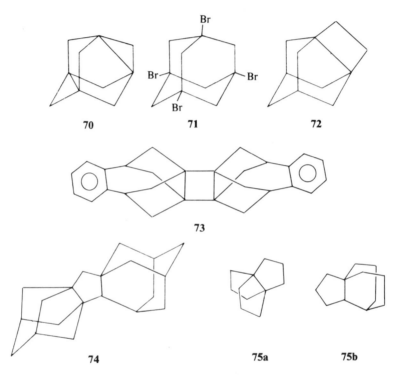

Compound 73 may be recognized as a dimer of a pyramidalized olefin (see Section 3.D.1), while 74 is inextricably related to Pt-catalyzed formation of adamantanes (see Section 4.A.2). It might appear that propellanes are as widespread as a clandestine World War II air force. Indeed, they are; quoting from Ginsburg's index,[13d] we find a collection of miscellaneous propellanes: apopseudoakuammigine, homoechiboline, villamine, etc.

[90] A. G. Yuchenko, A. T. Voroshchvenko, and F. V. Stepanov, *J. Org. Chem. USSR* 6, 188 (1970).

[91] R. Greenhouse, W. T. Borden, K. Hirotsu, and J. Clardy, *J. Am. Chem. Soc.* 99, 1664 (1977).

[92] W. Burns, M. A. McKervey, J. J. Rooney, N. G. Samman, J. Collins, P. von R. Schleyer and E. Osawa, *J. Chem. Soc., Chem. Commun.* p. 95 (1977).

While propellanes are multifarious, they are not nefarious. Indeed, species **50**, **51**, and **73** suggest types of winged apparitions different from the aerial strike force invoked above. These three species may be identified as compounds consisting of propellanes that share a ring with another propellane. We hesitate to call these compounds propellane dimers or bis- or double-propellanes. Instead, we recognize the structure shown in **76** as one of a butterfly.[93] As such, those species with a cyclobutane (cf. "cyclobuttane") ring in the center, such as **77–79**, are relabeled buttaflanes, while those with a

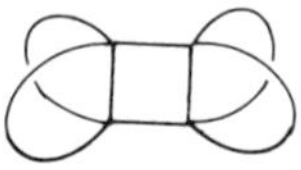

**76**

cyclopentane ring would be pentaflanes, etc. (Regrettably, species **74** is not a pentaflane.) To be specific, compound **50** is renamed [2.2.2.2]buttaflane (**73**), **51** is renamed 1,2:3,4:5,6:7,8-tetrabenzo[2.2.2.2]buttaflane (**78**), and **73** is

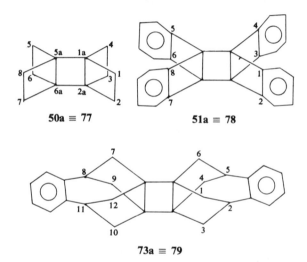

**50a ≡ 77**          **51a ≡ 78**

**73a ≡ 79**

renamed dibenzo[3.3.3.3]buttaflane (**79**). Humor aside, we seriously suggest the names [*m.n.p.q*]buttaflane and [*m.n.p,q*]pentaflane for **80** and **81**.

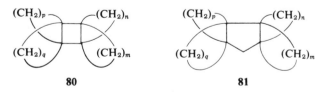

**80**          **81**

[93] J. S. Vincent and J. F. Liebman, personal observations.

We now turn to the so-called "paddlanes," [94] characterizable as having two quaternary carbons joined by four polymethylene chains of varying lengths, **82**. Few paddlanes are known, and before chronicling some species, we note

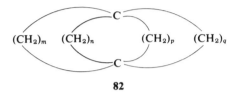

$(CH_2)_m$    $(CH_2)_n$         $(CH_2)_p$      $(CH_2)_q$

**82**

at this time that the four integers $m$, $n$, $p$, and $q$ do not uniquely define a paddlane. If all four numbers are different, then there are the distinct $[m.n.p.q]$ and $[m.n.q.p]$ isomers; if $m = n$ and $p \neq q$, $[m.m.p.q]$ and $[m.p.m.q]$ are still inequivalent; and finally, if $m = n$ and $p = q$, there is still the isomeric pair $[m.m.p.p]$ and $[m.p.m.p]$. This discussion is analogous, indeed equivalent, to that of the possible number of isomers of variously substituted *planar* methanes (see Section 6.C). The reader may recall from undergraduate organic chemistry that this tally count provides considerable support for the idea of tetrahedral carbon. Indeed, for small enough $[m.n.p.q]$paddlanes, the carbons at the bridgeheads may well be forced to be planar.[35,95] However, we remain too ignorant to anticipate the relative stability and interconversion rate of isomers (e.g., $[m.n.p.q] \rightleftharpoons [m.n.q.p]$, **83a** and **83b**, respectively) from the little that is known of experimental paddlane chemistry. Indeed, what constitutes a range of possible values for $m$, $n$, $p$, and $q$ remains unestablished.

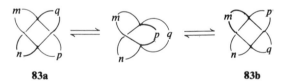

**83a**                                                      **83b**

Some literature paddlanes include derivatives of the [8.2.2.1] member (**84**,[95] **85**,[96] and **86**,[97]), the [12.2.2.1],[98] the [12.3.3.1], (**87**)[99] (but not yet the [12.2.2.2][94]), the [32.2.2.2],[94] and the [7.1.2.1] (dibenzoequinene) **88**.[100a,b] The "a" species are the paddlane frameworks, and we have chosen to draw

[94] E. Hahn, H. Bohm, and D. Ginsburg, *Tetrahedron Lett.* p. 507 (1973).
[95] R. Helder and H. Wynberg, *Tetrahedron Lett.* p. 4321 (1973).
[96] H. Hogeveen and T. B. Middelkoop, *Tetrahedron Lett.* p. 4325 (1973).
[97] R. Helder and H. Wynberg, *Tetrahedron Lett.* p. 4317 (1971).
[98] C. F. H. Allen, and J. A. Van Allen, *J. Org. Chem.* **18**, 882 (1953).
[99] H. Nozaki, T. Koyama, and T. Mori, *Tetrahedron* **25**, 5337 (1961).
[100a] H. H. Wasserman and P. M. Keehn, *J. Am. Chem. Soc.* **89**, 2770 (1967).
[100b] A. V. Fratini, *J. Am. Chem. Soc.* **90**, 1688 (1968).

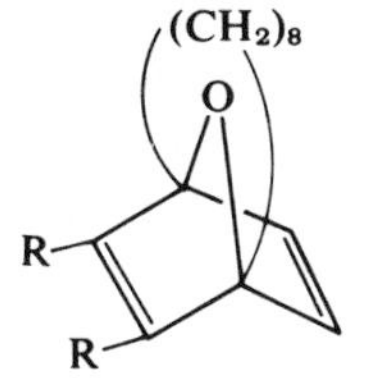

**84**, R = CN
**85**, R = COOC$_2$H$_5$

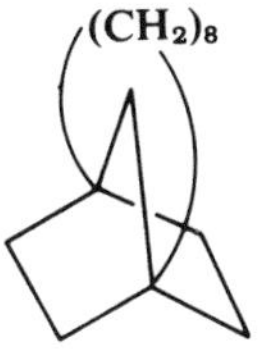

**84a (85a)**

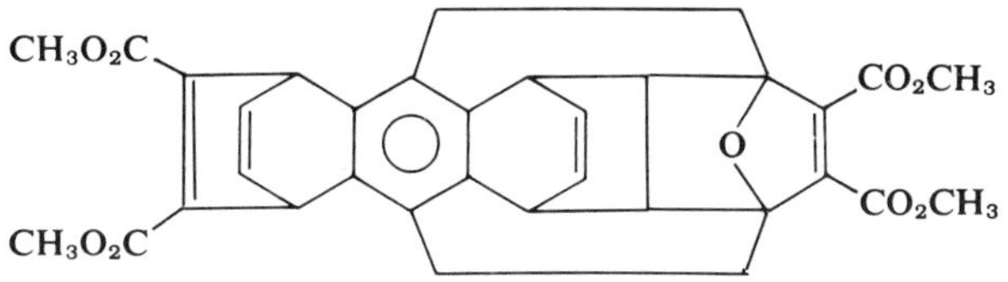

**86**

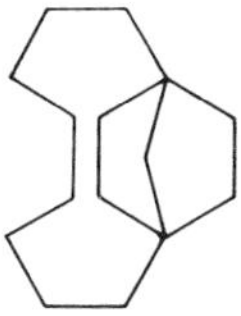

**86a**

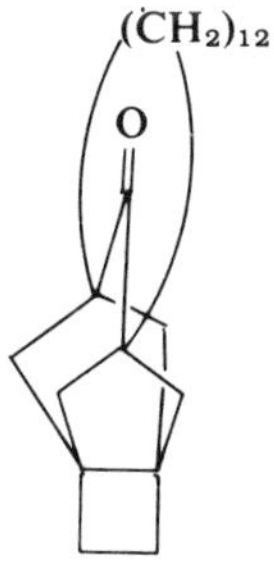

**87**

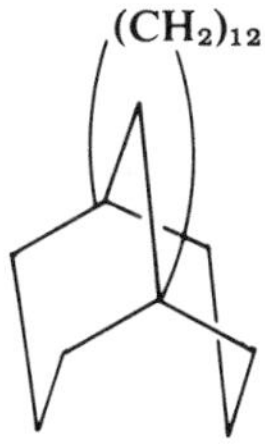

**87a**

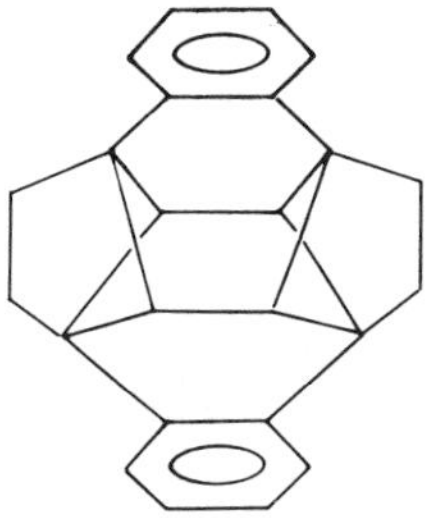

**88**

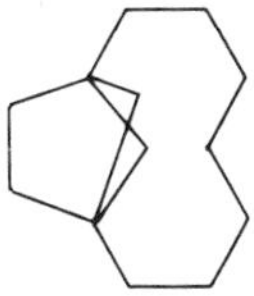

**88a**

only the paddlanes that are somewhat incognito so as to "disguise the limit." Earlier-cited failures include the [3.2.1.2] and ["2½.2.1.2"]paddlanes,[46] and we note a corresponding lack of success for a [3.2.2.1]derivative.[101] It might appear that a large value is needed for at least one of the four numbers $m$, $n$,

---

[101] A. P. Krapcho and R. C. H. Perths, *J. Chem. Soc., Chem. Commun.* p. 1615 (1968).

$p$, and $q$. However, there are so few examples of paddlanes that few regularities can be deduced. In particular, the [3.1.3.1] and [4.1.4.1]paddlanes, **89** and **90**, look essentially "normal" and appear considerably less strained than the isomeric [3.3.1.1] and [4.4.1.1] species **89a** and **90a**, as well as the larger [4.2.4.2]paddlane, **91**. Encouragingly, preliminary results[102] are suggestive of the formation of a [4.1.3.1]paddlane, **92**.

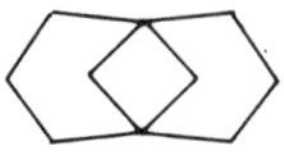

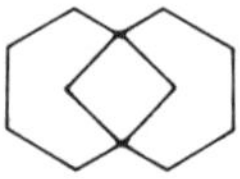

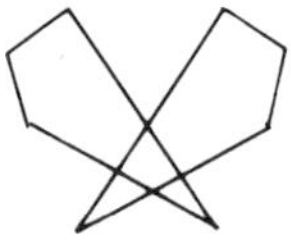

We now briefly mention the possibility[103a,b] of making acetylenic paddlanes (**93**) in which $m$, $n$, and $p$ have been chosen as 8, 10, and 12, respectively. These species are interesting for several reasons. First, it would appear that the central acetylenic unit would be extremely inert because of steric hindrance. Reactions with protons, solvated electrons, and photons (i.e., the photo-chemistry) of these species would be interesting. Indeed, paddlanes generally appear to be good species to hide reactive functionalities. (The subclass of compounds, **93**, has been renamed [$m.n.p$]keuthynes.[103a,b]) Second, these species may alternatively be described as distended propellanes with an effect-ive conjoining bond length of ca. 4.2 Å [$R(C\!-\!C \equiv C\!-\!C) \cong (1.5 + 1.2 + 1.5)$Å]. This suggests that there are relationships between paddlanes and propellanes other than the commonality of their parentage by Ginsburg and their structural novelties and pathologies. Perhaps the simplest interrelation

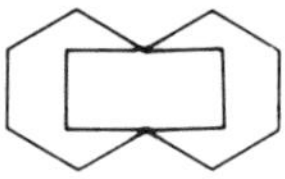

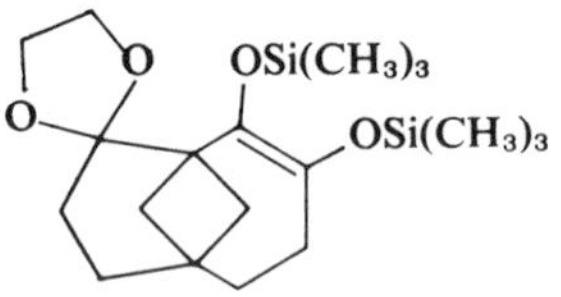

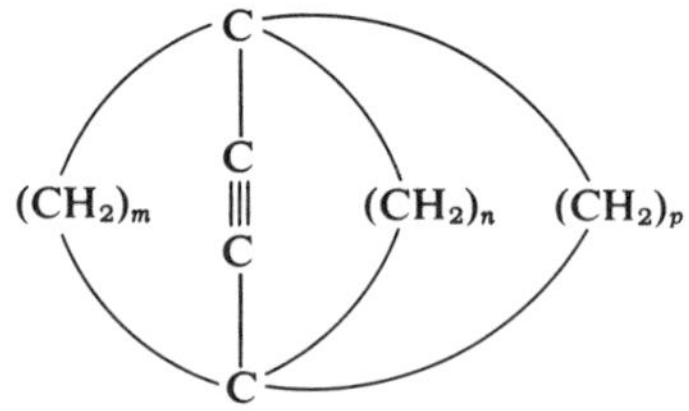

[102] Prof. Philip Warner, personal communication to Joel F. Liebman.

[103a] R. S. Macomber, *Am. Chem. Soc. Natl. Meet. 172nd, San Francisco, 1976* Org. Abstr. No. 240.

[103b] Prof. Roger S. Macomber, personal communication to Joel F. Liebman.

is the systematic "zeroing" of the four numbers that characterize paddlanes. We recognize that [m.n.p.0]paddlane is synonymous with [m.n.p]propellane. Likewise, a [m.n.0.0]paddlane or [m.n.0]propellane is identifiable as a bicyclo-[m.n.0]alkene where the bridgehead carbons are additionally joined by a $\pi$ bond (e.g., species **14** and **15** have $m = 5$, $n = 1$). Extending our analogies still further, we get cycloalkynes (see Section 3.G) and eventually arrive at $C_2$. However, this diatomic molecule does not have a C—C quadruple bond but is seemingly better described as having two $\pi$ bonds and no $\sigma$ bond.[104]

## 6.C  Planar Methane and Its Derivatives

Throughout all of our discussions, we have maintained the description of tetracoordinate carbon as essentially tetrahedral. We usually delegate either to history or the archives of undergraduate student error the idea of planar tetracoordinate carbon. As chroniclers of strained geometries, we may argue that this constitutes "merely" another distortion mode to be considered. And so we will consider it. Nonetheless, before commencing our admittedly brief discussion, a brief quote of Hoffmann's[105] is appropriate:

> Attempts to subvert something as basic to organic chemistry as the tetrahedral tetracoordinate carbon should perhaps be viewed as acts appropriately described by the Yiddish word *chutzpah* and/or the Greek *hubris*.

This psychological awareness did not dissuade him[105,106] or others[107–118]

[104] A. G. Gaydon, "Dissociation Energies and Spectra of Diatomic Molecules," 3rd Ed., pp. 234–236. Chapman & Hall, London, 1968.

[105] R. Hoffman, *Pure Appl. Chem.* **28**, 181 (1971).

[106] R. Hoffmann, R. W. Alder, and C. F. Wilcox, Jr., *J. Am. Chem. Soc.* **92**, 4992 (1970).

[107] G. Herzberg, "Molecular Spectra and Molecular Structure, III: Electronic Spectra and Electronic Structure of Polyatomic Molecules." Van Nostrand, Princeton, New Jersey, 1967.

[108] A. F. Saturno, *Theor. Chim. Acta* **28**, 273 (1967).

[109] H. J. Monkhorst, *J. Chem. Soc., Chem. Commun.* p. 1111 (1968).

[110a] B. M. Gimarc, *J. Am. Chem. Soc.* **93**, 593 (1971).

[110b] B. M. Gimarc, *Acc. Chem. Res.* **7**, 384 (1974).

[111] W. A. Lathan, W. J. Hehre, L. A. Curtiss, and J. A. Pople, *J. Am. Chem. Soc.* **93**, 6377 (1971).

[112] G. Olah and G. Klopman, *Chem. Phys. Lett.* **11**, 604 (1971).

[113a] S. Durmaz, J. N. Murrell, and J. B. Pedlay, *J. Chem. Soc., Chem. Commun.* p. 933 (1972).

[113b] J. N. Murrell, J. B. Pedlay, and S. Durmaz, *J. Chem. Soc., Faraday Trans. 2* **29**, 1370 (1973).

[114] R. A. Firestone, *J. Chem. Soc., Chem. Commun.* p. 163 (1973).

[115] R. J. Buenker and S. D. Peyerimhoff, *Chem. Rev.* **74**, 127 (1974).

[116] H. Wynberg and L. A. Halshof, *Tetrahedron* **30**, 593 (1971).

from theoretically studying the energetics, orbitals, and structures of such species. The energy difference between the square-planar and tetrahedral geometries for methane varies with the calculation, with the most rigorous[117a,b,118] calculation finding the latter more stable by ca. 160 kcal/mole. This value is in excess of the C—H bond strength,[119] so that planar methane itself may seemingly be disregarded.

Some care must be exercised, however, because square-planar $CH_4$ does not uniquely describe the electronic state of the species. (This is true with all formulas but in most cases we "know" the correct answer for at least the ground state.) In particular, there are several options, of which 94 and 95 are the most apparent. Species 94 has four "normal" C—H bonds and an

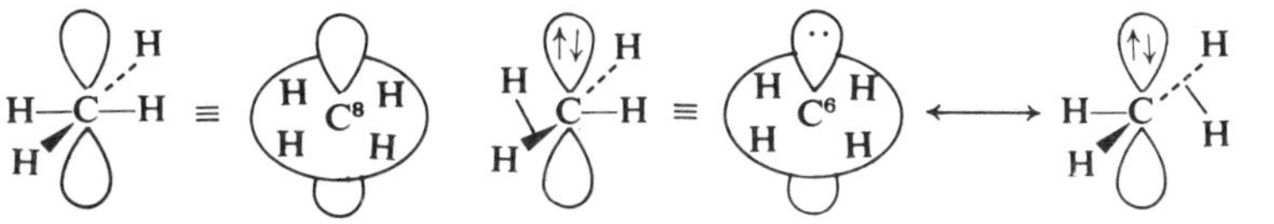

empty $p_z$ or $\pi$ orbital, while 95 has two such C—H bonds, one three-center/two-electron C—$H_2$ unit, and a doubly occupied $p_z$ or $\pi$ orbital, now perhaps better described as a lone pair. (95a and 95b are two resonance structures for this compound.) The above-cited calculations[117a,b,118] show that 95 is lower in energy for $CH_4$ but for $BH_4^-$ and $SiH_4$[118] the isoelectronic analogue of 94 is preferred. The lone pair in 95 is responsible for most of the interesting chemistry that follows. Some unlikely looking species that are calculated[116,118] to be more stable in the planar form than in the tetrahedral are the unsaturated if not aromatic 96, the cyclopropanoid 97 and 98, and acyclic carbenoid 99.

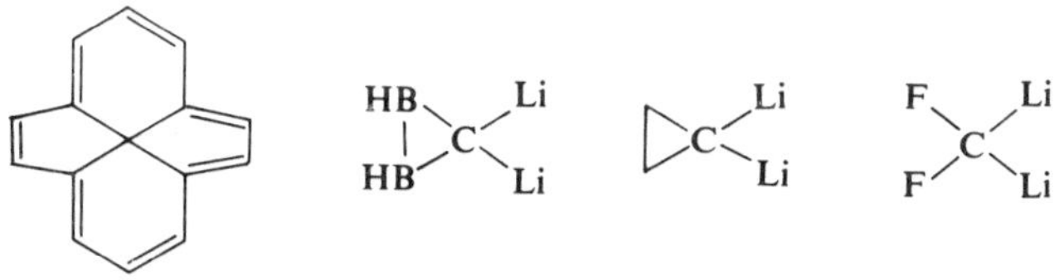

[117a] S. Palalikit, P. Hariharan, and I. Shavitt, unpublished results, *Can. Theor. Chem. Symp. 5th, Ottawa, 1974* (contributed paper).

[117b] Prof. Isaiah Shavitt, personal communication to Joel F. Liebman.

[118] J. B. Collins, J. D. Dill, E. D. Jemmis, Y. Apeloig, P. von R. Schleyer, R. Seeger, and J. A. Pople, *J. Am. Chem. Soc.* 93, 5419 (1976).

[119] B. deB. Darwent, *Natl. Stand. Ref. Data Ser. Natl. Bur. Stand.* NSRDS–NBS 31, (1970).

It is hard to consider such species in "conventional" strain-energy contexts. First of all, the multicenter bonds incumbent in all of these species are hard to define. (We have consciously tried to avoid discussion of species containing such bonds so as to avoid the "classical" vs. "nonclassical" controversy.) Second, general nonbenzenoid aromatic species such as **96** have rarely been considered in connection with strain, and so we laud a recent seemingly successful fusion[120] of molecular-mechanical and quantum-chemical calculations. In addition, **97–99** contain enough hetero atoms and substituents (i.e., atoms other than carbon and hydrogen) to confound most strain-energy analysis. Finally, should, say, **97** and **98** be isolable, it will be hard to readjust our thinking to affirm that these species are more stable than **97a** and **98a** and to consider the latter compounds as "strained."

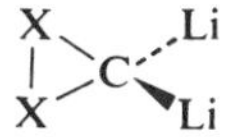

**97a** X = BH
**98a** X = CH$_2$

We now recall that the most rigorous calculations[117a,b,118] show that "square-pyramidal" CH$_4$, **100** ($\theta = 32°$), is more stable than square-planar **95** by ca. 26 kcal/mole. Calculations on substituted derivatives have not been performed. We may expect that a different set of stabilized compounds may well be found and that the relative stability of the square-planar and "square-pyramidal" forms depends on the substituents. We now take a brief theoretical excursion to explain this result. (If the reader wishes to omit this discussion, he

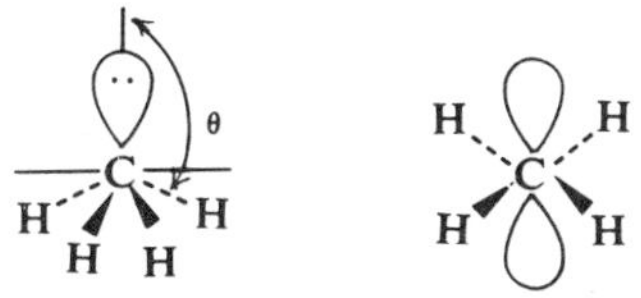

**100**                    **100a**

should proceed to the next paragraph.) The first explanation entails Nyholm–Gillespie or valence shell electron pair repulsion theory.[121] The square-planar ($D_{4h}$) form or isomer corresponds to a six-coordinate species in that it mimics an octahedral arrangement of groups. By contrast, the "square-pyramidal" ($C_{4v}$) form likewise corresponds to a five-coordinate geometry and indeed is a quite common arrangement for five groups. As five is closer to four (the normal coordination number) than six is, "square pyramidal" may be said to be more appropriate than square planar. The second explanation[110b] mimics the reason ammonia is pyramidal instead of planar: the highest occupied orbital in the

[120] J. Kao and N. L. Allinger, *J. Am. Chem. Soc.* **99**, 975 (1977).
[121] R. J. Gillespie, *J. Chem. Educ.* **40**, 295 (1963).

planar form is a $\pi$ orbital that is nonbonding with respect to the four hydrogens. When the molecule is bent from square planar to "square pyramidal," there is now positive overlap and so bonding with the four hydrogens. Accordingly, the molecule prefers a $C_{4v}$ geometry over $D_{4h}$. [We remind the reader that analogous orbital logic also shows why tetrahedral ($T_d$) geometry is preferred over $D_{4h}$ in the absence of any special electronic or steric effects.] The third approach recasts $CH_4$ as a complex of $CH_2$ and $H_2$, analogous to the earlier description of cyclopropane as a complex of $CH_2$ and ethylene (see Section 2.A). Although it has been outdated by other approaches to the structure of this three-membered ring, we nonetheless recall the connection with the Skell hypothesis of carbene additions to olefins (see Section 2.A). This suggests $CH_2 \cdot H_2$ is better described as **101** than as **102**, although both are poorer than **103**, i.e., tetrahedral methane. The following is a thermochemical

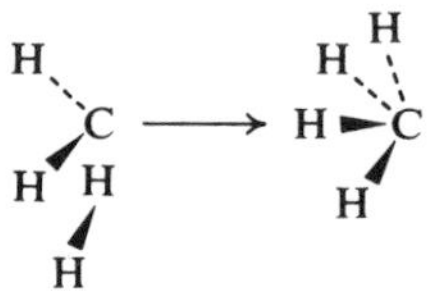

**101** ($C_{4v}$)

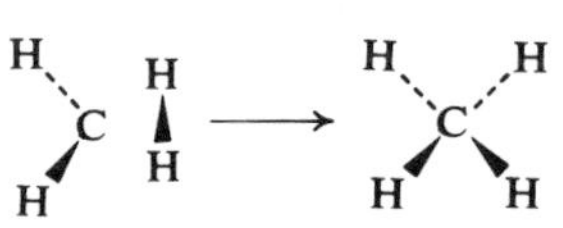

**102** ($D_{4h}$)

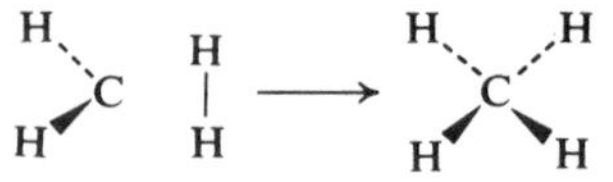

**103** ($T_d$)

cycle (Scheme 1) describing a complex of $CH_2$ and $H_2$ as nearly isoenergetic with the $C_{4v}$ "square-pyramidal" methane already discussed. (The numbers above the arrows are the respective heats of reaction in kcal/mole.)

$$CH_4\ (T_d) \xrightarrow{103} CH_3 + H \xrightarrow{113} {}^3CH_2 + H + H \xrightarrow{-104} {}^3CH_2 + H_2$$

$$CH_4\ (T_d) \xrightarrow{160} CH_4\ (D_{4h}) \qquad {}^3CH_2 + H_2 \xrightarrow{19.5} {}^1CH_2 + H_2$$

$$CH_4\ (D_{4h}) \xrightarrow{-26} CH_4\ (C_{4v}) \xleftarrow{+4} \underset{(C_{2v}\ CH_4)}{CH_2\text{---}H_2} \xleftarrow{-2} {}^1CH_2 + H_2$$

**Scheme 1**

The energetics for chipping off hydrogens from $CH_4$ to form sequentially $CH_3 + H$ and $CH_2 + 2H$, and for $H_2$ formation, were taken from Darwent;[119] the singlet–triplet energy difference for $CH_2$ is from Zittel *et al.*;[122] binding

[122] P. F. Zittel, G. B. Ellison, S. V. O'Neil, E. Herbst, W. C. Lineberger, and W. P. Reinhardt, *J. Am. Chem. Soc.* **98**, 3731 (1976).

energy of $CH_2$ and $H_2$ from Collins *et al.*,[123] where the corresponding energies were given for $H_2$ binding with LiH and $BH_3$; and the distortion energies of methane were taken from Palalikit *et al.*[117a] and Collins *et al.*[118] This analysis would also suggest that square-pyramidal tetracoordinate carbon compounds have a tendency to decompose into $CR_2$ and $R_2$. We note that this will depend on the substituents, much as substituents modify the energy difference of tetrahedral and square-planar geometries.

Electronically innocuous substituents, such as saturated organic groups, should not modify the relative stability of square-planar and "square-pyramidal" carbon. We thus believe that alicyclic hydrocarbons containing suitable quaternary carbons presumably have local geometries that mimic **100** rather than **100a**. It is precisely this type of compound for which strain-energy logic has been most commonly employed. This suggests reevaluation of some literature proposals involving square-planar carbon. In particular, we recall the proposed synthesis of "fenestrane," **104**,[124a,b] (tetracyclo[3.3.1.0.$^{3,9}$0$^{7,9}$]-nonane). We believe that the all-*cis* isomer **105** will be more stable. Indeed, excluding the four-membered rings, the latter species (redrawn as **105a**) does not look exceptionally strained. ([2.2.2.2]Paddlane[35] may be redrawn analogously as **106**, and appears "reasonable.")

| **104** | **105** | **105a** | **106** |

We may define a class of compounds [*m.n.p.q*]fenestranes, **107**, wherein $m, n, p$, and $q$ are the ring sizes. Accordingly, species **105** and **106** are [4.4.4.4]-fenestranes, and the literature precursor, **108**, is a [6.6.5.4]fenestrane derivative.

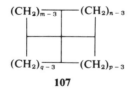

**107**

Few fenestranes have been reported. The parent [5.5.5.5]hydrocarbon is unknown to us, but a tetranitrogen derivative, **109** (1,4,7,10-tetraazatetracyclo-[3.3.1.0.$^{4,13}$0$^{10,13}$]tridecane) has been synthesized and was shown to "undergo a

[123] J. B. Collins, P. von R. Schleyer, J. S. Binkley, J. A. Pople, and L. Radom, *J. Am. Chem. Soc.* **98**, 3436 (1976).

[124a] V. Georgian and M. Saltzman, *Tetrahedron Lett.* p. 4315 (1972).

[124b] Prof. Vlasios Georgian and Martin Saltzman, personal communication to Joel F. Liebman.

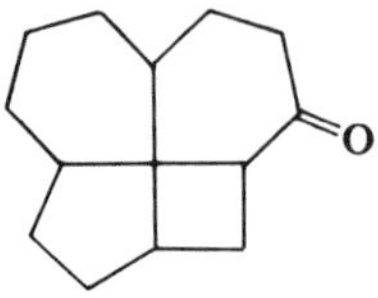

**108** 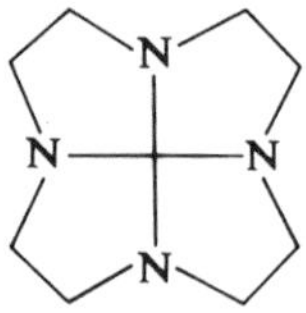            **109** 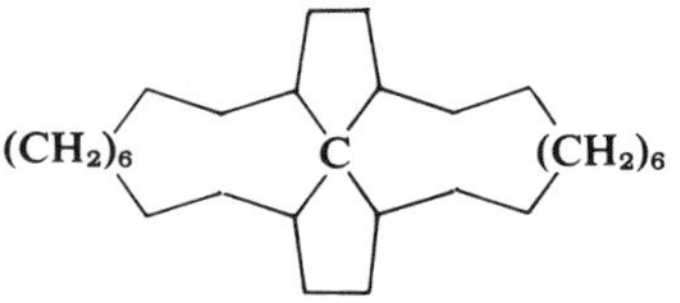                        **110**

remarkable series of degenerate prototropic and conformational equilibria."[125] We also identify [6.6]vespirene, **2**, as a [5.13.5.13]fenestrane derivative, **110**. (This comparison is useful insofar as racemization of vespirenes may be suggested to proceed through "square-pyramidal" intermediates instead of square-planar ones.) Finally, we note that the porphyrin $+Fe$, $+Mg$ and corrin $+Co$ "nuclei" of hemoglobin,[126a] chlorophyll[126b] and vitamin $B_{12}$[126c] may be identified as [6.6.6.6], [6.6.6.6], and [5.6.6.6]fenestrane derivatives, respectively. (We leave it to the reader to confirm this assignment and to provide any biochemical significance.) We may additionally define a hierarchy of structures where $C$—$CH_2$ bonds in fenestranes are sequentially cut and replaced by $C$—$H$ bonds. These [m], [m.n], and [m.n.p]fenestranes (which may be spoken of as broken windows) have precedent absent for the smaller [m.n.p.q]fenestranes. [m] and [m.n]fenestranes may be identified as cyclo-alkanes and spiroalkanes. [m.n.p]Fenestranes remain rare but include such species as **111** and **112**, a degradation product of **108** and precursor of the hoped-for planar **113**.[127]

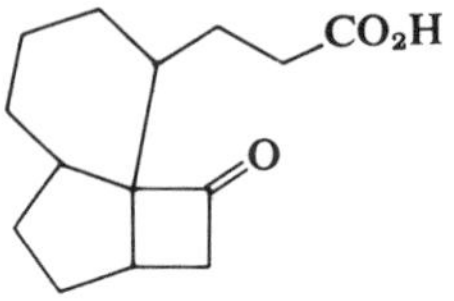

**111** 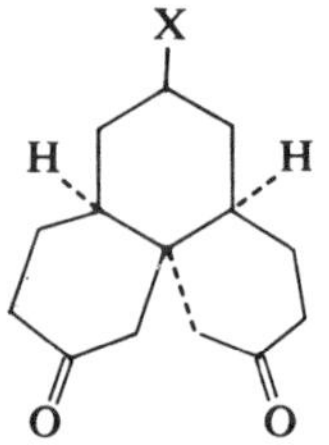                  **112** 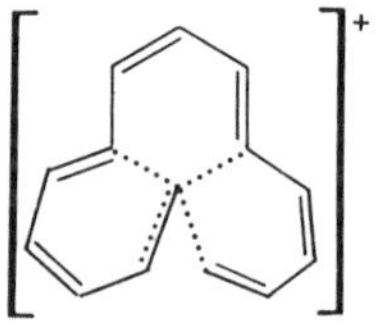                    **113**

Square-planar and "square-pyramidal" carbon-containing species are already extant in the experimental literature, although they are well hidden. (We remind the reader that Ginsburg[13d] noted this earlier, for propellanes.) For example, we recall an interesting study[128] of the reaction of alkali-metal

[125] J. E. Richman and H. E. Simmons, *Tetrahedron* **30**, 1769 (1974).

[126a] A. L. Lehninger, "Biochemistry," 2nd Ed., p. 491. Worth Publ., New York, 1975.

[126b] See Lehninger (Ref. 126a), p. 595.

[126c] See Lehninger (Ref. 126a), p. 349.

[127] C. E. Wilcox, Jr. and G. D. Graham, *J. Org. Chem.* **40**, 1974 (1975).

[128] P. Haberfield and R. B. Trattner, *J. Chem. Soc., Chem. Commun.* p. 1481 (1971).

2-naphtholates with acyl chlorides to form the naphthyl ester and acyl naphthol. The products are not surprising. However, from a thorough analysis of the data that considered ion aggregation, solvation, and the ambient nature of their nucleophile, these authors suggested a square-planar intermediate, **114**, as opposed to one that is essentially tetrahedral (**115**). Other more definitive but less "organic" examples are the carborane $B_5C_3H_7$

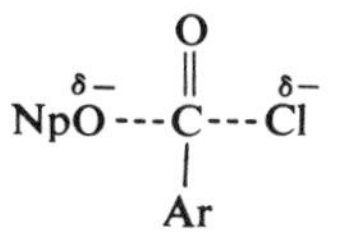

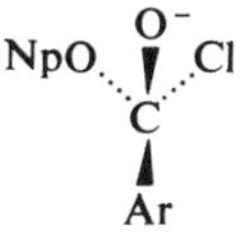

**114**            **115**

or $C(CH)_2(BH)_5$, **116**[129] with an "exopolyhedral" lone pair of electrons, and the iron carbide carbonyl, $Fe_5C(CO)_{15}$ or $C[Fe(CO)_3]_5$, **117**[130] with its "inverted square-pyramidal carbon." We conclude by recalling that in alicyclic species with tetracoordinate carbon having "square-pyramidal"

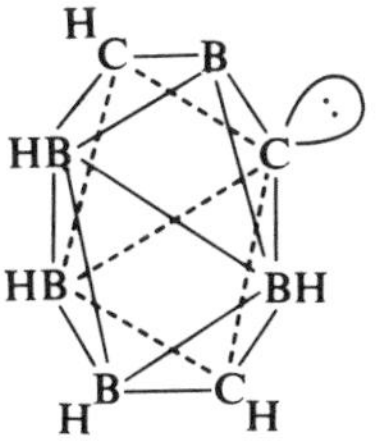

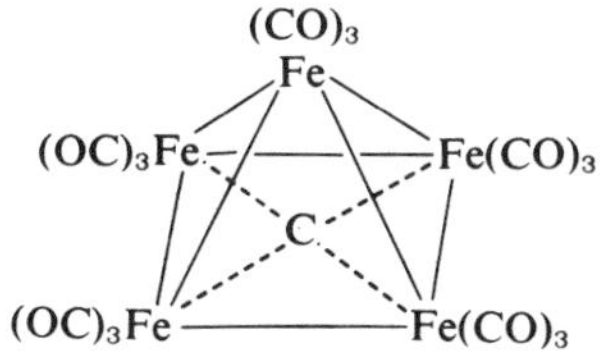

**116**            **117**

geometry, the lone pair is seemingly unused (cf. species **95** and **100**). This suggests the possibility of Lewis basicity and of pentacoordinate carbon, the latter being the topic of the next and final class of compounds to be discussed in this chapter on structural pathologies.

## 6.D   Five- and Six-Coordinate Carbon

Five-coordinate carbon has long been postulated in intermediate or transition states in $S_N2$,[131] $S_H2$,[132] and $S_E2$[133,134] reactions, and indeed they

[129] M. L. Thompson and R. N. Grimes, *J. Am. Chem. Soc.* **93**, 6677 (1971).

[130] E. H. Braye, L. E. Dahl, W. Hubel, and D. L. Wampler, *J. Am. Chem. Soc.* **84**, 4633 (1962).

[131] A. Streitwieser, Jr., "Solvolytic Displacement Reactions." McGraw-Hill, New York, 1962.

[132] K. U. Ingold and B. P. Roberts, "Free Radical Substitution Reactions." Wiley (Interscience), New York, 1971.

seem inherent in these bimolecular displacements. Ion triplets have also been suggested for $S_N2$[135] and $S_E2$[133] but even these would qualify as five-coordinate species. (We remind the reader that "covalent" and "ionic" are convenient words but are nonetheless artificial descriptions of molecular charge densities or distributions.) We recall that surprisingly stable species containing five- and/or six-coordinate carbon have been chronicled among the carboranes and other cluster compounds.[136-140] Among these compounds are to be found octahedral and icosahedral structures, the two Platonic solids or regular polyhedra omitted in Section 4.D.1. Octahedral species may have the higher-coordinated carbon interior or exterior to the polyhedron. We may identify species **116** and the isomeric carboranes **118** and **119**[141a,b] as having the carbons "exterior," while the ruthenium carbide carbonyl, **120**,[142,143] has carbon "interior." (For most of the compounds shown in this section, as well as **116** and **117** in the previous section, the lines connecting the various atoms need *not* be construed as bonds.) All the icosahedral structures, for example

---

[133] D. J. Cram, "Fundamentals of Carbanion Chemistry." Academic Press, New York, 1965.

[134] G. A. Olah, "Carbocations and Electrophilic Reactions." Wiley, New York, 1974.

[135] N. F. Raaen, T. Juhlke, F. J. Brown, and C. J. Collins, *J. Am. Chem. Soc.* **96**, 5928 (1974).

[136] R. N. Grimes, "Carboranes." Academic Press, New York, 1970.

[137a] G. B. Dunksgard and N. F. Hawthorne, *Acc. Chem. Res.* **6**, 124 (1973).

[137b] M. F. Hawthorne, K. R. Callahan, and R. J. Wiersma, *Tetrahedron* **30**, 1795 (1974).

[138a] E. L. Muetterties, *in* "Boron Hydride Chemistry" (E. L. Muetterties, ed.), Ch. 1. Academic Press, New York, 1975.

[138b] W. N. Lipscomb, *in* "Boron Hydride Chemistry" (E. L. Muetterties, ed.), Ch. 2. Academic Press, New York, 1975.

[138c] H. Beall, *in* "Boron Hydride Chemistry" (E. L. Muetterties, ed.), Ch. 9. Academic Press, New York, 1975.

[138d] T. Onak, *in* "Boron Hydride Chemistry" (E. L. Muetterties, ed.), Ch. 10. Academic Press, New York, 1975.

[138e] G. B. Dunks and M. F. Hawthorne, *in* "Boron Hydride Chemistry" (E. L. Muetterties, ed.), Ch. 11. Academic Press, New York, 1975.

[138f] P. A. Wegner, *in* "Boron Hydride Chemistry" (E. L. Muetterties, ed.), Ch. 12. Academic Press, New York, 1975.

[139a] K. Wade, *New Sci.* **62**, 615 (1974).

[139b] K. Wade, *Chem. Brit.* **11**, 177 (1975).

[139c] K. Wade, *Adv. Inorg. Nucl. Chem.* **18**, 1 (1976),

[140] R. E. Williams, *Adv. Inorg. Nucl. Chem.* **18**, 67 (1976).

[141a] R. A. Beaudet and R. L. Poynteur, *J. Chem. Phys.* **53**, 1899 (1970).

[141b] G. L. McKown and R. A. Beaudet, *Inorg. Chem.* **10**, 1350 (1971).

[142] B. F. G. Johnson, R. D. Johnston, and J. Lewis, *J. Chem. Soc. A* p. 2865 (1968).

[143] A. Sirigu, M. Bianchi, and E. Benedetti, *J. Chem. Soc., Chem. Commun.* p. 596 (1969).

**118a** **118b** **118c** **119a** **119b**

**119c** **120a** **120b**

the isomeric carboranes **121–123**,[144a,b,145] have exterior carbons. Indeed, there are so few cases of "interior" carbons that it is probably safe to drop the adjectives "interior" and "exterior" in the discussion that follows.

**121a** **121b** **122** **123a** **123b**

Although we argue that "inorganic" and "organic" are antiquated and artificial adjectives, we must admit that species **116–123** have too few carbons to be discussed further in our text owing to limitations of space and continuity with what else we have written. (The reader will note that compound **117** has 16 carbons, while **120** has 18, but in each case no more than one carbon is at all "organic.") One can hypothesize the existence of octahedral $(CH)_6$ (**124**)

[144a] J. A. Potenza and W. N. Lipscomb, *Inorg. Chem.* **5**, 1471 (1966).
[144b] H. Beall and W. N. Lipscomb, *Inorg. Chem.* **6**, 874 (1967).
[145] R. K. Bohm and M. D. Bohn, *Inorg. Chem.* **10**, 350 (1971).

and icosahedral $(CH)_{12}$ (125) and present them as possibly new members of the six- and twelve-carbon variety of valence isomers in the $(CH)_n$ series (see Section 4.D.2). Explicit quantum-chemical calculations on 124 show it

124a                          124b                          125

to be unstable relative to benzene,[146] and indeed, its dianion $(CH)_6{}^{-2}$ has been suggested[147] to be unstable relative to two cyclopropenyl anion molecules, despite their antiaromaticity (see Section 3.D.3). We may additionally inquire whether 124 has any unpaired electrons, as the presence thereof would suggest high reactivity accompanying low stability. The above calculations[146,147] are moot. However, if we trust the isoelectronic principle, then the orbital-energy pattern for $(CH)_6$ should be the same as that for $(BH)_6{}^{-6}$. Calculations on this latter anion are unreported. However, theoretical studies on $(BH)_6{}^{-2}$ show[148] the lowest unoccupied molecular orbital to be of $f_{2u}$ ($t_{2u}$) symmetry (i.e., triply degenerate). Accordingly, by reasoning based on Hund's rule and the Jahn–Teller principle analogous to octahedral metal complexes,[149a] octahedral $(CH)_6$ (124) should have two unpaired electrons but still must deform from its highly symmetric geometry. Figure 6.1 shows the interconversion of 124 to benzene (126) and to prismane (126a). (We recognize the former process as an interconversion of a "closo" to "arachno" geometry,[139a–c,140] where "closo" means a totally triangulated polyhedron or deltahedron, "arachno" means two nontriangular faces, and for completeness, "nido" means but one nontriangular face. The latter process is a "polytopal rearrangement" of the "six-atom" group.[138a]

CNDO studies on the closed-shell singlet icosahedral 125 show it to be ca. 600 kcal/mole less stable than two benzenes [corresponding calculations find octahedral $(CH)_6$ to be over 150 kcal/mole less stable than benzene].[149b] For $(CH)_{12}$ strict icosahedral geometry is not expected, as corresponding orbital analysis derived from the orbital energy levels of $(BH)_{12}{}^{-2}$ shows[148] an analogous distortion when we add ten electrons. (While the lowest unoccupied

[146] M. D. Newton, J. M. Schulman, and M. M. Manus, *J. Am. Chem. Soc.* **96**, 17 (1974).

[147] K. van der Meer and J. J. C. Mulder, *Theor. Chim. Acta* **41**, 183 (1976).

[148] R. Hoffmann and W. N. Lipscomb, *J. Chem. Phys.* **36**, 2179 (1962).

[149a] F. A. Cotton and G. Wilkinson, "Advanced Inorganic Chemistry," 3rd Ed., Sect. 20–3. Wiley (Interscience), New York, 1972.

[149b] R. M. Cooper and J. F. Liebman, unpublished results.

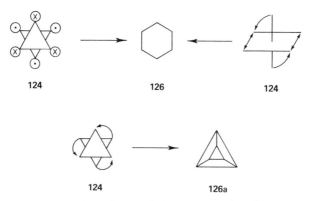

**Figure 6.1.** Interconversion of octahedral $(CH)_6$, **124**, into benzene, **126**, and prismane, **126a**, where the unmarked vertices are CH groups. The arrows show the required atomic motion and we remind the reader that · and ×, the arrow tip and tail, respectively, represent motion up and down relative to the plane of the paper.

molecular orbital of the boron species is "but" fourfold degenerate ($g_g$ or $u_g$), the next higher orbital is of fivefold degeneracy ($h_u$ or $v_u$). (Figure 6.2 interrelates **125** and an unbridged paracyclophane or benzene dimer). We wish to emphasize that these multiplicity and distortion conclusions are not a consequence of the high energy of the orbitals of interest but merely of their degeneracy and symmetries. More precisely, the conclusions are independent of the exotic higher coordination numbers for carbon atoms in these species.

Turning from the "inorganic" **116–123** and the hypothetical **124** and **125**, we now chronicle existent "organic" compounds with analogous structural features. In particular, we cite the "pyramidal" cations[150] **127** and **128**, which may be seen to mimic the top halves of **124** and **125**, respectively. Both species have been of intense interest, with the former characterized better by both

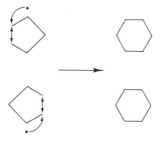

**Figure 6.2.** Interconversion of icosahedral $(CH)_{12}$ into an unbridged paracyclophane or benzene dimer. The arrows show the required atomic motion.

[150] H. Hogeveen and P. W. Kwant, *Acc. Chem. Res.* **8**, 413 (1976).

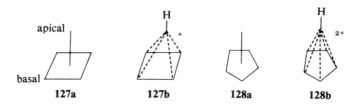

| 127a | 127b | 128a | 128b |

theory[151–155] and experiment.[150] Indeed, species **127** is one of those lovely but unlikely species that first was predicted theoretically[151,152] and soon afterward characterized experimentally.[156a,b,157] "Isoelectronic" comparisons with boron hydrides[139a–c,140,151,152] and with other "cluster compounds"[139a–c,140] have proved highly useful in understanding **127** as well as numerous members of these other classes of compounds. In particular, we may cite the square-pyramidal geometry of pentaborane(9), **129**[158a,b] and cyclobutadiene iron tricarbonyl, **130**.[159] Fragmentation into $(CH)^+$ and the antiaromatic $(CH)_4$[152,153,154b,156a,b,157] has provided additional insights and comparisons. For example, this logic suggests $(CH)_6^{2+}$ can be reformulated as a complex of $(CH)^+$ and the antiaromatic $(CH)_5^+$.[160a,b,161] Correspondingly, the trigonal pyramidal $(CH)_4$, or tetrahedrane, "can be viewed as the covalent representation of the product from $HC:^+$ and the antiaromatic cyclopropenyl anion," and thereby one can deduce from "the simplest Hückel model that the stability order will be $(CH)_6^{2+} > (CH)_5^+ > (CH)_4$."[162a,b] This returns us to four-coordinate carbon and normal chemical

[151] R. E. Williams, *Inorg. Chem.* **10**, 210 (1971).

[152] W.-D. Stohrer and R. Hoffmann, *J. Am. Chem. Soc.* **90**, 1661 (1972).

[153] S. Yoneda and Z. Yoshida, *Chem. Lett.* p. 607 (1972).

[154a] H. Kollmar, H. O. Smith, and P. von R. Schleyer, *J. Am. Chem. Soc.* **95**, 5834 (1973).

[154b] W. J. Hehre and P. von R. Schleyer, *J. Am. Chem. Soc.* **95**, 5838 (1973).

[155] M. J. S. Dewar and R. C. Haddon, *J. Am. Chem. Soc.* **95**, 5836 (1973); 255 (1974).

[156a] S. Masamune, M. Sakai, and H. Ona, *J. Am. Chem. Soc.* **94**, 8955 (1972).

[156b] S. Masamune, M. Sakai, H. Ona, and A. L. Jones, *J. Am. Chem. Soc.* **94**, 8956 (1972).

[157] H. Hart and M. Kuzuya, *J. Am. Chem. Soc.* **94**, 8958 (1972).

[158a] W. J. Dulmage and W. N. Lipscomb, *Acta Crystallogr.* **5**, 260 (1952).

[158b] K. Hedberg, M. E. Jones, and V. Schomaker, *Proc. Natl. Acad. Sci. U.S.A.* **38**, 679 (1952).

[159] A. C. Villa, L. Coghi, A. G. Manfredotti, and C. Guastini, *Acta Crystallogr., Sect. B* **30**, 2101 (1974).

[160a] H. Hogeveen and P. W. Kwant, *Tetrahedron Lett.* p. 1655 (1973).

[160b] H. Hogeveen and P. W. Kwant, *J. Am. Chem. Soc.* **96**, 1965 (1974).

[161] H. T. Jonkman and W. C. Nieuwport, *Tetrahedron Lett.* p. 1671 (1973).

[162a] H. Hart and M. Kueuza, *J. Am. Chem. Soc.* **96**, 6436 (1974), footnote 6.

[162b] Personal communication from Prof. Harold Hart, Prof. James F. Harrison, and Dr. Masayumi Kuzuya to Joel F. Liebman and Arthur Greenberg.

structures and descriptions, although tetrahedrane would scarcely be considered normal in any other context.

The reader may have noticed that we referred to the "lovely" $(CH)_5^+$ and the "antiaromatic" $(CH)_5^+$, the square-pyramidal cation **127** and a destabilized cyclopentadienyl cation, **131**. What is the preferred structure of $(CH)_5^+$ ? We recall that cyclobutadiene is more stable than tetrahedrane (see Section 3.C), and so it is legitimate to inquire about the relative stability of these five-carbon ions. We also know to be wary when comparing isoelectronic all-carbon and all-boron cluster species. For example, bicyclobutane **(132)**[163]

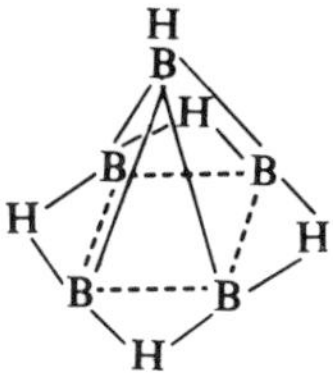

|     |     |     |     |
| --- | --- | --- | --- |
| **129** | **130** | **131s** | **131t** |

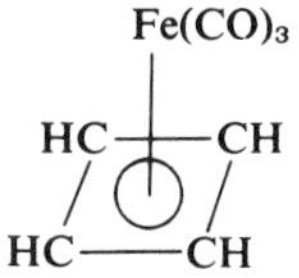

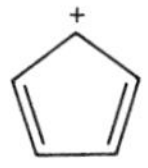

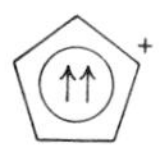

and tetraborane(10), **133**[164] are isoelectronic "nido" species with similar geometries. However, bicyclobutane is experimentally less stable than 1,3-butadiene, while for the corresponding boron case, tetraborane(10) is more stable than the unknown bisdiboranyl, **134**. Are ions **127** (and **128**) "merely"

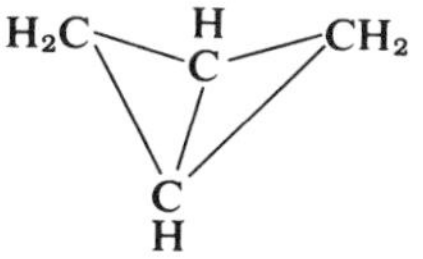

|     |     |
| --- | --- |
| **132** | **133** |

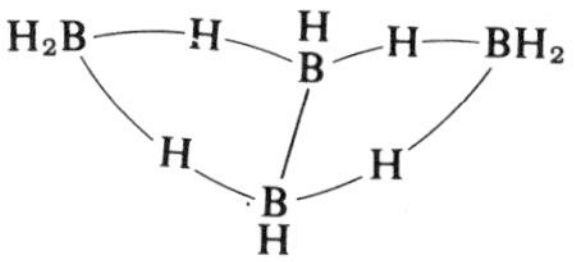

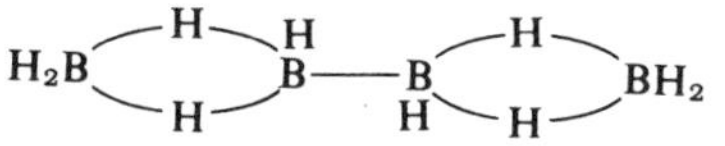

**134**

theoreticians' ideas of chemical "reality"? Experimentally, the relative stability of the five-carbon ions depends on the substituent. In accord with the most rigorous calculations,[154a,b,155] the unsubstituted species is a pentagonal triplet, **131t**.[165] Likewise, the pentachloro and pentaphenyl derivatives

[163] K. W. Cox, M. D. Harmony, G. Nelson, and K. B. Wiberg, *J. Chem. Phys.* **50**, 1976 (1968).

[164] E. B. Moore, R. E. Dickerson, and W. N. Lipscomb, *J. Chem. Phys.* **27**, 209 (1957).

[165] M. Saunders, R. Berger, A. Jaffe, J. M. McBride, J. O'Niell, R. Breslow, J. M. Hoffman, Jr., C. Perchonock, E. Wasserman, R. S. Hutton, and V. J. Kuck, *J. Am. Chem. Soc.* **95**, 3017 (1973).

are also pentagonal, but while the former is a triplet[166a,b] like the parent, the pentaphenyl[166c,167a,b] (and other pentaaryl derivatives)[167a,b] is slightly stabilized as singlets ($\leq 1$ kcal/mole; cf. **131s**). Aliphatic derivatives such as the dimethyl **135**[156a,b,168a,c] and bishomo compounds[157,168a–170b] such as **136** (R = H,[168a–c,169] R = CH$_3$[170a,b]) have spectroscopic properties consistent with the square-pyramidal structure. (For the bishomo compounds, there

**135**

**136a** R = H
**136b** R = CH$_3$

additionally seems to be some disparity between theory and experiment with regard to stabilization of pyramidal cations by alkyl[171a] (and aryl[171b]) groups, but here other rearrangement processes (and steric inhibition of resonance) are additionally possible. It is additionally found that "O$^-$"[152] and other anionic substituents such as "CH$_2^-$" stabilize the pentagonal singlet structure **131s**—we know of no case where cyclopentadienones, **137**,[172,173] or fulvenes, **138**,[173,174a,b] exist in the pyramidal form. A useful generalization is that

[166a] R. Breslow, R. Hill, and E. Wasserman, *J. Am. Chem. Soc.* **86**, 5349 (1964).

[166b] R. Breslow, H. W. Chang, R. Hill, and E. Wasserman, *J. Am. Chem. Soc.* **89**, 1112 (1967).

[166c] R. Breslow, H. W. Chang, and W. A. Yager, *J. Am. Chem. Soc.* **85**, 2033 (1963).

[167a] W. Brosser, H. Kurreck, and P. Siegle, *Chem. Ber.* **100**, 78 (1967).

[167b] W. Brosser, P. Siegle, and H. Kurreck, *Chem. Ber.* **101**, 39 (1969).

[168a] S. Masamune, M. Sakai, A. V. Kemp-Jones, H. Ona, A. Venot, and T. Nakashina, *Angew. Chem., Int. Ed. Engl.* **12**, 769 (1973).

[168b] A. V. Kemp-Jones, N. Nakamura, and S. Masamune, *J. Chem. Soc., Chem. Commun.* p. 1001 (1974).

[168c] A. V. Kemp-Jones, N. Nakamura, and S. Masamune, *J. Chem. Soc., Chem. Commun.* p. 109 (1974).

[169] R. M. Coates and K. Yano, *Tetrahedron Lett.* p. 2289 (1972).

[170a] H. Hart and M. Kuzuya, *J. Am. Chem. Soc.* **94**, 8958 (1972); **95**, 4096 (1973); **96**, 6436 (1974).

[170b] H. Hart and M. Kuzuya, *Tetrahedron Lett.* p. 4123 (1973).

[171a] H. Hart and M. Kuzuya, *J. Am. Chem. Soc.* **97**, 2459 (1975).

[171b] H. Hart and M. Kuzuya, *J. Am. Chem. Soc.* **97**, 2450 (1975).

[172] M. A. Ogliaruso, M. G. Romanelli, and E. I. Becker, *Chem. Rev.* **65**, 261 (1965).

[173] P. J. Garratt, "Aromaticity," Ch. 6. McGraw-Hill, New York, 1972.

[174a] E. D. Bergmann, *Chem. Rev.* **68**, 41 (1968).

[174b] P. Yates, *Adv. Alicyclic Chem.* **2**, 60 (1968).

137 X = O
138 X = CR$_2$

substituents with either unpaired or $\pi$ electrons seem to stabilize the planar cyclopentadienyl form.

Analogous findings apply to the $(CH)_6^{2+}$ ion and its derivatives. The hexamethyl derivatives,[160a,b,161] like numerous boranes and cluster compounds,[136,140] have the pentagonal–pyramidal geometry given in structure **128**. For example, the carborane $(CH)_4(BH)_2$[141b,175,176] has the related structure, **140**. However, the difluoro analogue of this species, $(CH)_4(BF)_2$, is a planar hexagon, **141**.[177] The structure of this latter species should remind the reader of $p$-benzoquinone. Indeed, substituents such as $O^-$ and $CH_2^-$ on **128** revert the pyramid to planarity—we know of no quinone or quinone methide[178a–g] with a pyramidal geometry. The hexachloro dication is a planar triplet,[179] while the seemingly isoelectronic hexa-$O^-$ derivative of $(CH)_6^{2+}$ [i.e., the tetraanion of 2,3,5,6-tetrahydroxy-$p$-benzoquinone (rhodizonate ion) **142**] is the corresponding planar singlet.[180a,b] Indeed, the above-cited hexamethyl

[175] I. A. McNeil, K. L. Gallaher, F. R. Scholer, and S. H. Bauer, *Inorg. Chem.* **12**, 2108 (1973).

[176] V. S. Mastryukov, O. V. Dorofeeva, L. V. Vilkov, A. F. Zhigach, V. T. Laptev, and A. B. Petrunin, *J. Chem. Soc., Chem. Commun.* p. 276 (1973).

[177] P. L. Timms, *J. Am. Chem. Soc.* **90**, 4584 (1968).

[178a] G. J. Gleicher, *in* "The Chemistry of the Quinonoid Compounds" (S. Patai, ed.), Part 1, Ch. 1. Wiley, New York, 1974.

[178b] J. Bernstein, M. D. Cohen, and L. Leiserowitz, *in* "The Chemistry of the Quinonoid Compounds" (S. Patai, ed.), Part 1, Ch. 2. Wiley, New York, 1974.

[178c] K.-P. Zeller, *in* "The Chemistry of the Quinonoid Compounds" (S. Patai, ed.), Part 1, Ch. 5. Wiley, New York, 1974.

[178d] H. W. Moore and R. J. Wikhom, *in* "The Chemistry of the Quinonoid Compounds" (S. Patai, ed.), Part 1, Ch. 8. Wiley, New York, 1974.

[178e] J. H. Fendler and E. J. Fendler, *in* "The Chemistry of the Quinonoid Compounds" (S. Patai, ed.), Part 2, Ch. 10. Wiley, New York, 1974.

[178f] P. Hodge, *in* "The Chemistry of the Quinonoid Compounds" (S. Patai, ed.), Part 2, Ch. 11. Wiley, New York, 1974.

[178g] H.-U. Wagner and R. Gompper, *in* "The Chemistry of the Quinonoid Compounds" (S. Patai, ed.), Part 2, Ch. 18. Wiley, New York, 1974.

[179] E. Wasserman, R. S. Hutton, V. J. Kuck, and E. A. Chandross, *J. Am. Chem. Soc.* **96**, 1965 (1974).

[180a] R. West and J. Niu, *in* "Nonbenzenoid Aromatics" (J. P. Snyder, ed.), Vol. 1, pp. 311–345. Academic Press, New York, 1969.

[180b] R. West and J. Niu, *in* "The Chemistry of the Carbonyl Group" (J. Zabricky, ed.), Vol. 2, pp. 241–275. Wiley (Interscience), New York, 1970.

compound is seemingly unique among "organic" derivatives in being of pentagonal–pyramidal geometry.

139                    140                    141                    142

The possibility of $CH^+$ interacting with other planar $\pi$ species is a reasonable idea. However, few of these species have the maximal symmetry anticipated for the product. For example, with cyclopentadiene anion, the product with $CH^+$ is seemingly benzvalene, **143**, and not a $(CH)_6$ valence isomer with $C_{5v}$ or five-fold symmetry, **144** (see Section 5.B.2). Likewise, $CH^+$ "reacts" with benzene to form norbornadienyl cation, **145**, calculationally[181] and experimentally[182] shown to be more stable than the earlier-suggested[183] $C_{6v}$ pyramidal cation, **146**. Analogues of this latter pyramid have been offered for species with two fewer electrons,[150] and we note that $Sn^{2+}$ and $Pb^{2+}$, isoelectronic with $SnH^{3+}$, $PbH^{3+}$, and $CH^{3+}$) do form such a complex with benzene, **147**.[184a,b] All these products are consistent with the qualitative,

143          144          145          146          147a M = Sn
                                                    147b M = Pb

theoretical "method of structural fragments."[185a,b] We find that this method predicts that $CH^+$ will "react" with ethylene to form both allyl and cyclopropyl cations, and with trimethylenemethane, **148** to form a rather unique representation (**149**) of bicyclo[1.1.1]pentyl cation, **150**. While we may

[181] C. Cone, M. J. S. Dewar, and D. Landman, *J. Am. Chem. Soc.* **99**, 372 (1977).

[182] P. R. Story, L. C. Snyder, D. C. Douglas, E. W. Anderson, and R. L. Kornegay, *J. Am. Chem. Soc.* **85**, 3630 (1963).

[183] S. Winstein and C. Odronneau, *J. Am. Chem. Soc.* **82**, 2084 (1960).

[184a] $Sn^{+2}$: H. Luth and E. L. Amma, *J. Am. Chem. Soc.* **91**, 7515 (1969).

[184b] $Pb^{+2}$: A. G. Gash, P. F. Rodesiler, and E. L. Amma, *Inorg. Chem.* **13**, 2429 (1974).

[185a] J. F. Liebman, *J. Am. Chem. Soc.* **96**, 3053 (1974).

[185b] Robert M. Cooper, Stephen D. Frans, Robin L. Kellert, Robert D. White, and Joel F. Liebman, unpublished results.

inquire if **149** is at all reasonable, this ion mimics the earlier discussed tri-methylenemethane iron tricarbonyl derivative **6**. While we remember that this extremely strained bridgehead carbonium ion is readily formed by

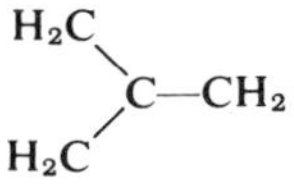

|  **148**  |  **149**  |  **150**  |

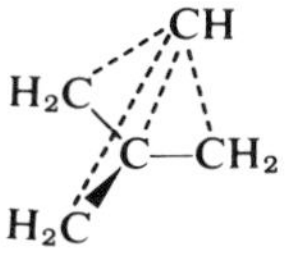

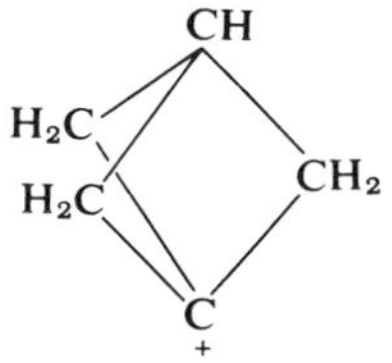

solvolysis (see Section 3.B.2), it is seemingly not produced by deamination of aminospiropentane.[186] It is thus interesting, if not encouraging, that our study of pyramidal cations and five- and six-coordinated carbon in general has taken us on a guided tour through antiaromatic species, valence isomers, and now seemingly "simple" bicyclic derivatives of well-defined and long known hydrocarbons.

[186] M.-J. Chang and J. J. Gajewski, unpublished results, personal communication to Joel F. Liebman.

# Addendum

A new compendium of thermochemical data,[1] updating the Cox and Pilcher book, has recently appeared. The inclusion of lower-order torsional barriers,[2,3] coupled with "smaller, softer" hydrogen and "larger, softer" carbon atoms, has supplied Allinger's 1977 force field (MMII) with exceedingly accurate predictive value for $\Delta H_f^0$ of diverse hydrocarbons.[4] It also appears[4] to obviate the need for postulating the "gauche hydrogen effect," a concept which had received criticism.[5] The new force field yields an $\Delta H_f^0$ of +22.17 kcal/mole for dodecahedrane,[6] a value which is about midway between those previously calculated using the Schleyer (EAS) and Allinger (MMI) force fields. The Ermer–Lifson force field has been used to successfully calculate the $C_1C_2C_3$ angle (128°) in endo,endo-tetracyclo[6.2.1.1.$^{3,6}0^{2,7}$]dodecane (**149**, p. 219) and the $C_2C_3C_4$ angle (139°) in 2,2,5,5-tetramethyl-7,8:9,10-dibenzobicyclo[4.2.2]dec-3-ene, both of which exceed the corresponding ideal (unstrained) angles by almost 20°.[7a,b] A new force field for alkanes and

[1] J. B. Pedley and J. Rylance, *Sussex-N.P.L. Computer Analysed Thermochemical Data: Organic and Organometallic Compounds*, University of Sussex, Sussex, 1977.

[2] L. S. Bartell, *J. Am. Chem. Soc.*, **99**, 3279 (1977).

[3] N. L. Allinger, D. Hindman, and H. Honig, *J. Amer. Chem. Soc.*, **99**, 3282 (1977).

[4] N. L. Allinger, *J. Am. Chem. Soc.*, **99**, 8127 (1977).

[5] E. Osawa, J. B. Collins, and P. von R. Schleyer, *Tetrahedron*, **33**, 2667 (1977).

[6] Prof. Norman L. Allinger, personal communication to Arthur Greenberg.

[7a] O. Ermer, *Angew. Chem., Int. Ed. Engl.*, **16**, 798 (1977).

[7b] O. Ermer, *Angew. Chem., Int. Ed. Engl.*, **16**, 658 (1977).

nonconjugated olefins that very accurately calculates $\Delta H_f^0$ for a variety of compounds, including certain "problem cases," has been published.[8] For example, cyclodecane, in agreement with the results of experimentation, as well as the Ermer–Lifson force field, but in contrast to earlier EAS and MMI results, has been found to be most stable in the boat–chair–boat conformation. Furthermore, the 2,4-ethanonoradamantane molecule is calculated to be more stable (by only 0.5 kcal/mole) than the 2,8-isomer, a finding which is in qualitative agreement at least with experiment (both EAS and MMI found these isomers to be isoenergetic). However, there is still poor agreement (error of 6 kcal/mole) between the calculated and experimental values for manxane. The rotational barriers about the central single bonds in bitryptycyls are very high (in one case, exceeding 54 kcal/mole[9]); perhaps calculations of approximate transition-state geometries will allow a check on the energy function of an exceedingly long carbon–carbon bond. A strain energy of 35.5 kcal/mole has been calculated for the newly prepared *trans*-1-methyl-2-adamantylidene-1-methyladamantane.[10]

ESR studies indicate that the singly-occupied orbital of the bridgehead tryptycyl radical is an sp[8.6] hybrid. Orthogonality considerations lead to the conclusion that this orbital forms a 102.5° angle with the three bonding orbitals at the bridgehead carbon. Since the enforced geometry is virtually tetrahedral, bent bonds (ca.7° bending at carbon) are the result.[11] The tentative conclusion has been reached that 1-methylcyclobutyl cation contains an sp[3]-hybridized tricoordinate carbon. This finding, which contradicts the findings of sophisticated calculations, is rationalized by the dominance of strain over intrinsic hybridization in this species.[12] Upon photolysis, bridged bicyclobutanes such as **60** (p. 87) apparently do not pass through tetrahedranes or 2,4-bicyclobutanediyls en route to cyclobutadienes.[13]

1,5,9-Cyclododecatriyne (**237**, p. 138) is transformed at 650°C into the previously unknown parent molecule [6]radialene (**105**, p. 104; only substituted derivatives, including the hexamethyl compound, had previously been reported).[14] New evidence supporting the intermediacy of butalene

---

[8] D. N. J. White and M. J. Bovill, *J. Chem. Soc. Perkin II*, 1610 (1977).

[9] L. H. Schwartz, C. Koukotas, and C.-S. Yu, *J. Am. Chem. Soc.*, **99**, 7710 (1977).

[10] D. Lenoir and R. Frank, *Tetrahedron Lett.*, 53 (1978).

[11] C. L. Reichel and J. M. McBride, *J. Am. Chem. Soc.*, **99**, 6758 (1977).

[12] R. P. Kirchen and T. S. Sorensen, *J. Am. Chem. Soc.*, **99**, 6687 (1977).

[13] G. Maier, H. P. Reisenauer, and H.-A. Freitag, *Tetrahedron Lett.*, 121 (1978).

[14] A. J. Barkovich, E. S. Strauss, and K. P. C. Vollhardt, *J. Am. Chem. Soc.*, **99**, 8321 (1977).

has been published (**95**, p. 100).[15] Cyclobutenylidene, formed via the reaction of atomic carbon with cyclobutenone, rearranges to vinyl-acetylene through a bicyclobutene-like transition state ( $\Delta H_f^0$ = 121 kcal/mole, MINDO/3).[16] A crystal structure of *syn*-2,2'-bifenchylidene E (**125**, p. 110) shows there to be a normal-length double bond and a surprisingly small (ca. 12°) twist in this highly crowded olefin.[17] Dissymmetric tricyclo[3.3.3.0$^{2,6}$]undec-2(6)-ene (**181**, p. 126), which cannot dimerize by any $_\pi 2$ + $_\pi 2$ pathway, dimerizes nonetheless via an apparently concerted ene reaction to form only one of two possible diastereomers.[18] Bicyclo[4.2.1]non-1-en-8-one maintains some twisted overlap of its conjugated system.[19] 4,5-Benzo-1,2,4,6-cycloheptatetraene has been trapped by furan as well as by styrene (the benzene ring helps to stabilize this allene relative to the isomeric cycloheptatrienylidene).[20] 1-Aza-1,2,4,6-cycloheptatetraene has been observed in an argon matrix, and is a key intermediate in the photolytic reaction of phenyl azide with diethylamine.[21]

Two newly synthesized species pair aromaticity versus electrostatic strain. Cycloheptatriene trianion has been observed by nmr, but this aromatic system tends to be easily oxidized.[22] Dicyclooctatetraeno(1,2:4,5) benzene dianion, avoiding a structure having one nonplanar 8-membered ring fused to a planar $14\pi$ system, apparently adopts the nonaromatic planar $20\pi$ structure that more effectively delocalizes charge.[23] The corresponding tetraanion has also been prepared.[23]

[5]Metacyclophane has been obtained, quite unexpectedly, as a stable by-product (!) of the synthesis of its transient isomer, [5]paracyclophane. Its spectral properties are consistent with the presence of a highly distorted benzene ring and an extremely rigid pentamethylene chain. At 150°C, the molecule rearranges to form indane, thus suffering the same eventual fate as the paracyclophane.[24] Comparison of the UV spectral

[15] R. Breslow and P. L. Khana, *Tetrahedron Lett.*, 3429 (1977).

[16] S. F. Dyer, S. Kammula, and P. B. Shevlin, *J. Am. Chem. Soc.*, **99**, 8104 (1977).

[17] T. Pilati and M. Simonetta, *J. Chem. Soc. Perkin II*, 1435 (1977).

[18] R. Greenhouse, W. T. Borden, T. Ravindranathan, K. Hirotsu, and J. Clardy, *J. Am. Chem. Soc.*, **99**, 6955 (1977).

[19] W. Carruthers and A. Orridge, *J. Chem. Soc. Perkin I*, 2411 (1977).

[20] E. E. Waali, J. M. Lewis, D. E. Lee, E. W. Allen, III, and A. K. Chappell, *J. Org. Chem.*, **42**, 3460 (1977).

[21] O. L. Chapman and J.-P. Le Roux, *J. Am. Chem. Soc.*, **100**, 282 (1978).

[22] J. J. Bahl, R. B. Bates, W. A. Beavers, and C. R. Launer, *J. Am. Chem. Soc.*, **99**, 6126 (1977).

[23] L. A. Paquette, G. D. Ewing, S. Traynor, and J. M. Gardlik, *J. Am. Chem. Soc.*, **99**, 6114 (1977).

[24] J. W. van Straten, W. H. de Wolf, and F. Bickelhaupt, *Tetrahedron Lett.*, 4667 (1977).

properties of [5]metacyclophane with those reported for **327** (p. 166) fuels our skepticism over the formation of the latter, particularly under the high-temperature conditions cited. However, in light of the stability of [5]metacyclophane (as well as **339–343**, p. 170), we wonder if **327** is synthesizable under suitable conditions. The cyclophane structure **326** on p. 166 was apparently incorrectly assigned to a partially closed reaction product.[25] A crystal structure has been published for [2.2.2.2](1,2,4,5)cyclophane (**322**, p. 165).[26] The middle ring of triply-stacked paracyclophane is known to have a twist-boat shape, and has been accorded a greater experimental strain energy than the other two benzene rings, which are themselves boat-shaped.[27]

The x-ray structure of *anti*-tetramantane (**16**, p. 187) has been reported.[28] The first simple derivative of tetracyclo[4.2.1.0.$^{2,8}$0$^{5,7}$]non-3-ene has been observed, and has been found to be extremely reactive.[29] A stable $Fe(CO)_3$ complex of the transient molecule tricyclo[6.2.2.0$^{2,7}$]-dodeca-3,5,9,11-tetraene (the Diels–Alder dimer adduct of benzene) has been isolated.[30] The hydrocarbon, while somewhat more stable thermodynamically than the *anti*-(2+2)-benzene dimer (**114**, p. 214), is less stable kinetically, since it has a "thermally-allowed" route to benzene. Tetracyclo[4.4.0.0.$^{2,4}$0$^{3,5}$]deca-7,9-diene, a new $(CH)_{10}$, does not rearrange thermally to its as-yet-unknown isomer **112** (p. 213).[31]

The norbornadienyl–quadricyclanyl excited singlet and triplet interconversion surfaces have been entered on each "side" by means of photolyses of the corresponding azo precursors. The results suggest that the excited singlet surface has a minimum that delivers diradicals preferentially to norbornadiene (N), and that the triplet surface possesses a minimum that delivers diradicals preferentially to quadricyclane (Q). The results are employed to explain the photochemistry and thermochemistry of N–Q interconversion.[32] Thermolysis of a deuterated Dewar benzene, in conjunction with other experimental findings, leads to the conclusion that at least 93% of the aromatization occurs via a formally "forbidden" $_\sigma 2_s$ + $_\pi 2_s$ transformation.[33] [Turro had earlier concluded (Section 5.B.2) that at least 0.02% of the reaction passed through triplet benzene.] Thus, despite

[25] J. A. Gladysz, J. G. Fulcher, S. J. Lee, and A. B. Bocarsley, *Tetrahedron Lett.*, 3421 (1977).

[26] A. W. Hanson, *Acta Crystallogr.*, **B33**, 2003 (1977).

[27] K. Nishiyama, M. Sakiyama, and S. Seki, *Tetrahedron Lett.*, 3739 (1977).

[28] P. J. Roberts and G. Ferguson, *Acta Crystallogr.*, **B33**, 2335 (1977).

[29] R. D. Miller and D. L. Dulce, *Tetrahedron Lett.*, 3329 (1977).

[30] W. Grimme and E. Schneider, *Angew. Chem., Int. Ed. Engl.*, **16**, 717 (1977).

[31] M. Christl, H.-J. Lüddeke, A. Nagyrevi-Neppel, and G. Freitag, *Chem. Ber.*, **110**, 3745 (1977).

[32] N. J. Turro, W. R. Cherry, M. F. Mirbach, and M. J. Mirbach, *J. Am. Chem. Soc.*, **99**, 7388 (1977).

[33] M. J. Goldstein and R. S. Leight, *J. Am. Chem. Soc.*, **99**, 8112 (1977).

its high exothermicity, the aromatization of Dewar benzene is seen to be not very different from the isomerization of bicyclo[2.2.0]hex-2-ene.[33] Intermediacy of *trans*-Dewar benzene in the aromatization of the *cis*-isomer is dismissed on energetic grounds.[33] This all-but-forgotten $(CH)_6$, which is calculated (MNDO) to be almost 150 kcal/mole less stable than benzene, has a calculated aromatization activation energy of 5 kcal/mole.[34] It will be a brave chemist indeed who invests grant money in the search for this molecule.

Careful exclusion of oxygen during catalyst preparation is now known to preclude the retrocarbene reaction of scheme 8 (Chapter 5).[35a] Arguments against the metallocarbene–metallocyclobutane mechanism for olefin metathesis (schemes 6–8, Chapter 5) have appeared, the most powerful being the nonobservation of cyclopropanes under the usual reaction conditions. It is argued that the (gas-phase) equilibrium constant predicts significant cyclopropane concentrations, and since cyclopropane appears to be comparable in reactivity to ethylene in these reactions, the reported nonobservation of cyclopropanes appears to violate microscopic reversibility.[35b] However, although cyclopropane itself is more soluble than ethylene in solvents such as cyclohexane,[36] we hesitate to apply gas-phase equilibrium constants to condensed-phase equilibria, especially in cases in which the free energy differences are slight.

The first stable thiiranones have been reported, and the structure of a sterically-hindered derivative structure has been determined by x-ray.[37] Evidence for the intermediacy of an oxasilacyclopropane has also been reported.[38] Some spectroscopically-derived structural data have been published for dioxirane (**139,** p. 286), an important intermediate in the ozonolysis of ethylene.[39a,b] Mass spectral studies, supported by calculations, indicate that cyclic $C_2H_5S+$ is considerably more stable than cyclic $C_2H_5O+$, and is also more stable than its acyclic isomers, in contrast to the situation for the oxygen analogue[40] (contrast with the transformation of **149** to **150,** p. 291). This is in line with the reduced strain of thiirane relative to oxirane (pp. 281, 283). The x-ray structure for 1-methyl-2,3-di-*tert*-butylthiirenium tetrafluoroborate, the first stable thiirenium salt, has

[34] M. J. S. Dewar, G. P. Ford, and H. S. Rzepa, *J. Chem. Soc. Chem. Comm.*, 728 (1977).

[35a] Prof. Paul G. Gassman, personal communication to Arthur Greenberg.

[35b] F. D. Mango, *J. Am. Chem. Soc.*, **99,** 6117 (1977).

[36] J. H. Hildebrand, J. M. Prausnitz, and R. L. Scott, "Regular and Related Solutions." Van Nostrand, New York, 1970. See Appendix 3.

[37] E. Schaumann and U. Behrens, *Angew. Chem., Int. Ed. Engl.*, **16,** 722 (1977).

[38] W. Ando, M. Ikeno, and A. Sekiguchi, *J. Am. Chem. Soc.*, **99,** 6447 (1977).

[39a] F. J. Lovas and R. D. Suenram, *Chem. Phys. Lett.*, **51,** 453 (1977).

[39b] R. I. Martinez, R. E. Huie, and J. T. Herron, *Chem. Phys. Lett.*, **51,** 457 (1977).

[40] B. van de Graaf and F. W. McLafferty, *J. Am. Chem. Soc.*, **99,** 6806 (1977).

been published.[41] A *tetrakis*(trifluoromethyl)"Dewar" pyrrole has been isolated[42] (the corresponding "Dewar" thiophene derivative is also known; p. 336). Nonplanarity of the dioxetane ring in **183** (p. 303) somewhat relieves the nonbonded hydrogen repulsions in this molecule.[43] A thermostable imino-bridged triaza-$\lambda^5$-phosphetine is the first bicyclic molecule containing a P=N structural element and is an inorganic analogue of the unknown bicyclo[1.1.1]pentene system.[44]

1,1-Dibromo-*trans*-2,2-diphenylcyclopropanes have ring geometries virtually identical to that of cyclopropane. The authors reporting these geometries summarize current difficulties in interpreting substituted cyclopropane structures.[45] Substituted bicyclobutanes have been prepared via oxidation of highly conjugated 1,3-dimethylenecyclobutanes.[46a] The enhanced stability observed for 1,3-dipyridiniumbicyclobutanes[46b] is reminiscent of the thermodynamic stabilization noted (p. 327) for 1-cyanobicyclobutane. Furthermore, the anticipated increase in the ring-puckering angle[46b] is consistent with the large puckering angle found in 1,3-dicyanobicyclobutane.[47] And, in a close encounter with dodecahedrane, the molecule $C_{16}$-hexaquinacene [$(CH)_{16}$, i.e., dodecahedrane minus a "$(CH)_4$ isobutanoid cap"] has been synthesized.[48]

The calculations by Dill and Schleyer on tetralithiotetrahedrane (p. 338) suggested studies of the photochemical dimerization of dilithioacetylene. This has now been done in liquid ammonia and a new product obtained which has a single $C_{13}$ nmr peak, formula $C_4Li_4$ (field ionization mass spectrometry), and forms acetylene almost quantitatively upon hydrolysis. The evidence is employed to suggest that the crystalline substance, stable at $-20°C$, is tetralithiotetrahedrane.[49]

The synthesis of *stable* (mp 138°C) tetra-*tert*-butyltetrahedrane ($H^7$ nmr: $\delta 1.21$; $C^{13}$ nmr: $\delta 10.20, 28.32, 32.26$; molecular mass = 276) via photochemical decomposition of tetra-*tert*-butylcyclopentadienone at $-80°C$ has been submitted for publication.[50]

[41] R. Destro, T. Pilati, and M. Simonetta, *J. Chem. Soc. Chem. Comm.*, 576 (1977).

[42] Y. Kobayashi, I. Kumadaki, A. Ohsawa, and A. Ando, *J. Am. Chem. Soc.*, **99**, 7350 (1977).

[43] J. Hess and A. Vos, *Acta Crystallogr.*, **B33**, 3527 (1977).

[44] E. Niecke and H.-G. Schäfer, *Angew. Chem., Int. Ed. Engl.*, **16**, 783 (1977).

[45] M. E. Jason and J. A. Ibers, *J. Am. Chem. Soc.*, **99**, 6012 (1977).

[46a] M. Horner and S. Hünig, *J. Am. Chem. Soc.*, **99**, 6120 (1977).

[46b] M. Horner and S. Hünig, *J. Am. Chem. Soc.*, **99**, 6122 (1977).

[47] P. L. Johnson and J. P. Schaefer, *J. Org. Chem.*, **37**, 2762 (1972).

[48] L. A. Paquette, R. A. Snow, J. L. Muthard, and T. Cynkowski, *J. Am. Chem. Soc.*, **100**, 1600 (1978).

[49] G. Rauscher, T. Clark, T. Poppinger, and P. von R. Schleyer, *J. Am. Chem. Soc.*, in press. Prof. Paul von R. Schleyer, personal communication to Joel F. Liebman.

[50] G. Maier, S. Pfriem, U. Schäfer, and R. Matush, submitted to *Angew. Chem., Int. Ed. Engl.* We thank Prof. Emile H. White for communicating this information.

# Index

# ORGANIC CHEMISTRY
## A SERIES OF MONOGRAPHS

EDITOR

### HARRY H. WASSERMAN

*Department of Chemistry*
*Yale University*
*New Haven, Connecticut*

1. Wolfgang Kirmse. CARBENE CHEMISTRY, 1964; 2nd Edition, 1971

2. Brandes H. Smith. BRIDGED AROMATIC COMPOUNDS, 1964

3. Michael Hanack. CONFORMATION THEORY, 1965

4. Donald J. Cram. FUNDAMENTALS OF CARBANION CHEMISTRY, 1965

5. Kenneth B. Wiberg (Editor). OXIDATION IN ORGANIC CHEMISTRY, PART A, 1965; Walter S. Trahanovsky (Editor). OXIDATION IN ORGANIC CHEMISTRY, PART B, 1973; PART C, 1978

6. R. F. Hudson. STRUCTURE AND MECHANISM IN ORGANO-PHOSPHORUS CHEMISTRY, 1965

7. A. William Johnson. YLID CHEMISTRY, 1966

8. Jan Hamer (Editor). 1,4-CYCLOADDITION REACTIONS, 1967

9. Henri Ulrich. CYCLOADDITION REACTIONS OF HETEROCUMULENES, 1967

10. M. P. Cava and M. J. Mitchell. CYCLOBUTADIENE AND RELATED COMPOUNDS, 1967

11. Reinhard W. Hoffmann. DEHYDROBENZENE AND CYCLOALKYNES, 1967

12. Stanley R. Sandler and Wolf Karo. ORGANIC FUNCTIONAL GROUP PREPARATIONS, VOLUME I, 1968; VOLUME II, 1971; VOLUME III, 1972

13. Robert J. Cotter and Markus Matzner. RING-FORMING POLYMERIZATIONS, PART A, 1969; PART B, 1; B, 2, 1972

14. R. H. DeWolfe, CARBOXYLIC ORTHO ACID DERIVATIVES, 1970

15. R. Foster. ORGANIC CHARGE-TRANSFER COMPLEXES, 1969

16. James P. Snyder (Editor). NONBENZENOID AROMATICS, VOLUME I, 1969; VOLUME II, 1971

17. C. H. Rochester. ACIDITY FUNCTIONS, 1970

18. Richard J. Sundberg. THE CHEMISTRY OF INDOLES, 1970

19. A. R. Katritzky and J. M. Lagowski. CHEMISTRY OF THE HETEROCYCLIC N-OXIDES, 1970

20. Ivar Ugi (Editor). ISONITRILE CHEMISTRY, 1971

21. G. Chiurdoglu (Editor). CONFORMATIONAL ANALYSIS, 1971

22. Gottfried Schill. CATENANES, ROTAXANES, AND KNOTS, 1971

23. M. Liler. REACTION MECHANISMS IN SULPHURIC ACID AND OTHER STRONG ACID SOLUTIONS, 1971

24. J. B. Stothers. CARBON-13 NMR SPECTROSCOPY, 1972

25. Maurice Shamma. THE ISOQUINOLINE ALKALOIDS: CHEMISTRY AND PHARMACOLOGY, 1972

26. Samuel P. McManus (Editor). ORGANIC REACTIVE INTERMEDIATES, 1973

27. H. C. Van der Plas. RING TRANSFORMATIONS OF HETEROCYCLES, VOLUMES 1 AND 2, 1973

28. Paul N. Rylander. ORGANIC SYNTHESES WITH NOBLE CATALYSTS, 1973

29. Stanley R. Sandler and Wolf Karo. POLYMER SYNTHESES, VOLUME I, 1974; VOLUME II, 1977

30. Robert T. Blickenstaff, Anil C. Ghosh, and Gordon C. Wolf. TOTAL SYNTHESIS OF STEROIDS, 1974

31. Barry M. Trost and Lawrence S. Melvin, Jr. SULFUR YLIDES: EMERGING SYNTHETIC INTERMEDIATES, 1975

32. Sidney D. Ross, Manuel Finkelstein, and Eric J. Rudd. ANODIC OXIDATION, 1975

33. Howard Alper (Editor). TRANSITION METAL ORGANOMETALLICS IN ORGANIC SYNTHESIS, VOLUME I, 1976; VOLUME II, 1978

34. R. A. Jones and G. P. Bean. THE CHEMISTRY OF PYRROLES, 1976

35. Alan P. Marchand and Roland E. Lehr (Editors). PERICYCLIC REACTIONS, VOLUME I, 1977; VOLUME II, 1977

36. Pierre Crabbé (Editor). PROSTAGLANDIN RESEARCH, 1977

37. Eric Block. REACTIONS OF ORGANOSULFUR COMPOUNDS, 1978

38. Arthur Greenberg and Joel F. Liebman, STRAINED ORGANIC MOLECULES, 1978

39. Philip S. Bailey. OZONATION IN ORGANIC CHEMISTRY, VOL. I, 1978

40. Harry H. Wasserman and Robert W. Murray (Editors). SINGLET OXYGEN, 1978

A
B 8
C 9
D 0
E 1
F 2
G 3
H 4
I 5
J 6